CASUAL CALCULUS

3 Volumes

A Friendly Student Companion

CASUAL CALCULUS

3 Volumes

A Friendly Student Companion

Kenneth H Luther

Valparaiso University, USA

World Scientific

NEW JERSEY • LONDON • SINGAPORE • BEIJING • SHANGHAI • HONG KONG • TAIPEI • CHENNAI • TOKYO

Published by

World Scientific Publishing Co. Pte. Ltd.
5 Toh Tuck Link, Singapore 596224
USA office: 27 Warren Street, Suite 401-402, Hackensack, NJ 07601
UK office: 57 Shelton Street, Covent Garden, London WC2H 9HE

British Library Cataloguing-in-Publication Data
A catalogue record for this book is available from the British Library.

CASUAL CALCULUS: A FRIENDLY STUDENT COMPANION
(In 3 Volumes)

ISBN 978-981-124-263-2 (set_hardcover)
ISBN 978-981-124-264-9 (set_paperback)
ISBN 978-981-124-265-6 (set_ebook for institutions)
ISBN 978-981-124-266-3 (set_ebook for individuals)

ISBN 978-981-122-392-1 (vol. 1_hardcover)
ISBN 978-981-122-488-1 (vol. 1_paperback)
ISBN 978-981-122-393-8 (vol. 1_ebook for institutions)
ISBN 978-981-122-394-5 (vol. 1_ebook for individuals)

ISBN 978-981-124-197-0 (vol. 2_hardcover)
ISBN 978-981-124-198-7 (vol. 2_paperback)
ISBN 978-981-124-199-4 (vol. 2_ebook for institutions)
ISBN 978-981-124-211-3 (vol. 2_ebook for individuals)

ISBN 978-981-122-395-2 (vol. 3_hardcover)
ISBN 978-981-122-489-8 (vol. 3_paperback)
ISBN 978-981-122-396-9 (vol. 3_ebook for institutions)
ISBN 978-981-122-397-6 (vol. 3_ebook for individuals)

For any available supplementary material, please visit
https://www.worldscientific.com/worldscibooks/10.1142/11927#t=suppl

Printed in Singapore

The content from my favorite semester of Calculus is dedicated to my favorite people: Kathy, Melody, and Joy.

Preface

Welcome back! I hope that you are visiting Casual Calculus — Volume 3 because you found Volume 1 and / or 2 helpful.

The Preface of Volume 1 gave the set-up for the work as a whole, across three volumes, so I won't repeat it here. I will just reiterate some of the organizational items in case some readers are jumping directly into this Volume 3.

The large structure is:

- Volume 1 contains Chapters 1–6, which correspond to a standard first semester of Calculus, ending with the Fundamental Theorem of Calculus.
- Volume 2 contains Chapters 7–12, which go with a standard second semester of single-variable Calculus.
- Volume 3 contains Chapters 13–18, which match what is often Calculus 3 (Multivariable Calculus).

The section-by-section set up of this book is as follows: Each topic of content begins with a narrative section that leads you through the main ideas and presents examples along the way. After each Example is a "You Try It" problem that's very similar to the example. What I hope you do, as you're reading, is stop after an Example you think you understood, and immediately try your hand at the associated YTI problem. The solutions to all YTI problems are at the end of the very section they're shown in — so, if you think you've succeeded at the YTI problem, go check the solution to be sure. Or, if you get stuck on the YTI problem, then go look at its solution to get a hint. Once you've completed a section, you'll see the YTI problems collected, along with a set of Practice Problems and Challenge Problems. The Practice Problems should be similar to the YTI Problems, but you're getting them all at once, and so you don't necessarily know which specific technique to use or which Example to follow — you have to think about it! The solutions to the Practice Problems are at the back end of the book. So while they're all available, they are more physically separated from the section they come from; the idea is that you might be inclined to rely on them a bit less, although they are still there when you need them. Then finally, the Challenge Problems are a bit tougher than the others, and you can use those to see if you're successfully synthesizing the ideas you've seen in their section. Solutions to Challenge Problems come after those to the Practice Problems.

At many locations in the text, I will pose some "Food For Thought" (FFT) based on an open question left unanswered. These little puzzlers are bracketed by the symbol (it's supposed to be a fork, plate, and knife).

To keep you focused on problem solving, some derivations or other more theoretical discussions are held off until the end of a section. I have always been a fan of heavy metal, and I see jumping into a derivation or proof as the mathematical version of jumping into a mosh pit: you're mostly there to sing along, but every once in a while, you have to wade in and get a bit bruised. So, each of these more mathematically violent discussions are set off with a subheading of, "Into the Pit!!"

The content here in Volume 3 is by far my favorite to teach, out of all three semesters of Calculus. I hope you enjoy it, too!

Contents

Chapter 13

Mathematical Deja Vu

13.1 Life in Three Dimensions

Introduction

Calculus I was boring. All your work was done in the land of two dimensions; you had one independent variable and one dependent variable, and only two axes on a graph. Ha, that's kid stuff. Now it's time for grown-up math! Here you start on your journey into the land of three dimensions. Many of the concepts you'll encounter at first will be familiar — things you learned to do in 2D now get extended to 3D, there's just "more". Points have three coordinates instead of two. Graphs require three axes, not two. Functions can have more than one independent variable. We'll eventually see partial derivatives and double (and triple!) integrals.

There will be some significant changes and additions along the way. For example, life in three dimensions is made more fruitful with the use of vectors, vector functions, and vector fields. Sure, those things are used in two dimensions as well, but working in higher dimensions causes us to sigh and realize we need more robust tools: rather than worrying about whether we're in two, three, or even more dimensions, can we develop mathematical tools that work the same way regardless of how many dimensions we have?

To reach that point, we'll start off getting comfortable with the 3D rectangular (Cartesian) coordinate system.

The 3D Rectangular Coordinate System

Based on your experience with the 2D coordinate system, some extensions of familiar concepts into 3D should not be painful:

- In 2D, we had two coordinate axes, usually called x and y; the x-axis displayed the only independent variable (input), and the y-axis displayed the dependent variable (output). In 3D we have three coordinate axes, usually called x, y and z; x and y are both independent variables and z is the dependent variable. (There is still only one dependent variable, that will always be true no matter how many dimensions we use!)

  Figure 13.1 shows a 3D coordinate system. I can't say this is THE 3D coordinate system, because there are several options for how the axes are presented. Some people like to have the positive x-axis pointing out of the page, with the positive y-axis going eastward on the page. Personally, I prefer the x-axis to be oriented rightwards, with the positive y-axis pointing into the page. This is like taking a 2D coordinate system on the page and then just tipping it over so that it falls into the page. It's what helps me visualize things best. If you like other configurations, by all means, use them — but to repeat what you've been told since middle school: *label the dang axes*!

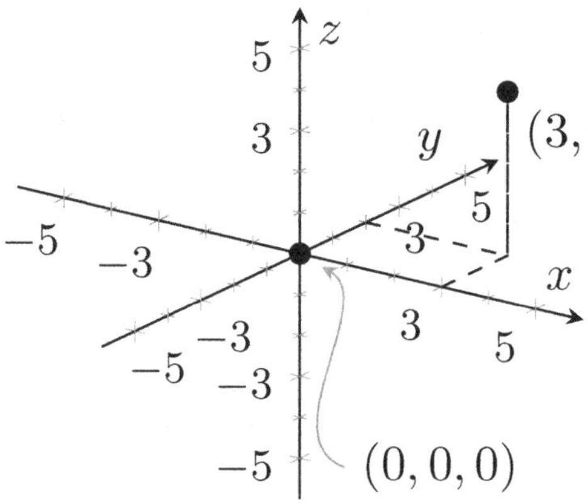

Fig. 13.1 The 3D rectangular coordinate system.

- In 2D, we had one coordinate plane. It was the xy-plane. In 3D, we have three coordinate planes: the xy-plane, the yz-plane, and the xz-plane.

The planes intersect along the coordinate axes. The origin is the one point that all three planes have in common. (Remember this later on when we ask if it's possible for three planes to intersect in only one single point.)
- In 2D, space was divided into four quadrants. In 3D, space is divided into octants. Can you guess how many octants there are? (Hint: it's implied in the name *octant.*)
- In 2D, each point was described uniquely by two coordinates: in rectangular coordinates these were x and y, and in polar coordinates they were r and θ. In 3D, each point is described by three coordinates; the rectangular coordinate names are (unsurprisingly) x, y and z; eventually, we'll see other 3D coordinates systems too.

Plotting points in 3D is rather simple. Use coordinates the same way you do in 2D. For example, to locate the point $(3, 2, 4)$, you start at the origin and then:

- Move along the positive x-axis 3 units.
- Move parallel to the positive y-axis a distance of 2 units (remain in the xy-plane).
- Move parallel to the positive z-axis a distance of 4 units. This point now "floats in space".

(The point $(3, 2, 4)$ is shown in Fig. 13.1.) Negative coordinates are interpreted just as in 2D — you move in the opposite direction of the appropriate positive axis.

You Try It

(1) Which axes do the points $(1, 0, 0)$, $(0, 1, 0)$ and $(0, 0, 1)$ lie on?
(2) Which coordinate planes do the points $(1, 1, 0)$, $(0, 1, 1)$ and $(1, 0, 1)$ lie on?

You Try It

(3) An observer is at the 3D coordinate origin, and can pivot East / West to look in the positive / negative x direction; North / South to look in the positive / negative y direction, and Up / Down to look in the positive / negative z direction. How does the observer orient to see the point $(-1, 4, 2)$? $(3, -3, -5)$? $(-3, -4, -2)$?

You Try It

(4) In 2D, it takes two points to uniquely define a line. In 3D, how many points do you think it takes to uniquely define a line?

Descriptions of Regions, Curves, and ...?

Here is one tidbit that might actually come as a surprise:

- In 2D, the equation of a line is, in standard form, $ax + by + c = 0$. You might think, then, that the equation of a line in 3D would be $ax + by + cz + d = 0$. But it's not. That's the equation of a plane! The equation of a line is usually presented in parametric form. We'll see plenty of lines and planes quite soon.

We must be able to describe regions and curves in 3D. Perhaps there are also new, other things that we didn't encounter in 2D that we will need to describe. Here are some examples of familiar themes:

- The equation $x = 1$ describes the set of all points with an x-coordinate of 1. In 2D, this is a line, parallel to the y-axis. In 3D, though, the equation $x = 1$ now describes an entire plane! All the points that have an x-coordinate of 1 comprise a plane parallel to the yz-plane. In general, anything in 3D of the form $x = a$, $y = b$ or $z = c$ is a plane, parallel to one of the coordinate planes. Figure 13.2 shows a scattering of points which all have $x = 1$ in common.

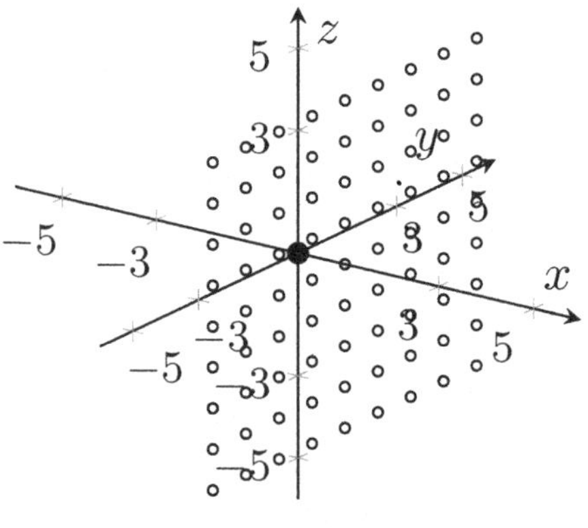

Fig. 13.2 Points on the 3D plane $x = 1$.

- The inequality $y \geq 2$ describes the set of all points with a y coordinate greater than or equal to 2. In 2D, this is a portion of the coordinate plane, the entire "half-plane" above (and including) the horizontal line $y = 2$. In 3D, this inequality describes and entire portion of 3D space — it's a "half-space" consisting of all points "above" (and including) the *plane* $y = 2$. I had to put "above" in quotes because the actual physical orientation of this half-space depends on how your axes are oriented on your paper.

You Try It

(5) What region is described by the expression $x > 3$ in 2D and 3D?

Here are a couple of things that are new or different:

- In 2D, curves (graphs of functions?) wiggle through the xy-plane. In 3D, curves can be restricted to one of the coordinate planes, or can wiggle through space. Usually, curves that wiggle through space are described with parametric equations (coming soon!).
- In 2D, we only plot points and lines / curves. In 3D, there's something new! Consider that in 2D, the equation $x^2 + y^2 = 1$ describes the unit circle. What do you think this equation describes in 3D? Hint: It's not a circle. Or, if we extend a familiar equation from $x^2 + y^2 = 1$ to $x^2 + y^2 + z^2 = 1$, what do we have now? Hint: This is not a circle, either. When graphed in 3D, the equations $x^2 + y^2 = 1$ and $x^2 + y^2 + z^2 = 1$ produce entire *surfaces* (not just curves). We'll look in detail at three-dimensional surfaces very soon, however the latter of those two expressions can be understood with the information in this very next section.

The Distance Formula

One of the most important tools in any coordinate system is the computation of distance. You should recall that in 2D, the distance between two (rectangular) points (x_1, y_1) and (x_2, y_2) is

$$d = \sqrt{(x_2 - x_1)^2 + (y_2 - y_1)^2}$$

You might imagine that the extension of this to 3D is simple, and you'd be right: in 3D, the distance between two points (x_1, y_1, z_1) and (x_2, y_2, z_2) is

$$d = \sqrt{(x_2 - x_1)^2 + (y_2 - y_1)^2 + (z_2 - z_1)^2}$$

If you prefer this without the radical, the same relationship can be written

$$d^2 = (x_2 - x_1)^2 + (y_2 - y_1)^2 + (z_2 - z_1)^2$$

In fact, you should start learning to recognize the form of the right hand side, so that when you see something similar, you can start to think about what the expression might mean in terms of distance. For example, if you don't already know what the expression $x^2 + y^2 + z^2 = 1$ creates in 3D, match it to the 3D distance formula ... if it helps, expand the expression to $(x-0)^2+(y-0)^2+(z-0)^2$. We can see that the expression $x^2+y^2+z^2 = 1$ describes all points (x, y, z) whose distance from the point (0,0,0) is 1. What is that collection of points? It's a sphere! Specifically, it's the sphere of radius 1 centered at the origin, also known as the *unit sphere*.

You Try It

(6) What region is described by the expression $x^2 + y^2 + z^2 > 1$?

The things you can do with the 3D distance formula are similar to things you can do in 2D.

EX 1 What is the distance between the points $(1, 2, 1)$ and $(-2, 4, -1)$?

We have

$$d = \sqrt{(-2-1)^2 + (4-2)^2 + (-1-1)^2} = \sqrt{9+4+4} = \sqrt{17} \quad \blacksquare$$

You Try It

(7) Is the triangle defined by the points $P(-2, 4, 0)$, $Q(1, 2, -1)$, and $R(-1, 1, 2)$ an equilateral triangle?

Life in Three Dimensions — Problem List

Life in Three Dimensions — You Try It

These appeared above; solutions begin on the next page.

(1) Which axes do the points $(1,0,0)$, $(0,1,0)$ and $(0,0,1)$ lie on?
(2) Which coordinate planes do the points $(1,1,0)$, $(0,1,1)$ and $(1,0,1)$ lie on?
(3) An observer is at the 3D coordinate origin, and can pivot East / West to look in the positive / negative x direction; North / South to look in the positive / negative y direction, and Up / Down to look in the positive / negative z direction. How does the observer orient to see the point $(-1,4,2)$? $(3,-3,-5)$? $(-3,-4,-2)$?
(4) In 2D, it takes two points to uniquely define a line. In 3D, how many points do you think it takes to uniquely define a line?
(5) What region is described by the expression $x > 3$ in 2D and 3D?
(6) What region is described by the expression $x^2 + y^2 + z^2 > 1$?
(7) Is the triangle defined by the points $P(-2,4,0)$, $Q(1,2,-1)$, and $R(-1,1,2)$ an equilateral triangle?

Life in Three Dimensions — Practice Problems

Try these as you get the hang of the You Try It problems. Solutions to these problems are available in Sec. B.1.1.

(1) What region is described by the expression $y \geq 0$ in 2D and 3D?
(2) What region is described by the expression $1 \leq x^2 + y^2 + z^2 \leq 25$
(3) Is the triangle defined by the points $P(1,1,0)$, $Q(2,4,1)$, and $R(-1,-1,3)$ a right triangle? (Hint: Remember the Pythagorean Theorem.)

Life in Three Dimensions — Challenge Problems

Try these problems to test your skills with the ideas in this section. Solutions to these problems are available in Sec. C.1.1.

(1) What is described by the expression $x = y$ in 2D and 3D?
(2) What is described by the expression $x^2 + y^2 = 1$ in 2D and 3D?
(3) Is the triangle defined by the points $A(1,2,-3)$, $B(3,4,-2)$, and $C(3,-2,1)$ an isoceles triangle?

Life in Three Dimensions — You Try It — Solved

(1) Which axes do the points $(1, 0, 0)$, $(0, 1, 0)$ and $(0, 0, 1)$ lie on?

☐ These lie on the x-, y-, and z-axes, respectively. ■

(2) Which coordinate planes do the points $(1, 1, 0)$, $(0, 1, 1)$ and $(1, 0, 1)$ lie on?

☐ These lie in the xy-, yz-, and xz-planes, respectively. ■

(3) An observer is at the 3D coordinate origin, and can pivot East / West to look in the positive / negative x direction; North / South to look in the positive / negative y direction, and Up / Down to look in the positive / negative z direction. How does the observer orient to see the point $(-1, 4, 2)$? $(3, -3, -5)$? $(-3, -4, -2)$?

☐ To see the indicated points, the observer must be oriented in the following directions:

- for $(-1, 4, 2)$: West, North, and Up.
- for $(3, -3, -5)$: East, South, and Down.
- for $(-3, -4, -2)$: West, South, and Down.

Figure 13.3 shows all three points. ■

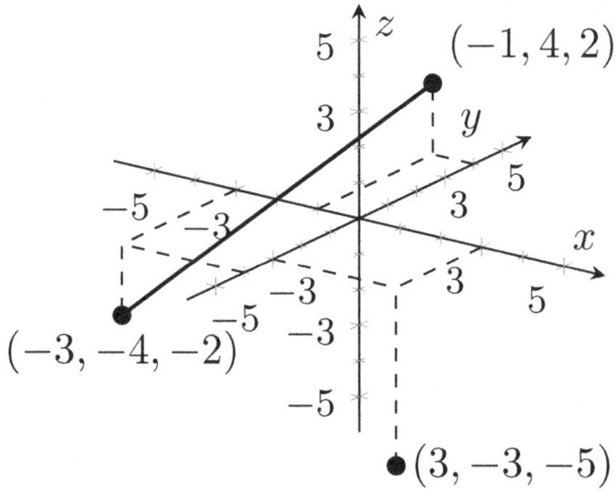

Fig. 13.3 Three points and a line segment between two of them.

(4) In 2D, it takes two points to uniquely define a line. In 3D, how many points do you think it takes to uniquely define a line?

□ In 3D, we still only need two points to define a line; the line defines the path of shorted distance between the two points. And, two points still define a line in 3D. But that line is shared by an infinite number of planes! Can you picture that? Figure 13.3 shows the line between $(-1, 4, 2)$ and $(-3, -4, -2)$. ■

(5) What region is described by the expression $x > 3$ in 2D and 3D?

□ The 3D region $x > 3$ is everywhere to the "right" of (but not including) the plane $x = 3$. This would be called a "semi infinite half space". ■

(6) What region is described by the expression $x^2 + y^2 + z^2 > 1$?

□ This region is everything outside the unit sphere (the sphere of radius 1 centered at the origin). ■

(7) Is the triangle defined by the points $P(-2, 4, 0)$, $Q(1, 2, -1)$, and $R(-1, 1, 2)$ an equilateral triangle?

□ The triangle connecting the points is equilateral if the lengths of the sides are the same:

$$|PQ| = \sqrt{(1+2)^2 + (2-4)^2 + (-1-0)^2} = \sqrt{14}$$
$$|QR| = \sqrt{(-1-1)^2 + (1-2)^2 + (2+1)^2} = \sqrt{14}$$
$$|PR| = \sqrt{(-1+2)^2 + (1-4)^2 + (2-0)^2} = \sqrt{14}$$

The lengths of all three sides are the same, so the triangle is an equilateral triangle. ■

13.2 Multivariable Functions

Introduction

In Calculus I, functions only had 2 variables: one independent and one dependent. But who can be satisfied with only two variables? Not us! We can allow as many independent variables as we want, and in "real life", most quantities are determined by more than one thing. The volume of a cylinder depends on its radius *and* height. The temperature outside depends on (among other things) your latitude, longitude, and altitude. The distance a well hit baseball travels depends on bat speed, pitch speed, angle of impact, wind resistance, and more.

This is a good news / bad news situation. The good news is that we're going to learn about functions that have more than two variables. The bad news is that when we have two independent variables, it can become difficult to plot the function, and with two independent variables and one dependent variable, we have now used up all the possible axes in three-dimensional space. The worse news is that if we have three or more independent variables, it is not possible to plot the function in a standard coordinate system, although there may be ad-hoc methods of visualization.

In this section and the next, we'll learn the basics of these functions (domain, range, etc), review functions in parametric form, see several common types and shapes, and find alternate ways to plot & visualize functions. Once we learn these things, then we'll go through the same sequence of topics as in Calc 1: limits, continuity, derivatives, and integrals. Luckily, most topics will contain strange twists — that's how it stays interesting, after all.

Notation, Domain, and Range

We describe multivariable functions using the same functional notation as before, just with more variables. We can have $f(x)$, $f(x, y)$, $f(x, y, z)$, $h(p, q, w, y)$, etc. A function of the form $z = f(x, y)$ is the most complicated we can get and still be able to graph it. Evaluation is indicated using this notation, too: plug in whatever sits in the variable's spot in the $f(x, y, \ldots)$ notation.

EX 1 If $f(x, y) = x^2 + y^3$, what are $f(1, 1)$, $f(x, x)$ and $f(x^2, y^2)$?

$$\begin{aligned} f(1,1) &= (1)^2 + (1)^3 = 2 \\ f(x,x) &= (x)^2 + (x)^3 = x^2 + x^3 \\ f(x^2, y^2) &= (x^2)^2 + (y^2)^3 = x^4 + y^6 \quad \blacksquare \end{aligned}$$

Some other notation we'll use is $\mathbb{R}$, $\mathbb{R}^2$, $\mathbb{R}^3$, etc. It gets old writing "two-dimensional space" or "three-dimensional space" over and over, and so we use the symbol $\mathbb{R}$ for "the set of all real numbers" accompanied by an exponent denoting the dimension of the space we're in. Our two-dimensional coordinate system is $\mathbb{R}^2$, and we are now starting to explore the three dimensional coordinate system $\mathbb{R}^3$.

The domain of a function retains the same general meaning it had before: it is the collection of values that are allowed to be used for the variables, such that the function remains defined. The difference now is that the physical location of the domain gets more complicated. Consider the function $f(x, y) = \sqrt{x} + \sqrt{y}$. Here, we know that we're allowed to use any $x \geq 0$ and any $y \geq 0$. Where are all those points located? In the first quadrant of the xy-plane. In $\mathbb{R}^2$, the domain of a function was part of the x-axis; in $\mathbb{R}^3$, domains are entire regions of the xy-plane.

The range of a function also retains the same general meaning it had before: it is the collection of values that are possible outputs of the function. The range of a function $f(x, y)$ will be part of (or all of) the z-axis. The range of $z = f(x, y) = \sqrt{x} + \sqrt{y}$ is the set of all values $z \geq 0$.

While ranges remain restricted to the possible values of the single dependent variable, regardless of how many independent variables there are, the domain of a function gets more complicated as the number of independent variables increases.

EX 2 Describe the domain and range of $f(x, y, z) = \sqrt{x^2 + y^2 + z^2 - 1}$.

We need to collect all possible values of x, y, and z for which $f(x, y, z)$ is still defined; this means we need $x^2 + y^2 + z^2 - 1 \geq 0$, or in a better form, we need $x^2 + y^2 + z^2 \geq 1$. The collection of points $x^2 + y^2 + z^2 = 1$ is the unit sphere in $\mathbb{R}^3$, and so $x^2 + y^2 + z^2 \geq 1$ is everything outside and on the unit sphere. If we name the dependent variable $w = f(x, y, z)$, then we can describe the range as $w \geq 0$, since we'll always get positive numbers or zero out of the function. ■

You Try It

(1) For $f(x,y) = x^2 e^{3xy}$, find $f(2,0)$ and $f(3x,2y)$, and describe the domain and range of $f(x,y)$.
(2) Describe the domain and range of $f(x,y) = \sqrt{x+y}$.
(3) Describe the domain and range of $f(x,y,z) = \sqrt{1-x^2-y^2-z^2}$.

Parametric Equations and Lines in 3D

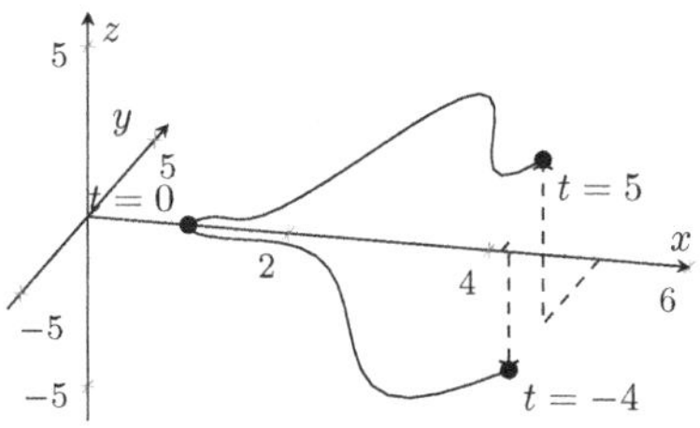

Fig. 13.4 The parametric curve $x = \sqrt{t^2+1}$, $y = t\cos(2t)$, $z = t^3/(t^2+1)$ for $-4 \leq t \leq 5$.

Functions do not have to be described as $z = f(x,y)$. Another way to assemble a function is to select a parameter, like t, that determines ALL of the other variables in the function. You have already done this in 2D (see Chapter 12), and you should recall, for example, that the equations $x = \cos t, y = \sin t, 0 \leq t \leq 2\pi$ describe the unit circle. As t varies from $t = 0$ to $t = 2\pi$ we create x and y values that become the points (x,y) on the curve. (If that doesn't sound familiar, go review!)

This idea extends to three (or more) dimensions in two ways:

(1) We can use a single parameter, say t, to determine any number of other variables. Here is a perfectly good set of parametric equations that form a curve in three-space:

$$x(t) = \sqrt{t^2+1}\,;\; y(t) = t\cos(2t)\,;\; z(t) = \frac{t^3}{t^2+1} \quad ; \quad -4 \leq t \leq 5$$

As t varies from $t = -4$ to $t = 5$, we create a collection of points (x,y,z) that form the curve. Parametric equations like this —

which have one parameter — form curves. Figure 13.4 shows this particular parametric curve, with a few representative points.

(2) We can use more than one parameter to determine our variables. Here is a perfectly good set of parametric equations in which our variables x and y depend on TWO parameters, say u and v:

$$x(u,v) = u^2 + v^2\,;\; y(u,v) = \frac{u}{v} \quad ; \quad 0 \le u \le 1, 1 \le v \le 3$$

It turns out parametric constructions like this — with two parameters — define surfaces, and we'll take a look at parametric surfaces later.

You Try It

(4) As mentioned above, the parametric equations $x = \cos t, y = \sin t$ form the unit circle. With no restrictions on t, so that t can be any real number, we trace the unit circle over and over. What do we get if we extend these equations to the 3D parametric curve $x = \cos t, y = \sin t, z = t$?

When studying two-dimensional forms, we noted that although the equation $ax+by+c = 0$ describes a line in 2D, the equation $ax+by+cz+d = 0$ does NOT describe a line in 3D, but rather an entire plane. Rather, lines in 3D are usually described in parametric form:

Useful Fact 13.1. *Parametric equations of the form*

$$x(t) = a + bt\,;\; y(t) = c + dt\,;\; z(t) = m + nt \quad ; \quad t_1 \le t \le t_2$$

form a straight line segment in $\mathbb{R}^3$. *When* t *is unrestricted, we get an infinite line.*

EX 3 Find the equations of the line segment starting at the point $(1, 1, 1)$ and ending at $(2, 3, 4)$.

Right now, we have no methodical way of finding the line segment's equations, we just have to use our imagination. At some point, we have to decide on the window of t values to apply — when in doubt, we can always start with $t = 0$ to $t = 1$ then adapt if a more clever arrangement is found. Here, since the x values we want to span start at 1 and end at 2, we can write $x = 1 + t$ for $0 \le t \le 1$. Similarly, we need to have y start at $y = 1$ and go two units to $y = 3$, and have z start at $z = 1$ and go three units to $z = 4$. So how about this, displayed in Fig. 13.5:

$$x(t) = 1 + t\,;\; y(t) = 1 + 2t\,;\; z(t) = 1 + 3t \quad ; \quad 0 \le t \le 1 \quad \blacksquare$$

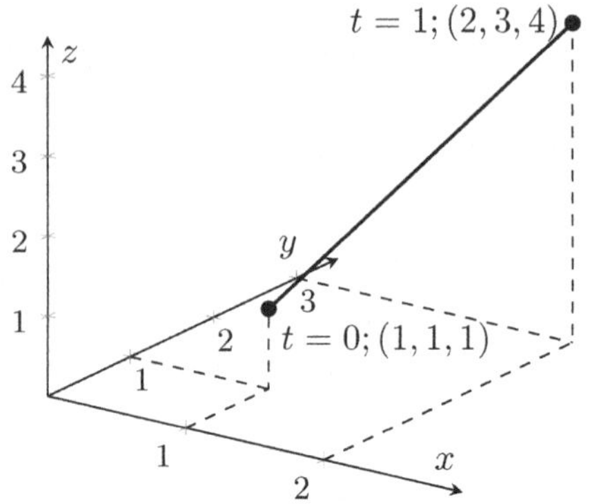

Fig. 13.5 The line segment $x = 1 + t, y = 1 + 2t, z = 1 + 3t$ for $0 \leq t \leq 1$.

Parametric descriptions of curves are not unique; for example, the same line segment can be described as:

$$x(t) = 1 + \frac{t}{2}\,;\; y(t) = 1 + t\,;\; z(t) = 1 + \frac{3}{2}t \qquad ; \qquad 0 \leq t \leq 2$$

Also note that with parametric equations, we get *direction*. These sets of equations describe the line segment *from* $(1, 1, 1)$ *to* $(2, 3, 4)$. What if we wanted to go the other way, from $(2, 3, 4)$ to $(1, 1, 1)$? Then we could use

$$x(t) = 2 - t\,;\; y(t) = 3 - 2t\,;\; z(t) = 4 - 3t \qquad ; \qquad 0 \leq t \leq 1$$

You Try It

(5) Find parametric equations form the line segment starting at the point $(0, -2, 5)$ and ending at $(\pi, -1, 5)$. Then give equations for the reverse segment, starting at $(\pi, -1, 5)$ and ending at $(0, -2, 5)$.

Here's a twist on lines that differs from what you know about lines in 2D. In the xy-plane, if lines are not parallel, then they must intersect. In 3D, that's not true. Lines which are not parallel do not have to intersect — which is excellent news for people traveling in airplanes. In 3D, lines can be parallel, perpendicular, skew, or none of the above. Skew lines are lines that are not parallel but also do not intersect. It will be easier to figure out whether lines are parallel or perpendicular later on, when we're working with vectors. But we can learn a few things from the parametric equations of lines, such as ...

EX 4 Two lines in $\mathbb{R}^3$ are given by the parametric equations

$$x = t\,,\ y = 2 + t\,,\ z = 1 - 2t \quad \text{and} \quad x = 2s\,,\ y = -1 + 3s\,,\ z = \frac{s}{2}$$

Do these lines intersect?

In order to intersect, the lines must share a point. That is, there must be a pair of parameter values s and t that create the same coordinates. For the x-coordinates to be equal, we'd need $t = 2s$. To have the y-coordinates be the same, we'd need $2 + t = -1 + 3s$; but to have this happen at the same values that make the x-coordinates the same, i.e. $t = 2s$, we'd need $2 + 2s = -1 + 3s$, or $s = 3$ (and so also $t = 6$). So, with $t = 6$ and $s = 3$, we have the same x and y coordinates. At these values the z-coordinate of the first line is $z = 1 - 2(6) = -11$, and the z-coordinate of the second line is $z = 3/2$. So we cannot make all three coordinates the same, and the lines do not share a point.

A briefer way to describe this is: we can always make the x and y coordinates the same by solving two equations in two unknowns, $t = 2s$ and $2 + t = -1 + 3s$. Having done that, will the resulting s and t values also make the z-coordinates equal? Try it out! ■

You Try It

(6) Do the lines $x = t - 1, y = 2t, z = 1 - t$ and $x = -2 + s, y = -1 + s, z = 2s$ share a point?

Multivariable Functions — Problem List

Multivariable Functions — You Try It

These appeared above; solutions begin on the next page.

(1) For $f(x, y) = x^2 e^{3xy}$, find $f(2, 0)$ and $f(3x, 2y)$, and describe the domain and range of $f(x, y)$.
(2) Describe the domain and range of $f(x, y) = \sqrt{x + y}$.
(3) Describe the domain and range of $f(x, y, z) = \sqrt{1 - x^2 - y^2 - z^2}$.
(4) As mentioned above, the parametric equations $x = \cos t, y = \sin t$ form the unit circle. With no restrictions on t, so that t can be any real number, we trace the unit circle over and over. What do we get if we extend these equations to the 3D parametric curve $x = \cos t, y = \sin t, z = t$?
(5) Find parametric equations form the line segment starting at the point $(0, -2, 5)$ and ending at $(\pi, -1, 5)$. Then give equations for the reverse segment, starting at $(\pi, -1, 5)$ and ending at $(0, -2, 5)$.
(6) Do the lines $x = t - 1, y = 2t, z = 1 - t$ and $x = -2 + s, y = -1 + s, z = 2s$ share a point?

Multivariable Functions — Practice Problems

Try these as you get the hang of the You Try It problems. Solutions to these problems are available in Sec. B.1.2.

(1) For $f(x, y) = \ln(x + y - 1)$, find $f(e, 1)$ and $f(4x, x + 1)$, and describe the domain and range of $f(x, y)$.
(2) Describe the domain and range of $f(x, y) = \sqrt[3]{x} + \sqrt[4]{y}$.
(3) Describe the domain and range of $f(x, y, z) = \ln(16 - 4x^2 - 4y^2 - z^2)$.
(4) Find the equations of the line segment starting at the point $(2, 5, -1)$ and ending at $(3, 3, 0)$.
(5) Describe the curve given by the parametric equations $x = 3 - t, y = 1 + 2t, z = -t$ for $-1 \leq t \leq 1$.
(6) Do the lines $x = 1 + 2t, y = 2 - t, z = -t$ and $x = -s, y = 1 - 2s, z = 1 + s$ share a point?

Multivariable Functions — Challenge Problems

Try these problems to test your skills with the ideas in this section. Solutions to these problems are available in Sec. C.1.2.

(1) Describe the domain and range of $f(x, y) = \sqrt{y - x}\ln(y + x)$; include a sketch of the domain.
(2) Give TWO different sets of parametric equations that produce the line segment starting at $(1, 1, 2)$ and ending at $(-2, 0, 2)$.
(3) Do the lines $x = t, y = t, z = t$ and $x = 2 - s$, $y = -1 + 2s$, $z = (s + 1)/2$ share a point?

Multivariable Functions — You Try It — Solved

---

(1) For $f(x, y) = x^2e^{3xy}$, find $f(2, 0)$ and $f(3x, 2y)$, and describe the domain and range of $f(x, y)$.

□ We have $f(2, 0) = 2^2e^{3\cdot 2\cdot 0} = 4$ and $f(3x, 2y) = (3x)^2e^{3\cdot 3x\cdot 2y} = 9x^2e^{18xy}$. Since we can use all real numbers for both x and y in this function, its domain is all of $\mathbb{R}^2$, i.e. the entire xy-plane. Only positive numbers and zero can result from this function, so the range is $z \geq 0$ (having named $z = f(x, y)$). ■

---

(2) Describe the domain and range of $f(x, y) = \sqrt{x + y}$.

□ We must have $x + y \geq 0$, i.e. $y \geq -x$. This is everything on and above the line $y = -x$ in the xy-plane. ■

---

(3) Describe the domain and range of $f(x, y, z) = \sqrt{1 - x^2 - y^2 - z^2}$.

□ We must have $1 - x^2 - y^2 - z^2 \geq 0$, i.e. $x^2 + y^2 + z^2 \leq 1$. This is the surface and interior of the unit sphere centered at the origin. ■

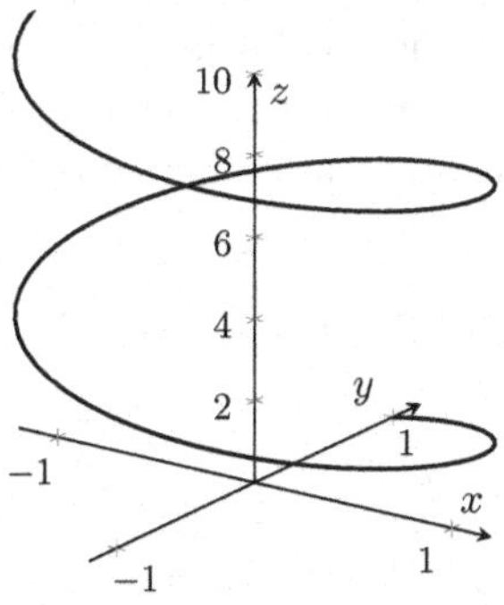

Fig. 13.6 The helix $x = \cos t, y = \sin t, z = t$.

(4) As mentioned above, the parametric equations $x = \cos t, y = \sin t$ form the unit circle. With no restrictions on t, so that t can be any real number, we trace the unit circle over and over. What do we get if we extend these equations to the 3D parametric curve $x = \cos t, y = \sin t, z = t$?

□ These parametric equations generate a *helix*, like DNA. In the xy-plane, the points rotate around the unit circle, but z is increasing as t increases. So we have circular cross sections continually elevating. Figure 13.6 shows this parametric curve. Do you know the direction of the curve? ■

(5) Find parametric equations form the line segment starting at the point $(0, -2, 5)$ and ending at $(\pi, -1, 5)$. Then give equations for the reverse segment, starting at $(\pi, -1, 5)$ and ending at $(0, -2, 5)$.

□ For the line going from $(0, -2, 5)$ to $(\pi, -1, 5)$, how about:

$$x = \pi t, y = -2 + t, z = 5 \qquad \text{for} \qquad 0 \leq t \leq 1$$

Then for the reverse, from $(\pi, -1, 5)$ to $(0, -2, 5)$, how about:

$$x = \pi - \pi t, y = -1 - t, z = 5 \qquad \text{for} \qquad 0 \leq t \leq 1$$ ■

(6) Do the lines $x = t-1, y = 2t, z = 1-t$ and $x = -2+s, y = -1+s, z = 2s$ share a point?

□ We can make the x and y coordinates match by forcing $t-1 = -2+s$ and $2t = -1+s$. From the first equation, we have $t = -1+s$. Plugging into the second, we get $2(-1+s) = -1+s$, or $s = 1$ ... and so also $t = 0$. Therefore, $t = 0$ and $s = 1$ make the x and y coordinates the same. Passing these values to the z-coordinates we get (1) $z = 1 - 0 = 1$ and (2) $z = 2(1) = 2$. The z-coordinate can not be made the same while the x and y coordinates are also the same. The lines do not share a point. ■

13.3 Three-Dimensional Surfaces

Introduction

Having encountered simple regions and lines in 3D, you're ready for more general three-dimensional surfaces. We'll start with the common ones, and then look at some ways to make their visualization a bit easier.

Upscaling of Equations

In Sec. 13.1, we asked what happens to the graphical "thing" described by $x^2+y^2=1$ when we upscale it from being considered as a 2D equation to a 3D equation. Note that there are two ways we can consider this upscaling:

(1) Keep the equation exactly as it is ($x^2+y^2=1$), and ask what it now represents in 3D.
(2) Extend the equation by introducing a term that seems appropriate for z ($x^2+y^2+z^2=1$), and ask what we now get in 3D.

Remember that an equation of a graphical thing tells you what all the points on it have in common.

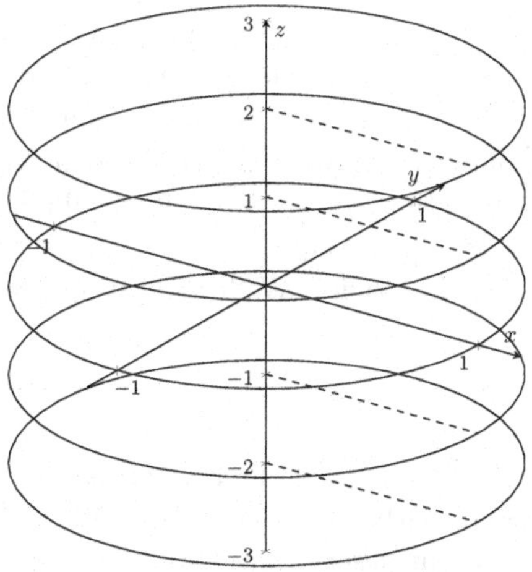

Fig. 13.7 Traces of $x^2+y^2=1$ for $z=-2,-1,0,1,2$.

Let's scale up $x^2+y^2=1$ using the first method: we'll keep the equation exactly as it is, but consider it as a 3D equation. The variable z does not appear in the equation, and yet the equation now represents a three-dimensional collection of points. Well, if z does not appear, then z is said to be *unrestricted* — that is, it can take on any value. But regardless of what z value we consider, the x and y coordinates of the points getting collected by the equation still have to obey the rule $x^2 + y^2 = 1$. In 2D, this rule gives it's a circle since it collects all points one unit from the origin. But in 3D, we now say, "The variable z can be anything, but for any one value of z we fix, we must collect all the points whose (x, y, z) coordinates obey $x^2 + y^2 = 1$". And so, at any z-coordinate, we must see a cross section in the x & y directions that is a circle of radius 1. When all the results from all values of z are assembled, we get a cylinder. Figure 13.7 shows several of these circular cross sections for different values of z; if you imagine now filling in the figure with all the other such cross sections for all possible values of z, you will (hopefully) imagine the formation of an entire cylinder.

Note that since we have three variables to play with, we can also create the similar equations $y^2 + z^2 = 1$ and $x^2 + z^2 = 1$. These are also infinitely long cylinders of radius 1; the equation $y^2 + z^2 = 1$ is a cylinder centered on the x-axis, and $x^2 + z^2 = 1$ is a cylinder centered on the y-axis. Make sure you understand why.

Now let's scale up $x^2 + y^2 = 1$ using the second method: we will add to the equation a term for z that continues the pattern already displayed by x and y, and get $x^2 + y^2 + z^2 = 1$. All points collected by this equation are the points that obey the stated rule, so (recognizing that this is just the distance formula in action) we are looking for all points whose distance from the origin, $x^2 + y^2 + z^2$, is equal to 1 — that is, we have the sphere of radius 1 centered at the origin. We'll look at spheres in general down below.

You Try It

(1) What do $y = x^2$ and $x = z^2$ look like in 3D?
(2) Describe the 3D surface $y^2 + 4z^2 = 4$. Give identifying information that supports your description.

Planes

We have already considered the "upscaling" of another 2D equation, $ax + by + c = 0$. Remember that in 2D this represents a line. In this case, both methods of upscaling result in a plane — in 3D, the specific equation $ax + by + c = 0$, and also the more generally extended $ax + by + cz + d = 0$ represent planes. The difference is that a plane of the form $ax + by + c = 0$ will be parallel to the z-axis, but $ax + by + cz + d = 0$ can be any plane. Do you see why?

If you're still reading a couple of chapters after this, where we introduce vectors, we'll revisit the subject of planes and how to describe them very efficiently. For now, we'll just consider some basics to get started, since we'll need to know about planes in general quite often.

One thing we often need to know about a plane is how it is oriented. A quick way to determine the orientation of a plane is to ask about its three intercepts (x, y, and z). This is easy to do.

EX 1 What are the x, y, and z intercepts of the plane $3x - 2y + z - 6 = 0$?

The x intercept occurs where the y and z coordinates are both 0. In this case, the equation gives $3x - 6 = 0$, or $x = 2$. So the x intercept is the point $(2, 0, 0)$. The y intercept occurs where the x and z coordinates are both 0. In this case, the equation gives $-2y - 6 = 0$, or $y = -3$. So the y intercept is the point $(0, -3, 0)$. The z intercept occurs where the x and y coordinates are both 0. In this case, the equation gives $z - 6 = 0$, or $z = 6$. So the z intercept is the point $(0, 0, 6)$. Knowing that this plane must hit these three points, we can get a good idea of its orientation. Figure 13.8 shows these three points of intersection and an indication of how the plane slopes through the first octant. ■

Just like it takes two points to uniquely define a line in 2D, it also only takes two points to uniquely define a line in 3D, and it takes 3 points to uniquely define a plane. For example, if you know the three intercepts of a plane, you can determine its equation.

EX 2 Find the equation of the plane whose x, y, and z intercepts are $(-1, 0, 0)$, $(0, 2, 0)$ and $(0, 0, 4)$.

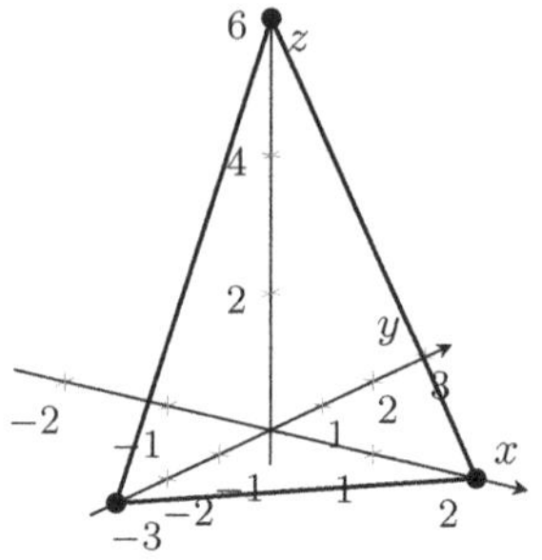

Fig. 13.8 Axis intercepts of the plane $3x - 2y + z - 6 = 0$.

Each point must obey the rule stated in the equation of the plane. In general, the equation of a plane is $ax + by + cz + d = 0$. Plugging each point into this general form and simplifying, we get:

$$a(-1) + b(0) + c(0) + d = 0 \quad \to \quad -a + d = 0$$
$$a(0) + b(2) + c(0) + d = 0 \quad \to \quad 2b + d = 0$$
$$a(0) + b(0) + c(4) + d = 0 \quad \to \quad 4c + d = 0$$

Here we have 3 equations, but 4 unknowns. That sounds bad, but it's actually good! The extra unknown is unrestricted, and can be anything. Let's let d be the "extra" unknown, and choose it to be a convenient value — such as 4. Then:

- from $-a + d = 0$, we get $-a + 4 = 0$, or $a = 4$.
- from $2b + d = 0$, we get $2b + 4 = 0$, or $b = -2$.
- from $4c + d = 0$, we get $4c + 4 = 0$, or $c = -1$.

These values create the equation of this plane: $4x - 2y - z + 4 = 0$. It's common to rewrite this as $z = 4x - 2y + 4$. FFT: Would we get an equally valid equation of the plane if we selected a different value of d to use to start things off, such as $d = 12$? Why or why not? ■

Note that the three points on the plane you know do not have to be intercepts, and we do exactly the same thing to find the equation of the plane:

EX 3 Find the equation of the plane containing the points $(1, 1, 2)$, $(0, -2, 1)$ and $(2, -1, 2)$.

Each point must obey the rule stated in the equation of the plane. Plugging each point into the general form for a plane and simplifying, we get:

$$\begin{aligned} a(1) + b(1) + c(2) + d = 0 \quad &\rightarrow \quad a + b + 2c + d = 0 \\ a(0) + b(-2) + c(1) + d = 0 \quad &\rightarrow \quad -2b + c + d = 0 \\ a(2) + b(-1) + c(2) + d = 0 \quad &\rightarrow \quad 2a - b + 2c + d = 0 \end{aligned}$$

Again, we have 3 equations, but 4 unknowns. Let's let d be the "extra" unknown, and choose it to be a convenient value — such as -2. With $d = -2$, we then get the three equations

$$\begin{aligned} a + b + 2c &= 2 \\ -2b + c &= 2 \\ 2a - b + 2c &= 2 \end{aligned}$$

Do whatever it is you know to do to solve a system of equations like this. You should get $a = -4/7$, $b = -2/7$, and $c = 10/7$, so that the equation of the plane becomes

$$-\frac{4}{7}x - \frac{2}{7}y + \frac{10}{7}z - 2 = 0$$

or multiplying by 7 and dividing by 2,

$$-2x - y + 5z - 7 = 0$$

We might also rewrite this as

$$z = \frac{2}{5}x + \frac{1}{5}y + \frac{7}{5} \quad \blacksquare$$

You Try It

(3) What are the x, y, and z intercepts of the plane $x+3y-2z-10 = 0$?

(4) Find the equation of the plane containing the points $(-1,-1,1)$, $(1,0,3)$ and $(2,2,2)$.

Spheres

You already know that the equation $x^2 + y^2 + z^2 = 1$ describes the unit sphere, i.e. the sphere of radius 1 centered at the origin. We could generalize this to $x^2 + y^2 + z^2 = r^2$, which describes a sphere of radius r centered at the origin. That's easy. But think back to your experience with circles. What if a circle is not centered at the origin? It's equation looks like $(x - x_0)^2 + (y - y_0)^2 = r^2$, and this is a circle of radius r with center (x_0, y_0).

So what if we wanted a sphere with a center at any point (x_0, y_0, z_0)? You might guess that its equation would be $(x-x_0)^2+(y-y_0)^2+(z-z_0)^2=r^2$ ... and you'd be right!

Now here's the annoying part. Remember that in 2D, the equation of a circle doesn't always come in that form. Often, it comes in "standard" form, $x^2+y^2+cx+dy+f=0$. We know that this is a circle, but we don't know its radius or its center yet. In order to find that center and radius, we had to .... sigh ... complete the square on the x and y terms. The same thing happens with spheres now.

Useful Fact 13.2. *The standard form of the equation of a sphere in three dimensions is:*

$$x^2+y^2+z^2+cx+dy+fz+g=0$$

By completing the square on x, y, and z terms, we can rewrite this as

$$(x-x_0)^2+(y-y_0)^2+(z-z_0)^2=r^2$$

which displays the center $(x_0, y_0, z_0$ and radius r.

EX 4 Find the center and radius of the sphere $x^2+y^2+z^2-2x+4y=0$.

Remember that to complete the square, we group terms of the same variable, like this:

$$(x^2-2x)+(y^2+4y)+(z^2)=0$$

Then we take half of the coefficient of each linear term if present, square it, and add it to both sides of the equation:

$$(x^2-2x+1)+(y^2+4y+4)+(z^2)=1+4$$

By doing so, we've generated perfect squares:

$$(x-1)^2+(y+2)^2+(z-0)^2=5$$

And now we know that the sphere has center $(1, -2, 0)$ and radius $\sqrt{5}$. ■

You Try It

(5) Find the center and radius of the sphere $x^2+y^2+z^2-6x+4y-2z=11$.

Quadric Surfaces

In the land of two dimensions, the equation $ax^2 + by^2 + cx + dy + f = 0$ can generate several types of curves: ellipses, parabolas, and hyperbolas (remember, circles are special ellipses). It all depends on the size and sign of the coefficients. Remember that if $a = b$, and both are positive, you'll probably get a sphere. If a and b are opposite in sign, you can get a hyperbola. If a or b is zero, you'll get a parabola, and so on. These are all called *conic sections*, since each shape can be formed by passing a plane through a full cone (with both halves) either straight or at an angle. Appendix A.1 has a quick summary of the common quadric surfaces. When you look there, don't freak out about things you have not encountered yet, such as the parametric representations of the surfaces. That will come in time.

When we generalize that concept to 3D, we get *quadric surfaces.* They are generated by the equation $ax^2 + by^2 + cz^2 + dx + fy + gz + h = 0$ (just take the general conic section equation and toss in the z's). Just like with conic sections, these surfaces are formed by various combinations of size and sign of the coefficients a, b, and c in that standard form. There are many more types of quadric surfaces than there are conic sections, and Appendix A.2 has descriptions of several of the common ones. There are ellipsoids, hyperboloids, paraboloids, hyperbolic paraboloids, hyperboloids of one and two sheets, elliptic paraboloids, etc etc etc. While you can get by memorizing the conic sections and their properties, it is fruitless to attempt that with quadric surfaces. That would make your brain hurt. So instead, we learn how to determine properties and appearances of quadric surface by visualize their cross sections, or *traces.* It's kind of like taking an x-ray of a 3D surface.

Like conic sections, quadric surfaces can be aligned along any of the axes, or rotated away from an axis. While a sphere looks the same with respect to any axis, you can imagine a cylinder whose centerline is the x-axis looks different than one whose centerline is the y-axis, or the z-axis. For our purposes, alignment along different axes is the worst variations we will have — we won't see quadric surfaces oriented in random directions. I'm sorry for the disappointment.

Cross Sections and Traces

Let's pick a random quadric surface, $x^2 - y^2 + z^2 + 4 = 0$. We know this isn't a sphere because the coefficient of y^2 is negative. So what is it? We can visualize it via cross sections, which are formed by taking *traces* in certain planes. Once we resolve cross sections in each coordinate direction, we should have a pretty good idea of the overall structure of the surface.

- The intersection of this surface with the plane $z = 1$, called the *trace* of the surface in that plane, is found by plugging in $z = 1$: we get $x^2 - y^2 + 5 = 0$. Since we are good with conic sections, we know that this equation gives a hyperbola, so we say that the trace (cross section) of $x^2 - y^2 + z^2 + 4 = 0$ in the plane $z = 1$ is the hyperbola $x^2 - y^2 + 5 = 0$. (Fig. 13.9 shows this trace.) Perhaps you can also realize that not matter what plane $z = c$ we choose, cross sections of this surface will be hyperbolas.
- If we fix a value of x, thus finding the trace in some plane $x = c$, we'll get a hyperbola in the yz direction. Do you see why? The equation will looks like $z^2 - y^2 + (4 + c^2) = 0$. (Fig. 13.9 shows the trace of this surface in the plane $x = 2$.)

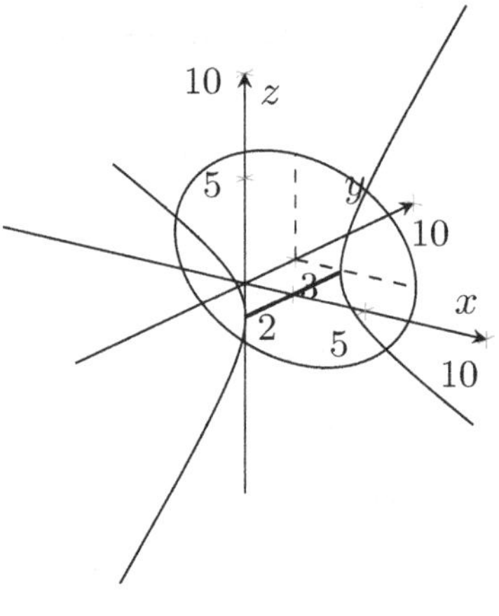

Fig. 13.9 Traces of $x^2 - y^2 + z^2 + 4 = 0$ in planes $x = 2$ and $y = 3$.

- If we fix a value of y, thus finding the trace in some plane $y = d$, we'll see circular cross sections in the xz direction, since the equation will now look like $x^2 + z^2 + (4 - d^2) = 0$. (Fig. 13.9 shows the trace of this surface in the plane $y = 3$.)

Finally, there's one extra thing you may have noticed: since we can rearrange the equation to get $x^2 + z^2 = y^2 - 4$; this equation will only be valid when the right hand side is non-negative (do you see why?), so we must have $y^2 \geq 4$. Therefore, this surface simply does not exist for values of y between -2 and 2.

To summarize, we have a surface that has hyperbolic cross sections in two directions, circular cross sections in the other, and is split into two pieces (since nothing's there for $-2 < y < 2$). If you go back to the link containing all the pictures, I hope you'll agree that the only shape following these properties is a hyperboloid of two sheets. Further, we know it opens (is symmetric) around the y-axis, since the y-axis is where the big gap is.

Other quadric surfaces might have easier traces (cross sections). For example, the traces of $x^2 + 2y^2 + z^2 - 1 = 0$ are ellipses in planes $x = c$ or $z = c$ and are circles in planes $y = c$. (Be sure you see why.) And therefore, by either just good personal visualization, or by searching the options in Appendix A.2, we determine that this quadric surface is an ellipsoid.

In general, there's no one specific system for investigating traces, you just have to be clever. Try finding traces in the coordinate planes. Find out if traces of the surface don't exist for certain values of a variables (thus showing that the surface has more than just one piece).

EX 5 Identify the surface $x^2 = 2y^2 + 3z^2$ using traces. Provide at least one trace parallel to each each of the xy-, yz-, and xz-planes.

- In the coordinate planes $y = 0$ and $z = 0$, the traces are lines. Do you see why?
- In the coordinate plane $x = 0$ (the yz-plane), this surface only exists at the point (0,0,0).
- In any plane $x = c$, traces look like $2y^2 + 3z^2 = C$, which are ellipses.
- In any plane $y = c$ or $z = c$, traces look like $x^2 - 3z^2 = C$ or $x^2 - 2y^2 = C$, which are hyperbolas.

Figures 13.10 and 13.11 show some vertical and horizontal traces for this surface. By either good personal visualization, or by searching the options in Appendix A.2, we determine that this quadric surface can only be a cone, opening along the x-axis. ■

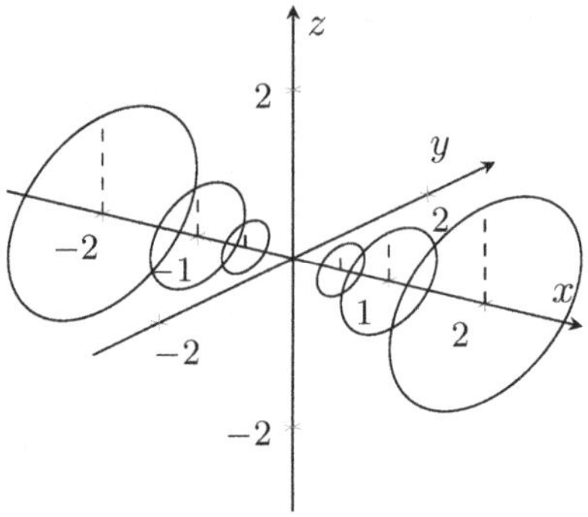

Fig. 13.10 Traces of $x^2 = 2y^2 + 3z^2$ in planes $x = -2, -1, -0.5, 0.5, 1, 2$.

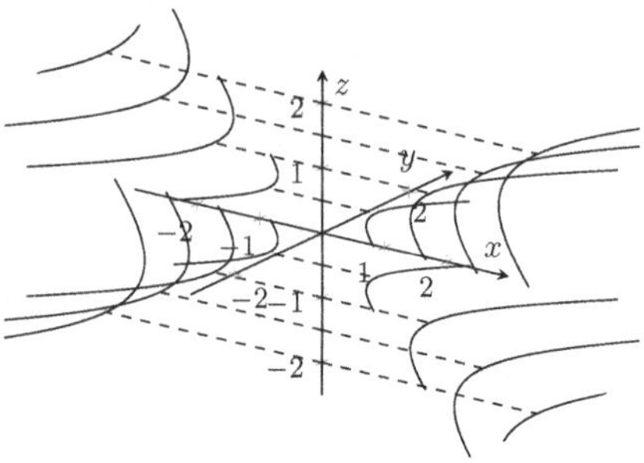

Fig. 13.11 Traces of $x^2 = 2y^2 + 3z^2$ in planes $z = -2, -1.5, \ldots, 2$.

You Try It

(6) Identify the surface $4x^2 + 9y^2 + 36z^2 = 36$ using traces. Provide at least one trace parallel to each each of the xy-, yz-, and xz-planes.

Fig. 13.12 Sample topographic contours from a USGS topographic map.

Contour Plots

If you're like me, your skills at drawing three-dimensional objects are limited. We are lucky when we have a function whose graph becomes part of a quadric surface, but that is not always the case. The surface described by just some random function $z = f(x, y)$ that shows up on your doorstep can be much more complicated than a simple quadric surface. Fortunately, we have a method by which we can represent the dynamics of a general 3D function in the 2D plane: contour plots.

Whether you know it or not, you have been looking at contour plots all your life. If you have ever seen a topographic map, a weather map with isobars, or a bathymetric map (any boaters out there?) then you have seen a contour map. A topographic map is a contour plot reflecting elevation as a function of latitude and longitude. We connect all the points on a 2D map that share some common elevation (z value). If you've ever hiked using a topo map as a guide, you know that tightly packed elevation contours mean either a steep ascent or descent. Figure 13.12 shows a portion of a topographic map (elevation in feet) from a topographic map.[1] Did you know that the United States Geological Survey has such topographic maps at different scales for all areas in the US? You should look for a map containing your home town! You might be surprised by the topography in the area where you live.

On many weather maps you've seen all your life, contours connect points on a 2D map that share the same value of some property; when the data displayed is air pressure, the contours are called *isobars*; the more tightly packed the isobars, the stronger the wind will be, since the density of isobars is related to the pressure gradient. Surely you've seen a temperature plot that shows temperature as a function of position on a map (where red represents hotter and blue represents cooler) — sometimes contour plots have colormaps associated with them, and sometimes the contours are just curvy lines. Either way, we have a two-dimensional display of three-dimensional data.

Figure 13.13 shows a display of magnetic declination around the globe; the three-dimensional globe (sorry, flat-earthers) is projected onto a two-dimensional region, and then within that region, we see contours representing magnetic declination — which is a measure of the angular difference between geographic north and magnetic north.[2]

[1] `https://pubs.usgs.gov/gip/topomapping/topo.html`
[2] U.S. Geological Survey Open-File Report 99-0418, by Saltus *et al.*

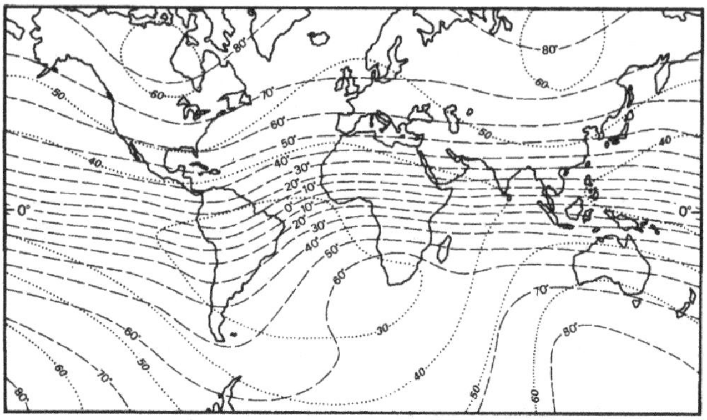

Fig. 13.13 A contour map showing magnetic declination (USGS).

Here is a very generic and dry approach to contour plots (sorry). We take a function $z = f(x, y)$ and ask: what set of (x, y) values produce the value $z = 1$? Or $z = 2$? Or any other chosen value of z? We identify all those points in the xy-plane and sketch all those points together; the resulting contour gets labelled by the associated value of z. When we generate such contours for several values of z and play connect the dots on the resulting points in the plane, we generate a nest of contours which together become the contour plot.

As an easy example, consider the paraboloid $z = x^2 + y^2$ (if you want to know what this looks like, think about traces). The collection of (x, y) values that yield $z = 1$ are the points satisfying $x^2 + y^2 = 1$, or the unit circle. So the *contour* corresponding to $z = 1$ is the unit circle. You can probably see that the contour corresponding to any $z = c$ is the circle $x^2 + y^2 = c$. A contour plot of this function would consist of nested circles, each circle labeled with the appropriate z value. Slightly more interesting might be the contour plot for $z = 2x^2 + 3y^2$, in which contours are ellipses. Figure 13.14 shows a contour plot for that function.

You should be able to visualize the contour plot associated with any quadric surface. I'll describe the way I see it, and hope it makes sense. Imagine the full 3D quadric surface as an inflated balloon sitting on a piece of paper, and pretend you have a paintbrush. Trace with your brush several solid lines around the outside of the quadric surface, each at a constant

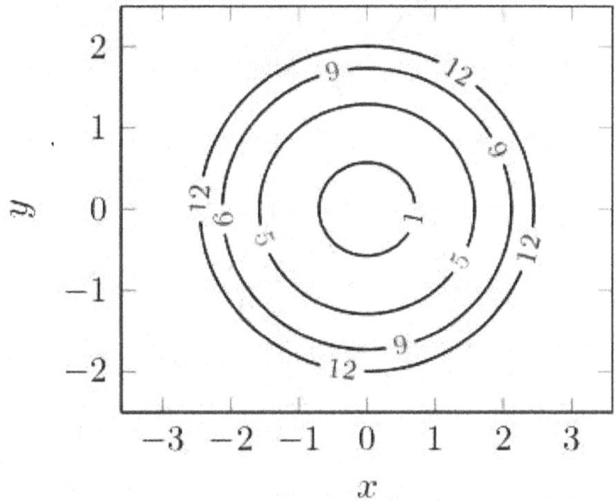

Fig. 13.14 Contour plot for $z = 2x^2 + 3y^2$ (for $z = 1, 5, 9, 12$).

elevation on the surface (at $z = 0$, $z = 1$, etc.). Then squish the surface flat. The pattern all the wet paint from the surface will make on the piece of paper is the contour map. Some generalities about contour maps are:

- The more tightly packed the lines, the steeper the surface
- If you move in the direction of increasing contour level values, you are moving uphill on the surface
- If you move in the direction of decreasing contour level values, you are moving downhill on the surface

EX 6 Does the contour plot in Fig. 13.15 represent $z = x + \cos(y)$ or $z = y + \cos(x)$? Why? (And how can we tell if there are no labels? What the heck!)

The two candidate functions would produce contours from the equations $x + \cos(y) = c$ or $y + \cos(x) = c$, respectively (where c is a constant). If we rearrange these equations, they are either $x = c - \cos(y)$ or $y = c - \cos(x)$. Remembering that $\cos(x)$ and $\cos(y)$ are always between -1 and 1, then contours of the first candidate function would oscillate around vertical lines $x = c$ by an amount between -1 and 1 that varies in the y-direction. Similarly, contours of the second candidate function would oscillate around horizontal lines $y = c$ by an amount between -1 and 1 that varies in the

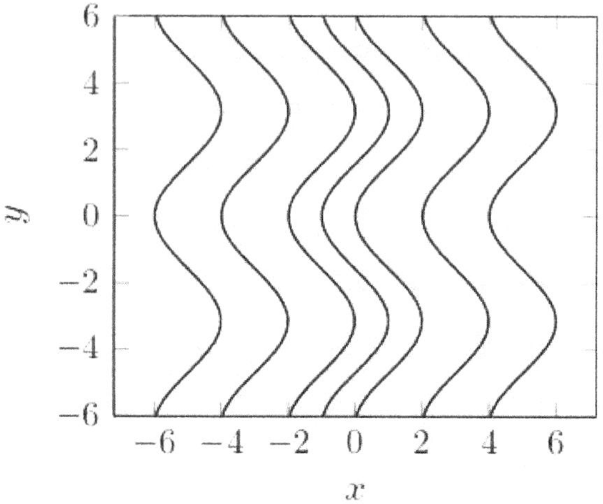

Fig. 13.15 Contours for $z = x + \cos(y)$ or $z = y + \cos(x)$?

x-direction. The contour map itself matches the first description, so the contours are for the function $z = x + \cos(y)$. ■

You Try It

(7) Does the contour plot in Fig. 13.16 represent $z = x^2 + y^2$ or $z = x^2 - y^2$? Why? (This figure is at the end of the section.)

Level Curves and Level Surfaces

Oh, no, another topic! Well, not really. A level curve of a function $f(x, y)$ is the collection of all members from its domain that produce the same output. For a function $f(x, y)$, a level curve is a contour. If $f(x, y) = x^2 + 2y^2$, then the level curve corresponding to $c = 2$ is the ellipse $x^2 + 2y^2 = 2$. The reason we're looking at this alternate terminology is to prepare us for later on, when we'll use the idea of a *level surface*. A level surface of a function $f(x, y, z)$ is the collection of all members from its domain that produce the same output.

EX 7 Describe the level surfaces of the function $w = x^2 + y^2 + z^2$ for the values $w = -1, 0, 1, 100$. Can you make an observation about what all the level surfaces of this function have in common?

First, note that the function $w = f(x, y, z)$ itself cannot be directly visualized, we would need 4 axes. The best we can do is say, "well, let's just look at all the points (x, y, z) that give a specific value of w.

- There is no level surface for $w = -1$ since there are no points (x, y, z) in $\mathbb{R}^3$ that yield $x^2 + y^2 + z^2 = -1$.
- The level surface for $w = 0$ is a single point — more specifically, the origin; we are looking for all points (x, y, z) in $\mathbb{R}^3$ which yield $x^2 + y^2 + z^2 = 0$. There is only one such point, $(x, y, z) = (0, 0, 0)$.
- The level surface for $w = 1$ is the unit sphere, $x^2 + y^2 + z^2 = 1$.
- The level surface for $w = 100$ is the sphere of radius 10 centered at the origin, $x^2 + y^2 + z^2 = 100$.
- Every level surface of $w = f(x, y, z)$ is a sphere centered at the origin. ■

We get level curves from functions $z = f(x, y)$ by collecting the points (x, y) from $\mathbb{R}^2$ that yield a specific (constant) value of z. We get level surfaces from functions $w = f(x, y, z)$ by collecting the points (x, y, z) from $\mathbb{R}^3$ that yield a specific (constant) value of z. Is there a level *thing* of some sort for functions $y = f(x)$?

You Try It

(8) Describe the following level curves:

- the level curve of $z = x^2 + 2y^2$ corresponding to $z = 3$.
- the level curve of $t = q - 4r$ corresponding to $t = 7$.
- the level curve of $x_3 = -3x_1^2 + x_2$ corresponding to $x_3 = 2$.

Three-Dimensional Surfaces — Problem List

Three-Dimensional Surfaces — You Try It

These appeared above; solutions begin on the next page.

(1) What do $y = x^2$ and $x = z^2$ look like in 3D?
(2) Describe the 3D surface $y^2 + 4z^2 = 4$. Give identifying information that supports your description.
(3) What are the x, y, and z intercepts of the plane $x + 3y - 2z - 10 = 0$?
(4) Find the equation of the plane containing the points $(-1, -1, 1)$, $(1, 0, 3)$ and $(2, 2, 2)$.
(5) Find the center and radius of the sphere $x^2 + y^2 + z^2 - 6x + 4y - 2z = 11$.
(6) Identify the surface $4x^2 + 9y^2 + 36z^2 = 36$ using traces. Provide at least one trace parallel to each each of the xy-, yz-, and xz-planes.
(7) Does the contour plot in Fig. 13.16 represent $z = x^2 + y^2$ or $z = x^2 - y^2$? Why?
(8) Describe the following level curves:
    - the level curve of $z = x^2 + 2y^2$ corresponding to $z = 3$.
    - the level curve of $t = q - 4r$ corresponding to $t = 7$.
    - the level curve of $x_3 = -3x_1^2 + x_2$ corresponding to $x_3 = 2$.

Three-Dimensional Surfaces — Practice Problems

Try these as you get the hang of the You Try It problems. Solutions to these problems are available in Sec. B.1.3.

(1) Find the equation of the plane containing the points $(1, 0, 1)$, $(0, 1, 1)$ and $(1, 1, 0)$.
(2) Identify the surface $x^2 + y^2 + z^2 = 4x - 2y$. Give at least two pieces of identifying information that distinguishes this surface from others of the same type.
(3) Find the equation of a sphere that has center $(3, 8, 1)$ and passes through the point $(4, 3, -1)$.
(4) Describe the 3D surface $x^2 - y^2 = 1$. Give identifying information that supports your description.
(5) Identify the surface $z = x^2 - y^2$ using traces. Provide at least one trace parallel to each each of the xy-, yz-, and xz-planes.
(6) Complete the squares on $4x^2 + y^2 + 4z^2 - 4y - 24z + 36 = 0$ and identify it using traces.

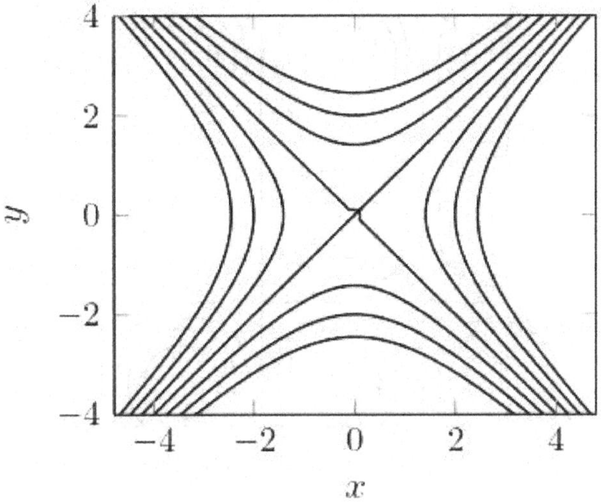

Fig. 13.16 Contours for $z = x^2 + y^2$ or $z = x^2 - y^2$?

(7) Identify the surface $z^2 = 4x^2 + 9y^2 + 36$ using traces. Provide at least one trace parallel to each each of the xy-, yz-, and xz-planes.
(8) Does the contour plot in Fig. 13.17 represent $z = \sin(x)\cos(y)$ or $z = \sin(y)\cos(x)$? Why?
(9) Present the level curves of $z = 3x^2 - 2y^2$ for $= -2, -1, 0, 1, 2$ as a 2D contour plot.

Three-Dimensional Surfaces — Challenge Problems

Try these problems to test your skills with the ideas in this section. Solutions to these problems are available in Sec. C.1.3.

(1) A plane shares x and y intercepts with the plane $2x + y - 3z - 4 = 0$ but has its own z intercept of $(0, 0, -2)$. What is the equation of this plane?
(2) Identify the surface $4x^2 + 4y^2 + 4z^2 - 8x + 16y = 1$. Give at least two pieces of identifying information that distinguishes this surface from others of the same type.
(3) Identify the surface $25y^2 + z^2 = 100 + 4x^2$ using traces. Provide at least one trace parallel to each each of the xy-, yz-, and xz-planes.

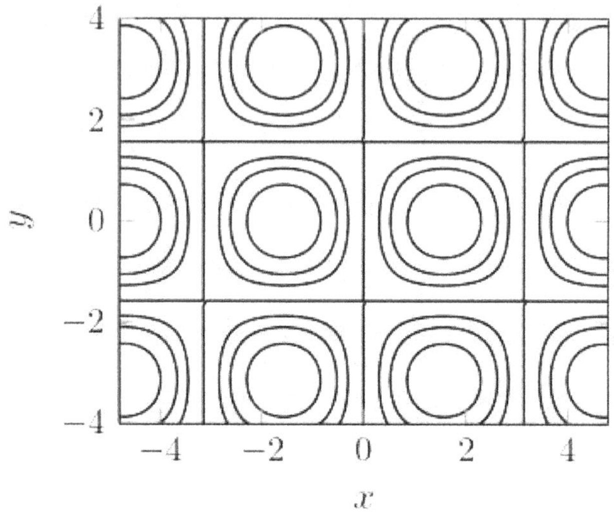

Fig. 13.17 Contours for $z = \sin x \cos y$ or $z = \sin y \cos x$?

(4) *(Bonus! This problem will connect to others in later sections.)* Find the level surface of $w = 2x^2 + y^2 + z$ corresponding to $w = 4$, and present that level surface as a contour plot showing the values $z = 0, 1, 2, 3$. (Thus, from a 4D hypersurface, we generate a 3D level surface, and for that we show several 2D level curves!)

Three-Dimensional Surfaces — You Try It — Solved

(1) What do $y = x^2$ and $x = z^2$ look like in 3D?

☐ In 2D, $y = x^2$ is a parabola. When we expand to 3D, the z coordinate is now unspecified and unrestricted. We can picture the parabola $y = x^2$ elongated along the z-axis. This is called a parabolic cylinder.

In the 2D xz-plane, $z = y^2$ is a parabola. When we expand to 3D, the x-coordinate is now unspecified and unrestricted. We get a parabolic cylinder elongated along the x-axis.

Use tech to plot these and see what they look like. ■

(2) Describe the 3D surface $y^2 + 4z^2 = 4$. Give identifying information that supports your description.

☐ In the 2D yz-plane, this is an ellipse. When we expand to 3D, the x-coordinate is now unspecified and unrestricted. We can picture the 2D ellipse $y^2 + 4z^2 = 4$ elongated along the x-axis. This is called an elliptic cylinder. ■

(3) What are the x, y, and z intercepts of the plane $x + 3y - 2z - 10 = 0$?

☐ The x intercept occurs where the y and z coordinates are both 0. In this case, the equation gives $x - 10 = 0$, or $x = 10$. The y intercept occurs where the x and z coordinates are both 0. In this case, the equation gives $3y - 10 = 0$, or $y = 10/3$. The z intercept occurs where the x and y coordinates are both 0. In this case, the equation gives $-2z - 10 = 0$, or $z = -5$. So the intercepts are $(10, 0, 0)$, $(0, 10/3, 0)$ and $(0, 0, -5)$. ■

(4) Find the equation of the plane containing the points $(-1, -1, 1)$, $(1, 0, 3)$ and $(2, 2, 2)$.

☐ Plugging each point into the general form for a plane and simplifying, we get:

$$\begin{aligned} a(-1) + b(-1) + c(1) + d = 0 \quad &\rightarrow \quad -a - b + c + d = 0 \\ a(1) + b(0) + c(3) + d = 0 \quad &\rightarrow \quad a + 3c + d = 0 \\ a(2) + b(2) + c(2) + d = 0 \quad &\rightarrow \quad 2a + 2b + 2c + d = 0 \end{aligned}$$

Again, we have 3 equations, but 4 unknowns. Let's let d be the "extra" unknown, and choose it to be a convenient value — such as -2. With $d = -2$, we then get the three equations

$$\begin{aligned} -a - b + c &= 2 \\ a + 3c &= 2 \\ 2a + 2b + 2c &= 2 \end{aligned}$$

The solution to this system is $a = -5/2$, $b = 2$, and $c = 3/2$, so that the equation of the plane becomes

$$-\frac{5}{2}x + 2y + \frac{3}{2}z - 2 = 0$$

or multiplying by 2,

$$-5x + 4y + 3z - 4 = 0$$

This can also be written as

$$z = \frac{5}{3}x - \frac{4}{3}y + \frac{4}{3} \quad ■$$

---

(5) Identify the surface $x^2 + y^2 + z^2 - 6x + 4y - 2z = 11$. Give at least two pieces of identifying information that distinguishes this surface from others of the same type.

□ To completing the square on the given equation, we have to group the variables and include "clever" terms on both sides:

$$\begin{aligned} &(x^2 - 6x + (-3)^2) + (y^2 + 4y + (2)^2) + (z^2 - 2z + (-1)^2) \\ &\qquad = 11 + (-3)^2 + (2)^2 + (-1)^2 \end{aligned}$$

which then allows clean-up and factoring,

$$\begin{aligned} (x^2 - 6x + 9) + (y^2 + 4y + 4) + (z^2 - 2z + 1) &= 11 + 9 + 4 + 1 \\ (x-3)^2 + (y+2)^2 + (z-1)^2 &= 25 \end{aligned}$$

So this is a sphere with center $(3, -2, 1)$ and radius 5. ■

---

(6) Identify the surface $4x^2 + 9y^2 + 36z^2 = 36$ using traces. Provide at least one trace parallel to each each of the xy-, yz-, and xz-planes.

□ In the plane $x = 0$, we get the trace $9y^2 + 36z^2 = 36$, which is an ellipse. In the plane $y = 0$, we get the trace $4x^2 + 36z^2 = 36$, another ellipse. The trace in the plane $z = 0$ is also an ellipse. In any plane $x = c$, we get

$$\begin{aligned} 4c^2 + 9y^2 + 36z^2 &= 36 \\ 9y^2 + 36z^2 &= 36 - 4c^2 \end{aligned}$$

which is an ellipse only defined for $|c| \le 3$. The trace in the plane $y = c$ looks like

$$\begin{aligned} 4x^2 + 9c^2 + 36z^2 &= 36 \\ 4x^2 + 36z^2 &= 36 - 9c^2 \end{aligned}$$

which is an ellipse only defined for $|c| \le 2$. The trace in the plane $z = c$ looks like

$$\begin{aligned} 4x^2 + 9y^2 + 36c^2 &= 36 \\ 4x^2 + 9y^2 &= 36 - 36c^2 \end{aligned}$$

which is an ellipse only defined for $|c| \le 1$. Since all traces are ellipses, this is an ellipsoid. ■

---

(7) Does the contour plot in Fig. 13.18 represent $z = x^2+y^2$ or $z = x^2-y^2$? Why?

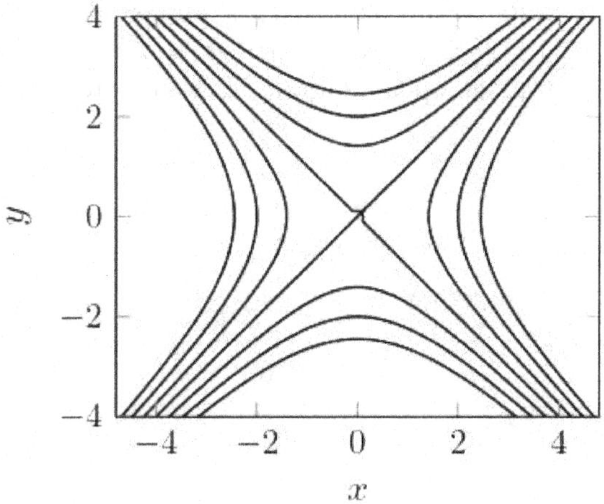

Fig. 13.18 Contours for $z = x^2 + y^2$ or $z = x^2 - y^2$?

□ Contours of the function $z = x^2 + y^2$ are of the form $x^2 + y^2 = c$, i.e. circles. Contours of the function $z = x^2 - y^2$ are of the form $x^2 - y^2 = c$, i.e. hyperbolas. The contour plot shows hyperbolic contours, so the plot represents $z = x^2 - y^2$. Labels of the contours have been added here. ■

(8) Describe the following level curves:

- the level curve of $z = x^2 + 2y^2$ corresponding to $z = 3$.
- the level curve of $t = q - 4r$ corresponding to $t = 7$.
- the level curve of $x_3 = -3x_1^2 + x_2$ corresponding to $x_3 = 2$.

□ The level curve of $z = x^2 + 2y^2$ corresponding to $z = 3$ is the collection of all points (x, y) in $\mathbb{R}^2$ that yield $x^2 + 2y^2 = 3$. This is an ellipse, going through $(\pm\sqrt{3}, 0)$ and $(0, \pm\sqrt{3/2})$.

The level curve of $t = q - 4r$ corresponding to $t = 7$ is the collection of all points (q, r) from $\mathbb{R}^2$ (still a 2D plane, just with different variable names — let's assume assume q is horizontal, r is vertical) that yield $q - 4r = 7$. This is a line, with intercepts $(q, r) = (7, 0)$ and $(0, -7/4)$.

The level curve of $x_3 = -3x_1^2 + x_2$ corresponding to $x_3 = 2$ is the collection of all points (q, r) from $\mathbb{R}^2$ (still a 2D plane, just with different variable names — let's assume assume x_1 is horizontal, x_2 is vertical) that yield $-3x_1^2 + x_2 = 2$. Since we can arrange this as $x_2 = 2 + 3x_1^2$, we are looking at a parabola, opening upwards, with vertex $(x_1, x_2) = (0, 2)$.

■

13.4 Limits of Multivariable Functions

Introduction

Now that you are (hopefully) somewhat comfortable with the basics of multivariable functions, we can proceed through the same sequence of concepts that we went through earlier for single variable functions. You may not have realized it at the time, but the entirety of calculus is based on the idea of a limit. Derivatives and definite integrals are both defined using limits. So without further ado, let's examine limits of multivariable functions.

Notation, Meaning, and the Easy Ones

For a single variable function $f(x)$, we use the notation

$$\lim_{x \to a} f(x) = L$$

This means that as a the independent variable x (in the domain of $f(x)$) approaches a target value a, the value of the function (in the range) is approaching a corresponding target L. When the limit is easily evaluated and there are no surprises, the problem is boring. All the fun stuff happens in the strange cases where there are asymptotes, infinities, or other pathologies in the function.

Let's upgrade this idea for a function of two independent variables, $z = f(x, y)$. If we take the notation away, the idea of a limit does not change: as a member of the domain of the function approaches some target value, the value of the function is approaching its own target. But when we take a closer look, things start to get very interesting than before. For one thing, the domain of a function of two variables is part to all of the xy-plane, so a member of the domain is a point (x, y), not just x or y alone. Therefore, we are interested in what happens to the function's output value z as (x, y) approaches some target, say (a, b), and we write

$$\lim_{(x,y) \to (a,b)} f(x, y) = L$$

In the simplest cases, this still holds no surprises. For example, without any further coaching, it probably makes sense to you that

$$\lim_{(x,y) \to (1,2)} x^2 + y^2 = 5$$

or that

$$\lim_{(x,y) \to (\pi/2,0)} \frac{\cos y}{\sin x} = 1$$

When you tackled single variable limits for the first time, you learned all sorts of limit rules like

$$(a)\ \lim_{x \to a} (c \cdot f(x)) = c \lim_{x \to a} f(x)$$

$$(b)\ \lim_{x \to a} (f(x) \pm g(x)) = \lim_{x \to a} f(x) + \lim_{x \to a} g(x)$$

$$(c)\ \lim_{x \to a} f(x) \cdot g(x) = \lim_{x \to a} f(x) \cdot \lim_{x \to a} f(x)$$

$$(d)\ \lim_{x \to a} x^n = \left(\lim_{x \to a} x\right)^n$$

$$(e)\ \lim_{x \to a} \frac{f(x)}{g(x)} = \frac{\lim_{x \to a} f(x)}{\lim_{x \to a} g(x)} \text{ as long as } \lim_{x \to a} g(x) \neq 0$$

We can expect the multivariable analogs to hold as well, but note that these are *organizational* rules of limits: they show how we can simplify or rearrange limits, not what the actual value of any one limit is. For limit rules that actually demonstrate how to predict limit values, we have to pick those up individually as we look into specific functions.

One helpful rule from Chapter 2 that delivered actual values for limits said that if $f(x)$ is a polynomial or rational function and $x = a$ is in the domain of $f(x)$, then

$$\lim_{x \to a} f(x) = f(a)$$

In other words, the limit of such a function at a given target point is just what we get when we plug $x = a$ in to the function. The multivariable analog of this continues to hold, and in fact, we'll be quite cavalier in its implementation. We'll even extend it to more general functions by posing this rule of thumb: "If you have a pretty good guess as to what a limit is, just by examining the value of the function at the limit point, that guess is likely correct". It is rarely technically proper to claim that you have evaluated a limit by "plugging in a value". But we're all friends here, nod nod wink wink, and so we can propose this mathematical agreement among friends:

Useful Fact 13.3.

- *(Single Variable) If you want to evaluate the limit of $f(x)$ as x approaches a, just test the target value $x = a$ in the function $f(x)$. If no problems occur, and you get a perfectly good answer L, then odds are L is the limit you're looking for.*

- *(Multivariable) If you want to evaluate the limit of $f(x,y)$ as (x,y) approaches (a,b), just test the target value(s) $(x,y)=(a,b)$ in the function $f(x,y)$. If no problems occur, and you get a perfectly good answer L, then odds are L is the limit you're looking for.*

Put another way, when functions behave themselves, their limits are not surprising.

EX 1 Investigate the limit $\lim_{(x,y)\to(1,2)} \dfrac{x+y}{x^2+y^2}$.

We should not evaluate limits by plugging in values. But just between friends, let's try it anyway. At $(x,y)=(1,2)$,

$$\frac{x+y}{x^2+y^2}=\frac{1+2}{1^2+2^2}=\frac{3}{5}$$

Since the function is equal to $3/5$ at $(x,y)=(1,2)$, chances are that as (x,y) *approaches* $(1,2)$, our function is approaching $3/5$. Let's take a leap and predict that

$$\lim_{(x,y)\to(1,1)} \frac{x+y}{x^2+y^2}=\frac{3}{5} \quad \blacksquare$$

You Try It

(1) Investigate the limit $\lim_{(x,y)\to(5,2)} (x^3-y^2+\sqrt{x-2y})$.

Would an increase in the number of variables have any significant effect on what we do?

You Try It

(2) Investigate the limit $\lim_{(x,y,z)\to(1,-1,1)} \dfrac{xy}{1+z}$.

These have been examples of the easy limits. They are no fun. Let's get to the interesting stuff.

The Interesting Limits and Path Independence

As in the single variable case, the more interesting limit problems are the ones in which strange things happen at or near the limit point — such as when indeterminate forms like $0/0$ turn up.

The danger of using our little mathematical agreement between friends to investigate limits is that we can't turn it around and use it to conclude opposite outcomes. In other words, suppose we are given a function $f(x)$ or $f(x, y)$, and we start investigating the limit by just plugging in the target point $x = a$ or $(x, y) = (a, b)$. If we get a good value back, that value is likely the correct limit value. But if we FAIL to get a good value back, we *cannot* conclude that the limit does not exit.

Let me repeat that again, because this is one of the most common sources of error by students at this stage of calculus: *you cannot say a limit doesn't exist just because things go bad when you plug in the limit point.* Failure of a function to exist at a limit's target point simply means you have to take a closer look at what's going on. Do you remember this limit?

$$\lim_{x \to 0} \frac{\sin x}{x}$$

If we are on autopilot, we might conclude that since the denominator of this function is zero at $x = 0$ — and so the function is not defined at $x = 0$ — then the limit must not exist. But that's false! This is actually a perfectly good limit, and it is equal to 1.

One of the first strategies to employ when you encounter a function which appears to be undefined at the limit point is to see if you can algebraically simplify the function:

EX 2 Investigate the limit $\displaystyle\lim_{(x,y)\to(1,1)} \frac{x-y}{x^2-y^2}$.

If we try to just plug in $(x, y) = (1, 1)$ we get $0/0$. This does not mean the limit doesn't exist yet. If we simplify the function, we can actually see that

$$\lim_{(x,y)\to(1,1)} \frac{x-y}{x^2-y^2} = \lim_{(x,y)\to(1,1)} \frac{x-y}{(x+y)(x-y)} = \lim_{(x,y)\to(1,1)} \frac{1}{x+y} = \frac{1}{2} \quad \blacksquare$$

You Try It

(3) Investigate the limit $\displaystyle\lim_{(x,y)\to(0,0)} \frac{x^2+y^2}{\sqrt{x^2+y^2+1}-1}$. (Hint: Rationalize the denominator ... come on now, you're reading about multivariable calculus, you can't let a phrase like "rationalize the denominator" intimidate you!)

But algebraic simplification won't always solve our problems. Another strategy to apply is to consider the idea of *path independence.* Think back to the idea of left hand and right hand limits for single variable functions. As $x \to a$, there are only two ways x can approach a: from the left or from the right. And for a limit of some function $f(x)$ to exist as $x \to a$, the left and right hand limits must agree. This is why, for example, the limit

$$\lim_{x \to 0} \frac{|x|}{x}$$

does NOT exist: the left hand limit is -1 and the right hand limit is 1.

The idea of left and right hand limits does scale up to multivariable functions, but it gets a lot worse! One the real line, there are only two directions in which x can approach a: from the left, and from the right. But suppose we have a multivariable limit in which the target point is $(x, y) \to (1, 2)$. This approach to $(1, 2)$ happens in the xy-plane. From how many directions can (x, y) approach $(1, 2)$? Here are some possibilities:

- (x, y) can approach $(1, 2)$ along the vertical line $x = 1$, from above or below.
- (x, y) can approach $(1, 2)$ along the horizontal line $y = 2$, from the left or the right.
- (x, y) can approach $(1, 2)$ along the line $y = 2x$, up from the lower left or down from the upper right.
- (x, y) can approach $(1, 2)$ along the curve $y = 2 - 2(x - 1)^2$, from the left or the right. (Paths of approach do not have to be straight lines!)

(These paths are included in Fig. 13.19, with others.) Untimately, there are *infinitely many* directions of approach to the point $(1, 2)$ in the xy-plane! And in order for a limit

$$\lim_{(x,y) \to (1,2)} f(x, y)$$

to exist, we have to get the same value of the limit *from every single possible direction of approach.* Now, how feasible is it to check a limit along an infinite number of paths of approach? Not very. It's impossible to exhaustively check all possible directions of approach of (x, y) to (a, b). We actually never show that a limit exists by investigating different paths of approach. Rather, this idea is used in the reverse sense: we can show a limit does NOT exist by finding two directions of approach to our limit point that produce different limits.

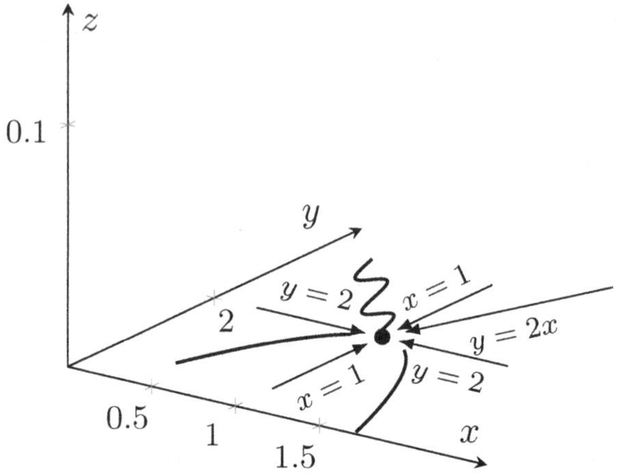

Fig. 13.19 Paths of approach to the point $(1, 2)$ in the xy-plane.

Before solving entire limit problems this way, we have to know how to check a limit along a specific path of approach. Here's how that works:

(1) Choose a path of approach to a limit point (a, b). Often we start with a the vertical line $x = a$ or the horizontal line $y = b$.
(2) Restrict the function to values on that path alone by plugging in the expression describing the path to the function; this reduces the function to a single variable. For example, plugging in $x = a$ reduces $f(x, y)$ to a single variable function $f(a, y)$.
(3) Investigate the limit of this single variable function using all tools available for such limits — including L-Hopital's Rule, if needed!

EX 3 Find the limit of $f(x, y) = \dfrac{xy}{x^2 + y^2}$ along these paths of approach to $(0, 0)$: $x = 0$ and $y = x$.

Along the path $x = 0$, the function $f(x, y)$ reduces to

$$f(x, y) = f(0, y) = \frac{0 \cdot y}{0^2 + y^2} = \frac{0}{0 + y^2} = 0$$

Since the function collapses to the constant value 0 at every point on the path $x = 0$, then the limit of $f(x, y)$ as $(x, y) \to (0, 0)$ on the path $x = 0$ is 0. Along the path $y = x$, the function $f(x, y)$ reduces to

$$f(x, y) = f(x, x) = \frac{x \cdot x}{x^2 + x^2} = \frac{x^2}{x^2 + x^2} = \frac{x^2}{2x^2} = \frac{1}{2}$$

Since the function collapses to the constant value $1/2$ at every point on the path $y = x$, then the limit of $f(x, y)$ as $(x, y) \to (0, 0)$ on the path $y = x$ is $1/2$. ∎

You Try It

(4) Find the limit of $f(x, y) = \sin(x + y)/(x + y)$ along these paths of approach to $(0, 0)$: $y = 0$ and $y = x$. Do we know the limit of $f(x, y)$ as $(x, y) \to (0, 0)$ from these two results?

So, suppose you're investigating $\lim\limits_{(x,y)\to(a,b)} f(x, y)$, and $f(x, y)$ happens to be undefined at (a, b). By now you know you cannot simply say the limit does not exist, and move on to the next problem. But maybe you're suspicious that the limit really doesn't exist — here's what you can do:

- Select one path of approach for (x, y) to approach (a, b) — perhaps the vertical line $x = a$ or horizontal line $y = b$. Determine the limit on this specific path.
- Select another path and repeat. If the new limit value is different than the first one, you're done — there are two paths of approach that give different limits, and the limit does not exist. If the limit is the same as before, then repeat again with a new path.
- Repeat until you get contradictory limit values or you just plain give up.

Let me repeat something important. Path testing CANNOT be used to claim that a limit exists. You cannot test, say, 3 paths, get a limit value $L = 2$ on all of them, and then conclude that the limit really is 2. You can only use path testing to demonstrate that a limit does NOT exist.

Knowing which paths are the most strategic to use is a matter of experience and intuition. There are no magic path generators which will give you the best paths each time.

EX 4 Investigate the limit $\lim\limits_{(x,y)\to(0,0)} \dfrac{xy}{x^2 + y^2}$.

We cannot evaluate the limit directly; if we try the point $(0, 0)$ we fail to get a result since the function becomes $0/0$ there. So now we'll hope the limit does NOT exist. We show above in EX 3 that along the path $x = 0$; we get:

$$\lim_{(x,y)\to(0,0)} \frac{xy}{x^2+y^2} = \lim_{y\to 0} \frac{(0)y}{(0)^2+y^2} = \lim_{y\to 0}(0) = 0$$

along the path $y = x$ we get

$$\lim_{(x,y)\to(0,0)} \frac{xy}{x^2+y^2} = \lim_{x\to 0} \frac{x(x)}{x^2+(x)^2} = \lim_{x\to 0} \frac{x^2}{2x^2} = \lim_{x\to 0} \frac{1}{2} = \frac{1}{2}$$

Since we have found two paths of approach to the limit point $(0,0)$ that give different limit values, we can conclude that this limit does not exist. Figure 13.20 shows $f(x,y)$ near the origin. The graph has quite a snarl near $(0,0)$, like someone poked a pencil in the (x,y) plane at the origin and then twisted. But, you can see a "ridge" on the surface at an elevation of $z = 1/2$; this is over the line $y = x$. ■

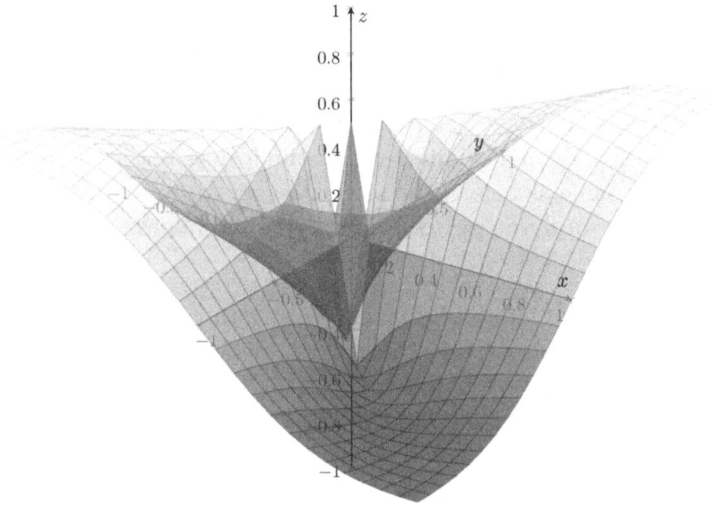

Fig. 13.20 The function $f(x,y) = xy/(x^2+y^2)$.

You Try It

(5) Investigate the limit $\displaystyle\lim_{(x,y)\to(0,0)} \frac{x^2}{x^2+y^2}$.

(6) Investigate the limit $\displaystyle\lim_{(x,y)\to(0,0)} \frac{xy^2}{x^2+y^4}$.

NOTE: When picking paths to test, make sure your path actually goes to the point you want it to. For example, we cannot use the path $y = 1/x$ to test a limit as $(x, y) \to (0, 0)$. Make sure you see why!

Continuity

The concepts of limits and continuity go hand in hand. Loosely speaking, a function is continuous anywhere that the value of the limit as $(x, y) \to (a, b)$ is the value the function really hits at $(x, y) = (a, b)$. This way, there are no breaks, holes, gaps, etc at $(x, y) = (a, b)$. This line of thinking masks some complications, though. For good old fashioned $y = f(x)$ types of functions, we introduced the idea of continuity from the left and the right; on an interval $[a, b]$, $f(x)$ was continuous from, say, the left if

$$\lim_{x \to b^-} f(x) = f(b)$$

and then we would include $x = b$ in the interval of continuity. That's why we say $f(x) = \sqrt{1 - x^2}$ is continuous on the closed interval $[-1, 1]$: the function is not defined for $x > 1$, but continuity from the left at $x = 1$ is good enough to say the function is "continuous" there ... and similarly for $x = -1$. When we scale up to domains of functions $f(x, y)$ that are regions of the 2D xy-plane, things along the edge of a finite domain are more complicated than in a scenario of just left and right sided limits. How would we assess continuity at, say, the edge of a circular domain such as for $f(x, y) = \sqrt{1 - x^2 - y^2}$? If you want to explore this in detail, take a course in Advanced Calculus or Real Analysis (and it's often the second semester of either when multivariable functions are opened up again). For our purposes, we can simply state the following:

Definition 13.1. *A function $f(x, y)$ is continuous at (a, b) if* $\lim_{(x,y) \to (a,b)} f(x, y) = f(a, b)$.

At this basic level, points of continuity of a function can be seen as points from the domain at which no strange things happen.

EX 5 Where is the function $f(x, y) = \dfrac{\sin(xy)}{e^x - y}$ continuous?

The domain of this function does not include any points from the curve $y = e^x$; with that, we can say that all points in the xy-plane not on that curve are points of continuity of the function. ∎

You Try It

(6) Where is the function $\ln(x^2 + y^2 - 4)$ continuous?

Into the Pit!!

The Formal Definition of Limits

Here is an extra portion for those of you who may have enjoyed Sec. 2.5 (Volume 1) and the peek behind the curtain of formal definitions of limits which quantify all the loose language of the function "getting closer" to its limit as the input value(s) "get closer" to their target. We will not do any problem solving, but rather will just update the language for those who are interested.

In the single variable scenario, the formal definition looked like this:

Definition 13.2. $\lim_{x \to a} f(x) = L$ *if, given any* $\epsilon > 0$ *there exists a* $\delta > 0$ *such that* $|f(x) - L| < \epsilon$ *whenever* $0 < |x - a| < \delta$.

The quantity $|f(x) - L|$ measures how close $f(x)$ is getting to its supposed limit, while $|x - a|$ measures how close the input variable x is getting to its target. Overall, the definition requires that we can link together the measures of "closeness", so that we can answer a challenge like, "Guarantee that we can get $f(x)$ within 0.01 of its limit value" by answering with, "Okay, as long as you choose x to be within 0.005 of its target, then $f(x)$ is guaranteed to be within 0.01 of the limit value." When we can discover the link between these two small windows ($[L - \epsilon, L + \epsilon]$ on the y-axis and $[a - \delta, a + \delta]$ on the x-axis), regardless of how small ϵ is, we have conclusively proven the limit.

When we move to the multivariable case $z = f(x, y)$, then the idea of the input variable(s) getting closer to their target is more complicated: we have $(x, y) \to (a, b)$, which takes place anywhere in a neighborhood around the point (a, b) in the xy-plane. A measure of "how close" (x, y) is to (a, b) relies on establishing a circle around (a, b), because — as we know — (x, y) can approach (a, b) from any direction.

And so, the new and improved formal definition of a (finite) limit in 3D looks like this:

Definition 13.3. $\lim_{(x,y)\to(a,b)} f(x,y) = L$ *if, given any* $\epsilon > 0$ *there exists a* $\delta > 0$ *such that* $|f(x) - L| < \epsilon$ *whenever* $0 < \sqrt{(x-a)^2 + (y-b)^2} < \delta$.

An illustration of this definition in action is shown in Fig. 13.21, which involves the limit L of $f(x,y) = 1+xy$ as (x,y) approaches $(3,2)$. We know (or, without rigid proof, we're *pretty sure*) that this limit value is $L = 7$. The figure shows a circle of radius 0.5 around $(3,2)$ in the xy-plane; we can see that as long as input to $f(x,y)$ comes from within that circle, the resulting values of $f(x,y)$ are certain to be within some ϵ of $L = 7$. This value happens to be $\epsilon \approx 1.7$ in this case, and the image demonstrates the statement, "Given $\epsilon = 1.7$, then there is a corresponding value $\delta = 0.5$ such that $|f(x) - L| < \epsilon$ whenever $0 < \sqrt{(x-3)^2 + (y-2)^2} < \delta$." The formal definition of the limit holds for this one particular pairing of ϵ and δ. That's great for an illustration, but it's not the whole story.

A complete proof that our limit is correct requires that we generate a general link between δ and ϵ for *any* epsilon; that way, we can "squeeze" the window around $L = 7$, and correspondingly squeeze the window around $(3,2)$ — so that no matter how small ϵ gets, we can always find a δ such that $|f(x) - L| < \epsilon$ whenever $0 < \sqrt{(x-a)^2 + (y-b)^2} < \delta$. Establishing this general argument locks in the proof that the limit is correct.

Fighting through this kind of argument for multivariable functions is not a trivial task; if these sorts of deeper mathematical details are intriguing to you, then you definitely should plan to take a course in Advanced Calculus or Real Analysis.

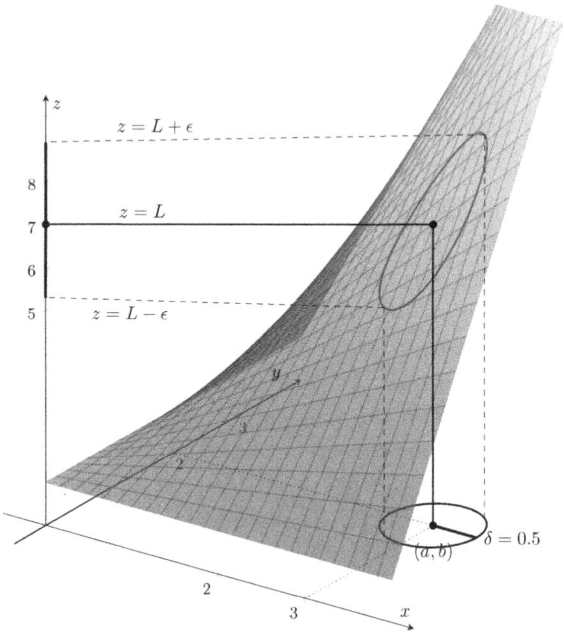

Fig. 13.21 The $\epsilon - \delta$ frame for a limit with $f(x, y) = 1 + xy$.

Limits of Multivariable Functions – Problem List

Limits of Multivariable Functions — You Try It

These appeared above; solutions begin on the next page.

(1) Investigate the limit $\lim_{(x,y)\to(5,2)} (x^3 - y^2 + \sqrt{x - 2y})$.

(2) Investigate the limit $\lim_{(x,y,z)\to(1,-1,1)} \dfrac{xy}{1+z}$.

(3) Investigate the limit $\lim_{(x,y)\to(0,0)} \dfrac{x^2 + y^2}{\sqrt{x^2 + y^2 + 1} - 1}$. (Hint: Rationalize the denominator.)

(4) Find the limit of $f(x, y) = \dfrac{\sin(x + y)}{x + y}$ along these paths of approach to $(0, 0)$: $y = 0$ and $y = x$. Do we know the limit of $f(x, y)$ as $(x, y) \to (0, 0)$ from these two results?

(5) Investigate the limit $\displaystyle\lim_{(x,y)\to(0,0)} \frac{x^2}{x^2+y^2}$.

(6) Investigate the limit $\displaystyle\lim_{(x,y)\to(0,0)} \frac{xy^2}{x^2+y^4}$.

(7) Where is the function $\ln(x^2+y^2-4)$ continuous?

Limits of Multivariable Functions — Practice Problems

Try these as you get the hang of the You Try It problems. Solutions to these problems are available in Sec. B.1.4.

(1) Investigate the limit $\displaystyle\lim_{(x,y)\to(0,0)} \frac{xy}{\cos(xy)}$.

(2) Investigate the limit $\displaystyle\lim_{(x,y)\to(0,0)} \frac{x^2+\sin^2 y}{2x^2+y^2}$.

(3) Investigate the limit $\displaystyle\lim_{(x,y)\to(0,0)} \frac{xy^4}{x^2+y^8}$.

(4) Where is the function $f(x,y)=(x-y)/(1+x^2+y^2)$ continuous?

Limits of Multivariable Functions — Challenge Problems

Try these problems to test your skills with the ideas in this section. Solutions to these problems are available in Sec. C.1.4.

(1) Investigate the limit $\displaystyle\lim_{(x,y)\to(0,0)} \frac{6x^3y}{2x^4+y^4}$.

(2) Investigate the limit $\displaystyle\lim_{(x,y)\to(0,0)} \frac{x^4-y^4}{x^2+y^2}$.

(3) Where is the function $f(x,y,z)=\sqrt{1-x^2-y^2-z^2}$ continuous?

Limits of Multivariable Functions — You Try It — Solved

---

(1) Investigate the limit $\lim_{(x,y)\to(5,2)} x^3 - y^2 + \sqrt{x-2y}$.

□ We do not evaluate limits by plugging in values. No, sir! But let's try it anyway. At $(x, y) = (5, 2)$,

$$x^3 - y^2 + \sqrt{x-2y} = 5^3 - 2^2 + \sqrt{5-4} = 125 - 4 + 1 = 122$$

Since the function is equal to 122 at $(x, y) = (5, 2)$, chances are that as (x, y) *approaches* $(5, 2)$, our function is approaching 122. ■

---

(2) Investigate the limit $\lim_{(x,y,z)\to(1,-1,1)} \frac{xy}{1+z}$.

□ We do not evaluate limits by plugging in values. No way, no how. But let's try it anyway. At $(x, y, z) = (1, -1, 1)$,

$$\frac{xy}{1+z} = \frac{(1)(-1)}{1+1} = -\frac{1}{2}$$

Since the function is equal to $-1/2$ at $(x, y, z) = (1, -1, 1)$, chances are that as (x, y, z) *approaches* $(1, -1, 1)$, our function is approaching $-1/2$. ■

---

(3) Investigate the limit $\lim_{(x,y)\to(0,0)} \frac{x^2+y^2}{\sqrt{x^2+y^2+1}-1}$. (Hint: Rationalize the denominator.)

□ Remember that whole "rationalizing the denominator" business from when you did single variable limits? Consider that

$$\begin{aligned}\frac{x^2+y^2}{\sqrt{x^2+y^2+1}-1} &= \frac{x^2+y^2}{\sqrt{x^2+y^2+1}-1} \cdot \frac{\sqrt{x^2+y^2+1}+1}{\sqrt{x^2+y^2+1}+1} \\ &= \frac{(x^2+y^2)(\sqrt{x^2+y^2+1}+1)}{(x^2+y^2+1)-1} \\ &= \frac{(x^2+y^2)(\sqrt{x^2+y^2+1}+1)}{x^2+y^2} = \sqrt{x^2+y^2+1}+1\end{aligned}$$

so

$$\lim_{(x,y)\to(0,0)} \frac{x^2+y^2}{\sqrt{x^2+y^2+1}-1} = \lim_{(x,y)\to(0,0)} (\sqrt{x^2+y^2+1}+1) = 2$$ ■

(4) Find the limit of $f(x,y) = \dfrac{\sin(x+y)}{x+y}$ along these paths of approach to $(0,0)$: $y = 0$ and $y = x$. Do we know the limit of $f(x,y)$ as $(x,y) \to (0,0)$ from these two results?

□ Along $y = 0$, the function reduces to

$$f(x,0) = \frac{\sin(x+0)}{x+0} = \frac{\sin(x)}{x}$$

and we know that $\lim\limits_{x\to 0} \dfrac{\sin(x)}{x} = 1$.

Along $y = x$, the function reduces to

$$f(x,x) = \frac{\sin(x+x)}{x+x} = \frac{\sin(2x)}{2x}$$

and we will also have $\lim\limits_{x\to 0} \dfrac{\sin(2x)}{2x} = 1$.

Since we found a limit of $L = 1$ on both paths, the overall limit is $L = 1$, right? ... Wrong! We can make no conclusions about the limit with these results. It might be 1, but it might not be. To conclude that the limit is 1, we need to see the limit be 1 on every possible path of approach to $(0,0)$. We now have $\infty - 2$ paths left to check. ■

(5) Investigate the limit $\lim\limits_{(x,y)\to(0,0)} \dfrac{x^2}{x^2+y^2}$.

□ Along the path $x = 0$,

$$\lim_{(x,y)\to(0,0)} \frac{x^2}{x^2+y^2} = \lim_{(x,y)\to(0,0)} \frac{0}{0+y^2} = 0$$

Along the path $y = 0$,

$$\lim_{(x,y)\to(0,0)} \frac{x^2}{x^2+y^2} = \lim_{(x,y)\to(0,0)} \frac{x^2}{x^2} = 1$$

So we get two different limits on two different paths, and the limit does not exist. ■

(6) Investigate the limit $\lim\limits_{(x,y)\to(0,0)} \dfrac{xy^2}{x^2+y^4}$.

□ Along the path $x = 0$,

$$\lim_{(x,y)\to(0,0)} \frac{xy^2}{x^2 + y^4} = \lim_{(x,y)\to(0,0)} \frac{0}{0 + y^4} = 0$$

Along the path $y = 0$, we'll get $L = 0$ too. How about along the path $x = y^2$?

$$\lim_{(x,y)\to(0,0)} \frac{xy^2}{x^2 + y^4} = \lim_{(x,y)\to(0,0)} \frac{(y^2)y^2}{(y^2)^2 + y^4} = \lim_{(x,y)\to(0,0)} \frac{y^4}{2y^4} = \frac{1}{2}$$

So we get two different limits on two different paths, and the limit does not exist. ■

---

(7) Where is the function $\ln(x^2 + y^2 - 4)$ continuous?

□ The domain of this function is all points such that $x^2 + y^2 - 4 > 0$, i.e. everywhere that $x^2 + y^2 > 4$. This is everywhere outside the circle of radius 2 centered at the origin. Within this domain, all points (x,y) are points of continuity. ■

13.5 Partial Derivatives

Introduction

Ah, life was good back in the days when functions had only one first derivative, df/dx or f', and only one second derivative, d^2f/dx^2 or f''. Now that we are considering functions of more than one variable, the number of possible derivatives really starts to stack up. The number of first derivatives of a multivariable function is the same as the number of independent variables; for example, $f(x, y)$ will have two first derivatives. The number of second derivatives is the number of possible pair-wise combinations of independent variables; $f(x, y)$ would have four second derivatives, corresponding to the combinations xx, xy, yx, yy. A function of three variables has three first derivatives, nine second derivatives, and 27 third derivatives. I'll bet you are pretty excited now!

The Definition of a Partial Derivative

Here is the formal definition of a derivative in the old, boring single variable setting:

$$\frac{df}{dx} = f'(x) = \lim_{h \to 0} \frac{f(x+h) - f(x)}{h}$$

Try to remember the interpretation of this: df/dx measures the instantaneous change in output of the function caused by the instantaneous change in the independent variable at a certain value. As we pass some $x = a$, how fast is the function changing? This is called the rate of change of f with respect to x.

In the new multivariable setting, we want to measure the same thing: the instantaneous change in output of the function caused by the instantaneous change in *one* of the independent variables at a certain value. We still want to examine this change one variable at a time. That's why, as stated above, the function $f(x, y)$ will have two first derivatives: the rate of change with respect to x, and the rate of change with respect to y. Each variable contributes a part of the total change in f at a given point, say $(x, y) = (a, b)$, and so the first derivatives are now called *partial derivatives.* Based on the fact that we're still measuring essentially the same thing, a rate of change of the function with respect to a variable, these new formal definitions of first partial derivatives should make some sense:

Definition 13.4. *For a function $f(x,y)$, the first partial derivative with respect to x is defined by*

$$\frac{\partial f}{\partial x} = f_x(x,y) = \lim_{h\to 0} \frac{f(x+h,y) - f(x,y)}{h}$$

and the first partial derivative with respect to y is defined by

$$\frac{\partial f}{\partial y} = f_y(x,y) = \lim_{h\to 0} \frac{f(x,y+h) - f(x,y)}{h}$$

Notation: Remember that in single variable derivatives, we had two notations to describe the derivative of f with respect to x: df/dx and f'. In the multivariable setting there are also two notations, shown in the definitions above. *Operator notation* uses the curly-d symbol ∂ to indicate a partial derivative. Subscript notation shows a which variable is active during a derivative: for example, f_x represents a derivative with respect to x (a prime, f', is not sufficient anymore).

We don't really use these limit definitions to compute partial derivatives, but as long as you're going to be computing partial derivatives, you might as well know what they mean!

FFT: This following attempted definition would be wrong:

$$\lim_{h\to 0} \frac{f(x+h,y+h) - f(x,y)}{h}$$

Do you know why?

Can you construct the formal definition of the first derivative with respect to y of the function $u = f(x,y,z,w)$?

Computing First Order Partial Derivatives

Recall from dealing with single variable functions $f(x)$ that we can compute either the numerical value of a derivative at a given point, or we can compute a derivative *function* f_x that will give us the derivative values at all useable points. For example, if $f(x) = x^2$, then the derivative *function* is $df/dx = 2x$ whereas the *value* of the derivative at $x = 2$ is 4. This same idea holds true in the multivariable setting as well. We can compute derivative *values* or derivative *functions*.

We find partial derivatives by treating one variable as constant and doing the derivative with respect to the other variable, using all the familiar single variable derivative rules. For example, to compute $f_x(x, y)$ we treat y as constant and do the derivative with respect to x.

EX 1 Find all first order derivatives of $f(x, y) = x^5 + x^2y^3 - 7xy + 6x - y + 1$.

Since there are two independent variables, x and y, there will be two first order partial derivatives, $\partial f/\partial x$ and $\partial f/\partial y$ (also known as f_x and f_y). To find f_x, we treat y as a constant and do the "usual" derivative operations with respect to x, so we get:

$$f_x(x, y) = 5x^4 + (2x)(y^3) - 7y + 6$$

The y^3 in the second term is temporarily treated as a multiplicative constant on the x^2, and so it doesn't change. In the term $7xy$, the 7 and the y are treated as multiplicative constants — essentially, the term is $(7y)x$ and so the derivative with respect to x is $7y$. The final chunk $-y + 1$ is considered constant, and so its derivative is 0. Similarly, to find f_y, we treat x as a constant and do the "usual" derivative stuff with respect to y, so we get:

$$f_y(x, y) = 3x^2y^2 - 7x - 1$$

The term x^5 is temporarily treated as a constant, so its derivative is 0. The same is true for $6x$. The x^2 in the term x^2y^3 is a temporary multiplicative constant, and so the derivative of that term is $x^2(3y^2)$ which rearranges to $3x^2y^2$. ■

The tracking of variables gets a little trickier when the product, quotient, or chain rules are involved.

EX 2 Find all first order derivatives of $f(x, y) = xe^{xy}$.

We'll have two first partial derivatives. Note that when we take the derivative with respect to x, there's a product rule waiting to happen, and a chain rule within one term of the product rule:

$$\begin{aligned}\frac{\partial f}{\partial x} &= \left(\frac{\partial}{\partial x}x\right)(e^{xy}) + x \cdot \left(\frac{\partial}{\partial x}e^{xy}\right)\\ &= (1)e^{xy} + x\left(e^{xy}\frac{\partial}{\partial x}(xy)\right)\\ &= e^{xy} + x\left(e^{xy}(y)\right) = (1 + xy)e^{xy}\end{aligned}$$

The partial derivative with respect to y is a bit easier, but there's a chain rule again:

$$\frac{\partial f}{\partial y} = xe^{xy} \cdot \frac{\partial}{\partial y}(xy) = xe^{xy} \cdot x = x^2 e^{xy} \quad \blacksquare$$

You also have to be able to do derivatives for variables other than the usual x and y, and/or for functions that have more than two independent variables.

EX 3 Find all first order derivatives of $f(x, y, z) = x^2 + xyz + xy + yz$.

There are three first order partial derivatives; they are

$$f_x = 2x + yz + y \qquad f_y = xz + x + z \qquad f_z = xy + y \quad \blacksquare$$

EX 4 Find all first order derivatives of $z = \sin\alpha\cos\beta$.

There are two first order first partial derivatives, with respect to α and β. In both notations, they are:

$$\frac{\partial z}{\partial \alpha} = z_\alpha = \cos\alpha\cos\beta$$

$$\frac{\partial z}{\partial \beta} = z_\beta = \sin\alpha(-\sin\beta) = -\sin\alpha\sin\beta \quad \blacksquare$$

You Try It

(1) Find both first order derivatives of $z = xe^{3y}$.
(2) Find all three first order derivatives of $f(x, y, z) = xy^2z^3 + 3yz$.

To compute a specific derivative *value*, the easiest thing to do is get the appropriate derivative function, and then plug in the point in question.

EX 5 If $f(x, y) = xe^{xy}$, what is $f_x(1, 0)$?

We found the partial derivative f_x in EX 2 above, and it was $f_x(x, y) = (1 + xy)e^{xy}$. So $f_x(1, 0) = (1 + 1 \cdot 0)e^{1 \cdot 0} = (1)e^0 = 1$. $\blacksquare$

You Try It

(3) If $f(x, y) = \sqrt{x^2 + y^2}$, what is $f_x(3, 4)$?

Higher Order Derivatives

You should have noted by now that the partial derivatives of a function are themselves functions of all the variables in the original. If we start with $f(x, y)$ then both first partial derivatives are functions of x and y, too, and they are noted $f_x(x, y)$ and $f_y(x, y)$. Since each first derivative is a function of both x and y, then each has its own pair of derivatives. And you probably recall from Calc I that a derivative of a first derivative is called a second derivative. Therefore, a function of two variables x and y has four possible second partial derivatives:

(1) $f_x(x, y)$ has a derivative with respect to x.
(2) $f_x(x, y)$ has a derivative with respect to y.
(3) $f_y(x, y)$ has a derivative with respect to x.
(4) $f_y(x, y)$ has a derivative with respect to y.

(How many second derivatives would $f(x, y, z)$ have?)

You have to be alert to the differences in notation for second partial derivatives and higher depending on whether you are using subscript or operator notation. In subscript notation, we just list the derivatives in the order they happen, left to right. The derivative f_{xy} represents Case 2 above — we find the first partial derivative with respect to x then find the partial derivative with respect to y of that. However, in operator notation (the notation with the ∂ symbols), we'd write the explicit sequence of derivative actions, and its more compressed form, as

$$\frac{\partial}{\partial y}\left(\frac{\partial f}{\partial x}\right) \quad \rightarrow \quad \frac{\partial^2 f}{\partial y \partial x}$$

So, we have two versions of the same derivative,

$$f_{xy} \quad \text{or} \quad \frac{\partial^2 f}{\partial y \partial x}$$

Note that the order of the x and y is different! You have to be very aware of which notation you're using, and be alert to the order of derivative operations that are indicated.

A relief valve in the complexity of higher order derivatives is:

Theorem 13.1. ***Clairaut's Theorem:*** *If $f(x, y)$ has continuous second order partial derivatives at and around a point (a, b), then $f_{xy}(a, b) = f_{yx}(a, b)$.*

The informal interpretation is that unless $f(x, y)$ is an oddball function, then the mixed partial derivatives are going to be the same.

EX 6 If $f(x, y) = \sin(x^2 + y^2)$, find all second order partial derivatives and check for consistency with Clairaut's Theorem.

We start with the first derivatives

$$f_x = 2x\cos(x^2 + y^2) \qquad ; \qquad f_y = 2y\cos(x^2 + y^2)$$

and then can compute second derivatives from these. First, we take the derivative with respect to x of the first derivative with respect to x. Watch out for the product rule!

$$\begin{aligned} f_{xx} = \frac{\partial^2 f}{\partial x^2} &= 2\cos(x^2 + y^2) + 2x\left(-2x\sin(x^2 + y^2)\right) \\ &= 2\cos(x^2 + y^2) - 4x^2\sin(x^2 + y^2) \end{aligned}$$

Next, we take the derivative with respect to y of the first derivative with respect to x:

$$f_{xy} = \frac{\partial^2 f}{\partial x \partial y} = 2x\left(-2y\sin(x^2 + y^2)\right) = -4xy\sin(x^2 + y^2)$$

The derivative with respect to x of the first derivative with respect to y is:

$$f_{yx} = \frac{\partial^2 f}{\partial y \partial x} = 2y\left(-2x\sin(x^2 + y^2)\right) = -4xy\sin(x^2 + y^2)$$

And finally, the derivative with respect to y of the first derivative with respect to y is (product rule alert!):

$$\begin{aligned} f_{yy} = \frac{\partial^2 f}{\partial y^2} &= 2\cos(x^2 + y^2) + 2y\left(-2y\sin(x^2 + y^2)\right) \\ &= 2\cos(x^2 + y^2) - 4y^2\sin(x^2 + y^2) \end{aligned}$$

Since $f_{xy} = f_{yx}$, these results are consistent with Clairaut's Theorem — as they should be, since the second partials are continuous everywhere. ■

EX 7 If $f(x, y) = x^{11} + xy^2$, what are $f_{xxx}(2, 5)$ and $\partial^{12} f / \partial x^{12}$?

First, we need the third partial derivative with respect to x. Going down the line, we have

$$\begin{aligned} f_x &= 11x^{10} + y^2 \\ f_{xx} &= 11 \cdot 10x^9 = 110x^9 \\ f_{xxx} &= 110 \cdot 9x^8 = 990x^8 \end{aligned}$$

and therefore $f_{xxx}(2,5) = 990(2)^8 = 253,440$. The derivative $\partial^{12}f/\partial x^{12}$ represents the twelfth derivative with respect to x. But each derivative knocks down the power of of x^n by 1. Since we start with x^{11}, then after 12 derivatives there will be nothing left. So $\partial^{12}f/\partial x^{12} = 0$. ∎

You Try It

(4) Find all second order derivatives of $f(x,y) = x^4 - 3x^2y^3$.
(5) If $f(x,y,z) = 3xy^4 + x^3y^2$, what are f_{xxy} and f_{yyy}?
(6) If $u = e^{r\theta}\sin\theta$, what is $u_{\theta rr}$?

Laplace's Equation

Laplace's Equation is one of the most important equations in upper level applied mathematics. It shows up in fluid flow problems, gravitation problems, electrostatics problems, meteorology, and so forth. Its development requires a combination of rates of change with conservation principles. of Laplace's Equation for a function $f(x,y)$ is

$$f_{xx} + f_{yy} = 0$$

A function which satisfies Laplace's Equation can be called a *potential* function. (You have probably heard of gravitational potential and electrostatic potential — that carry over in terminology is not accidental.) We'll come back to potential functions in general later on, for now we're just identifying this important equation.[3]

EX 8 Does $u = \sqrt{x^2+y^2}$ satisfy Laplace's Equation?

Apart from the change in name of the function to u, all we need to do is find out if $u_{xx} + u_{yy} = 0$. Getting those derivatives, we have (the quotient rule details in the second derivatives are omitted):

$$u_x = \frac{x}{\sqrt{x^2+y^2}} \quad \rightarrow \quad u_{xx} = \frac{y^2}{(x^2+y^2)^{3/2}}$$

$$u_y = \frac{y}{\sqrt{x^2+y^2}} \quad \rightarrow \quad u_{yy} = \frac{x^2}{(x^2+y^2)^{3/2}}$$

The function does not satisfy Laplace's Equation because,

$$u_{xx} + u_{yy} = \frac{x^2+y^2}{(x^2+y^2)^{3/2}} = \frac{1}{\sqrt{x^2+y^2}} \neq 0 \quad ∎$$

[3]If you find Laplace's Equation and potential functions really interesting, be sure to enroll in a course about Partial Differential Equations!

You Try It

(7) Does $u = x^2 + y^2$ satisfy Laplace's Equation?

Partial Derivatives — Problem List

Partial Derivatives — You Try It

These appeared above; solutions begin on the next page.

(1) Find both first order derivatives of $z = xe^{3y}$.
(2) Find all three first order derivatives of $f(x, y, z) = xy^2z^3 + 3yz$.
(3) If $f(x, y) = \sqrt{x^2 + y^2}$, what is $f_x(3, 4)$?
(4) Find all second order derivatives of $f(x, y) = x^4 - 3x^2y^3$.
(5) If $f(x, y, z) = 3xy^4 + x^3y^2$, what are f_{xxy} and f_{yyy}?
(6) If $u = e^{r\theta} \sin\theta$, what is $u_{\theta rr}$?
(7) Does $u = x^2 + y^2$ satisfy Laplace's Equation?

Partial Derivatives — Practice Problems

Try these as you get the hang of the You Try It problems. Solutions to these problems are available in Sec. B.1.5.

(1) Find both first order derivatives of $z = y \ln x$.
(2) Find both first order derivatives of $u = te^{w/t}$.
(3) Find all first order derivatives of $f(x, y, z) = x^2e^{yz}$.
(4) Find all first order derivatives of $w = \sqrt{r^2 + s^2 + t^2}$.
(5) If $f(x, y) = \sin(2x + 3y)$, what is $f_y(-6, 4)$?
(6) Find all second order derivatives of $f(x, y) = \ln(3x + 5y)$.
(7) If $f(r, s, t) = r \ln(rs^2t^3)$, what are f_{rss} and f_{rst}?
(8) Does $u = x^2 - y^2$ satisfy Laplace's Equation?
(9) For $z = 3x^2 - 2y^2$, compute $z_x(1, 0)$, $z_y(1, 0)$, $z_x(1, 1)$, $z_y(1, 1)$, $z_x(0, 1)$ and $z_y(0, 1)$. In PP 9 of Sec. 13.3, you presented some level curves of this function. On which level curves (specified by associated value of z) will we find the points $(1, 0)$, $(1, 1)$, and $(0, 1)$?

Partial Derivatives — Challenge Problems

Try these problems to test your skills with the ideas in this section. Solutions to these problems are available in Sec. C.1.5.

(1) Find both first order derivatives of $f(s,t) = \dfrac{st}{s^2+t^2}$.
(2) Find all second order derivatives of $u = e^{-s}\sin t$.
(3) Does $u = \ln\sqrt{x^2+y^2}$ satisfy Laplace's Equation?
(4) *(Bonus! This problem connects to CP 4 in Sec. 13.3 and will be used later, too.)* For $w = 2x^2 + y^2 + z$, compute $w_x(1,1,1)$, $w_y(1,1,1)$, and $w_z(1,1,1)$. In CP 4 of Sec. 13.3, you presented some level curves associated with this level surface. Can you find where the point $(1,1,1)$ is indicated on the plot of those level curves?

Partial Derivatives — You Try It — Solved

---

(1) Find both first order derivatives of $z = xe^{3y}$.

☐ For $z = xe^{3y}$, we get

$$\frac{\partial z}{\partial x} = e^{3y} \qquad \text{and} \qquad \frac{\partial z}{\partial y} = 3xe^{3y} \quad ■$$

---

(2) Find all three first order derivatives of $f(x,y,z) = xy^2z^3 + 3yz$.

☐ For $f(x,y,z) = xy^2z^3 + 3yz$, we get

$$f_x = y^2z^3 \qquad f_y = 2xyz^3 + 3z \qquad f_z = 3xy^2z^2 + 3y \quad ■$$

---

(3) If $f(x,y) = \sqrt{x^2+y^2}$, what is $f_x(3,4)$?

☐ Find the first derivative f_x then plug in $(3,4)$:

$$f_x = \frac{x}{\sqrt{x^2+y^2}} \qquad \rightarrow \qquad f_x(3,4) = \frac{3}{\sqrt{3^2+4^2}} = \frac{3}{5} \quad ■$$

---

(4) Find all second order derivatives of $f(x,y) = x^4 - 3x^2y^3$.

☐ With this function, we have these first derivatives:

$$f_x = 4x^3 - 6xy^3 \qquad \text{and} \qquad f_y = -9x^2y^2$$

so that

$$f_{xx} = 12x^2 - 6y^3\,;\; f_{xy} = f_{yx} = -18xy^2\,;\; f_{yy} = -18x^2y \quad ■$$

---

(5) If $f(x,y,z) = 3xy^4 + x^3y^2$, what are f_{xxy} and f_{yyy}?

☐ For the first one,

$$f_x = 3y^4 + 3x^2y^2 \quad \rightarrow \quad f_{xx} = 6xy^2 \quad \rightarrow \quad f_{xxy} = 12xy$$

For the second one,

$$f_y = 12xy^3 + 2x^3y \quad \rightarrow \quad f_{yy} = 36xy^2 + 2x^3 \quad \rightarrow \quad f_{yyy} = 72xy \quad ■$$

(6) If $u = e^{r\theta} \sin\theta$, what is $u_{\theta rr}$?

☐ This is a third order derivative and there will be a product rules involved in the derivative with respect to θ:

$$\begin{aligned}
\frac{\partial u}{\partial \theta} &= re^{r\theta} \sin\theta + e^{r\theta} \cos\theta \\
&= e^{r\theta}(r\sin\theta + \cos\theta) \\
\frac{\partial^2 u}{\partial r \partial \theta} = \frac{\partial}{\partial r}\frac{\partial u}{\partial \theta} &= \theta e^{r\theta}(r\sin\theta + \cos\theta) + e^{r\theta}(\sin\theta) \\
&= e^{r\theta}(\theta r\sin\theta + \theta\cos\theta + \sin\theta) \\
\frac{\partial^3 u}{\partial^2 r \partial \theta} = \frac{\partial}{\partial r}\frac{\partial^2 u}{\partial r \partial \theta} &= \theta e^{r\theta}(\theta r\sin\theta + \theta\cos\theta + \sin\theta) + e^{r\theta}(\theta\sin\theta) \\
&= \theta e^{r\theta}(\theta r\sin\theta + \theta\cos\theta + 2\sin\theta) \quad \blacksquare
\end{aligned}$$

(7) Does $u = x^2 + y^2$ satisfy Laplace's Equation?

☐ We have $u_{xx} = 2$ and $u_{yy} = 2$, so $u_{xx} + uyy \neq 0$, and the function does not satisfy Laplace's equation. ■

Chapter 14

Can We All Agree That Just One Variable Was Lame?

14.1 The Chain Rule

Introduction

Here's a question that seems relatively harmless:

- The length and width of a rectangle are increasing at rates of 1 cm/sec and 1.5 cm/sec, respectively. How fast is the total area of the rectangle increasing at the instant when the length and width are 5 cm and 6 cm?

Now how hard can a problem about a rectangle be? Well, first let's think about an answer that might be intuitive, and yet wrong. Since $A = LW$, then if L changes at 1 cm/sec and W changes at 1.5 cm/sec, it almost makes sense that the area would be changing at $(1\,cm/sec)(1.5\,cm/sec) = 1.5\,cm^2/sec$. But that's wrong! For one thing, did you notice the units don't work out? Units are usually not something we just plop on at the end, they need to develop organically and properly through the mathematics. But further, here is a quick list of all the issues surrounding this question, which together make it more complicated than you'd think!

- We all know that for a rectangle, $A = LW$, and so area depends on length and width.
- But L and W are *changing*, and therefore their values depend on time t.
- So ultimately, the value of area depends on time t.
- The statement "length increases at 1 cm/sec" provides the value $dL/dt = 1\,\frac{cm}{sec}$.
- The statement "width increases at 1 cm/sec" provides a value of dW/dt.

- The question "how fast is the area increasing" is requesting a value of dA/dt.

Do you see the issue? We want a value of the derivative dA/dt. But the formula we have only links A to L and W, so there is no formula that links A to t. How in the world can we find the derivative dA/dt?

There is a chain of dependencies from t to L and W, and then from L and W to A. Effectively, the independent variable is t, and the dependent variable is A. L and W are just intermediate variables that allow A to talk to t. As you should remember from your journey through single variable Calculus, the method we use to find derivatives of a dependent variable with respect to an independent variable by going through intermediate variables is called the *chain rule*.

The Chain Rule (Non) Formula

Suppose we have a function z that depends on variables x and y, which in turn depend on yet another variable t. The notation soup for this situation is $z = f(x(t), y(t))$, and we are likely to be interested in the derivative dz/dt. But we also likely do not have an equation directly relating z to t. So how do we find that derivative? The Chain Rule!

The single variable scenario has, say, y depending on t through the intermediate variable x, and in that case, the applicable chain rule expression is

$$\frac{dy}{dt} = \frac{dy}{dx} \cdot \frac{dx}{dt} \tag{14.1}$$

Now suppose we have $z = f(x(t), y(t))$, which means z depends on t through the intermediate variables x and y. Think of all the derivatives implicit in this relationship. Let's say z is tied to x and y via $z = x^2 + y^2$. Then the derivatives we have available from this are

$$\frac{\partial z}{\partial x} = 2x \qquad \text{and} \qquad \frac{\partial z}{\partial y} = 2y$$

And let's say x and y are tied to t via $x = 2t$ and $y = e^t$. Then we also have the derivatives

$$\frac{\partial x}{\partial t} = 2 \qquad \text{and} \qquad \frac{\partial y}{\partial t} = e^t$$

The derivative we *really* want is $\partial z/\partial t$, but there's no formula that directly links z to t. (Please ignore that we could, in this case, plug in the expressions for x and y to get $z = (2t)^2 + (e^t)^2$ ... that's not always possible.) We can scale up the form of Eq. (14.1) to get:

$$\frac{\partial z}{\partial t} = \frac{\partial f}{\partial x}\frac{\partial x}{\partial t} + \frac{\partial f}{\partial y}\frac{\partial y}{\partial t} \tag{14.2}$$

and we can now fill it in with the derivatives themselves:

$$\frac{\partial f}{\partial t} = (2x)(2) + (2y)e^t = 4x + y2e^t$$

And that's the derivative we want! Technically, we should replace x and y by their expressions with respect to t, to get

$$\frac{\partial f}{\partial t} = 4(2t) + (e^t)(e^t) = 8t + 2e^{2t}$$

but in many cases this would get too messy, and so we'll often omit that final step. Also, very often, we have *values* of variables and derivatives to put in place, so putting too much effort into the cosmetic appearance of the derivative can be counterproductive. We just need to remember that x and y depend on t and so the derivative shown in the second to last step is a function of t, as desired. And, of course, read any instructions that are given.

Equation (14.2) presents a case where the use of either style of derivative notation (single vs. partial) can be correct. Certainly, the derivatives of z with respect to x and y must be shown as partial derivatives with ∂. But it would not be incorrect to write the derivatives of z, x, and y with single derivative notation, because each of z, x, and y ultimately depend only on t. That is, the alternate form

$$z't = \frac{\partial f}{\partial x}x't + \frac{\partial f}{\partial y}y't$$

is also fine. But, I recommend being in the habit of using the partial symbols anyway because we're going to find situations where all of the derivatives are partial derivatives.

There is no one magic chain rule formula that applies to all problems, we have to generate a new one each time we're faced with a new set of variable names and dependencies. The general strategy is to create an expression for the derivative we want such that

- there is one contribution from each intermediate variable
- each contribution is made of derivatives that can be found from the given equations
- each contribution simplifies to the derivative you want

EX 1 If $z = f(x(t), y(t)) = x\ln(x+2y)$ with $x(t) = \sin t$ and $y(t) = \cos t$, what is $\frac{\partial z}{\partial t}$?

Here are the derivatives we know about already from the equations:

$$\frac{\partial z}{\partial x} = \ln(x+2y) + \frac{x}{x+2y} \quad ; \quad \frac{\partial z}{\partial y} = \frac{2x}{x+2y}$$

$$\frac{\partial x}{\partial t} = \cos t \quad ; \quad \frac{\partial y}{\partial t} = -\sin t$$

There are two intermediate variables, x and y. Our chain rule expression will be

$$\frac{\partial z}{\partial t} = \frac{\partial f}{\partial x}\frac{\partial x}{\partial t} + \frac{\partial f}{\partial y}\frac{\partial y}{\partial t}$$

This follows the guidelines above: one contribution from each intermediate variable, all the derivatives are known, and each contribution (pair) reduces to $\frac{\partial z}{\partial t}$. Filling in the derivatives gives:

$$\frac{\partial z}{\partial t} = \left(\ln(x+2y) + \frac{x}{x+2y}\right)(\cos t) - \frac{2x}{x+2y}(\sin t)$$

This is a case when replacement of x and y with their expressions in terms of t would get really messy, so we'll leave it as-is. ■

You Try It

(1) If $z = f(x(t), y(t)) = x^2y + xy^2$ with $x(t) = 2 + t^4$ and $y(t) = 1 - t^3$, what is $\frac{\partial z}{\partial t}$?

More and Different Variables

Variations on chain rule problems arise when you have more than two intermediate variables, or even more than one independent variable. We apply the same ideas, and we can create chain rule expressions all day without even having any specific functions in mind, just the variable dependencies!

EX 2 If $z = (u(s), v(s), w(s))$, write a chain rule expression for $\dfrac{\partial z}{\partial s}$.

Since z depends on u, w, v and each of those depends on s, we have one independent variable (s) but three intermediate variables. So we'll have three contributions to the overall derivative, arranged this way:

$$\frac{\partial z}{\partial s} = \frac{\partial z}{\partial u}\frac{\partial u}{\partial s} + \frac{\partial z}{\partial v}\frac{\partial v}{\partial s} + \frac{\partial z}{\partial w}\frac{\partial w}{\partial s} \quad \blacksquare$$

What if the intermediate variables in a function themselves depend on more than one variable?

EX 3 If $f = f(x(s,t), y(s,t))$, write a chain rule expression for all possible first derivatives of f with respect to the independent variables.

The functional notation tells us that f depends on x and y, but x and y in turn depend on s and t. So, ultimately, the independent variables are s and t, and x and y are the intermediate variables. There are two derivatives of f with respect to the independent variables, $\partial f/\partial s$ and $\partial f/\partial t$ and those derivatives can be assembled like this:

$$\frac{\partial f}{\partial s} = \frac{\partial f}{\partial x}\frac{\partial x}{\partial s} + \frac{\partial f}{\partial y}\frac{\partial y}{\partial s}$$
$$\frac{\partial f}{\partial t} = \frac{\partial f}{\partial x}\frac{\partial x}{\partial t} + \frac{\partial f}{\partial y}\frac{\partial y}{\partial t} \quad \blacksquare$$

You Try It

(2) If $u = f(x, y)$ with $x = x(r, s, t)$ and $y = y(r, s, t)$, write the chain rule expressions for all possible first derivatives of u with respect to the three independent variables.

(3) If $z = f(x(s,t), y(s,t)) = x^2 + xy + y^2$ with $x(s,t) = s + t$ and $y(s,t) = st$, what are $\partial z/\partial s$ and $\partial z/\partial t$?

To see if you're catching on, here's one more example with an excessive number of variables.

EX 4 Suppose $w = f(x, y, z, t)$ and $x = x(u, v), y = y(u, v), z = z(u, v), t = t(u, v)$. Ultimately, how many first derivatives of w are there with respect to the independent variables? Write the chain rule expression for one of them.

There are two independent variables, u and v and four intermediate variables: x, y, z, t. So there are two derivatives of w that can be computed;

the chain rule expression for either must contain contributions from all four intermediate variables. One of the derivatives is:

$$\frac{\partial w}{\partial u} = \frac{\partial w}{\partial x}\frac{\partial x}{\partial u} + \frac{\partial w}{\partial y}\frac{\partial y}{\partial u} + \frac{\partial w}{\partial z}\frac{\partial z}{\partial u} + \frac{\partial w}{\partial t}\frac{\partial t}{\partial u} \quad \blacksquare$$

I hope you realize by now how essential it is that you understand functional notation, so that when you see an expression like

$$w = f(x(u,v), y(u,v), z(u,v), t(u,v))$$

you will understand immediately what all the variable dependencies are.

So far, our examples have been of creating formal chain rule expressions or general derivative formulas. But very often, we will be given *values* of variables and derivatives that need to be put together to generate the *value* of some overall derivative. In this case, there are two choices:

(1) Build the most general chain rule expression for the derivative that's needed, and only at the very end plug in any numerical data.
(2) Plug in numerical data at the appropriate point in the development of the chain rule formula.

The second option is often much cleaner, but also more fraught with opportunity for error — such as plugging in a numerical value too soon.

EX 5 Suppose $w = xe^{y/z}$ with $x = t^2, y = 1 - t, z = 1 + 2t$. What is $\frac{\partial w}{\partial t}$ when $t = 1$?

We see there is only one independent variable, t, and three intermediate variables x, y, z. So the chain rule expression for our derivative is:

$$\frac{\partial w}{\partial t} = \frac{\partial w}{\partial x}\frac{\partial x}{\partial t} + \frac{\partial w}{\partial y}\frac{\partial y}{\partial t} + \frac{\partial w}{\partial z}\frac{\partial z}{\partial t}$$

From our given equation relating w to x, y, and z, we can immediately compute these derivatives:

$$\frac{\partial w}{\partial x} = e^{y/z} \quad ; \quad \frac{\partial w}{\partial y} = \frac{x}{z}e^{y/z} \quad ; \quad \frac{\partial w}{\partial z} = -\frac{xy}{z^2}e^{y/z}$$

Let's now follow option (2) above, and start using our numerical data right now, rather than carrying around these general expressions. When $t = 1$, we know from the other equations that $x = 1$, $y = 0$ and $z = 3$, and so these partial derivatives become:

$$\frac{\partial w}{\partial x} = 1 \quad ; \quad \frac{\partial w}{\partial y} = \frac{1}{3} \quad ; \quad \frac{\partial w}{\partial z} = 0$$

Similarly, from the equations relating x, y, and z to t, we can find

$$\frac{\partial x}{\partial t} = 2t \quad ; \quad \frac{\partial y}{\partial t} = -1 \quad ; \quad \frac{\partial z}{\partial t} = 2$$

and so when $t = 1$, we have

$$\frac{\partial x}{\partial t} = 2 \quad ; \quad \frac{\partial y}{\partial t} = -1 \quad ; \quad \frac{\partial z}{\partial t} = 2$$

Now we have values for all six derivatives that appear in our chain rule expression, and we can plug them in to find that when $t = 1$,

$$\frac{\partial w}{\partial t} = (1)(2) + \frac{1}{3}(-1) + (0)(2) = \frac{5}{3} \quad \blacksquare$$

You Try It

(4) If $z = x^2 + xy^3$ where $x = uv^2 + w^3$ and $y = u + ve^w$, find the values of all possible first derivatives of z with respect to the three independent variables when $u = 2, v = 1, w = 0$.

And just for fun, let's answer the question that started this whole topic — and pay close attention to units along the way.

EX 6 The length and width of a rectangle are increasing at rates of 1 cm/sec and 1.5 cm/sec, respectively. How fast is the total area of the rectangle increasing at the instant when the length and width are 5 cm and 6 cm?

The area of a rectangle is $A = LW$; when the length and width are changing, they are both functions of time, t. Therefore, area A is a function of time t, with intermediate variables L and W. The proper chain rule expression for this situation is:

$$\frac{\partial A}{\partial t} = \frac{\partial A}{\partial L}\frac{\partial L}{\partial t} + \frac{\partial A}{\partial W}\frac{\partial W}{\partial t}$$

Note that both length and width contribute to the overall change in area. Now from the equation $A = LW$, we know that

$$\frac{\partial A}{\partial L} = W \quad \text{and} \quad \frac{\partial A}{\partial W} = L$$

The data in the problem tells us we are interested in the rate of change of area when $L = 5\,cm$ and $W = 6\,cm$, and with these values we have

$$\frac{\partial A}{\partial L} = 6 \qquad \text{and} \qquad \frac{\partial A}{\partial W} = 5$$

Let's pause and not forget that all the quantities we're throwing around have units. Since area (and so also ∂A) has units of cm^2, and L (and so also ∂L) has units of cm, then the units of $\partial A/\partial L$ are $cm^2/cm = cm$. Similarly, the units of $\partial A/\partial W$ are cm. So the new and improved versions of these derivatives are

$$\frac{\partial A}{\partial L} = 6\,cm \qquad \text{and} \qquad \frac{\partial A}{\partial W} = 5\,cm$$

Now let's carry on. We are given the derivative values

$$\frac{\partial L}{\partial t} = 1\,\frac{cm}{sec} \qquad \text{and} \qquad \frac{\partial W}{\partial t} = 1.5\,\frac{cm}{sec}$$

and so we have the values of all 4 derivatives in our chain rule expression. Therefore,

$$\frac{\partial A}{\partial t} = (6\,cm)\left(1\,\frac{cm}{sec}\right) + (5\,cm)\left(1.5\,\frac{cm}{sec}\right) = 13.5\,\frac{cm^2}{sec}$$

The area is changing at a rate of $13.5\,\frac{cm^2}{sec}$ at the instant in question. ■

The answer to EX 6 came with units. There are two ways that can happen — one correct, and one incorrect.

(1) We can start off the problem saying, "Well, we are looking for rate of change of area, so the units ought to be square centimeters per second" ... and so just plop those units on the final number that we get at the end.
(2) We can carry our units along at each step of the problem, and allow our chain rule formula to generate the final units organically.

Can you guess which is the wrong way to do it? Please pay close attention to the last few lines of EX 6, where units put into our chain rule expression blended together to create the units we need. That is how it's supposed to work. You don't just plop units onto your final result; rather, units should develop naturally though your chain rule. So, you should know what units to expect before you even start working, and **if your chain rule expression does not generate those units, your chain rule expression is wrong.**

The Chain Rule — Problem List

The Chain Rule — You Try It

These appeared above; solutions begin on the next page.

(1) If $z = f(x(t), y(t)) = x^2y + xy^2$ with $x(t) = 2 + t^4$ and $y(t) = 1 - t^3$, what is $\frac{\partial z}{\partial t}$?
(2) If $u = f(x, y)$ with $x = x(r, s, t)$ and $y = y(r, s, t)$, write the chain rule expressions for all possible first derivatives of u with respect to the three independent variables.
(3) If $z = f(x(s, t), y(s, t)) = x^2 + xy + y^2$ with $x(s, t) = s + t$ and $y(s, t) = st$, what are $\partial z/\partial s$ and $\partial z/\partial t$?
(4) If $z = x^2 + xy^3$ where $x = uv^2 + w^3$ and $y = u + ve^w$, find the values of all possible first derivatives of z with respect to the three independent variables when $u = 2, v = 1, w = 0$.

The Chain Rule — Practice Problems

Try these as you get the hang of the You Try It problems. Solutions to these problems are available in Sec. B.2.1.

(1) If $z = f(x(t), y(t)) = \sqrt{x^2 + y^2}$ with $x(t) = e^{2t}$ and $y(t) = e^{-2t}$, what is $\frac{\partial z}{\partial t}$?
(2) If $z = f(x(s, t), y(s, t)) = e^{xy} \tan y$ with $x(s, t) = s + 2t$ and $y(s, t) = \frac{s}{t}$, find both $\partial z/\partial s$ and $\partial z/\partial t$.
(3) If $w = w(x, y, z)$ with $x = x(t, u)$, $y = y(t, u)$, and $z = z(t, u)$, write chain rule expressions for all possible first derivatives of w with respect to the independent variables.
(4) If $u = \sqrt{r^2 + s^2}$ where $r = y + x \cos t$ and $s = x + y \sin t$, find the values of all possible first derivatives of u with respect to the independent variables when $x = 1, y = 2, t = 0$.
(5) If $z = f(\alpha(s, t), \beta(s, t)) = \sin \alpha \tan \beta$ with $\alpha(s, t) = 3s + t$ and $\beta(s, t) = s - t$, find the first derivatives of z with respect to the independent variables.
(6) If $u = u(s, t)$ with $s = s(w, x, y, z)$ and $t = t(w, x, y, z)$, write chain rule expressions for all possible first derivatives of u with respect to the independent variables.

(7) If $R = \ln(u^2 + v^2 + w^2)$ where $u = x + 2y$, $v = 2x - y$, and $w = 2xy$, find the values of all possible first derivatives of R with respect to x and y when $x = 1, y = 1$.

The Chain Rule — Challenge Problems

Try these problems to test your skills with the ideas in this section. Solutions to these problems are available in Sec. C.2.1.

(1) If $z = f(\alpha(s,t), \beta(s,t)) = \ln \alpha \cos \beta$ with $\alpha(s,t) = 2s + t$ and $\beta(s,t) = s - 2t$, find the first derivatives of z with respect to the independent variables s and t. Write your final expressions in terms of s and t.
(2) If $w = w(s,t)$ with $s = s(x,y,z,p)$ and $t = t(x,y,z,p)$, write chain rule expressions for all possible first derivatives of w with respect to the four independent variables.
(3) If $T = \cos(x^2 + y^2 + z^2)$ where $x = u + v$, $y = 2u - v$, and $z = 3uv^2$, find the values of all possible first derivatives of T with respect to u and v when $u = 1, v = 2$.

The Chain Rule — You Try It — Solved

(1) If $z = f(x(t), y(t)) = x^2y + xy^2$ with $x(t) = 2 + t^4$ and $y(t) = 1 - t^3$, what is $\partial z/\partial t$?

☐ z is a function of x and y, and then both x and y are functions of t, so the chain rule expression is:

$$\begin{aligned}\frac{\partial z}{\partial t} = \frac{\partial z}{\partial x}\frac{\partial x}{\partial t} + \frac{\partial z}{\partial y}\frac{\partial y}{\partial t} &= (2xy + y^2)(4t^3) + (x^2 + 2xy)(-3t^2)\\ &= (2xy + y^2)(4t^3) - (x^2 + 2xy)(3t^2)\end{aligned}$$

Since x and y are functions of t, then ultimately $\partial z/\partial t$ is a function of t. ■

(2) If $u = f(x,y)$ with $x = x(r,s,t)$ and $y = y(r,s,t)$, write the chain rule expressions for all possible first derivatives of u with respect to the three independent variables.

☐ There will be three first derivatives of u; these derivatives are with respect to r, s, t going through the intermediate variables of x, y. The chain rule formulations are:

$$\frac{\partial u}{\partial r} = \frac{\partial u}{\partial x}\frac{\partial x}{\partial r} + \frac{\partial u}{\partial y}\frac{\partial y}{\partial r}$$
$$\frac{\partial u}{\partial s} = \frac{\partial u}{\partial x}\frac{\partial x}{\partial s} + \frac{\partial u}{\partial y}\frac{\partial y}{\partial s}$$
$$\frac{\partial u}{\partial t} = \frac{\partial u}{\partial x}\frac{\partial x}{\partial t} + \frac{\partial u}{\partial y}\frac{\partial y}{\partial t} \quad \blacksquare$$

---

(3) If $z = f(x(s,t), y(s,t)) = x^2 + xy + y^2$ with $x(s,t) = s+t$ and $y(s,t) = st$, what are $\partial z/\partial s$ and $\partial z/\partial t$?

□ z is a function of x and y, and then both x and y are functions of both s and t, so there are two chain rule expressions:

$$\frac{\partial z}{\partial s} = \frac{\partial z}{\partial x}\frac{\partial x}{\partial s} + \frac{\partial z}{\partial y}\frac{\partial y}{\partial s} = (2x+y)(1)+(x+2y)(t) = (2x+y)+(x+2y)(t)$$

$$\frac{\partial z}{\partial t} = \frac{\partial z}{\partial x}\frac{\partial x}{\partial t} + \frac{\partial z}{\partial y}\frac{\partial y}{\partial t} = (2x+y)(1)+(x+2y)(s) = (2x+y)+(x+2y)(s)$$

Since x and y are functions of s and t, then ultimately $\partial z/\partial s$ and $\partial z/\partial t$ are functions of s and t. ■

---

(4) If $z = x^2 + xy^3$ where $x = uv^2 + w^3$ and $y = u + ve^w$, find the values of all possible first derivatives of z with respect to the three independent variables when $u = 2, v = 1, w = 0$.

□ There are three first partials with respect to u, v, w, going through the intermediate variables x and y. We want all three first derivatives when $u = 2, v = 1, w = 0$. First, note that for these values we have

$$x = 2(1)^2 + (0)^3 = 2$$
$$y = 2 + (1)e^0 = 3$$

So with these values,

$$\frac{\partial z}{\partial u} = \frac{\partial z}{\partial x}\frac{\partial x}{\partial u} + \frac{\partial z}{\partial y}\frac{\partial y}{\partial u} = (2x + y^3)(v^2) + (3xy^2)(1)$$
$$= (31)(1) + (54)(1) = 85$$

$$\frac{\partial z}{\partial v} = \frac{\partial z}{\partial x}\frac{\partial x}{\partial v} + \frac{\partial z}{\partial y}\frac{\partial y}{\partial v} = (2x + y^3)(2uv) + (3xy^2)(e^w)$$
$$= (31)(4) + (54)(1) = 178$$

$$\frac{\partial z}{\partial w} = \frac{\partial z}{\partial x}\frac{\partial x}{\partial w} + \frac{\partial z}{\partial y}\frac{\partial y}{\partial w} = (2x + y^3)(3w^2) + (3xy^2)(ve^w)$$
$$= (31)(0) + (54)(1) = 54 \quad \blacksquare$$

14.2 Optimization

Introduction

I'll be honest with you, the problems in this section will get lengthy. On the other hand, this section doesn't offer anything conceptually new. Think back to single variable Calculus: do you remember finding maximums and minimums? That's here. Remember critical points? Got 'em here, too. Do you remember the second derivative test? Guess what? That's here as well.

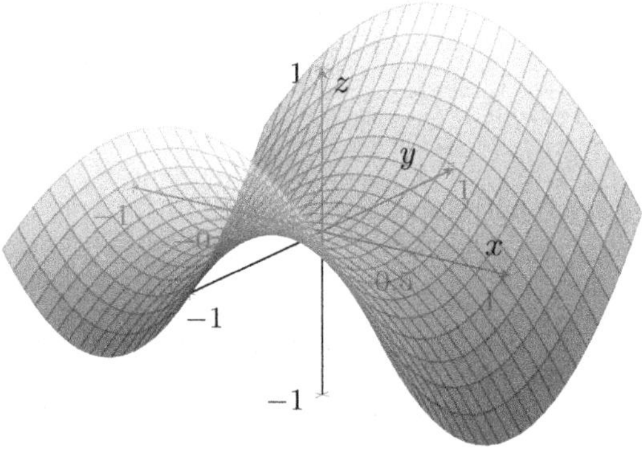

Fig. 14.1 The saddle point at the origin for the function $z = y^2 - x^2$.

This section extends everything you learned about graphical analysis of $f(x)$ via derivatives to graphical analysis of $f(x, y)$. You probably already have an intuitive sense of what a maximum or minimum on the graph of $f(x, y)$ might look like, but there are a couple of new subtleties involved:

- You should recall that the graph of $f(x) = x^3$ has a critical point at the origin (because $f'(x) = 0$ there), but that point is not a maximum or a minimum of the function; it was called a "level point," where the curve just rests for a moment before going on in the same direction. The new & improved version of this feature in three dimensions is called a *saddle point*, and name tells it all: just as the center of a saddle on a horse is a minimum point in the direction from the front to back of the horse, but is a maximum in the direction from the left to the right of the horse, so also a saddle point on a graph is a minimum in one direction, but a maximum in another — such as at the origin on the hyperbolic paraboloid $z = y^2 - x^2$ shown in Fig. 14.1. Also, if you like Pringles potato chips, then you enjoy the taste of saddle points.

- Maximums or minimums can occur at an infinite number of points. Consider the parabolic cylinder $z = 1 - x^2$ in three dimensions, shown in Fig. 14.2. Every point on this surface directly above the entire y-axis is a maximum of this function!

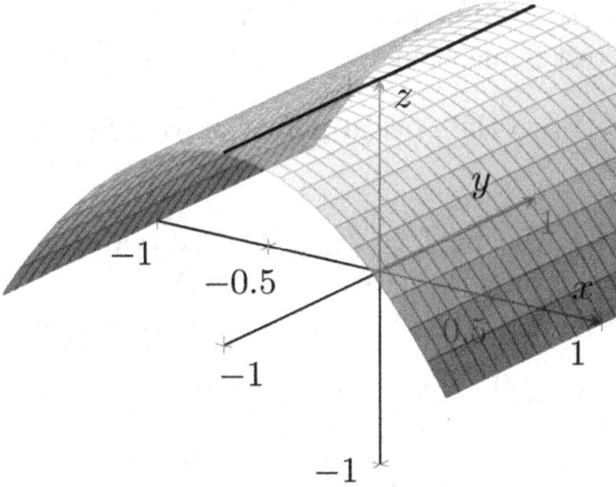

Fig. 14.2 The entire y-axis helping locate local maximums of $z = 1 - x^2$.

Critical Points and the Second Derivative Test

Just as a reminder, here's the old story: In single variable calculus, a critical point of $y = f(x)$ is a point where the derivative $f'(x)$ is either 0 or undefined. That critical point can locate any of the following: discontinuity, maximum, minimum, or level point. We distinguish them by graphing the function or by applying the second derivative test, which says: suppose $x = a$ is a critical point of $f(x)$ and $f'(a)$ exists, then

- if $f''(a) > 0$, there is a minimum at $(a, f(a))$
- if $dderf(a) < 0$, there is a maximum at $(a, f(a))$
- if $f''(a) = 0$, there is a level point at $(a, f(a))$

Here is the new & improved version for three dimensions:

Useful Fact 14.1. *The point (a, b) is a critical point of $f(x, y)$ if $f_x(a, b) = 0$* ***and*** *$f_y(a, b) = 0$,* ***or*** *if either doesn't exist. We distinguish critical points* *through a second derivative test that is a bit more complicated than the original. Let*

$$D = f_{xx}(a,b)f_{yy}(a,b) - [f_{xy}(a,b)]^2$$

Then,

- if $D > 0$ and $f_{xx}(a,b) > 0$, there is a minimum at $(a, b, f(a,b))$
- if $D > 0$ and $f_{xx}(a,b) < 0$, there is a maximum at $(a, b, f(a,b))$
- if $D < 0$, there is a saddle point at $(a, b, f(a,b))$
- if $D = 0$, the test is inconclusive

EX 1 Find and categorize the critical points of $f(x,y) = 2x^3 + xy^2 + 5x^2 + y^2$.

We have $f_x = 6x^2 + y^2 + 10x$ and $f_y = 2xy + 2y$. These are always defined, and so critical points will found where $f_x = 0$ AND $f_y = 0$. The latter is easier to deal with, so let's build and organize for solving $f_y = 0$:

$$f_y = 0 \quad \rightarrow \quad 2xy + 2y = 0$$
$$2y(x+1) = 0$$

So $f_y = 0$ when $x = -1$ or $y = 0$. Let's use these values, one at a time, and see what else it takes to make $f_x = 0$ simultaneously. We'll do $x = -1$ first, by building $f_x = 0$:

$$f_x = 0 \quad \rightarrow \quad 6x^2 + y^2 + 10x = 0 \tag{14.3}$$

and then plugging in $x = -1$,

$$6 + y^2 - 10 = 0 \quad \rightarrow \quad y^2 = 4$$

So already knowing that $x = -1$ will make $f_y = 0$, we've now found that either one of $y = \pm 2$ will also make $f_x = 0$. Putting $x = -1$ and $y = \pm 2$ together, we know that $(-1, -2)$ and $(-1, 2)$ are critical points.

Now we try $y = 0$ in (14.3):

$$6x^2 + (0)^2 + 10x = 0$$
$$2x(3x+5) = 0$$

So already knowing that $y = 0$ will make $f_y = 0$, we've now found that either one of $x = 0$ or $x = -5/3$ will simultaneously make $f_x = 0$. Putting $y = 0$ together with $x = 0$ and $x = -5/3$, we've found that $(0,0)$ and $(-5/3, 0)$ are also critical points.

Altogether, this surface has four critical points! To categorize them, we have to apply the new second derivative test from Useful Fact 14.1, and so we need some second derivatives. Continuing from $f_x = 6x^2 + y^2 + 10x$ and $f_y = 2xy + 2y$, we get:

$$f_{xx} = 12x + 10 \quad ; \quad f_{yy} = 2x + 2 \quad ; \quad f_{xy} = 2y$$

and then the quantity required for the second derivative test is:

$$D(x, y) = f_{xx} f_{yy} - [f_{xy}]^2 = (12x + 10)(2x + 2) - 4y^2$$

The second derivative test requires the value of both $D(x, y)$ and f_{xx} at each critical point. For our four critical points, they are:

- At $(-1, -2)$: $D(-1, -2) = -16$ and $f_{xx}(-1, -2) = -2$
- At $(-1, 2)$: $D(-1, 2) = -16$ and $f_{xx}(-1, 2) = -2$
- At $(0, 0)$: $D(0, 0) = 20$ and $f_{xx}(0, 0) = 10$
- At $(-5/3, 0)$: $D(-5/3, 0) = 40/3$ and $f_{xx} = (-5/3, 0) = -10$

Based on the signs of these values, comparison to Useful Fact 14.1 tells us that:

$A : (-1, -2)$ is a saddle point $C : (0, 0)$ is a minimum
$B : (-1, 2)$ is a saddle point $D : \left(-\frac{5}{3}, 0\right)$ is a maximum

Wow, that's a lot of work! Figure 14.3 shows this surface and the four critical points, labeled as A,B,C,D for the figure. ■

EX 2 How many critical points does $f(x, y) = e^x \cos(y)$ have?

We have $f_x = e^x \cos(y)$ and $f_y = -e^x \sin(y)$. Both of these will be zero when $\sin(y) = \cos(y) = 0$. (Remember that e^x is never 0.) But this never happens! So there are no critical points of this function. ■

You Try It

(1) Find and characterize the critical points of $f(x, y) = 9 - 2x + 4y - x^2 - 4y^2$.
(2) Find and characterize the critical points of $f(x, y) = 1 + 2xy - x^2 - y^2 = 1 - (x - y)^2$.

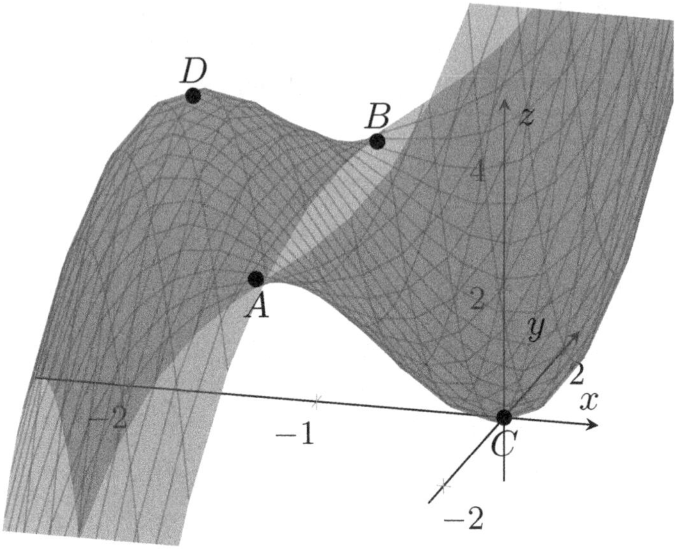

Fig. 14.3 Critical points of $z = 2x^3 + xy^2 + 5x^2 + y^2$.

Absolute Extremes

As we learned the last time we encountered critical points, they are often only *local* extremes — that is, the function may attains a value at that point which is lower or higher than anything else in its vicinity, but it may not be the biggest or smallest value the function hits over a larger interval. In order to to find *absolute* extremes over an interval $[c, d]$, we needed to test the function at the endpoints $x = c$ and $x = d$ to find if the value of f there is bigger or smaller than at any local extremes inside the interval.

Now that our playground is three-dimensional space, and points on our surfaces are associated with points in the xy-plane, we must search for absolute extremes over entire *regions* in the xy-plane; critical points will provide some local extremes, and these are then candidates to be absolute extremes over the specific region. And there is additional joy to be had, because the upgrade for examining endpoints of an interval for possible extremes, now we must look around the entire *boundary* of our region in the xy-plane. Altogether, we search for absolute extremes by:

- Find critical points on the interior of the given region.
- Pick a boundary of the given region and write its equation; plug that equation into $f(x, y)$ to restrict the function to that boundary; find other candidates for extremes there
- Repeat for all boundaries of the region
- Collect any corners of the region not already found as candidates
- Find the function's value at all candidate points (interior, boundary, corner) and compare them to discover the real extremes

Note that we don't need to apply the second derivative test in this process, all we have to do is identify possible extremes and get the function's value there.

EX 3 Find the absolute extremes of $f(x, y) = x^2 + y^2 + x^2y + 4$ over the rectangle $-1 \leq x \leq 1$, $-1 \leq y \leq 1$.

First, we find the critical points of this function to discover possible absolute extremes on the interior of the region. We have $f_x = 2x + 2xy$ and $f_y = 2y + x^2$. We'll get $f_x = 0$ only when $x = 0$ or $y = -1$. For $x = 0$, we'll also get $f_y = 0$ when $y = 0$, so $(0, 0)$ is a critical point. For $y = -1$, we'll also get $f_y = 0$ when $x = \pm\sqrt{2}$. But this is good news — these values are outside the region $-1 \leq x \leq 1$, so we don't care about them! The point $P_1(0, 0)$ is the only critical point on the interior of our domain. Now let's search the boundary of the region for possible extremes:

Let edge L1 be the left edge $x = -1$. On L1, the function reduces to $f(-1, y) = y^2 + y + 5$. This is now a single variable function, which has an extreme when $2y + 1 = 0$, or $y = -1/2$. So $P_2(-1, -1/2)$ is a possible absolute extreme.

Let edge L2 be the right edge $x = 1$. On L2, the function reduces to $f(1, y) = y^2 + y + 5$. This is now a single variable function, which has an extreme when $2y + 1 = 0$, or $y = -1/2$. So $P_3(1, -1/2)$ is a possible absolute extreme.

Let edge L3 be the lower edge $y = -1$. On L3, the function reduces to $f(x, -1) = 5$. This is constant, so there are no new possible extremes on this edge.

Let edge L4 be the upper edge $y = 1$. On L4, the function reduces to $f(x, 1) = 2x^2 + 5$. This is now a single variable function, which has an extreme when $4x = 0$, or $x = 0$. So $P_4(0, 1)$ is a possible absolute extreme.

The corners of this region are the points $P_5(-1, -1)$, $P_6(-1, 1)$, $P_7(1, -1)$, and $P_8(1, 1)$.

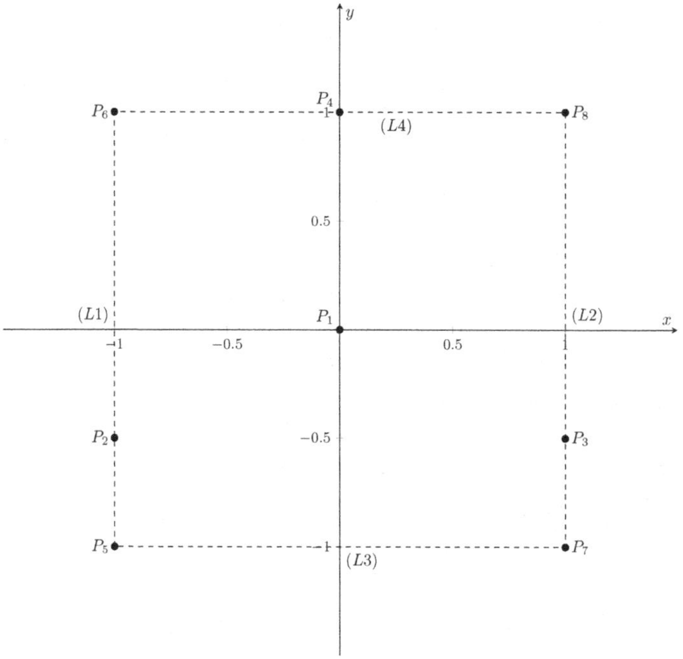

Fig. 14.4 Candidates for local extremes of $z = x^2 + y^2 + x^2y + 4$.

We now have a total of 8 candidates for absolute extremes: one critical point from the interior, three points on the boundaries of the region, and four corners. These have been named P_1 to P_8, and are shown in Fig. 14.4. To make determination on types of points, we have to check the function's value at each:

$$P_1 : f(0, 0) = 4$$
$$P_2 : f(-1, -1/2) = (1) + 1/4 + (1)(-1/2) + 4 = 21/4$$
$$P_3 : f(1, -1/2) = (1) + 1/4 + (1)(-1/2) + 4 = 21/4$$
$$P_4 : f(0, 1) = (0) + 1 + (0)(1) + 4 = 5$$

$$P_5 : f(-1,-1) = (1) + (1) + (1)(-1) + 4 = 5$$
$$P_6 : f(-1,1) = (1) + (1) + (1)(1) + 4 = 7$$
$$P_7 : f(1,-1) = (1) + (1) + (1)(-1) + 4 = 5$$
$$P_8 : f(1,1) = (1) + (1) + (1)(1) + 4 = 7$$

Comparing these values, we see an absolute minimum at $(0,0,4)$ and a pair of absolute maximums at $(-1,1,7)$ and $(1,1,7)$. ■

> **You Try It**
>
> (3) Find the absolute extremes of $f(x,y) = 1+4x-5y$ over the domain D that's a triangle with vertices $P(0,0)$, $Q(2,0)$, and $R(0,3)$.

Optimization Problems

Now that we can find extremes of multivariable functions, we can answer all sorts of problems that require identifying a maximum or minimum of such functions. These are called *optimization problems*. These problems often come with two important components; if you can identify them, you have a head start. These are the **objective function** and the **constraint**. The objective function is the quantity you are hoping to find a maximum or minimum for. Often, objective functions are broader than the given context calls for — such as being a function of three variables, when ultimately we want only two. The constraint is some sort of restriction on the given scenario, and the constraint usually allows us to focus the objective function more helpfully. These terms will make more sense after a couple of examples, I promise!

EX 4 What is the smallest distance from the point (1,1,1) to the paraboloid $z = x^2 + y^2 + 1$?

(Fig. 14.5 shows the point $(1,1,1)$, the given paraboloid, and a line from the point to the paraboloid where the distance is a minimum.) The opening phrase "What is the smallest distance from the point $(1,1,1)$..." tells us we are going to look for a minimum distance of some sort from that point. Therefore, our objective function will be related to the distance from the point $(1,1,1)$ to any other point (x,y,z); in 3D, that looks like:

$$d = \sqrt{(x-1)^2 + (y-1)^2 + (z-1)^2}$$

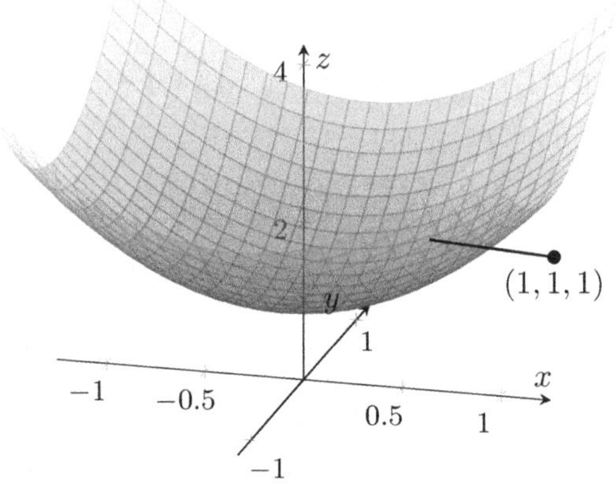

Fig. 14.5 Minimizing distance from $(1, 1, 1)$ to $z = x^2 + y^2 + 1$.

Now, strategy number one when dealing with minimizing a distance function is to note that whatever minimizes d^2 also minimizes d, so our objective function can be given as

$$d^2 = (x-1)^2 + (y-1)^2 + (z-1)^2$$

and now we've eliminated that pesky square root (which would make derivatives really awful). Note, though, that this is a function of three variables, and we don't quite know how to find a minimum of it. But, we don't want the distance from (1,1,1) to *any* point, we want the distance to the given paraboloid. That is our *constraint*: the points under consideration are not *any* points (x, y, z), but rather points of the specific form (x, y, x^2+y^2+1). We are *constrained* to those points in particular, and so we can restrict the formula for d^2 by replacing the generic z with $x^2 + y^2 + 1$:

$$d^2 = (x-1)^2 + (y-1)^2 + ((x^2 + y^2 + 1) - 1)^2$$

Now it's a function of two variables, and we're ready to roll. First, let's simplify it:

$$d^2 = (x-1)^2 + (y-1)^2 + (x^2 + y^2)^2$$

Next, treating the right hand side as the function $f(x, y)$ we want to minimize, let's get some derivatives:

$$f_x = 2(x-1) + 2(x^2 + y^2)(2x) = 2x - 2 + 4x^3 + 4xy^2$$
$$f_y = 2(y-1) + 2(x^2 + y^2)(2y) = 2y - 2 + 4x^2y + 4y^3$$

and since we need both $f_x = 0$ and $f_y = 0$, we need to solve the system of two equations in two unknowns

$$2x - 2 + 4x^3 + 4xy^2 = 0$$
$$2y - 2 + 4x^2y + 4y^3 = 0$$

I don't know about you, but I am not looking forward to that. Fortunately, we now have ready access to technical help. On-line resources such as Wolfram Alpha are ready and waiting to solve your two equations in two unknowns. From that source or any other you can access, we discover the solution to this pair of equations is $(x, y) = (1/2, 1/2)$. At these (x, y) coordinates, we get the distance between the paraboloid and $(1, 1, 1)$ by

$$d^2 = \left(\frac{1}{2} - 1\right)^2 + \left(\frac{1}{2} - 1\right)^2 + \left(\frac{1}{4} + \frac{1}{4}\right)^2 = \frac{1}{2} + \frac{1}{2} + \frac{1}{4} = \frac{3}{4}$$

and so the minimum distance is $d = \sqrt{5}/2$ (that is the distance of the line connecting the point to the paraboloid in Fig. 14.5). FFT: How do we know this is a minimum distance when we didn't do a second derivative test? Well, it certainly can't be a *maximum*, right? What is the maximum distance from the point $(1, 1, 1)$ to any point on the paraboloid $z = x^2 + y^2 + 1$? ■

You Try It

(4) Find the minimum distance d from the point $(2, 1, -1)$ to the plane $x + y - z = 1$.

EX 5 Suppose a rectangular box (with a lid) needs to hold a volume of $1\,m^3$, and the cardboard for the base and lid costs twice as much as the cardboard for the sides. What dimensions of the box will minimize the cost?

There are two basic geometric properties in play here, the volume of a box and the surface area of the box. Ultimately, we want to minimize a cost, but that cost will be based on the surface area. If the dimensions of the box are x, y, z (length, width, height), then — with a lid — the total surface area is $S = 2xy + 2yz + 2xz$. The base and lid are the xy's. But to adapt this to cost, we use the fact that the top and bottom are twice as expensive as the sides. Let's assume the price per unit area of the sides is p, so that total cost is

$$C = 2p(2xy) + p(2yz + 2xz) = 2p(2xy + yz + xz)$$

This cost function is our objective function. It shows the cost of any box in which the cost of the base and lid is twice as much as the cost of the sides. It's what we want to minimize, but currently it's a function of three variables. Fortunately, we have a constraint: we don't want just any old box, but rather, we are constrained to those boxes whose volumes are $1\,m^3$. The constraint $V = xyz = 1$ can be shuffled into $z = 1/(xy)$, and this will reduce the cost function from three variables to two:

$$C = 2p\left(2xy + y\cdot\frac{1}{xy} + x\cdot\frac{1}{xy}\right) = 2p\left(2xy + \frac{1}{x} + \frac{1}{y}\right)$$

Now we can proceed with the minimization. First, we get derivatives:

$$C_x = 2p\left(2y - \frac{1}{x^2}\right) \quad \text{and} \quad C_y = 2p\left(2x - \frac{1}{y^2}\right)$$

Next, we look for where $C_x = 0$ and $C_y = 0$:

$$C_x = 0 \quad \rightarrow \quad 2p\left(2y - \frac{1}{x^2}\right) = 0 \quad \rightarrow \quad 2yx^2 = 1$$

$$C_y = 0 \quad \rightarrow \quad 2p\left(2x - \frac{1}{y^2}\right) = 0 \quad \rightarrow \quad 2xy^2 = 1$$

We are seeking where both happen simultaneously, so let's merge them together (the choice of which to plug in to which is a toss-up). From the first, we have $y = 1/(2x^2)$, so the second becomes:

$$2x\left(\frac{1}{2x^2}\right)^2 = 1 \quad \rightarrow \quad x = \sqrt[3]{\frac{1}{2}}$$

So $C_y = 0$ when $x = \sqrt[3]{1/2}$, and then $C_x = 0$ at the same time when $y = 1/(2x^2)$, or

$$y = \frac{1}{2(\sqrt[3]{1/2})^2} = \sqrt[3]{\frac{1}{2}}$$

And when we have $x = y = \sqrt[3]{1/2}$, then the volume constraint $xyz = 1$ allows us to find the corresponding z:

$$\sqrt[3]{\frac{1}{2}}\cdot\sqrt[3]{\frac{1}{2}}\cdot z = 1$$
$$z = \sqrt[3]{2}\cdot\sqrt[3]{2} = \sqrt[3]{4}$$

Altogether, the dimensions that will minimize our cost are

$$(x, y, z) = \left(\sqrt[3]{\frac{1}{2}}, \sqrt[3]{\frac{1}{2}}, \sqrt[3]{4}\right)$$

(these are all in meters, per the original units). Note that the actual cost per unit area p never made a difference. FFT: How do we know we did not just find a maximum cost? Given our conditions, what is the maximum cost of a box? ■

You Try It

(5) Find the minimum surface area of a rectangular box (with no lid) that contains a volume of $32,000\ cm^3$.

Optimization — Problem List

Optimization — You Try It

These appeared above; solutions begin on the next page.

(1) Find and characterize the critical points of $f(x, y) = 9 - 2x + 4y - x^2 - 4y^2$.
(2) Find and characterize the critical points of $f(x, y) = 1 + 2xy - x^2 - y^2 = 1 - (x - y)^2$.
(3) Find the absolute extremes of $f(x, y) = 1 + 4x - 5y$ over the domain D that's a triangle with vertices $P(0, 0)$, $Q(2, 0)$, and $R(0, 3)$.
(4) Find the minimum distance d from the point $(2, 1, -1)$ to the plane $x + y - z = 1$.
(5) Find the minimum surface area of a rectangular box (with no lid) that contains a volume of $32,000\ cm^3$.

Optimization — Practice Problems

Try these as you get the hang of the You Try It problems. Solutions to these problems are available in Sec. B.2.2.

(1) Find and characterize the critical points of $f(x, y) = x^3y + 12x^2 - 8y$.
(2) Find and characterize the critical points of $f(x, y) = xy(1 - x - y) = xy - x^2y - xy^2$.
(3) Find the absolute extremes of $f(x, y) = 3 + xy - x - 2y$ over the domain D that's a triangle with vertices P(1,0), Q(5,0) and R(1,4).
(4) Find the minimum distance d from the point (1,2,3) to the plane $x - y + z = 4$ and the point on the plane where this distance occurs.
(5) What is the largest volume that can be contained in a rectangular box (with a lid!) that has a total surface area of $64\,cm^2$.

(6) Find and characterize the critical points of $f(x, y) = x^2 + y^2 + \dfrac{1}{x^2y^2}$.

(7) Find the absolute extremes of $f(x, y) = 4x + 6y - x^2 - y^2$ over the domain D with $0 \le x \le 4, 0 \le y \le 5$.

(8) What dimensions should we use for a rectangular aquarium of volume $12,000\ cm^3$ if we want to minimize its cost, given that the base costs 5 times

Optimization — Challenge Problems

Try these problems to test your skills with the ideas in this section. Solutions to these problems are available in Sec. C.2.2.

(1) Find and characterize the critical points of $f(x, y) = 8xy(x+y) + \sqrt{17}$.

(2) Find the absolute extremes of $f(x, y) = 3x^2 + 2xy + y^2$ over the domain D with $-2 \le x \le 2, 0 \le y \le 3$.

(3) Identify the coordinates of the point on the plane $2x - y + z = 16$ that is closest to the origin.

Optimization — You Try It — Solved

---

(1) Find and characterize the critical points of $f(x, y) = 1 + 2xy - x^2 - y^2 = 1 - (x - y)^2$.

□ We have that $f_x = 2y - 2x$ and $f_y = 2x - 2y$ and so $f_x = f_y = 0$ everywhere that $y = x$. Getting ready for the second derivative test, we have:

$$f_{xx} = -2 \to f_{xx}(x, x) = -2$$

$$f_{yy} = -2 \to f_{yy}(x, x) = -2$$

$$f_{xy} = 2 \to f_{xy}(x, x) = 2$$

$$\begin{aligned} D(x, x) &= f_{xx}(x, x) f_{yy}(x, x) - [f_{xy}(x, x)]^2 \\ &= (-2)(-2) - 2^2 = 0 \end{aligned}$$

Uh oh! Since $D = 0$, the second derivative test told us nothing. But we can still figure it out, because we're smart! Note that since $f(x, y) = 1 - (x - y)^2$, the function is equal to 1 whenever $y = x$ and is less than 1 everywhere else. So all points along $y = x$ are local maximums. ■

(2) Find the absolute extremes of $f(x, y) = 1 + 4x - 5y$ over the domain D that's a triangle with vertices $P(0, 0)$, $Q(2, 0)$, and $R(0, 3)$.

☐ Are there extremes over the interior of D? With $f_x = 4$ and $f_y = 5$, there are no points where $f_x = f_y = 0$, so there are no critical points. Extremes of this function will be on the boundary of D.

Let edge L1 be the line segment PR. On L1, $x = 0$ and the function reduces to $f(0, y) = 1 - 5y$. Since this is linear, there are no extremes along L1 itself.

Let edge L2 be the line segment PQ. On L2, $y = 0$ and the function reduces to $f(x, 0) = 1 + 4x$. Since this is linear, there are no extremes along L2 itself.

Let edge L3 be the line segment QR. This is the line $y = -3x/2 + 3$, and the function reduces to $f(x) = 23x/2 - 14$ here. Since this is also linear, there are no extremes along L3 itself.

The extremes must be at the vertices:

$$f(0, 0) = 1 \quad , \quad f(0, 3) = -14 \quad \text{and} \quad f(2, 0) = 9$$

So, the absolute minimum is $f(0, 3) = -14$ and the absolute maximum is $f(2, 0) = 9$. ■

(3) Find the minimum distance d from the point $(2, 1, -1)$ to the plane $x + y - z = 1$.

☐ To make the calculations easier, we'll minimize d^2 (whatever minimizes d^2 also minimizes d). The distance between (2,1,-1) and any point at all is given by:

$$d^2 = (x - 2)^2 + (y - 1)^2 + (z + 1)^2$$

This is our objective function which we want to minimize, but it has three variables. We need our constraint to, well, constrain it. We aren't interested in the distance from $(2, 1, -1)$ to any point in the universe. We are only interested in points on the plane $x + y - z = 1$, or rather, $z = x + y - 1$. This is our constraint, and it helps us mush our objective function down to having only two variables:

$$d^2 = (x - 2)^2 + (y - 1)^2 + (x + y)^2$$

Treating the right hand side as the function $f(x, y)$ we want to minimize, let's get some derivatives:

$$f_x = 2(x-2) + 2(x+y) = 4x + 2y - 4$$
$$f_y = 2(y-1) + 2(x+y) = 2x + 4y - 2$$

Our critical points come from there $f_x = 0$ and $f_y = 0$ simultaneously. The equation $f_x = 0$ reduces to $2x + y - 2 = 0$, and $f_y = 0$ reduces to $x + 2y - 1 = 0$. From the former, we have $y = 2 - 2x$; plugging that in to the latter gives

$$\begin{aligned} x + 2(2-2x) - 1 &= 0 \\ -3x + 3 &= 0 \\ x &= 1 \end{aligned}$$

and that value handed back to either $y = 2 - 2x$ gives us $y = 0$. Therefore $(x, y) = (1, 0)$ is a critical point. What kind of critical point is this? Let's keep going with second derivatives, and plan to use Useful Fact 14.1. With

$$f_{xx} = 4 \quad ; \quad f_{yy} = 4 \quad ; \quad f_{xy} = 2$$

we have

$$D(x, y) = f_{xx}f_{yy} - [f(x, y)]^2 = 12$$

Since $D(x, y) > 0$ and $f_{xx} > 0$ everywhere, any critical point is a local minimum, per Useful Fact 14.1. So $(x, y) = (1, 0)$ presents the minimum distance; the z-coordinate on the given plane there is $z = 1 - x - y = 0$ and so the minimum distance to $(2, 1, -1)$ happens at the point (1,0,0). The distance between the two points is $d = \sqrt{3}$. ■

---

(4) Find the minimum surface area of a rectangular box (with no lid) that contains a volume of $32,000\ cm^3$.

□ Let the dimensions of the box be x, y, z (length, width, height). We have no lid, so there is only one face of area xy and two of each other, xz and yz. The surface area of such a box is

$$S = xy + 2xz + 2yz$$

and this is our objective function, since we want to minimize this. It's currently a function of three variables, which is bad news. But, we don't want just any old box, we want a box that has a specific volume; this provides our constraint, we need to adhere to

$$xyz = 32000 \quad \text{or} \quad z = \frac{32000}{xy}$$

This constraint, when plugged in to the objective function, reduces the objective function to two variables:

$$\begin{aligned} S &= xy + 2x\left(\frac{32000}{xy}\right) + 2y\left(\frac{32000}{xy}\right) \\ &= xy + \frac{64000}{y} + \frac{64000}{x} \end{aligned}$$

Now we need derivatives:

$$S_x = y - \frac{64000}{x^2} \qquad \text{and} \qquad S_y = x - \frac{64000}{y^2}$$

Setting these equal to zero,

$$S_x = 0 \to y = \frac{64000}{x^2}$$

$$S_y = 0 \to x = \frac{64000}{y^2}$$

We need to merge these two conditions to find when both happen simultaneously. There is no one magic way to do this; how about we square the first, from $S_x = 0$:

$$y^2 = \frac{64000^2}{x^4}$$

then hand it to the second,

$$x = \frac{64000}{y^2} = 64000 \cdot \frac{x^4}{64000^2} = \frac{x^4}{64000}$$

Rearranged, we get $x^3 = 64000$, or $x = 40$. When $x = 40$, we have

$$y = \frac{64000}{x^2} = 40 \qquad \text{and} \qquad z = \frac{32000}{xy} = 20$$

Altogether, the surface area is minimized for dimensions $(x, y, z) = (40, 40, 20)$. ■

14.3 Double Integrals

Introduction

I'll bet you think back fondly on this expression:

$$\lim_{n\to\infty}\sum_{i=1}^{n} f(x_i^\star)\Delta x = \int_a^b f(x)\,dx$$

To the left of the equals sign is the Riemann Sum computation of the area under $f(x)$ over the interval $[a,b]$. To the right of the equals sign is the symbol we use to denote this value; this computation is called a definite integral. The Fundamental Theorem of Calculus gives us an easy way to compute the definite integral, but that theorem is not in itself the *definition* of a definite integral, this expression is. Well, if single integrals are fun to develop this way, imagine how much fun double integrals are.

Generalizing the Integration Process

The single integral definition above is formed by this (hopefully familiar) process, based on the fact that we want to compute the area under $f(x)$ over the interval $[a,b]$:

(1) Divide the interval $[a,b]$ into n partitions. Each partition has width Δx. This width represents the base of a rectangle.
(2) Within each partition i of the n total, select a representative point $x_i^\star$.
(3) Plug that value into the function, obtaining $f(x_i^\star)$; this value then acts as the height of a rectangle.
(4) Multiply this height $f(x_i^\star)$ by the width (Δx) to obtain the area $f(x_i^\star)\Delta x$ of our rectangle standing over the partition.
(5) Sum the resulting areas from all n partitions to get an *estimate* of the total area under $f(x)$,

$$A \approx \sum_{i=1}^{n} f(x_i^\star)\Delta x$$

(6) Recognize that our estimate of the area under $f(x)$ improves as the number of partitions increases, and ultimately the area is *exactly*

$$A = \lim_{n\to\infty}\sum_{i=1}^{n} f(x_i^\star)\Delta x$$

(7) Give this area the symbol $\displaystyle\int_a^b f(x)dx$.

Now we're going to upgrade from a single variable function $f(x)$ to a multivariable function $f(x, y)$ and try the same sort of thing. As a first step, let's rewrite all these items with the intention of making the process much more general (and thus more adaptable). For example, we've always known $[a, b]$ as the interval of integration, but more generally, it can be called a region of integration, which is carved out from the domain of the function. This removes any mention of the number of dimensions we're talking about. Also, rather than finding the "area" under $f(x)$, how about we just "obtain a measure of the region below $f(x)$". In a single integral, that measure is area — but what will it be when we consider $f(x, y)$? So here is the integration process, described in more general terms:

(1) Select a region of integration from the domain of the function.
(2) Partition the region of integration into tiny pieces. Each tiny piece, or *partition*,[1] will have its own size.
(3) Within each individual partition, select a representative point.
(4) Plug the coordinate(s) of that representative point into the function.
(5) Multiply the function's value at the representative point by the size of the partition.
(6) Sum these resulting contributions from all partitions.
(7) Find the limit of this sum as the number of partitions goes to ∞.
(8) Give this limit a name and symbol.

This now provides the integration process for any of $f(x)$, $f(x, y)$, or $f(x, y, z)$. We just have to figure out the specifics for each case, and see if there is a geometric interpretation for the calculation. With $f(x)$, you already know the geometric interpretation of the result of this partitioning process.

The Definition of a Double Integral

Let's go through the general integration process given above and figure out what each step really means for a function of two variables, $f(x, y)$. This process is not how we actually compute double integrals, but it will help you understand why double integrals give the results that they do.

[1]Yes, we are now using partition as a verb and a noun...

(1) Select a region of integration from the domain of the function.

The domain of a function $f(x, y)$ is part to all of the xy-plane. Therefore, our region of integration will be a portion of the xy-plane. For reasons that should become clear later, we'll assume that this region has at least two straight edges; that is, at least two boundaries of the region must be straight lines described by constant values of either x or y x or y values. Figures 14.6 and 14.7 show the two possible configurations: the region is either bounded by constant values of x ($x = a$ and $x = b$), or it's bounded by constant values of y ($y = c$ and $y = d$). Be careful that you understand these figures just show possible regions of integration from within the xy-plane; the function $f(x, y)$ that will be integrated over this region; is floating above (or below) these regions. A chosen region of integration is often generically named R (for region) or D (for domain).

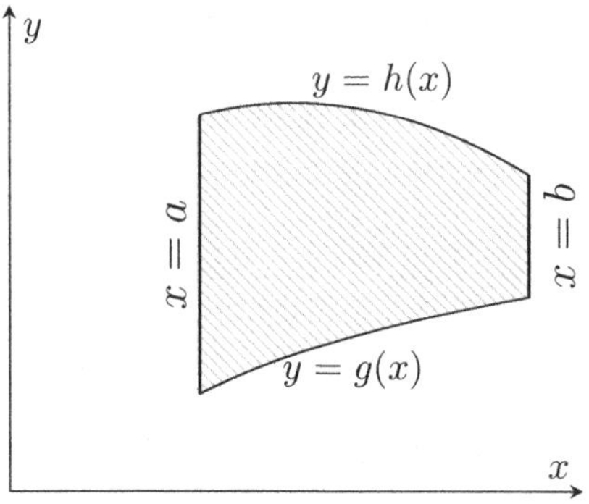

Fig. 14.6 2D region with vertical straight-line (constant) boundaries.

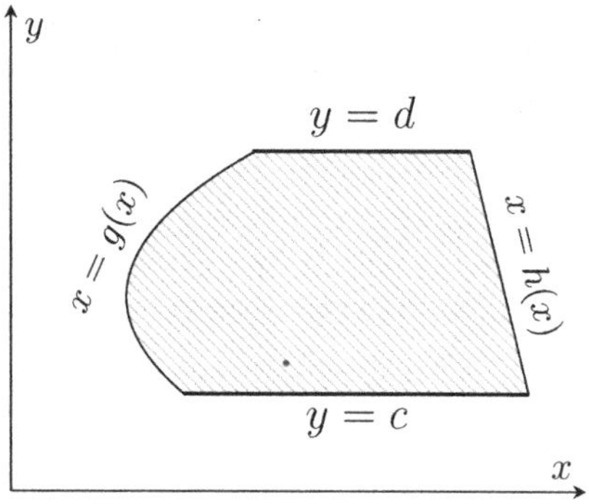

Fig. 14.7 2D region with horizontal straight-line (constant) boundaries.

(2) Divide the region of integration into partitions. Each partition may have its own size.

The region of integration R is part of the xy-plane. In single variable integrals, we partitioned our interval $[a, b]$ into pieces of size Δx. Now we partition in *both* the x and y directions, creating pieces of size Δx in one direction and Δy in the other. Thus, each partition is a tiny rectangle, and it has size (*area*) $\Delta A = \Delta x \Delta y$. Such a partition is indicated in Fig. 14.8. Let's say there are n divisions in the x-direction and m divisions in the y direction, and thus a total of nm small rectangles of size ΔA.

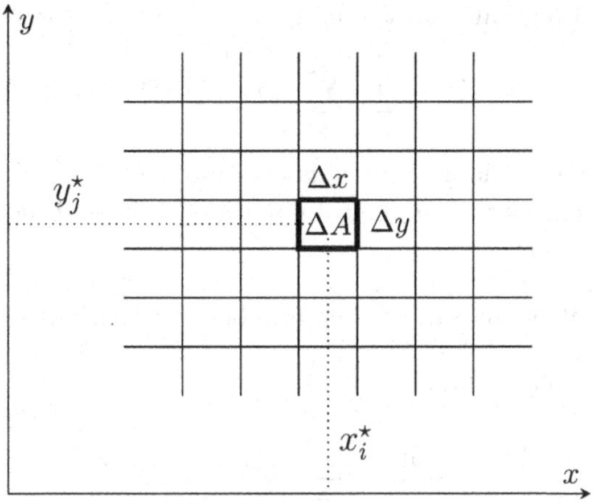

Fig. 14.8 Partitioning the xy-plane.

(3) Within each individual partition, select a representative point.

In the partition of size ΔA, we can select a representative point $(x_i^\star, y_j^\star)$. (The indices i and j simply locate it, for example the point in the 2nd x-column and 4th y-row would be (x_2, y_4).) Such representative point is shown in Fig. 14.8.

(4) Plug the coordinate(s) of that representative point into the function.

When we plug the representative point $(x_i^\star, y_j^\star)$ into our function, we get $f(x_i^\star, y_j^\star)$. Do you see what we're building yet? In a single integral, we built rectangles with width Δx and height $f(x_i^\star)$. Now, we're building standing rectangular columns; the base of each column has *area* ΔA, and now we've computed the *height* of the column. And therefore, when we do this:

(5) Multiply the function's value at the representative point by the size of the partition.

We've now found the *volume* of that standing column, $V_{ij} = f(x_i^\star, y_j^\star) \cdot \Delta A$. There are nm total standing columns, so when we do this:

(6) Sum these resulting contributions from all partitions.

We've found an estimate of the total volume underneath the function $f(x, y)$ over the region R. We have to sum over both counters i (in the x-direction) and j (in the y-direction) and therefore have a double sum:

$$V_{tot} \approx \sum_{j=1}^{m} \sum_{i=1}^{n} f(x_i^\star, y_j^\star)\Delta A$$

As in the single variable case, this estimate to the total volume under $f(x, y)$ will improve as the number of partitions in both directions go to ∞, and so when we do this:

(7) Find the limit of this sum as the number of partitions goes to ∞.

We get the true volume under $f(x, y)$:

$$V_{tot} = \lim_{n,m\to\infty} \sum_{j=1}^{m} \sum_{i=1}^{n} f(x_i^*, y_j^*)\Delta A$$

And there we have the Riemann Sum definition of the volume under $f(x, y)$ over a region of integration R.

(8) Give this limit a name and symbol.

This is a clunky thing to write all the time, so we'll give it this symbol:

$$\iint_R f(x, y)dA$$

and write the formal definition of a double integral as:

$$\iint_R f(x, y)dA = \lim_{n,m\to\infty} \sum_{i=j}^{m} \sum_{i=1}^{n} f(x_i^*, y_j^*)\Delta A$$

This notation now identifies the function being integrated and the region of integration R. We know that the result of this calculation will be the volume "under" $f(x, y)$ over the region R. (The word "under" has to be taken loosely, just like saying that single integrals give the area "under" a curve must be taken loosely, because the curve itself might be below the axis.)

The Computation of a Double Integral: Rectangular Regions

Fortunately, we don't use the Riemann Sum and limit definition of a double integral to evaluate the integral. Rather, we basically do the same thing as for single integrals — just twice! Let's start with the easiest case, when all the edges of the region of integration R are straight lines. That is, R is a rectangle defined by the statement $R = \{(x, y) : a \leq x \leq b, c \leq y \leq d\}$.

The computation of a double integral of $f(x,y)$ over such a rectangle is made possible by **Fubini's Theorem**, which says,

$$\iint_R f(x,y)\,dA = \int_a^b \int_c^d f(x,y)\,dydx$$

or equivalently

$$\iint_R f(x,y)\,dA = \int_c^d \int_a^b f(x,y)\,dxdy$$

These are "iterated integrals". The dA is exchanged for $dxdy$ or $dydx$, and the integral is solved inside-out using all the integral techniques you already know and love. Think of it as,

$$\int_a^b \int_c^d f(x,y)\,dydx = \int_a^b \left(\int_c^d f(x,y)dy \right) dx$$

When we are integrating with respect to x, we treat y as a constant, and vice versa. The inner integral should remove one variable completely, and then the outer integral removes the other variable and leaves only numbers. The final answer should be a number. The result represents the volume "under" $f(x,y)$, and so if you don't get a number, something is wrong.

EX 1 Evaluate $\displaystyle\int_1^3 \int_0^1 (1+4xy)\,dxdy$.

We start working on the inner integral; it's an integral with respect to x, and therefore we treat y as a constant. This inner integral should remove all references to x and leave only y's behind:

$$\int_0^1 (1+4xy)\,dx = (x+2x^2y)\Big|_0^1 = (1+2(1)^2y) - (0+2(0)^2y) = 1+2y$$

Now we pass this result to the outer integral; this outer integral will get rid of y and leave the final answer:

$$\int_1^3 \int_0^1 (1+4xy)\,dxdy = \int_1^3 (1+2y)dy = (y+y^2)\Big|_1^3 = (3+3^2)-(1+1^2) = 10$$

We can interpret this result as: The volume under the surface $z = 1+4xy$ over the region $R = \{(x,y) : 0 \le x \le 1, 1 \le y \le 3\}$ is 10. ■

We can also nest all the computations into one sequence; you can decide on your preference:

EX 2 Evaluate $\iint_R \frac{1+x^2}{1+y^2}\,dA$ over the region $R = \{(x, y) : 0 \le x \le 2, 0 \le y \le 1\}$.

We convert the boundaries of the region into limits of integration; make sure to match the right limits with the right variable! We can choose either $dA = dydx$ or $dA = dxdy$, it doesn't matter in this case. I'll show the latter:

$$\begin{aligned}
\int_0^1 \int_0^2 \frac{1+x^2}{1+y^2}\,dxdy &= \int_0^1 \left(\frac{1}{1+y^2} \left(x + \frac{x^3}{3} \right) \Big|_0^2 \right) dy \\
&= \int_0^1 \left(\frac{1}{1+y^2} \left(2 + \frac{2^3}{3} \right) \right) dy \\
&= \frac{14}{3} \int_0^1 \frac{1}{1+y^2}\,dy = \frac{14}{3} \tan^{-1}(y) \Big|_0^1 \\
&= \frac{14}{3} (\tan^{-1}(1) - \tan^{-1}(0)) = \frac{14}{3} \left(\frac{\pi}{4} - 0 \right) = \frac{7\pi}{6} \quad \blacksquare
\end{aligned}$$

You Try It

(1) Evaluate $\int_0^2 \int_0^{\pi/2} x \sin y \, dydx$.

(2) Evaluate $\iint_R (6x^2y^3 - 5y^4)\,dA$, where R is the region $\{(x, y) : 0 \le x \le 3, 0 \le y \le 1\}$.

(3) Find the volume under $3x + 2y + z = 12$ over the region $R = \{(x, y) : 0 \le x \le 1, -2 \le y \le 3\}$.

The Computation of a Double Integral: General Regions

Now we consider double integrations over a region R which has one or two boundaries that are not straight. Not much changes in the overall scheme of things, we still do a double integral as an iterated integral. But this time, one or more of the limits of integration will not be constant. Here are guidelines:

- The variable which has one or more non-constant limits must be integrated *first*, i.e. must be placed on the inner integral. In this case, the result of the inner integral might not be a number, but it should still completely remove the inner variable.

- At most two boundaries of the region can be identifiable as curves like $y = g(x)$ or $x = g(y)$, the rest must be lines such as $x = a$ or $y = c$. (Refer back to Figs. 14.6 and 14.7 if needed.)
- The outer integral must have constant endpoints, so that the final answer is a number. (Can you imagine what would happen if one of the outer integral's limits was not constant?)

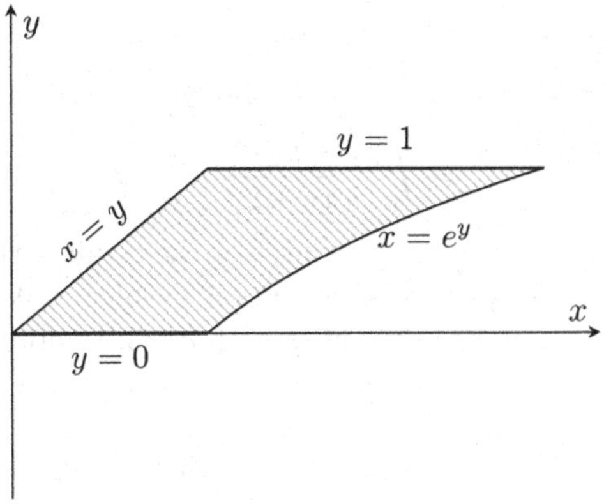

Fig. 14.9 Region between $x = y$, $x = e^y$, $y = 0$, $y = 1$ (w/ EX 3).

EX 3 Evaluate $\displaystyle\int_0^1 \int_y^{e^y} x\,dxdy$.

We can interpret this integral as seeking the volume under $f(x, y) = x$ over the region R bounded by the curves $x = y$ and $x = e^y$, and the lines $y = 0$ and $y = 1$. That region is shown in Fig. 14.9. Let's evaluate the inner integral by itself:

$$\int_y^{e^y} x\,dx = \frac{1}{2}x^2\bigg|_y^{e^y} = \frac{1}{2}\left((e^y)^2 - (y)^2\right) = \frac{1}{2}(e^{2y} - y^2)$$

Note that we have removed all instances of x, and are left with a function of y, which can now safely be passed to the outer integral:

$$\int_0^1 \int_y^{e^y} x\,dxdy = \int_0^1 \frac{1}{2}(e^{2y} - y^2)\,dy = \left(\frac{1}{4}e^{2y} - \frac{1}{6}y^3\right)\bigg|_0^1$$
$$= \left(\frac{1}{4}e^2 - \frac{1}{6}\right) - \frac{1}{4} = \frac{e^2}{4} - \frac{5}{12} \quad \blacksquare$$

EX 4 Evaluate $\iint_D \frac{4y}{x^3+2}\,dA$ over the region $D = \{(x,y) : 1 \le x \le 2, 0 \le y \le 2x\}$.

In this region, the variable x has constant limits, and the variable y has non-constant limits. So we must integrate with respect to y first. (Note that it would be impossible to integrate with respect to x first anyway!). To integrate with respect to y first, we set $dA = dydx$:

$$\int_1^2 \int_0^{2x} \frac{4y}{x^3+2}\,dydx = \int_1^2 \left(\frac{2y^2}{x^3+2}\bigg|_0^{2x} \right) dx$$
$$= \int_1^2 \left(\frac{2(2x)^2}{x^3+2} - 0 \right) dx = \int_1^2 \frac{8x^2}{x^3+2}\,dx$$

Now we pause for a moment and consider how the result of the inner integral has left us with a nice solvable substitution problem. We can choose to replace x^3+2 with u, and so also x^2dx with $du/3$, to obtain (don't forget to change the limits),

$$\int_1^2 \frac{8x^2}{x^3+2}\,dx = \int_3^{10} \frac{8}{3u}\,du = \frac{8}{3}\ln|u|\bigg|_3^{10} = \frac{8}{3}(\ln|10| - \ln|3|) = \frac{8}{3}\ln\frac{10}{3} \quad \blacksquare$$

You Try It

(4) Evaluate $\iint_D \frac{2y}{x^2+1}\,dA$, where D is the region $D = \{(x,y) : 0 \le x \le 1, 0 \le y \le \sqrt{x}\}$.

(5) Evaluate $\iint_D x\cos y\,dA$, where D is the region bounded by $y = 0, y = x^2$, and $x = 1$.

(6) Evaluate $\iint_D (x+2y)\,dA$, where D is the region between the curves $y = x$ and $y = x^4$.

The Order of Integration

Here are two things to know about order of integration:

(1) When the region of integration is a rectangle, i.e. all 4 limits of integration are constant, you can exchange the order of integration and get the same result. That is,

$$\int_a^b \int_c^d f(x,y)\,dydx = \int_c^d \int_a^b f(x,y)\,dxdy$$

This means that if it looks like the integration might be easier in one of the two possible orders, feel free to change the order of integration!

(2) You can also exchange (reverse) the order of integration for more general regions; that is, we can find two integrals that can produce the same result:

$$\int_a^b \int_{g(x)}^{h(x)} f(x,y)\,dydx = \int_c^d \int_{G(y)}^{H(y)} f(x,y)\,dxdy$$

as long as the limits of integration in each case delineate the same region of integration! After all, the double integral represents the volume under $f(x,y)$ over the region R, so as long as you describe the region R properly, it shouldn't matter which variable you take care of first. This is essential in the case that an integral cannot be done in one order; you have to be able to reverse the order of integration when needed.

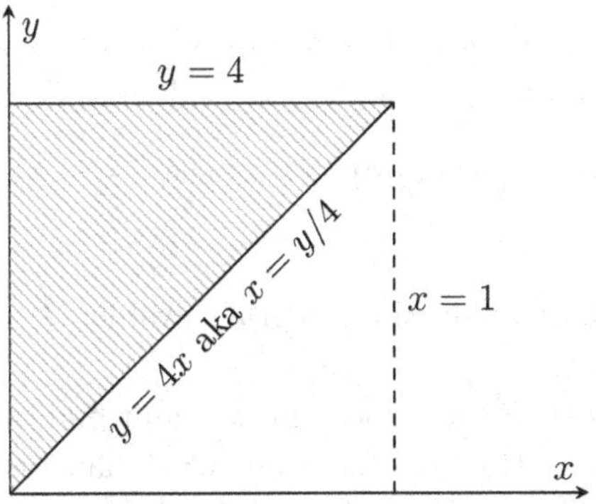

Fig. 14.10 Region between $y = 4x$ and $y = 4$ (w/ EX 5).

EX 5 Write the equivalent integral in which the order of integration is reversed: $\int_0^1 \int_{4x}^4 f(x,y)\,dydx$.

The limits of integration delineate the region R bounded by the lines $y = 4x$ and $y = 4$, and the lines $x = 0$ and $x = 1$. That region is shown in Fig. 14.10. Note that we can also describe this region as being the region between the lines $x = 0$ and $x = y/4$ and the lines $y = 0$ and $y = 4$ — it's the same region, with different description of the boundaries. Therefore, we can reverse the order of integration like this:

$$\int_0^1 \int_{4x}^4 f(x,y)\,dydx = \int_0^4 \int_0^{y/4} f(x,y)\,dxdy$$

Note that we didn't even need to supply a function $f(x, y)$, because the creation and reversal of limits of integration has absolutely nothing to do with the function being integrated! ■

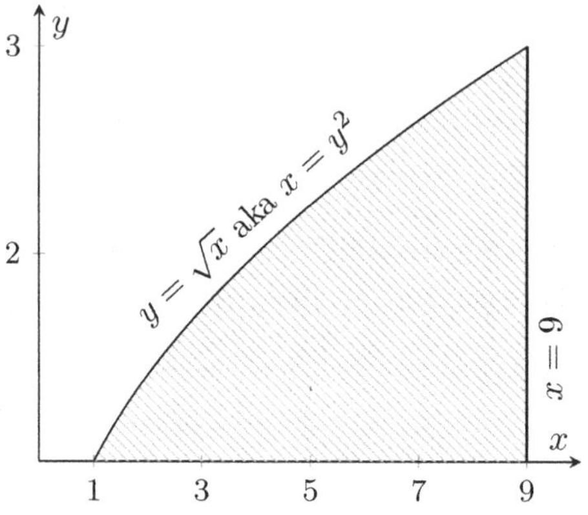

Fig. 14.11 Region between $x = y^2$ and $x = 9$ (w/ EX 6).

EX 6 Do something to make this integral solvable: $\int_0^3 \int_{y^2}^9 y\cos(x^2)\, dxdy$.

This integral is not solvable as-is because we can't find the inner antiderivative with respect to x. However, consider what happens if we reverse the order of integration. The region of integration is between the curves $x = y^2$ and $x = 9$ and the lines $y = 0$ and $y = 3$. That region is shown in Fig. 14.11. We can describe the same region as being between the curves $y = 0$ and $y = \sqrt{x}$ and the lines $x = 0$ and $y = 9$. Therefore, we can reverse the order like this:

$$\int_0^3 \int_{y^2}^9 y\cos(x^2)\, dxdy = \int_0^9 \int_0^{\sqrt{x}} y\cos(x^2)\, dydx$$

The new version of the integral is now easily solvable. Try it out if you don't believe me! ■

You Try It

(7) Write the equivalent integral in which the order of integration is reversed: $\int_0^1 \int_{3y}^3 e^{x^2}\, dxdy$.

Double Integrals — Problem List

Double Integrals — You Try It

These appeared above; solutions begin on the next page.

(1) Evaluate $\int_0^2 \int_0^{\pi/2} x \sin y \, dydx$.

(2) Evaluate $\iint_R (6x^2y^3 - 5y^4) dA$, where R is the region $\{(x, y) : 0 \leq x \leq 3, 0 \leq y \leq 1\}$.

(3) Find the volume under $3x + 2y + z = 12$ over the region $R = \{(x, y) : 0 \leq x \leq 1, -2 \leq y \leq 3\}$.

(4) Evaluate $\iint_D \frac{2y}{x^2+1} dA$, where D is the region $D = \{(x, y) : 0 \leq x \leq 1, 0 \leq y \leq \sqrt{x}\}$.

(5) Evaluate $\iint_D x \cos y \, dA$, where D is the region bounded by $y = 0, y = x^2$, and $x = 1$.

(6) Evaluate $\iint_D (x + 2y) \, dA$, where D is the region between the curves $y = x$ and $y = x^4$.

(7) Write the equivalent integral in which the order of integration is reversed: $\int_0^1 \int_{3y}^3 e^{x^2} \, dxdy$.

Double Integrals — Practice Problems

Try these as you get the hang of the You Try It problems. Solutions to these problems are available in Sec. B.2.3.

(1) Evaluate $\int_1^4 \int_0^2 x + \sqrt{y} \, dxdy$.

(2) Evaluate $\iint_R \frac{xy^2}{x^2+1} dA$, where R is the $\{(x, y) : 0 \leq x \leq 1, -3 \leq y \leq 3\}$.

(3) Find the volume under $z = 4 + x^2 - y^2$ over the region $R = \{(x, y) : -1 \leq x \leq 1, 0 \leq y \leq 2\}$.

(4) Evaluate $\iint_D e^{y^2} \, dA$, where D is the region $\{(x, y) : 0 \leq x \leq y, 0 \leq y \leq 1\}$.

(5) Evaluate $\iint_D (x+y)\,dA$, where D is between $y=\sqrt{x}$ and $y=x^2$.

(6) Evaluate $\iint_D (2x+y^2)\,dA$, where D is the region between $x=y^2$ and $x=y^3$.

(7) Write the equivalent integral in which the order of integration is reversed: $\int_0^1\int_{\sqrt{y}}^1 \sqrt{x^3+1}\,dxdy$.

(8) Evaluate $\iint_D xy^2\,dA$, where D is the region bounded by $x=0$ and $x=\sqrt{1-y^2}$.

(9) Find the volume under the paraboloid $z=x^2+3y^2$ over the region in the xy-plane bounded by the the lines $y=1, x=0$, and $y=x$.

(10) Evaluate $\int_0^1\int_{x^2}^1 x^3\sin(y^3)\,dydx$.

Double Integrals — Challenge Problems

Try these problems to test your skills with the ideas in this section. Solutions to these problems are available in Sec. C.2.3.

(1) Evaluate $\iint_D (1)\,dA$ where D is the region to the right of the parabola $x=y^2-1$ and to the left of the semicircle $x=\sqrt{1-y^2}$. Once you find the value of the integral, state what geometric measure you just calculated.

(2) Find the volume under the surface $z=2x+y^2$ over the region in the first quadrant bounded by $y=x^5$ and $y=x$.

(3) Reverse the order of integration of, and then evaluate, the following integral

$$\int_{-1}^0\int_{-\sqrt{y+1}}^{\sqrt{y+1}} y^2\,dxdy$$

Double Integrals — You Try It — Solved

---

(1) Evaluate $\int_0^2 \int_0^{\pi/2} x \sin y \, dy dx$.

$$\square \int_0^2 \int_0^{\pi/2} x \sin y \, dy dx = \int_0^2 \left[-x \cos y \Big|_0^{\pi/2} \right] dx$$
$$= \int_0^2 x \, dx = \left(\frac{1}{2} x^2 \right) \Big|_0^2 = 2 \quad \blacksquare$$

---

(2) Evaluate $\iint_R (6x^2y^3 - 5y^4) \, dA$ where R is the region $\{(x, y) : 0 \leq x \leq 3, 0 \leq y \leq 1\}$.

□ Setting the limits of integration,

$$\iint_R (6x^2y^3 - 5y^4) \, dA = \int_0^3 \int_0^1 (6x^2y^3 - 5y^4) \, dy dx$$
$$= \int_0^3 \left(\frac{3}{2} x^2 y^4 - y^5 \right) \Big|_0^1 dx = \int_0^3 \left(\frac{3}{2} x^2 - 1 \right) dx$$
$$= \left(\frac{1}{2} x^3 - x \right) \Big|_0^3 = \frac{27}{2} - 3 = \frac{21}{2} \quad \blacksquare$$

---

(3) Find the volume under $3x + 2y + z = 12$ over the region $R = \{(x, y) : 0 \leq x \leq 1, -2 \leq y \leq 3\}$.

□ We rewrite the function as $z = 12 - 3x - 2y$ and then

$$\int_{-2}^3 \int_0^1 (12 - 3x - 2y) \, dx dy = \int_{-2}^3 \left(12x - \frac{3}{2} x^2 - 2xy \right) \Big|_0^1 dy$$
$$= \int_{-2}^3 \left(\frac{21}{2} - 2y \right) dy = \left(\frac{21}{2} y - y^2 \right) \Big|_{-2}^3$$
$$= \frac{21}{2} (3 - (-2)) - (9 - 4) = \frac{95}{2} \quad \blacksquare$$

---

(4) Evaluate $\iint_D \frac{2y}{x^2 + 1} \, dA$ where D is the region $D = \{(x, y) : 0 \leq x \leq 1, 0 \leq y \leq \sqrt{x}\}$.

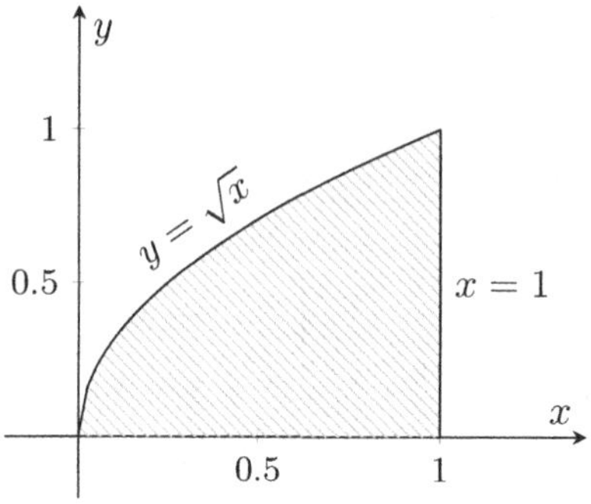

Fig. 14.12 Region of integration for YTI 4.

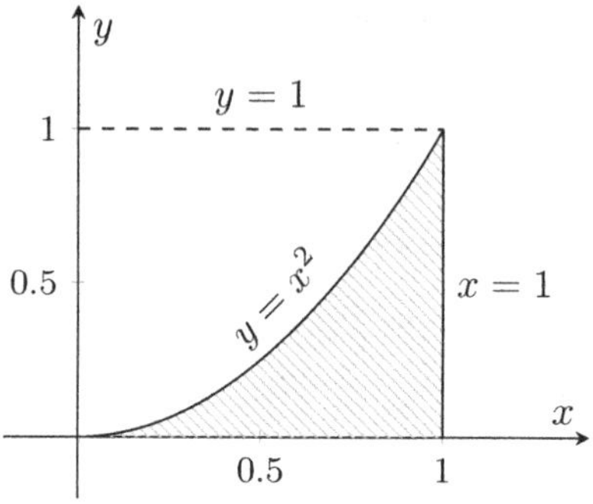

Fig. 14.13 Region of integration for YTI 5.

☐ The region is shown in Fig. 14.12. We have to integrate with respect to y first, since the boundaries of y are not constant:

$$\iint_D \frac{2y}{x^2+1}\,dA = \int_0^1\int_0^{\sqrt{x}} \frac{2y}{x^2+1}\,dydx = \int_0^1 \left(\frac{y^2}{x^2+1}\right)\Bigg|_0^{\sqrt{x}} dx$$
$$= \int_0^1 \frac{x}{x^2+1}\,dx = \frac{1}{2}\ln(x^2+1)\Big|_0^1 = \frac{\ln 2}{2} \quad ■$$

---

(5) Evaluate $\iint_D x\cos y dA$ where D is the region bounded by $y = 0, y = x^2$, and $x = 1$.

☐ The region D is also known as $\{(x, y) : 0 \le x \le 1, 0 \le y \le x^2\}$, i.e. the area under $y = x^2$ from $x = 0$ to $x = 1$. The region is shown in Fig. 14.13. We have to integrate with respect to y first, since the boundaries of y are not constant:

$$\iint_D x\cos y\,dA = \int_0^1\int_0^{x^2} x\cos y\,dydx = \int_0^1 (x\sin y)\Big|_0^{x^2} dx$$
$$= \int_0^1 x\sin(x^2)\,dx = -\frac{1}{2}\cos(x^2)\Big|_0^1 = \frac{1}{2}(1-\cos 1) \quad ■$$

---

(6) Evaluate $\iint_D (x + 2y)\,dA$, where D is the region between the curves $y = x$ and $y = x^4$.

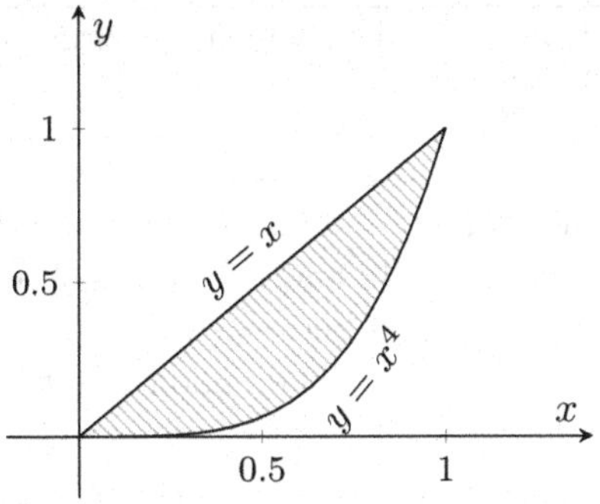

Fig. 14.14 Region of integration for YTI 6.

□ Since $y = x$ and $y = x^4$ intersect at (0,0) and (1,1) and $y = x^4$ is below the other on that interval, the region D bounded by them is the area under $y = x$ and over $y = x^4$ from $x = 0$ to $x = 1$. This region is shown in Fig. 14.14. To find the volume under the plane $x + 2y - z = 0$ over this region, we have to integrate $z = x + 2y$ with respect to y first, since the boundaries of y are not constant:

$$\iint_D (x + 2y)\, dA = \int_0^1 \int_{x^4}^{x} (x + 2y)\, dydx = \int_0^1 \left(xy + y^2\right)\Big|_{x^4}^{x} dx$$
$$= \int_0^1 \left(2x^2 - (x^5 + x^8)\right) dx = \left(\frac{2}{3}x^3 - \frac{1}{6}x^6 - \frac{1}{9}x^9\right)\Big|_0^1$$
$$= \frac{2}{3} - \frac{1}{6} - \frac{1}{9} = \frac{7}{18} \quad \blacksquare$$

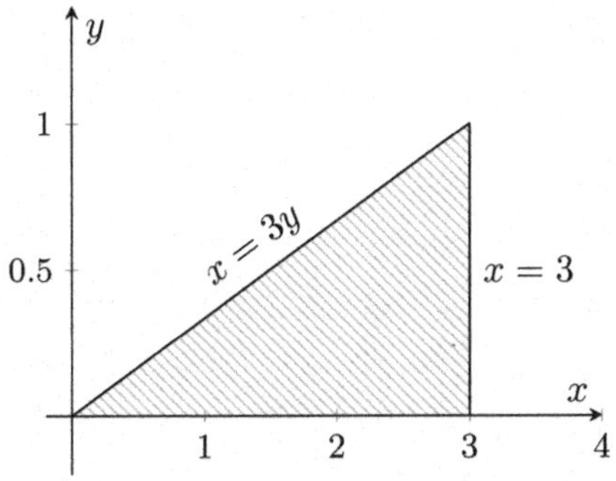

Fig. 14.15 Region of integration for YTI 7.

(7) Write the equivalent integral in which the order of integration is reversed: $\int_0^1 \int_{3y}^3 e^{x^2}\, dxdy$.

☐ We note that the limits of integration describe the triangular region to the right of the line $x = 3y$ and left of $x = 3$, from $y = 0$ to $y = 1$. This is the same as the region between above the line $y = 0$ and below $y = x/3$ from $x = 0$ to $x = 3$. The region is shown in Fig. 14.15. So, we have

$$\int_0^1 \int_{3y}^3 (x^2 + 3y^2)\, dxdy = \int_0^3 \int_0^{x/3} e^{x^2}\, dydx$$

This can now be solved, whereas the original could not. ■

14.4 Triple Integrals

Introduction

Single integrals are for functions like $y = f(x)$. Double integrals are for functions like $z = f(x, y)$. So, triple integrals are for functions like $w = f(x, y, z)$. The leap from single integrals to double integrals is an impressive one. You have to change the way you visualize problems and regions of integration. Once you've gotten comfortable with double integrals, triple integrals may not be as scary. In fact, without even seeing anything else, you might already have insight into the fundamental definition for triple integrals. Having started with the Riemann sum definition for single integrals,

$$\int_a^b f(x)\,dx = \lim_{n\to\infty} \sum_{i=1}^{n} f(x_i^*), \Delta x$$

and then scaling up to double integrals,

$$\iint_D f(x,y)\,dA = \lim_{n,m\to\infty} \sum_{i=j}^{m} \sum_{i=1}^{n} f(x_i^*, y_j^*)\,\Delta A$$

Can you guess how many summations will be involved for a triple integral? Better yet, can you guess the fundamental measurement of the partitioning? For single integrals, we used a bit of length, Δx; for double integrals, we used a bit of areal ΔA. What might we use for triple integrals? You can also probably guess how this pattern plays out to triple integrals:

$$\int_a^b f(x)\,dx \to \iint_D f(x,y)\,dA \ \to \ ???$$

Can you imagine what a triple integral actually computes? They are strange creatures. Consider this: it takes two axes to plot $y = f(x)$. It takes three axes to plot $z = f(x, y)$. How many axes does it take to plot $w = f(x, y, z)$? Four! Which means we're now integrating functions we can't even plot. But wait, it gets better!

Remember that a single integral gives the AREA under a function $f(x)$ over a region (interval) of integration. A double integral gives the VOLUME under a function $f(x, y)$ over a region (area) of integration.

So, naturally, a triple integral gives... ?? We sort of ran out of measurable quantities. What comes after area and volume? Beats me. So now we have a type of integral whose fundamental geometric interpretation is a bit dodgy.

Here's another fun fact about triple integrals: in single integrals, there was only one possible order of integration: dx. In double integrals, the area element could be written two ways: $dxdy$ or $dydx$. In triple integrals, we will have SIX different possible orderings of variables and their limits within the integral. (That sounds worse than it is, though, usually the order doesn't matter or there are clearly only one or two viable options.)

The Definition of a Triple Integral

Remember from the discussion of double integrals that we have a generalized integration process to build from:

(1) Select a region of integration from the domain of the function.
(2) Partition the region of integration into tiny pieces. Each tiny piece, or *partition*,[2] will have its own size.
(3) Within each individual partition, select a representative point.
(4) Plug the coordinate(s) of that representative point into the function.
(5) Multiply the function's value at the representative point by the size of the partition.
(6) Sum these resulting contributions from all partitions.
(7) Find the limit of this sum as the number of partitions goes to ∞
(8) Give this limit a name and symbol.

Let's go through this process for a function of three variables, $f(x, y, z)$, so that you know what a triple integral really is, then we'll see how to compute them.

(1) Select a region of integration from the domain of the function.

The domain of a function $f(x, y, z)$ is part to all of the three-dimensional coordinate system $\mathbb{R}^3$. Therefore, our region of integration will be a portion of $\mathbb{R}^3$. We will have to assume that a region of integration has at least two flat edges; in other words, at least two boundaries of the region will be planes described by constant x, y or z values. Having dealt with double integrals, you should understand the reason for this: since we progressively eliminate the variables through each step of integration, then the

[2]Yes, we are now using partition as a verb and a noun...

final (outermost) limits of integration must be constants to ensure we get a numerical value as the final result. Therefore, at least one variable must be bound by constants. Let's name our 3D regions of integration E (for, I suppose, "Ewww, this is terrible!").

(2) Divide the region of integration into partitions. Each partition will have its own size.

The region of integration E is a full three-dimensional solid. In single variable integrals, we partitioned our interval $[a, b]$ into pieces of size Δx. In double integrals, we partitioned our domain into pieces of size ΔA. Now we'll divide our solid of integration into little rectangular boxes of size (*volume*) ΔV. Each individual coordinate direction is partitioned as before, with lengths Δx, Δy, and Δz, so that $\Delta V = \Delta x \cdot \Delta y \cdot \Delta z$. Such an individual piece of the partition is called a *representative elementary volume.* (Imagine a large wedding cake being cut by a straight knife in all three directions; each individual piece is a representative elementary volume of the cake.)[3]

(3) Within each individual partition, select a representative point.

In the partition of size ΔV, we can select a representative point $(x_i^\star, y_j^\star, z_k^\star)$. Figure 14.16 shows a sample partition; we see a representative elementary volume with edges Δx, Δy, and Δz respectively, along with a representative point $(x_i^\star, y_j^\star, z_k^\star)$ chosen at the midpoint of each coordinate length.

(4) Plug the coordinate(s) of that representative point into the function.

When we plug the representative point $(x_i^\star, y_j^\star, z_k^\star)$ into our function, we get $f(x_i^\star, y_j^\star, z_k^\star)$.

(5) Multiply the function's value at the representative point by the size of the partition.

Here's where we start to lose our connection to known geometric measurements. In a single integral, we built rectangles, and found the area of each rectangle, $f(x_i^\star)\Delta x$. In double integrals, we built standing rectangular columns, and found the volume of each column, $f(x_i^\star, y_j^\star)\Delta A$. What the heck does $f(x_i^\star, y_j^\star, z_k^\star) \cdot \Delta V$ compute?? We don't really know, but that won't stop us from proceeding.

[3]But I don't recommend saying, "Thank you for my representative elementary volume!" because you may be asked to leave.

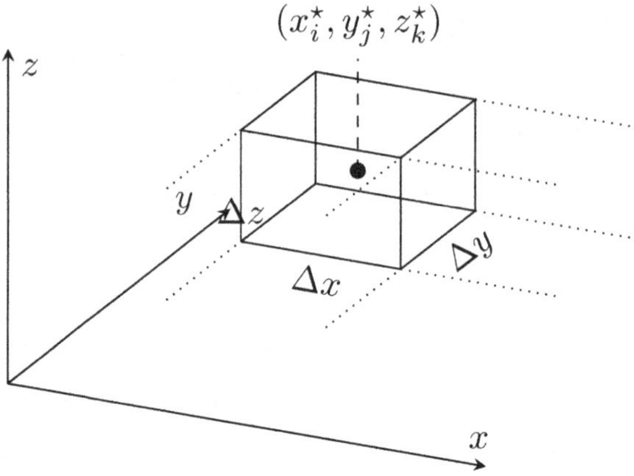

Fig. 14.16 Partitioning the 3D rectangular coordinate system.

(6) Sum these resulting contributions from all partitions.

This sum won't be an area A or volume V, so let's just call it T for total:

$$T \approx \sum_{k=1}^{p}\sum_{j=1}^{m}\sum_{i=1}^{n} f(x_i^\star, y_j^\star, z_k^\star)\Delta V$$

Now, whatever we're computing will undoubtedly improve as the number of partitions in all three directions go to ∞, so we can still do this:

(7) Find the limit of this sum as the number of partitions goes to ∞

$$T = \lim_{n,m,p\to\infty} \sum_{k=1}^{p}\sum_{j=1}^{m}\sum_{i=1}^{n} f(x_i^\star, y_j^\star, z_k^\star)\Delta V$$

And there we have the Riemann Sum definition of the triple integral of $f(x, y, z)$ over a 3D region of integration E.

(8) Give this limit a name and symbol.

Why ruin a good pattern? Let's write the formal definition of a triple integral as:

$$\iiint_E f(x, y, z)\, dV = \lim_{n,m,p\to\infty} \sum_{k=1}^{p}\sum_{j=1}^{m}\sum_{i=1}^{n} f(x_i^\star, y_j^\star, z_k^\star)\Delta V$$

The Computation of a Triple Integral

The process of computation of a triple integral is not that much different as for a double integral, there's just one more step. In fact, if the limits are all already given, you probably already know what to do:

EX 1 Evaluate $\displaystyle\int_0^1\int_0^z\int_0^y ze^{-y^2}\,dxdydz.$

We use Fubini's Theorem again, and work from the inside, outwards:

$$\begin{aligned}\int_0^1\int_0^z\int_0^y ze^{-y^2}\,dxdydz &= \int_0^1\int_0^z \left(xze^{-y^2}\right)\Big|_0^y\,dydz = \int_0^1\int_0^z yze^{-y^2}\,dydz\\ &= \int_0^1\left(-\frac{1}{2}ze^{-y^2}\right)\Big|_0^z\,dz = -\frac{1}{2}\int_0^1 \left(ze^{-z^2}-z\right)dz\\ &= -\frac{1}{2}\left[\left(-\frac{1}{2}e^{-z^2}\right)-\frac{1}{2}z^2\right]\Bigg|_0^1\\ &= \frac{1}{4}\left(e^{-z^2}+z^2\right)\Big|_0^1 = \frac{1}{4}(e^{-1}+1-1) = \frac{1}{4e} \quad \blacksquare\end{aligned}$$

You Try It

(1) Evaluate $\displaystyle\int_0^1\int_0^z\int_0^{x+z}(6xz)\,dydxdz.$

In the case that you need to decide on your own limits because the region of integration is described by its boundaries, we have similar general guidelines as for double integrals:

- At least one variable must have constant boundaries; the rest can be identifiable as surfaces like $z = f(x, y)$, $y = g(x, z)$ or $x = h(y, z)$.
- The variable which has constant limits must be integrated *last*, i.e. must be placed on the outer integral. The other two variables must be progressively eliminated from the inside, out.

These guidelines mean that the innermost variable can have limits that refer to the other two. But the limits on the middle integral can only refer to the outer variable. The innermost integral should eliminate all references to the inner variable. The middle integral should eliminate all references to itself. By the time we get to the outermost integral, only that final third variable should remain. See the example EX 1 above, since it followed the guidelines and worked out well. Can you identify which of the following are proper triple integrals that will result in a number?

$$(1)\ \int_0^y \int_0^2 \int_{x^2}^4 xyz\, dydxdz$$

$$(2)\ \int_0^1 \int_0^{x^2} \int_1^y xyz\, dxdydz$$

$$(3)\ \int_0^1 \int_0^1 \int_0^x xyz\, dxdydz$$

The answer is ... none of them! The final answer for the first will have y's in it. In the second integral, the inner variable is x, but that variable gets reintroduced in the limits of the middle variable, and those new x's will not go away. The same kind of thing happens in the third, too. Do you see why?

In all, the scheme for determining limits of integration (when they're not given explicitly) relies on your visualization of the 3D region of integration. One pair of boundaries will likely be surfaces, say $z_1 = f(x, y)$ and $z_2 = g(x, y)$, and these can be placed on the innermost integral as limits for z. (Often, one or both of these might even be constants.) What's left as the two inner integrals, then, forms a double integral; you can think of the triple integral as:

$$\iint_D \left(\int_{f(x,y)}^{g(x,y)} f(x, y, z)\, dz \right) dA$$

The trick in these problems isn't coming up with limits on z; they're often given, or easy to determine. Rather, the trick is to come up with the other limits. But the limits in the other directions x and y must cover the lateral extent of the full solid of integration, and I find there are three good ways to visualize this — you pick what works for you.

(1) The region in the xy-plane that must be described by the limits on x and y is the *shadow* of the full 3D solid of integration, as would be generated by a spotlight that's above the solid, pointing directly down.
(2) Imagine you are a tailor fitting the full 3D solid of integration for clothing. You wrap your tape measure around the top or bottom of solid, and slide it up and down to determine the shape and size of the cross section at all elevations. Whatever you found as the widest / largest

cross section anywhere on the solid is the shape you use to determine limits on x and y.

(3) (This is my favorite.) If you have seen the old British comedy show *Monty Python's Flying Circus*, then you are familiar with the animated "Monty Python Foot", which comes down from above and squishes other animated objects, usually at the end of the opening credits. If you imagine the Monty Python foot coming down from high on the z-axis and precisely squishing your solid of integration down flat into the xy-plane, what region would the *splat!* cover? A sphere would get smushed into a circle; an upright paraboloid would get smushed into a circle, but a paraboloid lying on its side would get smushed into a parabola. An upright hyperboloid of one sheet would get smushed into a circle. An ellipsoid whose long axis is parallel to the xy-plane would get smushed into an ellipse, but an ellipsoid whose long axis is parallel to the z-axis would get smushed into a circle. (Lots of 3D surfaces have circular 2D cross sections.)

But again, once you have pinned down limits on z and can visualize the region in the xy-plane covered by the solid, then you describe the bounds of this 2D region just like you would for a double integral — perhaps by bounding y with $y_1 = h_1(x)$ and $y_2 = h_2(x)$ and then x itself by constants, $x = a$ and $x = b$.

This entire discussion can get flipped on its side if it is more convenient to determine limits in, say, the y direction first — leaving the limits in x and z to be treated as a double integral.

No matter what else happens, the golden rule of setting limits is: make sure that a variable will completely disappear with successive integration from the inside out. If the inner-most integral goes with dz, then its limits can involve both x and y. But the middle integral can only involve one variable, and the outermost limits must both be constants.

EX 2 Evaluate $\iiint_E 6xy\,dV$ where E lies under the plane $z = 1 + x + y$ and above the region in the xy-plane bounded by $y = \sqrt{x}$, $y = 0$, and $x = 1$.

This is an irregularly shaped solid due to the base of the solid given by the x and y bounds. The top of the solid is the plane $z = 1 + x + y$; the bottom

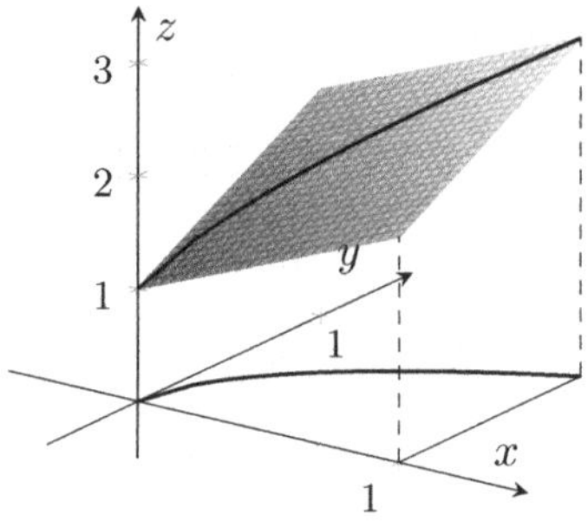

Fig. 14.17 The surface $z = 1 + x + y$.

of the solid must is inferred from the phrase "above the xy-plane" to be $z = 0$. Figure 14.17 shows the plane, the border curves in the xy-plane, and the trace of $y = \sqrt{x}$ in the plane. Since the limits on z refer to both x and y, then we must integrate z first. We don't really need to visualize what's happening down in the xy-plane since those limits are given explicitly: y is bound by $y - 0$ and $y = \sqrt{x}$, and x is then also bounded by $x = 0$ and $x = 1$. The integral we need is:

$$\iiint_E 6xy\, dV = \int_0^1 \int_0^{\sqrt{x}} \int_0^{1+x+y} 6xy\, dzdydx$$

and its evaluation proceeds like this:

$$\int_0^1 \int_0^{\sqrt{x}} \int_0^{1+x+y} 6xy\, dzdydx = \int_0^1 \int_0^{\sqrt{x}} 6xyz\Big|_0^{1+x+y} dydx$$

$$= \int_0^1 \int_0^{\sqrt{x}} 6xy(1 + x + y)\, dydx = \int_0^1 \int_0^{\sqrt{x}} 6xy + 6x^2y + 6xy^2\, dydx$$

$$= \int_0^1 3xy^2 + 3x^2y^2 + 2xy^3\Big|_0^{\sqrt{x}} dx = \int_0^1 3x(x) + 3x^2(x) + 2x(x^{3/2})\, dx$$

$$= \int_0^1 3x^2 + 3x^3 + 2x^{5/2}\, dx = x^3 + \frac{3}{4}x^4 + \frac{4}{7}x^{7/2}\Big|_0^1$$

$$= 1 + \frac{3}{4} + \frac{4}{7} = \frac{65}{28} \quad \blacksquare$$

You can see that evaluation of triple integrals is — even with "simple" limits — long and tedious. For me, once the integral is set up properly, the details of evaluation are secondary. The use of technology is encouraged for evaluation, and in all examples from here on, once the integral is presented, we will leap right to the final result. After all, there's only so much time in the day, and sadly, time for Calculus is often limited. In your allotted time, is it better to do two or integrals all the way out by hand, or set up six or eight integrals and use tech to evaluate them? I'll take the latter.

You Try It

(2) Evaluate $\iiint_E x^2 e^y \, dV$, where E is the region between $z = 1 - y^2$, $z = 0$, $x = -1$, and $x = 1$. (Once the integral is set up, you can use a CAS to evaluate it.)

OK, So They Do Compute Something!

I've been lying ... sort of ... as I've tried to lead you to believe that the result of a triple integral doesn't have any geometric meaning. Sometimes it does, as long as you integrate the proper function: $f(x, y, z) = 1$.

Let's think about this from the ground up. What do we get when we take a single integral of $f(x) = 1$ over the interval $[a, b]$?

$$\int_a^b (1)dx = x\Big|_a^b = b - a$$

You should recognize $b - a$ as the length of the interval $[a, b]$. So in general terms, the (single) integral of $f(x) = 1$ gave us a measure of the region of integration. Now how about the double integral of $f(x, y) = 1$ over a region R?

$$\iint_R (1)dA$$

From our general knowledge, this gives us the total volume underneath the plane $z = 1$ over the region of integration R. But the height of this region is 1, so if A is the area of the region of integration in the xy-plane, then the volume we're computing is $V = \text{area} \times \text{height} = A \cdot 1 = A$. In other words, the result of this double integral is equal to the area of the region of integration R. So in general terms, the integral of $f(x, y) = 1$ gave us a measure of the region of integration.

And so, now, what do you think we get from this integral?

$$\iiint_E (1)\, dV$$

We will get the volume of the solid of integration E! In other words, we can compute the volume of a 3D region bounded by pretty much any surfaces as long as we can describe that region using the limits of integration of a triple integral, and use those limits to integrate $f(x, y, z) = 1$.

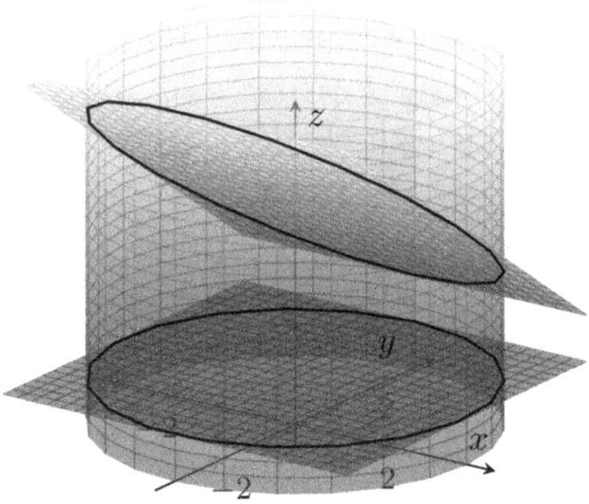

Fig. 14.18 Region between $x^2 + y^2 = 9$, $y + z = 5$, $z = 1$.

EX 3 Find the volume of the region enclosed by the cylinder $x^2 + y^2 = 9$ and the planes $y + z = 5$ and $z = 1$.

This cylinder has radius 3, and is centered around the z-axis. Figure 14.18 shows the cylinder and planes, with the intersections traced out. We enter the cylinder at the bottom through the flat plane $z = 1$ and exit its top through the slanted plane $y + z = 5$. Therefore, z is bounded by $z = 1$ and $z = 5 - y$. The 2D "shadow" of this solid down in the xy-plane is from the interior of the cylinder itself: $x^2 + y^2 = 9$; therefore, y is bounded by $\pm\sqrt{9 - x^2}$ and x is bounded by -3 and 3. (We progressively eliminate variables in the limits.) The triple integral needed to compute this volume is then

$$\int_{-3}^{3} \int_{-\sqrt{9-x^2}}^{\sqrt{9-x^2}} \int_{1}^{5-y} (1)\, dzdydx = 36\pi$$

(The value of the integral computed using tech.) ■

You Try It

(3) Find the volume of the tetrahedron (in the first octant) enclosed by the coordinate planes and the plane $2x + y + z = 4$. (Once the integral is set up, you can use tech to evaluate it.)

Triple Integrals — Problem List

Unless stated otherwise, once you have set up an integral, you can use tech to evaluate it.

Triple Integrals — You Try It

These appeared above; solutions begin on the next page.

(1) Evaluate $\int_0^1 \int_0^z \int_0^{x+z} (6xz)\, dydxdz$ by hand.

(2) Evaluate $\iiint_E x^2 e^y\, dV$, where E is the region between $z = 1 - y^2$, $z = 0$, $x = -1$, and $x = 1$.

(3) Find the volume of the tetrahedron (in the first octant) enclosed by the coordinate planes and the plane $2x + y + z = 4$.

Triple Integrals — Practice Problems

Try these as you get the hang of the You Try It problems. Solutions to these problems are available in Sec. B.2.4.

(1) Evaluate $\int_0^1 \int_x^{2x} \int_0^y (2xyz)\, dzdydx$ by hand.

(2) Evaluate $\iiint_E x\, dV$, where E is the region between the paraboloid $x = 4y^2 + 4z^2$ the plane $x = 4$.

(3) Find the the volume of the solid bounded by the parabolic cylinder $y = x^2$ and the planes $z = 0$, $z = 4$, and $y = 9$.

(4) Describe and sketch the 3D region whose volume is being computed in the following integral:

$$\int_{-3}^{3} \int_{-\sqrt{9-x^2}}^{\sqrt{9-x^2}} \int_{x^2+y^2-9}^{9-x^2-y^2} (1)\, dzdydx$$

(5) Evaluate (by hand) $\iiint_E yz\cos(x^5)\,dV$ where E is defined as the region $E = \{(x, y, z) : 0 \leq x \leq 1; 0 \leq y \leq x; x \leq z \leq 2x\}$.

(6) Find the volume of region between the paraboloid $x = y^2 + z^2$ and the plane $x = 16$.

Triple Integrals — Challenge Problems

Try these problems to test your skills with the ideas in this section. Solutions to these problems are available in Sec. C.2.4.

In these problems, once an integral is set up with all limits in place, you can use a CAS to evaluate it.

(1) Given a 3D region of integration E and a function $f(x, y, z)$ defined in that region, the average value of f over E is given by

$$f_{avg} = \frac{1}{V(E)} \iiint_E f(x, y, z)\,dV$$

where $V(E)$ is the volume of E. Find the average value of $f(x, y, z) = xz + 5z + 10$ over the region in the first octant between the plane $x + y + z = 5$ and the coordinate planes.

(2) Evaluate $\iiint_E y\ln(x) + z\,dV$, where E is defined as the region $E = \{(x, y, z) : 1 \leq x \leq e; 0 \leq y \leq \ln(x); 0 \leq z \leq 1\}$ using TWO equivalent orderings of integration. Obviously, you should get the same value from the integral with each ordering.

(3) Find the volume of region between the elliptic paraboloid $z = 2x^2 + y^2$ and the plane $z = 10$. (Use of tech for evaluation is highly recommended!)

Triple Integrals — You Try It — Solved

---

(1) Evaluate $\int_0^1 \int_0^z \int_0^{x+z} (6xz)\, dydxdz$ by hand.

☐ It's always nice when the integral comes with limits already in place!

$$\int_0^1 \int_0^z \int_0^{x+z} (6xz)\, dydxdz = \int_0^1 \int_0^z (6xzy)\Big|_0^{x+z} dxdz$$

$$= \int_0^1 \int_0^z 6xz(x+z)\, dxdz = \int_0^1 \int_0^z (6x^2z + 6xz^2)\, dxdz$$

$$= \int_0^1 (2x^3z + 3x^2z^2)\Big|_0^z dz = \int_0^1 5z^4 dz = 1 \quad \blacksquare$$

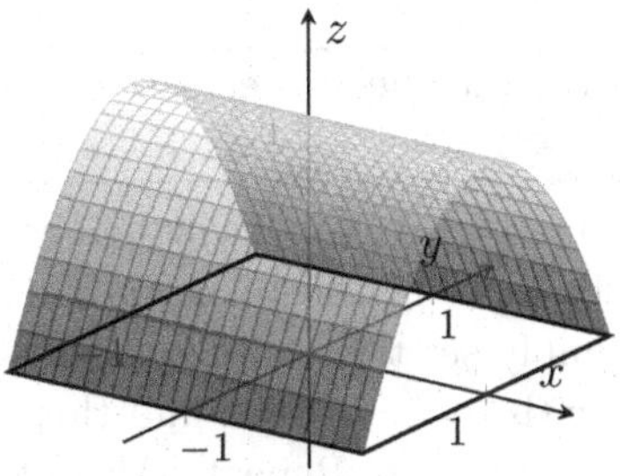

Fig. 14.19 Region between $z = 1 - y^2$, $x = \pm 1$, $z = 0$.

---

(2) Evaluate $\iiint_E x^2 e^y \, dV$ where E is the region between $z = 1 - y^2$, $z = 0$, $x = -1$, and $x = 1$.

☐ The roof of the region E is the parabolic cylinder $z = 1 - y^2$. It looks like a quonset hut. According to the x and y bounds, the floor of this region is the square with both x and y bounds of ± 1. Figure 14.19 shows the surface and the bounds in the xy-plane. We must set up limits of integration so that variables are progressively eliminated; the final limits must be constants. The variable z must go first, but then the order of x and y doesn't really matter. The default ordering is $dzdydx$, and there's no reason to deviate from that. So we have,

$$\iiint_E x^2 e^y \, dV = \int_{-1}^1 \int_{-1}^1 \int_0^{1-y^2} x^2 e^y \, dzdydx = \frac{8}{3e} \quad \blacksquare$$

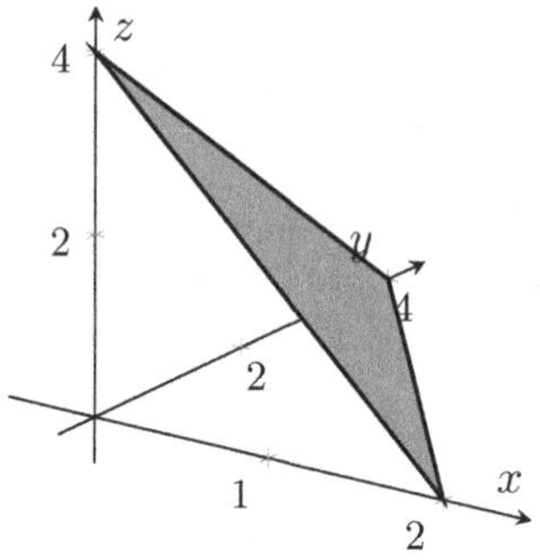

Fig. 14.20 The plane $2x+y+z=4$ in the first octant (w/ YTI 3).

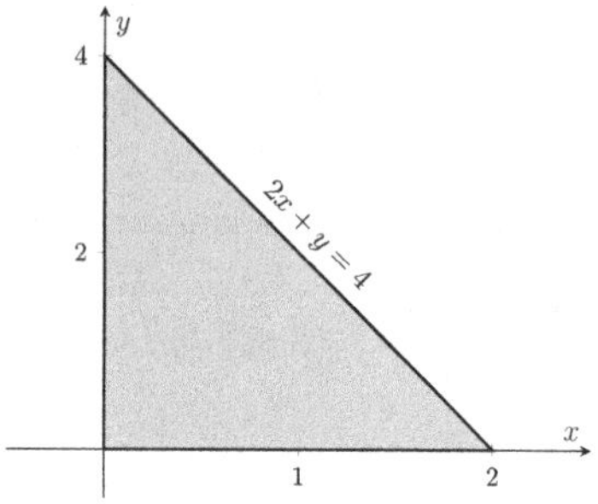

Fig. 14.21 2D region under $2x+y+z=4$ (w/ YTI 3).

---

(3) Find the volume of the tetrahedron enclosed by the coordinate planes and the plane $2x + y + z = 4$.

☐ The defining points of this solid will be the corners of the tetrahedron, i.e. the intercepts of the plane on the coordinate axes. To get these corners, simply set two coordinates to zero and solve for the third; for example, with $y = z = 0$ we find this tetrahedron intersects the x-axis at $2x = 4$ or $x = 2$. Repeat. Altogether, we find that this tetrahedron has corners $(2, 0, 0)$, $(0, 4, 0)$ and $(0, 0, 4)$. Then, we also know that in the xy-plane, the footing of this region is the triangle with coordinates $(2, 0)$, $(0, 4)$, and $(0, 0)$. Figure 14.20 shows this tetrahedron; Fig. 14.21 shows the extent of the region in the xy-plane which is used to determine limits on x and y.

In the z-direction, the tetrahedron goes from $z = 0$ up to the plane $z = 4 - 2x - y$. With z pinned down, we get limits on x and y by exploring the region of the xy-plane used by this tetrahedron.

The intersection of the tetrahedron with the xy-plane (i.e. z = 0) is the line $2x + y = 4$, i.e. $y = -2x + 4$. So limits on y are from $y = 0$ to $y = -2x + 4$. Now that z and y are pinned down, we need x. But we know that only $0 \leq x \leq 2$ is used.

We must order limits of integration so that variables are progressively eliminated; the final limits must be constants. So, z must go first, then

y and finally x: So,

$$V = \iiint E\, dV = \int_0^2 \int_0^{-2x+4} \int_0^{4-2x-y} dzdydz = \frac{16}{3}$$

FFT: A common error is to use the corners of the tetrahedron to immediately set bounds on the variables as $0 \leq x \leq 2$, $0 \leq y \leq 4$ and $0 \leq z \leq 4$. Why is this wrong? ■

Chapter 15

And Can We Agree Rectangular Coordinates Are Rather Dull?

15.1 Double Integrals in Polar Coordinates

Introduction

You may guess from the title of this chapter that polar coordinates are going to be involved. If you've been engaged with this content since Chapter 1, then you've seen polar coordinates. If you've jumped into this book at the start of the multivariable calculus chapters, you can backtrack to Chapter 12 to catch up on polar coordinates. If nothing else, you can get a lot of mileage out of the conversion equation $r^2 = x^2 + y^2$; draw that information onto a right triangle with inner angle θ, and you'll see everything you need to see about polar coordinates.

Double integrals are sometimes easier if we do them in polar coordinates instead of rectangular coordinates. So first, we need to see what that even means in the first place, and then find how to either set up a double integral from scratch in polar coordinates, or to convert one posed in rectangular coordinates into polar coordinates.

If changing coordinate systems in an integral sounds awful, be aware that you've already done this in a simpler setting. By now you should be well versed in "u-substitution"; in that process, you replace your x-coordinate with a new coordinate, called u. That is a change of coordinates in one dimension. Now we have two dimensions, but in one aspect, the conversion is even easier than in regular substitution. In substitution problems for single integrals, YOU had to decide on the right choice of conversion from x to u; in polar coordinates, the conversion is always done using the same equations — and either it works or it doesn't. There's less guesswork involved. An additional benefit is that, unlike in rectangular coordinates where there are two possible orders of integration for double integrals, once a double integral is set in polar coordinates, we usually maintain the same order of integration.

The problems given here are all problems that work well in polar coordinates. In later topics, you will encounter double integrals for which *you* will have to decide whether polar or rectangular coordinates are more convenient.

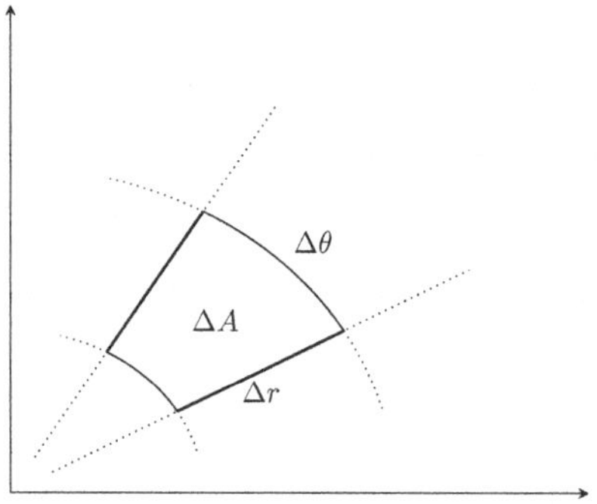

Fig. 15.1 Partitioning in polar coordinates.

When and How To Use Polar Coordinates

Some signs that you should consider using polar coordinates are:

(1) The problem explicitly tells you to use polar coordinates. (That one's sort of obvious, eh?)
(2) The region of integration is a circle or a portion of a circle. (Did you notice that when working in rectangular coordinates, a rectangular region of integration led us to limits of integration which were all constant? This is analogous to having constant limits in rectangular coordinates over a rectangular region. When our region of integration is circular, the limits of integration with respect to r and θ are constants.
(3) The function $f(x, y)$ being integrated contains the term $x^2 + y^2$, since that can be replaced by r^2.

In polar coordinates, a double integral has the form

$$\iint_R f(r,\theta)dA = \int_{\theta_1}^{\theta_2}\int_{r_1}^{r_2} f(r,\theta)rdrd\theta$$

That is,

- The area element dA becomes $rdrd\theta$ (see below for the reason why).
- Limits of integration are the bounds of r and θ which describe the region of integration.
- The function must be written in terms of r and θ.

It may be surprising that the area element in polar coordinates is not $dA = drd\theta$ but rather $dA = r\,dr\,d\theta$ (note the extra r). Here's why. If we discretize a region of integration with respect to r and θ in divisions of size Δr and $\Delta\theta$, we won't get rectangles but rather little radial sectors like the one shown in Fig. 15.1.

In rectangular coordinates, we'd write $\Delta A = \Delta y \Delta x$. But how do we write ΔA in polar coordinates? Note that the sector is *almost* a rectangle, and if it was, we'd have to come up with lengths of two sides to get the area. One side of the sector is already a length (Δr), but the other measure ($\Delta\theta$) is certainly not a length. Well, do you remember that old formula from geometry saying the length of a circular arc is $s\Delta\theta$, where s is the radius of the arc and $\Delta\theta$ is the angle swept out by that arc? If we adapt our notation to that idea, the other edge of the sector has length $r\Delta\theta$, and so the area of the sector is *approximately*

$$\Delta A \approx \Delta r(r\Delta\theta) = r\Delta r\Delta\theta$$

Then, when we go through the usual limiting process involved in the background of integration, this turns into $dA = rdrd\theta$.

Never forget that extra r, because if you do, your answers will be wrong!

Describing Polar Regions With r and θ

In order to generate correct limits of integration for double integrals in polar coordinates, we must be able to delineate regions in $\mathbb{R}^2$ by specifying bounds on r and θ. Here are some quick examples of doing that.

EX 1 Describe the following regions in polar coordinates:

(a) The upper semicircle of radius 4.

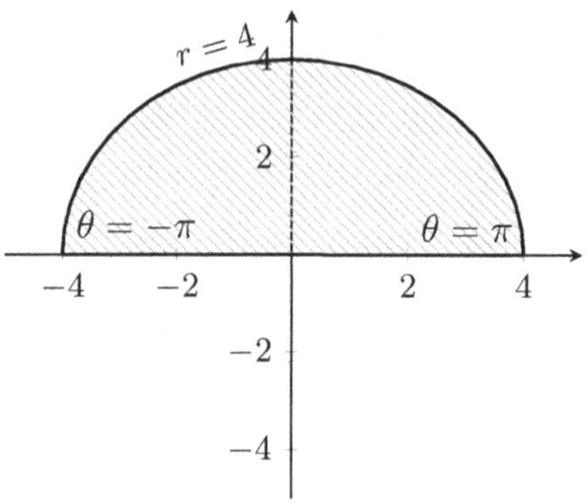

Fig. 15.2 The upper semicircle of radius 4.

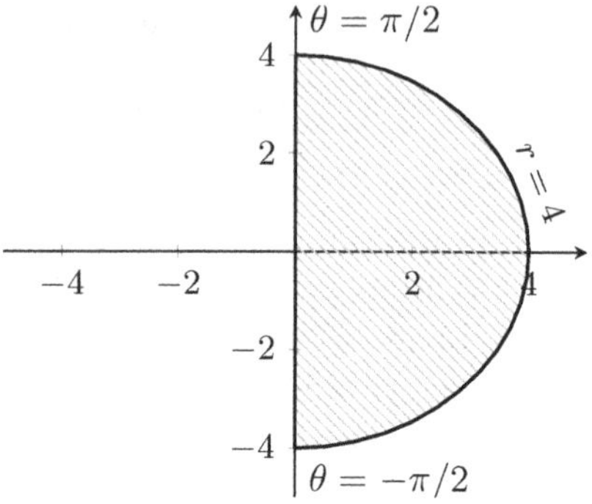

Fig. 15.3 The region between the right half of $x^2 + y^2 = 16$ and the y-axis.

(b) The region between the right half of $x^2 + y^2 = 16$ and the y-axis.
(c) The region $y = |x|$ and $x^2 + y^2 = 9$.
(d) The region above $y = -\sqrt{25 - x^2}$ and below the x-axis.
(e) One loop of the polar graph $r = \cos 2\theta$.

(a) The upper semicircle of radius 4, otherwise known as $y = \sqrt{16 - x^2}$, has polar bounds $0 \leq r \leq 4$, and $0 \leq \theta \leq \pi$, see Fig. 15.2.

(b) The region between the right half of $x^2 + y^2 = 16$ and the y-axis, otherwise known as $x = \sqrt{16 - y^2}$, is just a different portion of the same circle as the curve in (a), and has polar bounds $0 \leq r \leq 4$, and $-\pi/2 \leq \theta \leq \pi/2$ — see Fig. 15.3.

(c) The region between $y = |x|$ and $x^2 + y^2 = 9$ is shown in Fig. 15.4; this has polar bounds $0 \leq r \leq 3$ and $\pi/4 \leq \theta \leq 3\pi/44$.

(d) The region above $y = -\sqrt{25 - x^2}$ and below the x-axis is the lower half of a circle of radius 5 centered at the origin. Do we even need a picture for yet more parts of circles? This region uses $0 \leq r \leq 5$, and the angular bounds can be described in many ways, but usually either as $-\pi \leq \theta \leq 0$ or $\pi \leq \theta \leq 2\pi$, depending on how these bounds might be combined with others.

(e) You may remember $r = \cos 2\theta$ as a rose-petal shaped graph. We will describe the start and stop of a loop by two adjacent θ values that produce $r = 0$ (since the loop starts with $r = 0$ and then closes out when $r = 0$ again). We know that $\cos 2\theta = 0$ when $2\theta_1 = -\pi/2$ and then again at $2\theta_2 = \pi 2$. So a pair of θ values that start and stop a loop are $\theta_1 = -\pi/4$ and $\theta_2 = \pi/4$. The radial variable is bound by the function itself, as $r_1 = 0$ and $r_2 = \cos 2\theta$. A graph of this curve is shown in Fig. 15.5. ■

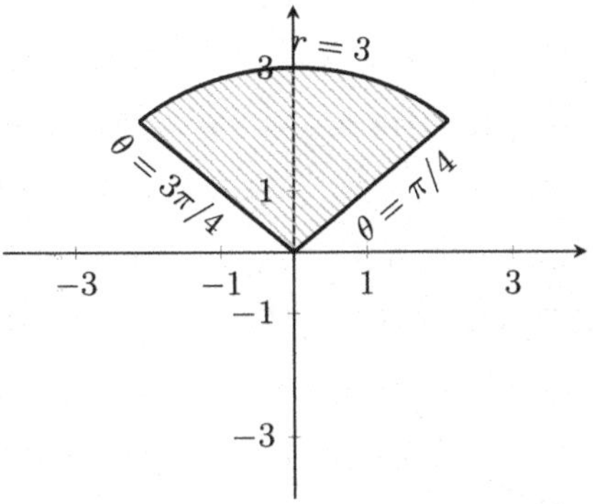

Fig. 15.4 The region between $y = |x|$ and $x^2 + y^2 = 9$.

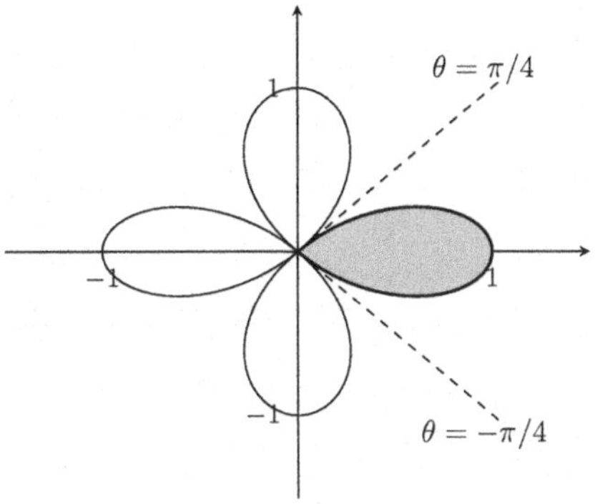

Fig. 15.5 One loop of the polar curve $r = \cos(2\theta)$.

You Try It

(1) Describe the following regions in polar coordinates:

(a) The region inside $x^2 + y^2 = 4$ and above $y = x$.

(b) The region between the circles $x^2 + y^2 = 4$ and $x^2 + y^2 = 9$, such that $y \geq 0$.

(c) The region between above the x-axis, inside $y = \sqrt{10 - x^2}$, and to the right of $y = -x$.

(d) One loop of the polar graph $r = 2 \sin 2\theta$.

Examples of Double Integrals in Polar Coordinates

Types of problems in which we might need polar coordinates include

- a generic integral which is easier in polar coordinates
- a problem requiring the area inside a polar curve

- the volume under a surface, when the volume integral $\iint_R f\,dA$ is easier to handle in polar coordinates

Here are examples of each type. First, let's see a problem which is just plain easier in polar coordinates.

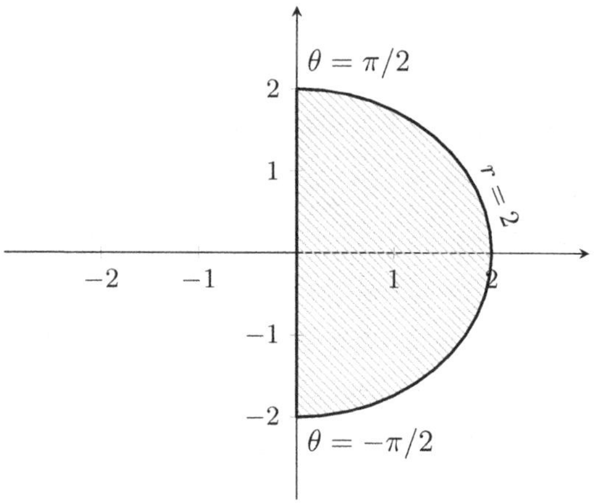

Fig. 15.6 Polar region between $x = \sqrt{4 - y^2}$ and the y-axis.

EX 2 Evaluate the integral $\iint_R e^{-x^2-y^2}\,dA$ where R is the region bounded by $x = \sqrt{4-y^2}$ and the y-axis.

This does not call for polar coordinates explicitly, but it will sure help. In fact, we can't even begin this problem in rectangular coordinates since we don't know the antiderivative of the function with respect to either x or y. Now, the region bounded by $x = \sqrt{4-y^2}$ and the y-axis is the inner half of the circle $x^2 + y^2 = 4$ to the right of the y-axis — see Fig. 15.6. We can describe this same region much easier in polar coordinates, as $0 \le r \le 2$ and $-\pi/2 \le \theta \le \pi/2$. In the function itself, we see the term $x^2 + y^2$ hiding, and so we can easily convert the function to polar coordinates:

$$e^{-x^2-y^2} = e^{-(x^2+y^2)} = e^{-r^2}$$

And so, remembering that $dA = r\,drd\theta$, the integral becomes

$$\iint_R e^{-x^2-y^2}\,dA = \int_{-\pi/2}^{\pi/2}\int_0^2 e^{-r^2}\,r\,drd\theta = \int_{-\pi/2}^{\pi/2}\left(-\frac{1}{2}e^{-4}+\frac{1}{2}\right)d\theta$$

$$= \frac{1}{2}\left((1-e^{-4})\theta\right)\Big|_{-\pi/2}^{\pi/2} = \frac{\pi}{2}(1-e^{-4})$$

Did you notice that the extra r in $dA = rdrd\theta$ is what made the integration possible because of the substitution necessary (but not shown, since we've done those a lot) in the second step? ■

You Try It

(2) Use polar coordinates to evaluate $\iint_R \cos(x^2+y^2)\,dA$ where R is the region above the x-axis and inside the circle $x^2+y^2=9$.

EX 3 Find the area of one loop of $r = \cos 2\theta$

Regardless of whether we are going to use rectangular or polar coordinates, we can find the area of a general 2D region R by doing the integral $\iint_R (1)dA$. This particular region suggests polar coordinates will be best. The bounds of this region were developed in EX 1 (see Fig. 15.5), and the resulting integral we need to solve is

$$\iint_R (1)\,dA = \int_{-\pi/4}^{\pi/4}\int_0^{\cos 2\theta}(1)\,r\,drd\theta = \int_{-\pi/4}^{\pi/4}\frac{1}{2}r^2\Big|_0^{\cos 2\theta}\,d\theta$$

$$= \frac{1}{2}\int_{-\pi/4}^{\pi/4}\cos^2(2\theta)\,d\theta = \frac{1}{2}\left(\frac{\theta}{2}+\frac{1}{8}\sin 4\theta\right)\Big|_{-\pi/4}^{\pi/4}$$

$$= \frac{1}{2}\left[\frac{\pi}{8}-\left(-\frac{\pi}{8}\right)\right] = \frac{\pi}{8}$$

and this is the area of one loop of the curve. Note that when we encountered the antiderivative of $\cos^2 2\theta$, we just relied on prior knowledge of that integral ... you can look it up if you need to. ■

You Try It

(3) Find the area of one loop of the curve $r = \cos 3\theta$.

How about a volume problem?

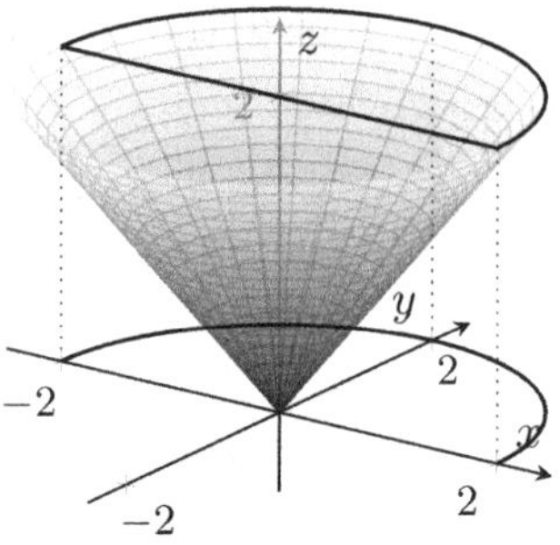

Fig. 15.7 The cone $z = \sqrt{x^2 + y^2}$; the upper half of $x^2 + y^2 = 4$.

EX 4 Find the volume under the cone $z = \sqrt{x^2 + y^2}$ and over the upper half of the circle $x^2 + y^2 = 4$.

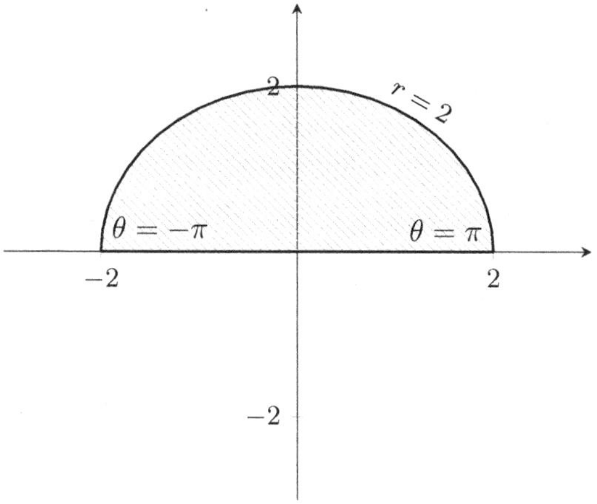

Fig. 15.8 The upper half of $x^2 + y^2 = 4$.

To find the volume under a 3D surface, we must evaluate the double integral $\iint_R z\,dA$, where z is the surface under which we're looking, and R is the region from the xy-plane over which the volume sits. Nothing changes here, except now we'll consider doing this problem in polar coordinates. Why? Because the region of integration is part of a circle and the function contains our friend x^2+y^2. In polar coordinates, the surface is $z = \sqrt{r^2} = r$.

The region of integration has bounds $0 \le r \le 2$ and $0 \le \theta \le \pi$. Figures 15.7 and 15.8 show the full surface as well as the corresponding polar region in the 2D plane. Therefore, the integral that gives the volume we want is:

$$\iint_R z\,dA = \int_0^\pi \int_0^2 (r)\, r\, dr d\theta = \int_0^\pi \int_0^2 r^2\, dr d\theta = \int_0^\pi \frac{1}{3} r^3 \bigg|_0^2 \, d\theta$$
$$= \int_0^\pi \left(\frac{8}{3}\right) d\theta = \frac{8}{3}\theta \bigg|_0^\pi = \frac{8\pi}{3}$$

so the volume we wanted is $V = 8\pi/3$. ■

You Try It

(4) Use polar coordinates to find the volume under the paraboloid $z = x^2 + y^2$ over the region $x^2 + y^2 \le 9$.

Finally, how about some practice converting a given integral? There are many cases where an integral posed in rectangular coordinates will be easier to solve after a conversion to polar coordinates.

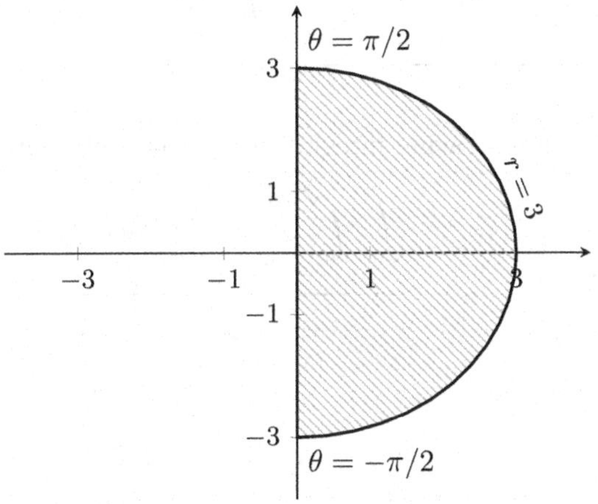

Fig. 15.9 The right half of $x^2 + y^2 = 9$.

EX 5 Convert the following integral into polar coordinates:

$$\int_0^3 \int_{-\sqrt{9-x^2}}^{\sqrt{9-x^2}} (x + y)\, dy dx$$

While the inner integral isn't bad, by the time we solved that integral and plugged in the inner limits of integration, the resulting outer integral would be horrid. This integral might be better posed in polar coordinates. The limits on y show that our region is bounded by $y_1 = -\sqrt{9-x^2}$ and $y_2 = \sqrt{9-x^2}$. These are the lower and upper halves of the circle $x^2+y^2=9$. But the x limits restrain x between 0 and 3. Therefore, our region of integration is only the right half of the circle $x^2+y^2=9$ — see Fig. 15.9. In polar coordinates, this region is described as $0 \leq r \leq 3$ and $-\pi/2 \leq \theta \leq \pi/2$. The function we're integrating is $x+y$, which we can't really do much with except use the conversion equations $x = r\cos\theta$ and $y = r\sin\theta$. Essentially, we're going to make the limits of integration easier but make the function a bit yuckier. It's a good trade. Finally, don't forget to change the area element from $dydx$ to $rdrd\theta$!

$$\int_0^3 \int_{-\sqrt{9-y^2}}^{\sqrt{9-y^2}} (x+y)\, dxdy = \int_{-\pi/2}^{\pi/2} \int_0^3 (r\cos\theta + r\sin\theta)\, r\, drd\theta$$

I suppose we should clean the integral up a bit, and get:

$$\int_0^3 \int_{-\sqrt{9-y^2}}^{\sqrt{9-y^2}} (x+y)\, dxdy = \int_{-\pi/2}^{\pi/2} \int_0^3 (\cos\theta + \sin\theta) r^2\, drd\theta \quad \blacksquare$$

You Try It

(5) Convert the following integral into polar coordinates:

$$\int_0^1 \int_0^{\sqrt{1-x^2}} e^{x^2+y^2}\, dydx$$

Double Integrals in Polar Coordinates — Problem List

Double Integrals in Polar Coordinates — You Try It

These appeared above; solutions begin on the next page.

(1) Describe the following regions in polar coordinates:
   (a) The region inside $x^2 + y^2 = 4$ and above $y = x$.
   (b) The region between the circles $x^2 + y^2 = 4$ and $x^2 + y^2 = 9$, such that $y \geq 0$.
   (c) The region between above the x-axis, inside $y = \sqrt{10 - x^2}$, and to the right of $y = -x$.
   (d) One loop of the polar graph $r = 2\sin 2\theta$.

(2) Use polar coordinates to evaluate $\iint_R \cos(x^2 + y^2)\, dA$, where R is the region above the x-axis and inside the circle $x^2 + y^2 = 9$.

(3) Find the area of one loop of the curve $r = \cos 3\theta$.

(4) Use polar coordinates to find the volume under (and outside of) $z = x^2 + y^2$ over the region $x^2 + y^2 \leq 9$.

(5) Convert the following integral into polar coordinates:

$$\int_0^1 \int_0^{\sqrt{1-x^2}} e^{x^2+y^2}\, dydx$$

Double Integrals in Polar Coordinates — Practice Problems

Try these as you get the hang of the You Try It problems. Solutions to these problems are available in Sec. B.3.1.

(1) Use polar coordinates to evaluate $\iint_R \sqrt{4 - x^2 - y^2}\, dA$, where R is the region $\{(x, y) : x^2 + y^2 \leq 4, x \geq 0\}$.

(2) Find the area inside the curve $r = 4 + 3\cos\theta$.

(3) Find the volume inside the sphere $x^2 + y^2 + z^2 = 16$ and outside the cylinder $x^2 + y^2 = 4$.

(4) Convert this integral into polar coordinates: $\int_{-a}^{a} \int_0^{\sqrt{a^2-y^2}} (x^2 + y^2)^{3/2}\, dxdy$.

(5) Find the area outside $r = 2$ and inside $r = 4\sin\theta$.

(6) Find the volume between the paraboloid $z = 10 - 3x^2 - 3y^2$ and the plane $z = 4$.

(7) Convert this integral into polar coordinates: $\int_0^2 \int_{-\sqrt{4-y^2}}^{\sqrt{4-y^2}} x^2y^2 \, dxdy.$

Double Integrals in Polar Coordinates — Challenge Problems

Try these problems to test your skills with the ideas in this section. Solutions to these problems are available in Sec. C.3.1.

Once these integrals are set up, you can use tech for the evaluation.

(1) Given a 2D region R and a function f defined in that region, the average value of f over R is given by

$$f_{avg} = \frac{1}{A(R)} \iint_R f \, dA$$

where $A(R)$ is the area of R. Find the average value of $f(x, y) = \sqrt{x^2 + y^2}$ over the region bounded by $r = 3\sin 2\theta$ between $\theta = 0$ and $\theta = \pi/2$.

(2) Use a double integral in polar coordinates to find the volume between the paraboloid $z = 16 - 2x^2 - 2y^2$ and the plane $z = 2$.

(3) Convert the following integral into polar coordinates and then evaluate:

$$\int_0^3 \int_{-\sqrt{9-y^2}}^{\sqrt{9-y^2}} (x^2 + y^2)^2 \, dxdy$$

Double Integrals in Polar— You Try It — Solved

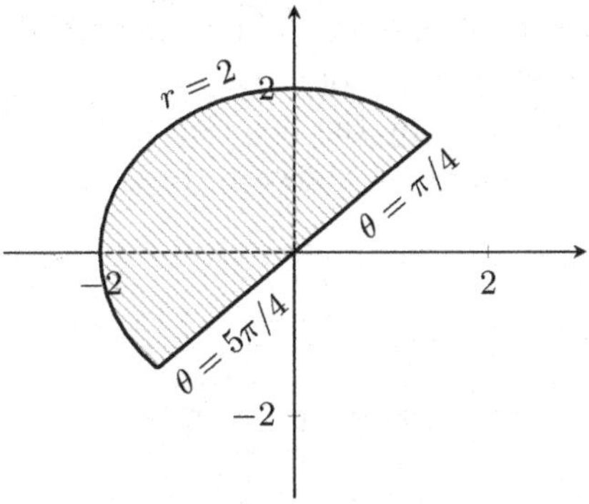

Fig. 15.10 The region inside $x^2 + y^2 = 4$ and above $y = x$.

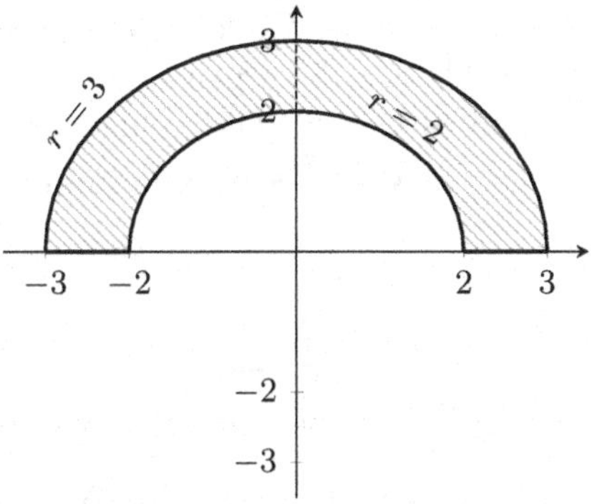

Fig. 15.11 Between $x^2 + y^2 = 4$ and $x^2 + y^2 = 9$, $y \geq 0$.

---

(1) Describe the following regions in polar coordinates:

(a) The region inside $x^2 + y^2 = 4$ and above $y = x$.

(b) The region between the circles $x^2 + y^2 = 4$ and $x^2 + y^2 = 9$, such that $y \geq 0$.

(c) The region between above the x-axis, inside $y = \sqrt{10 - x^2}$, and to the right of $y = -x$.

(d) One loop of the polar graph $r = 2\sin 2\theta$.

☐ (a) The region inside $x^2 + y^2 = 4$ and above $y = x$ is shown in Fig. 15.10. It has polar bounds $0 \leq r \leq 2$ and $\pi/4 \leq \theta \leq 5\pi/4$. When interpreting the description of a region, it's just as important to note what is not said as much as what is said. In this example, it's tempting to stop the region at the positive y-axis, since very often we work in the first quadrant. But no such restriction is explicitly given in the description of the region.

(b) The region between the circles $x^2 + y^2 = 4$ and $x^2 + y^2 = 9$, such that $y \geq 0$, is shown in Fig. 15.11. It has polar bounds $2 \leq r \leq 3$ and $0 \leq \theta \leq \pi$.

(c) The region between above the x-axis, inside $y = \sqrt{10 - x^2}$, and to the right of $y = -x$, is shown in Fig. 15.12. It has polar bounds $0 \leq r \leq \sqrt{10}$ and $0 \leq \theta \leq \pi/4$.

(d) To find one loop of the polar graph $r = 2\sin 2\theta$, we must find two adjacent θ values that produce $r = 0$ (since a loop starts and closes when $r = 0$). We know that $2\sin 2\theta = 0$ when $2\theta_1 = 0$ and then again at $2\theta_2 = \pi$. So a pair of θ values that start and stop a loop are $\theta_1 = 0$ and $\theta_2 = \pi/2$. The radius variable is bound by the function itself, as $r_1 = 0$ and $r_2 = 2\sin 2\theta$. A graph of this curve is shown in Fig. 15.13. ■

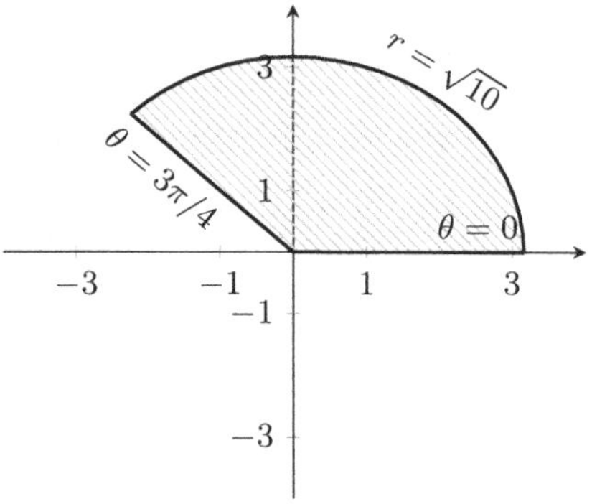

Fig. 15.12 Above the x-axis, inside $y = \sqrt{10 - x^2}$, right of $y = -x$.

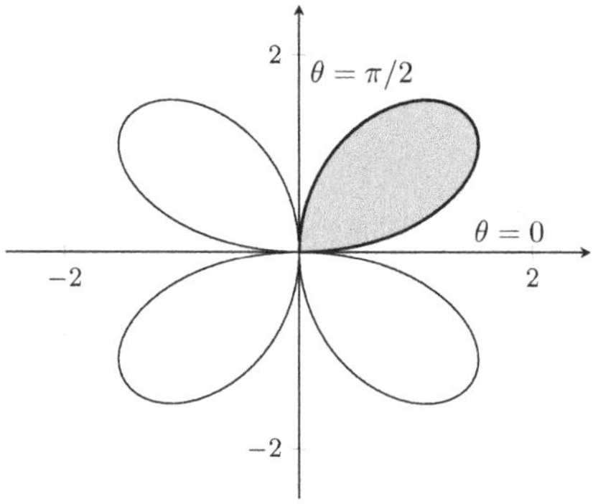

Fig. 15.13 One loop of the polar curve $r = 2\sin(2\theta)$.

---

(2) Use polar coordinates to evaluate $\iint_R \cos(x^2 + y^2)\, dA$, where R is the region above the x-axis and inside the circle $x^2 + y^2 = 9$.

□ I'm going to take a leap of faith that we don't need a figure for this one. The region R is swept out by the polar values $0 \le r \le 3$ and $0 \le \theta \le \pi$. Since $x^2 + y^2 = r^2$, and $dA = r dr d\theta$ in polar coordinates,

$$\iint_R \cos(x^2 + y^2)\, dA = \int_0^\pi \int_0^3 \cos(r^2) r\, dr d\theta$$

$$= \int_0^\pi \left(\frac{1}{2}\sin(r^2)\right)\Bigg|_0^3 d\theta = \int_0^\pi \frac{1}{2}\sin 9\, d\theta$$

$$= \left(\frac{1}{2}\sin 9\theta\right)\Bigg|_0^\pi = \frac{\pi}{2}\sin 9 \quad ■$$

(3) Find the area of one loop of the curve $r = \cos 3\theta$.

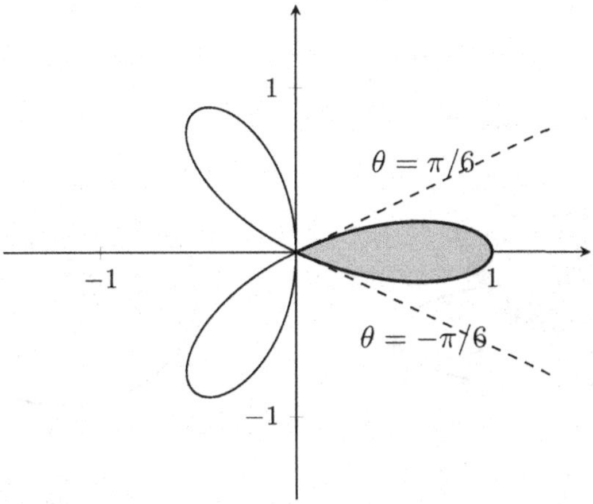

Fig. 15.14 One loop of $r = \cos(3\theta)$.

□ A graph of this curve is shown in Fig. 15.14. A single loop of the curve starts and stops at two consecutive locations where $r = 0$; this is when, for example, $3\theta = \pm\pi/2$, i.e. $\theta = \pm\pi/6$. The limits on r come from the equation of the curve itself. The double integral for the area starts off as

$$\iint_D dA = \int_{-\pi/6}^{\pi/6} \int_0^{\cos 3\theta} r dr\, d\theta = \int_{\pi/6}^{\pi/6} \left(\frac{1}{2}r^2\right) \Big|_0^{\cos 3\theta} d\theta$$
$$= \int_{\pi/6}^{\pi/6} \frac{1}{2} \cos^2 3\theta\, d\theta \cdots$$

Now a quick substitution of 3θ by u and so also $d\theta$ by $du/3$ (and remember that the substitution changes the limits, too!), we can continue,

$$\cdots = \int_{-\pi/2}^{\pi/2} \frac{1}{6} \cos^2 u\, du = \frac{1}{6}\left(\frac{1}{2}u + \frac{1}{4}\sin 2u\right)\Big|_{-\pi/2}^{\pi/2}$$

The contribution from $\sin 2u$ is zero at both endpoints, and so finally:

$$\cdots = \frac{1}{12}(u)\Big|_{-\pi/2}^{\pi/2} = \frac{\pi}{12}$$

(We can refer to prior knowledge of the antiderivative of $\cos^2 u$ here.)

■

(4) Use polar coordinates to find the volume under (and outside of) $z = x^2 + y^2$ over the region $x^2 + y^2 \leq 9$.

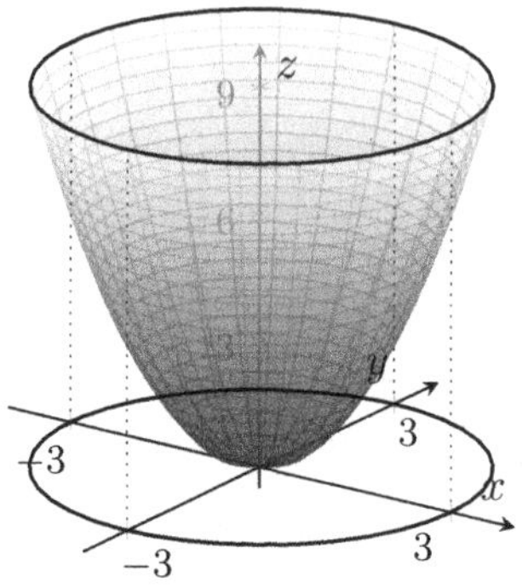

Fig. 15.15 Region under (outside of) $z = x^2 + y^2$, over $x^2 + y^2 \leq 9$.

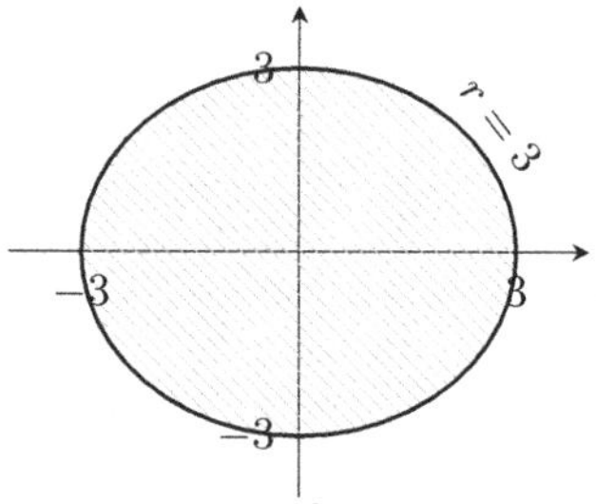

Fig. 15.16 Inside $x^2 + y^2 = 9$.

□ Figures 15.15 and 15.16 show the full surface as well as the corresponding polar region in the 2D plane. The stated domain is the area inside $r = 3$ for $0 \leq \theta \leq 2\pi$. So the volume under $z = x^2 + y^2$ i.e. $z = r^2$ over that domain is:

$$\iint_D r^2\,dA = \int_0^{2\pi}\int_0^3 r^2 \cdot r\,drd\theta = \int_0^{2\pi} \left(\frac{1}{4}r^4\right)\Big|_0^3 d\theta$$
$$= \int_0^{2\pi} \frac{81}{4}\,d\theta = \frac{81\pi}{2} \quad \blacksquare$$

(5) Convert the following integral into polar coordinates:

$$\int_0^1 \int_0^{\sqrt{1-x^2}} e^{x^2+y^2}\,dydx$$

□ The limits of integration describe the quarter of the unit circle in the first quadrant — see Fig. 15.17 — which is the region $0 \leq r \leq 1$ and $0 \leq \theta \leq \pi/2$. The overworked formula $x^2 + y^2 = r^2$ converts $e^{x^2+y^2}$ into e^{r^2}, and of course the area element $dydx$ become $r\,drd\theta$. Together,

$$\int_0^1 \int_0^{\sqrt{1-x^2}} e^{x^2+y^2}\,dydx = \int_0^{\pi/2}\int_0^1 e^{r^2} \cdot r\,drd\theta$$

Now the integral would be much easier to solve. ■

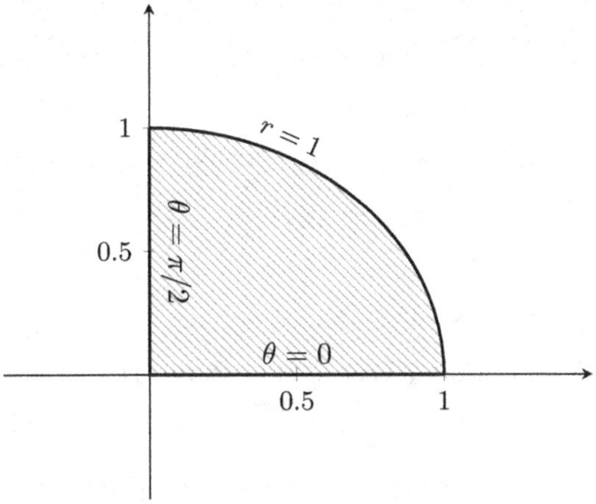

Fig. 15.17 The unit circle in the first quadrant.

15.2 Cylindrical and Spherical Coordinates

Introduction

Your entire Calculus life before Chapter 14 was spent with single integrals, and then — wham — you got hit with the one-two punch of double and then triple integrals.

Well, the same thing is about to happen with coordinate systems. You spent many years only knowing the rectangular coordinate system, and then only recently encountered the two-dimensional polar coordinate system — and now, wham, you're about to get hit with the one-two punch of two three-dimensional coordinate systems: cylindrical and spherical coordinates.

However, the thing to keep in mind is that even though they have scary names, the point of having access to cylindrical and spherical coordinates is to make your life *easier*. Really!

Cylindrical Coordinates

The good news is that if you get along with polar coordinates, then you'll be just fine with cylindrical coordinates, too. Cylindrical coordinates are just the polar coordinates r and θ with the cartesian z-coordinate tossed

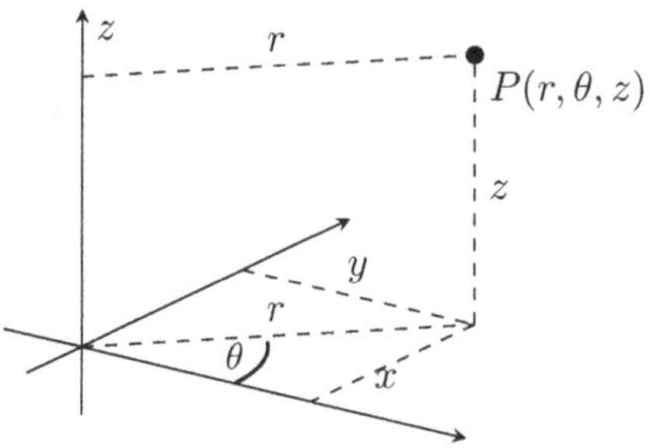

Fig. 15.18 A point's cylindrical coordinates (r, θ, z).

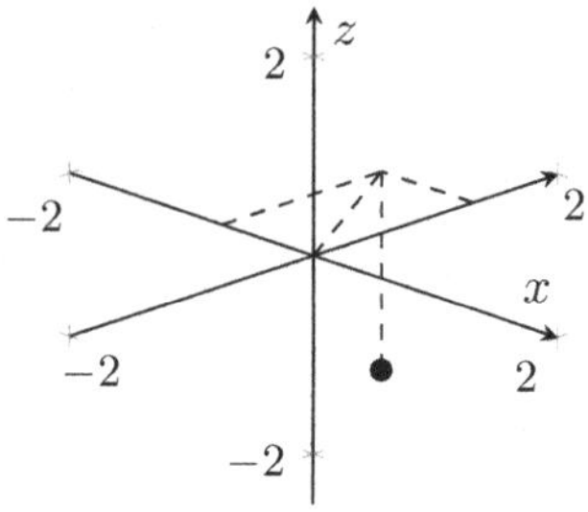

Fig. 15.19 This point's cylindrical coordinates are $(3/2, 2\pi/3, -2)$.

in. A point's cylindrical coordinates are presented as (r, θ, z), where r and θ are the same as in polar coordinates, and z is the same as in rectangular coordinates. In other words,

- r is the horizontal distance to the point as measured straight out from the z-axis. Equivalently, it's the distance of the point's shadow in the xy-plane to the origin $(0, 0)$.
- θ is the horizontal angle you have to turn, counterclockwise from the positive x-axis, to find the point's shadow in the xy-plane.
- z is the point's elevation

A general diagram of cylindrical coordinates is shown in Fig. 15.18, and the location of the point whose cylindrical coordinates are $(r, \theta, z) = (3/2, 2\pi/3, -2)$ are in Fig. 15.19.

The conversion equations we can use to go back and forth between cylindrical and rectangular coordinates are based on the polar-rectangular equations.

Useful Fact 15.1. *A point's rectangular coordinates* (x, y, z) *and cylindrical coordinates* (r, θ, z) *can be traded back and forth with these conversion equations:*

$$r^2 = x^2 + y^2 \quad ; \quad \tan\theta = \frac{y}{x} \quad ; \quad z = z \tag{15.1}$$

and

$$x = r\cos\theta \quad ; \quad y = r\sin\theta \quad ; \quad z = z \tag{15.2}$$

Since polar coordinates are most natural when describing something circular, you can probably imagine that cylindrical coordinates are most natural when describing ... something cylindrical? In fact, the equation $r = 1$ not only describes a circle in two dimensions (polar), it also describes a cylinder in three dimensions.

EX 1 Determine the cylindrical coordinates of the point whose rectangular coordinates are $(2\sqrt{3}, -2, -1)$.

Since the z-coordinate is the same in cylindrical and rectangular coordinates, we don't have to convert $z = -1$. We can convert $x = 2\sqrt{3}$ and $y = -2$ into cylindrical coordinates using (15.1):

$$r^2 = x^2 + y^2 = (2\sqrt{3})^2 + (2)^2 \quad \rightarrow \quad r = 4$$

$$\tan\theta = \frac{y}{x} = \frac{-2}{2\sqrt{3}} = -\frac{1}{\sqrt{3}} \quad \rightarrow \quad \theta = -\frac{\pi}{6}$$

The cylindrical coordinates of this point are $(r, \theta, z) = (4, -\pi/6, -1)$. ■

You Try It

(1) Locate the point whose cylindrical coordinates are $(r, \theta, z) = (2, \pi/4, 1)$ and determine its rectangular coordinates.
(2) What are the cylindrical coordinates of the point whose rectangular coordinates are $(1, -1, 4)$?

In addition to converting individual points with the conversion equations, we can also convert equations of surfaces. You already know how to do this, for example, by realizing that the circle whose equation is $x^2 + y^2 = 9$ in rectangular coordinates has the equation $r = 3$ in polar coordinates.

EX 2 Identify the surface and convert its rectangular equation into cylindrical form:

$$\text{(a) } z = 2x^2 + 2y^2 \quad , \quad \text{(b) } x^2 + y^2 + z^2 = 9 \quad , \quad \text{(c) } x = 3$$

□ (a) $z = 2x^2 + 2y^2$ is a paraboloid, and the equation can be written as $z = 2(x^2 + y^2)$, and so in cylindrical form we have $z = 2r^2$.

(b) $x^2 + y^2 + z^2 = 9$ is a sphere of radius 3, and in cylindrical form, it's $r^2 + z^2 = 9$. (Remember: this does NOT simplify to $r + z = 3$ no matter how much you want it to!)

(c) $x = 3$ is a plane, and all we can do is replace x with its direct equivalent to get $r\cos\theta = 3$.

Note that the two shapes with circular cross sections have nicer equations in cylindrical coordinates, while the surface which is a flat plane has a much better equation in rectangular coordinates. ■

You Try It

(3) What surface is described in cylindrical coordinates by the equation $r = 3$?
(4) Convert the equation $z = x^2/3 + y^2/2$ into cylindrical coordinates.

Spherical Coordinates

Spherical coordinates are useful when describing a 3D region that's shaped like .... well, I'll let you guess this one.

There are three coordinates of a point in the spherical system:

- ρ is the point's straight-line distance from the origin (This is the actual distance from the origin (0,0,0), not just distance outward from the z-axis, as r is in cylindrical coordinates.)
- θ is the horizontal angle you have to turn, counterclockwise from the positive x-axis, to find the point's image in the xy-plane. This is the same as θ in polar or cylindrical coordinates. θ usually ranges from 0 to 2π or $-\pi$ to π.
- ϕ is the point's azimuthal angle. That is, if you are standing at the origin looking straight overhead up the positive z-axis, ϕ is the angle you look down from overhead to get the point in sight. $\phi = 0$ is straight overhead, $\phi = \pi/2$ is straight out along the xy-plane, and $\phi = \pi$ is looking down through your feet. We would never use a ϕ value larger than π, because if the point is behind you, you're not going to find it by bending over and looking back up through your knees; rather, you're just going stay upright and turn around — and that move changes your θ coordinate instead.

A general diagram of spherical coordinates is shown in Fig. 15.20, and the location of the point whose cylindrical coordinates are $(r, \theta, z) = (2, \pi/4, 3\pi/4)$ are in Fig. 15.21.

When reading in other sources, be very alert to the naming of spherical coordinates. There is universal agreement about coordinate names in rectangular coordinates, but no so much in cylindrical and spherical coordinates. In some places, the names of the angular coordinates are swapped; the radial angle we're calling θ might be named ϕ instead, and the azimuthal angle we're calling ϕ might be θ. Sometimes the ordering is swapped so that even if they measure the same thing, the variables are ordered as (r, ϕ, θ). Some weirdos will even use ρ for the radial coordinate in polar and cylindrical coordinate, can you imagine that? Just be alert to notation whenever you switch from one source to another.

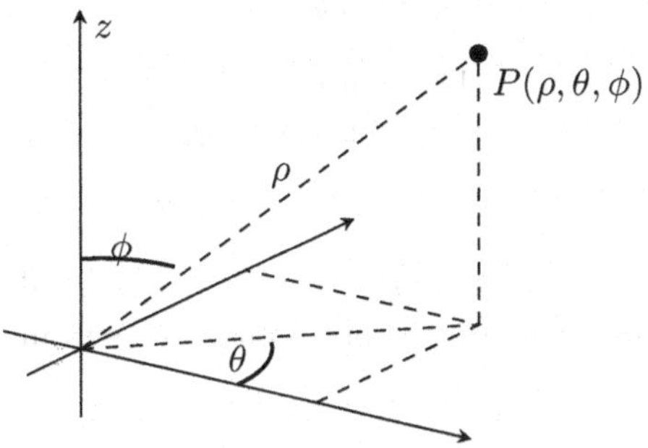

Fig. 15.20 A point's spherical coordinates (ρ, θ, ϕ).

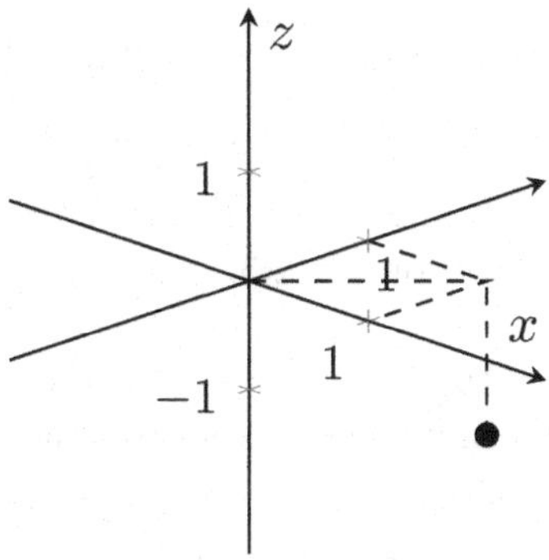

Fig. 15.21 This point's spherical coordinates are $(2, \pi/4, 3\pi/4)$.

Useful Fact 15.2. *A point's rectangular coordinates (x, y, z) and spherical coordinates (ρ, θ, ϕ) can be traded back and forth with these conversion equations:*

$$\rho^2 = x^2 + y^2 + z^2 \quad ; \quad \tan\theta = \frac{y}{x} \quad ; \quad \cos\phi = \frac{z}{\rho} \tag{15.3}$$

$$x = \rho\cos\theta\sin\phi \quad ; \quad y = \rho\sin\theta\sin\phi \quad ; \quad z = \rho\cos\phi \tag{15.4}$$

Conversion of coordinates of single points should be second nature for you by now, so just go ahead and give it a shot without prior coaching:

You Try It

(5) Locate the point whose spherical coordinates are $(\rho, \theta, \phi) = (1, \pi/6, \pi/6)$ and determine its rectangular coordinates.
(6) What are the spherical coordinates of the point whose rectangular coordinates are $(1, \sqrt{3}, 2\sqrt{3})$?

As above, we can also convert equations of surfaces in addition to individual points.

EX 3 Identify the surface and convert its rectangular equation into spherical form:

$$\text{(a) } 3x^2 + 3y^2 + 3z^2 = 9 \quad , \quad \text{(b) } z = \sqrt{x^2 + y^2} \quad , \quad \text{(c) } y = 1$$

□ (a) $3x^2 + 3y^2 + 3z^2 = 9$ is a sphere, and the equation can be written as $3(x^2 + y^2 + z^2) = 9$, i.e. $x^2 + y^2 + z^2 = 3$; so, in spherical form we have $\rho^2 = 3$ or $\rho = \sqrt{3}$.

(b) $z = \sqrt{x^2 + y^2}$ is the upper half of a cone, and to get it in spherical form, it is convenient to write it as $z^2 = x^2 + y^2$ and apply conversions from (15.3):

$$\begin{aligned}
z^2 &= x^2 + y^2 \\
(\rho\cos\phi)^2 &= (\rho\cos\theta\sin\phi)^2 + (\rho\sin\theta\sin\phi)^2 \\
\cos^2\phi &= \cos^2\theta\sin^2\phi + \sin^2\theta\sin^2\phi \\
\cos^2\phi &= \sin^2\phi(\cos^2\theta + \sin^2\theta) \\
\cos^2\phi &= \sin^2\phi \\
\sin\phi &= \cos\phi \quad \text{(ignoring the } \pm \text{ keeps the upper half of the cone only!)} \\
\tan\phi &= 1 \\
\phi &= \pi/4
\end{aligned}$$

So, note that this cone has a spherical equation of the form $\phi = c$, where c is a constant.

(c) $y = 1$ is a plane, and all we can do is replace y directly to get $\rho\sin\theta\sin\phi = 1$.

Like in EX 2, the shapes with circular or spherical characteristics have nicer equations in spherical coordinates, while the surface which is a flat plane still has a much better equation in rectangular coordinates. ■

You Try It

(7) What surface is described by the spherical equation $\rho \cos \phi = 2$?
(8) Convert the equation $z = x^2 + y^2$ into spherical coordinates.

You may notice that we do not go through direct conversion between cylindrical and spherical coordinates. That's because those conversions are messy (well, apart from θ being the same in each system) and if you really need to get from one to the other, you can always go through rectangular coordinates. So let's not clutter this up more than necessary.

Choosing Coordinates

The reason we want to be able to choose among coordinate systems is that some surfaces or regions are more easily and more naturally described in different systems. To see which system is best in which situations, consider what surfaces are found by setting each variable to a constant.

- In rectangular coordinates, any equation of the form $x = a$, $y = b$ or $z = c$ is a horizontal or vertical plane parallel to the yz-, xz- or xy-planes. Therefore, these coordinates are best for describing regions bounded by planes that are parallel to the coordinate planes.
- In cylindrical coordinates, any equation of the form $r = a$ is a cylinder. Any equation of the form $\theta = b$ is a vertical plane through the z-axis, but at an angle to the xz- or yz-planes. As in rectangular coordinates, an equation $z = c$ is a horizontal plane parallel to the xy-plane. Therefore, cylindrical coordinates may be best when describing a region that's cylindrical, or part of a vertical plane angled with respect to the rectangular coordinate planes.
- In spherical coordinates, any equation of the form $\rho = a$ is a sphere. Any equation of the form $\theta = b$ is a vertical plane through the z-axis, but at an angle to the xz- or yz-planes, as in cylindrical coordinates. Any equation of the form $\phi = c$ is a cone (see example EX 3 above). Therefore, spherical coordinates may be best when describing a region that's spherical or conical.

A surface that arises when we set a particular variable to a constant value is called an *isosurface*.[1]

[1]You may recognize "iso" from isobars, isoclines, and isotherms.

Several figures demonstrate isosurfaces that arise in conjunction with cylindrical and spherical coordinates:

- Fig. 15.22 shows a cylindrical isosurface $r = 2$.
- Fig. 15.23 shows a cylindrical and spherical isosurface $\theta = \pi/4$.
- Fig. 15.24 shows a spherical isosurface $\rho = 2$.
- Fig. 15.25 shows a spherical isosurface $\phi = \pi/4$.

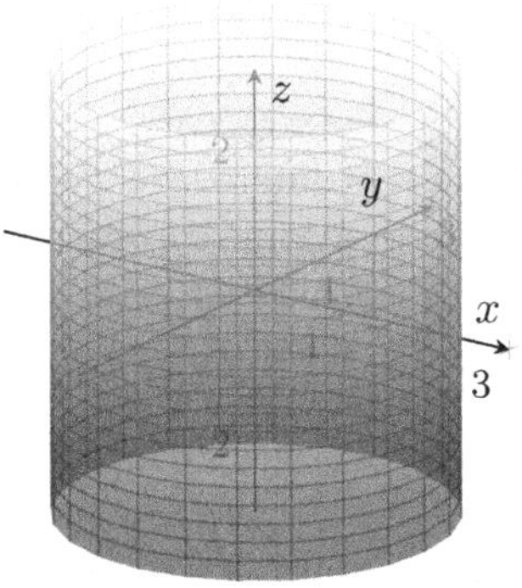

Fig. 15.22 Cylindrical isosurface $r = 2$.

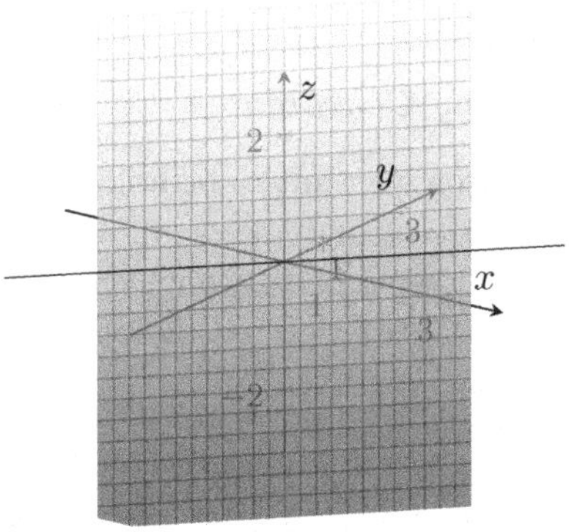

Fig. 15.23 Cylindrical isosurface $\theta = \pi/4$.

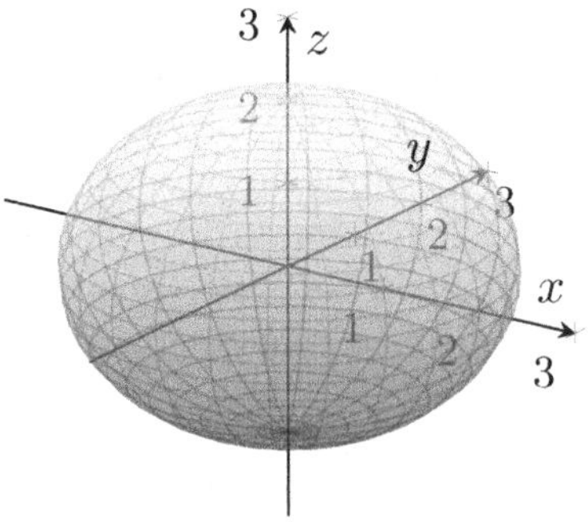

Fig. 15.24 Spherical isosurface $\rho = 2$.

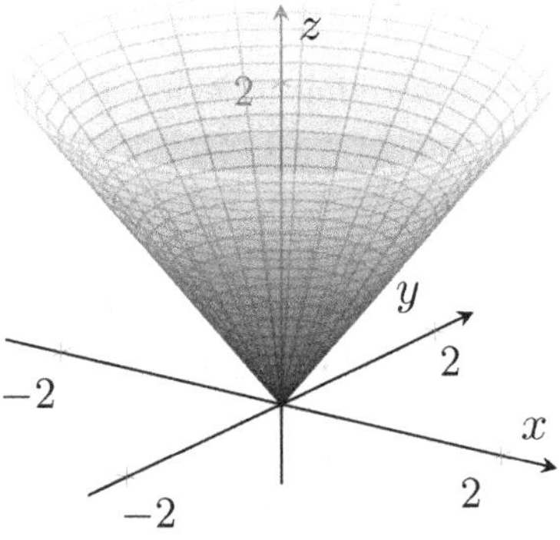

Fig. 15.25 Spherical isosurface $\phi = \pi/4$.

Describing Regions

I hate to be the bearer of bad news, but we are close to using cylindrical and spherical coordinates in the context of triple integrals. Much like with double integrals in polar coordinates, we will have to get good at delimiting 3D regions of integration by quantities which become our limits of integration. But again, remember that the entire purpose of having these other coordinate systems and your beck and call is to make your mathematical life easier! As an example, let's consider a scenario which seems like it would be quite benign — describing a spherical region of integration. In particular, let's think about a sphere of radius 3 centered at the origin.

In rectangular coordinates, it would be spectacularly **wrong** to assign our limits as $-3 \leq x \leq 3$, $-3 \leq y \leq 3$, and $-3 \leq z \leq 3$. These six bounds do make for very simple limits of integration, but the collection of these bounds describes a **cube**. Instead, we have to solve the equation of the sphere ($x^2 + y^2 + z^2 = 9$) for z and use those as our bounds on z:

$$-\sqrt{9 - x^2 - y^2} \leq z \leq \sqrt{9 - x^2 - y^2}$$

With limits on z determined, we next have to pick out the maximum expanse of this solid in the x and y directions; this maximum expanse is a circle of radius 3 (it's the trace of the sphere in the plane $z = 0$), i.e. $x^2 + y^2 = 3$. And from your experience with double integrals, you should recall that proper limits on this 2D region can be assigned as:

$$-\sqrt{9 - x^2} \leq y \leq \sqrt{9 - x^2} \qquad \text{and} \qquad -3 \leq x \leq 3$$

Let's say we're integrating some $f(x, y, z)$ over this solid region of integration; our triple integral would look like:

$$\int_{-3}^{3} \int_{-\sqrt{9-x^2}}^{\sqrt{9-x^2}} \int_{-\sqrt{9-x^2-y^2}}^{\sqrt{9-x^2-y^2}} f(x, y, z)\, dV$$

Now take a good hard look at that thing. It's horrible. And now, you should thank your lucky stars that we have other coordinate systems to use for this triple integral; in particular, you should be thankful for spherical coordinates, because when we use those and convert our function $f(x, y, z)$ to a function in spherical coordinates $f(\rho, \theta, \phi)$, then the triple integral will be this:

$$\int_0^3 \int_0^{\pi} \int_0^{2\pi} f(\rho, \theta, \phi)\, dV$$

That is so much better! And this becomes our game — to take a given region / solid of integration in $\mathbb{R}^3$, decide which coordinate system will provide the most benign limits of integration, and put that system to good use. In order to generate correct limits of integration for triple integrals in cylindrical and spherical coordinates, we must be able to delineate regions in $\mathbb{R}^3$ by specifying bounds on the appropriate coordinates. Here are some examples of that.

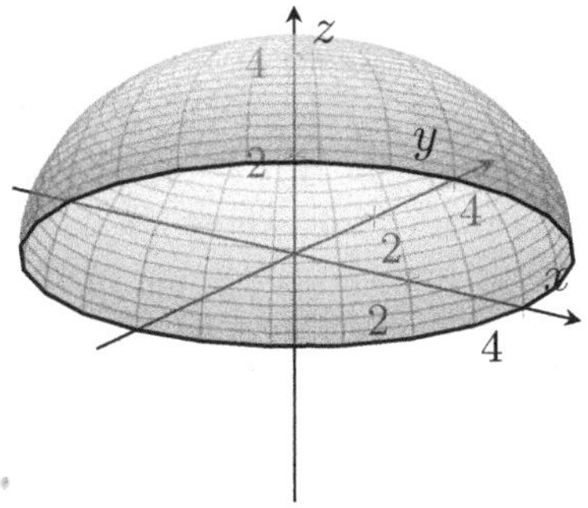

Fig. 15.26 Upper hemisphere of radius 4.

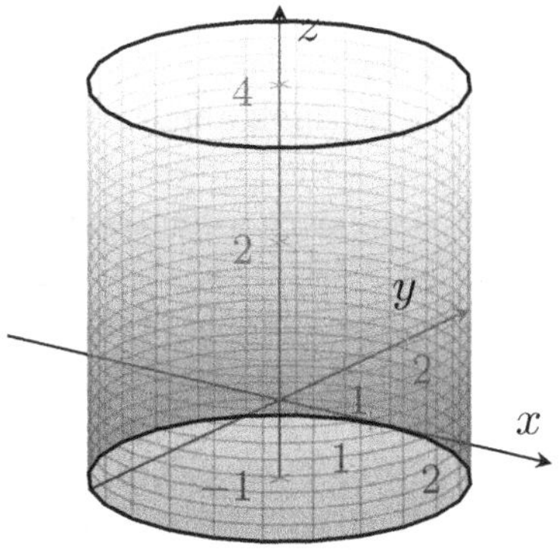

Fig. 15.27 $x^2 + y^2 = 3$ between $z = -1$, $z = 4$.

EX 4 Delineate the following regions by setting appropriate bounds on cylindrical coordinates:

(a) The upper hemisphere of radius 4.
(b) The region between $x^2 + y^2 = 3$, and the planes $z = -1$ and $z = 4$.
(c) The region under $z = 5 - x^2 - y^2$ and above the xy-plane.
(d) The enclosed region between $x = 5 - y^2 - z^2$ and the yz-plane.

(a) This region is shown in Fig. 15.26. The "floor" of the upper hemisphere of radius 4 is the xy-plane, i.e. $z = 0$. The "roof" is the hemisphere itself, $z = \sqrt{16 - x^2 - y^2}$. But in cylindrical coordinates, this is $z = \sqrt{16 - r^2}$. Radial limits for r and θ must cover the maximum extent of this solid found in the xy-plane, which is a circle of radius 4, and so we have $0 \le r \le 4$ and $0 \le \theta \le 2\pi$. Altogether,

$$0 \le \theta \le 2\pi \quad ; \quad 0 \le r \le 4 \quad ; \quad 0 \le z \le \sqrt{16 - r^2}$$

(b) This region is shown in Fig. 15.27. The region between $x^2 + y^2 = 3$ and the planes $z = -1$ and $z = 4$ has those values of z as constant limits on z. Then the circular cross section is described by polar coordinates as usual for a circle of radius $\sqrt{3}$; together,

$$0 \le \theta \le 2\pi \quad ; \quad 0 \le r \le \sqrt{3} \quad ; \quad -1 \le z \le 4$$

(It should be no surprise that all six limits in cylindrical coordinates are constants for a cylindrical region.

(c) This region is shown in Fig. 15.28. The region under $z = 5 - x^2 - y^2$ and above the xy-plane is like a parabolic tent. The "roof" is the paraboloid itself, which is $z = 5 - r^2$ in cylindrical coordinates. Radial limits for r and θ must cover the maximum extent of this solid found in the xy-plane, which comes from the intersection of the paraboloid with the xy-plane: $5 - x^2 - y^2 = 0$ becomes $x^2 + y^2 = 5$; it's is a circle of radius 5. Altogether,

$$0 \le \theta \le 2\pi \quad ; \quad 0 \le r \le \sqrt{5} \quad ; \quad 0 \le z \le 5 - r^2$$

(d) This region is shown in Fig. 15.29, and is is just the region from part (c) flipped on its side. In the vast majority of cases, we set up cylindrical coordinates so that z is the non-polar direction, but we can also choose either x or y to be the non-polar direction — so that r and θ take up the other two spots. That is, cylindrical coordinates can also be arranged as (x, r, θ) or (r, y, θ). When we look at the enclosed region between $x = 5 - y^2 - z^2$ and the yz-plane, we see that the traces of this region parallel the yz-plane are circular. In particular, the "floor" of this region is the intersection of $x = 5 - y^2 - z^2$ with $x = 0$, i.e. the circle $y^2 + z^2 = 5$. So in this case, we'd like the x direction to be the non-polar direction, and have r and θ active in the yz-plane's direction. We can describe the region as being bounded by points (x, r, θ) such that:

$$0 \le \theta \le 2\pi \quad ; \quad 0 \le r \le \sqrt{5} \quad ; \quad 0 \le x \le 5 - r^2 \quad \blacksquare$$

EX 5 With respect to spherical coordinates:

(a) Delineate the upper hemisphere of radius 4.
(b) Explain why it would be a *bad idea* to delineate the region between $x^2 + y^2 = 3$, and the planes $z = -1$ and $z = 4$, in spherical coordinates.
(c) Delineate the region above the cone $z^2 = x^2 + y^2$ and below the unit sphere.

(a) (See Fig. 15.26 again.) The full sphere of radius 4 is, in spherical coordinates, $\rho = 4$. We want the full sphere all the way around the

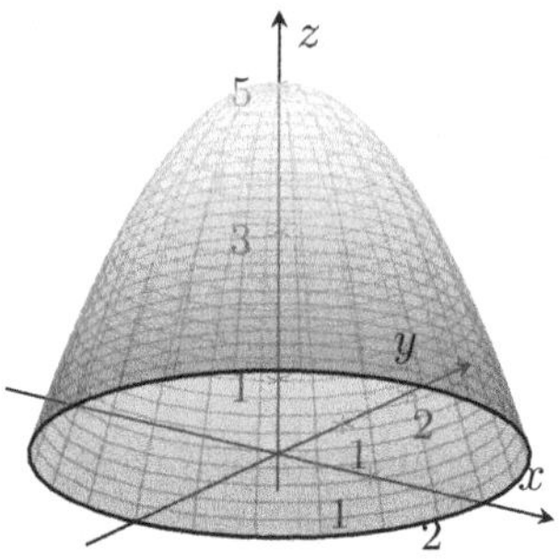

Fig. 15.28 Region under $z = 5 - x^2 - y^2$, above $z = 0$.

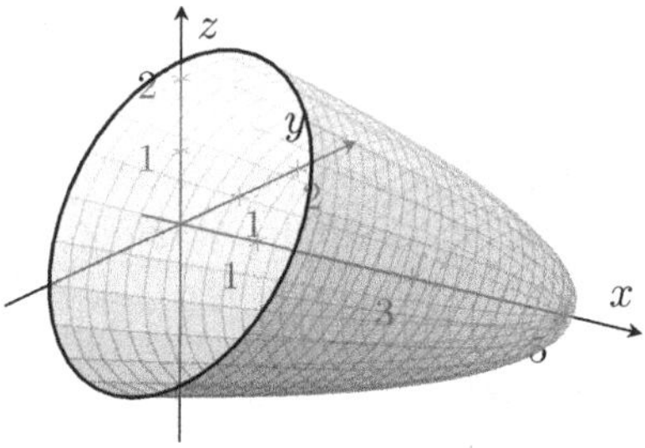

Fig. 15.29 Region between $x = 5 - y^2 - z^2$ and $x = 0$.

"equator", so we need $0 \leq \theta \leq 2\pi$. However, since we just want the upper hemisphere, it's like slicing off the lower half of a globe at the equator. We only want to keep points for which $0 \leq \phi \leq \pi/2$. Together,

$$0 \leq \rho \leq 4 \quad ; \quad 0 \leq \theta \leq 2\pi \quad ; \quad 0 \leq \phi \leq \frac{\pi}{2}$$

FFT: If we wanted the lower hemisphere of radius 4 instead, how would these limits change?

(b) (See Fig. 15.27 again.) The spherical coordinate ρ is the distance from the origin. If we take the cylinder $x^2 + y^2 = 3$ and give it a flat "roof" at $z = 4$, then points on that roof are not all the same distance from the origin; imagine a spotlight at the origin trying to track an ant that is walking all over this roof. The length of the beam of the spotlight would vary as the ant walked around. Because this distance is not constant, we will not get constant limits for ρ in this region. So spherical coordinates are a bad idea.

(c) As we move into setting up triple integrals in these other coordinate systems, we are going to be thinking of food a lot. The region above the cone $z^2 = x^2 + y^2$ and below the unit sphere ($\rho = 1$) can be seen in Fig. 15.30, and it looks like an ice cream cone! The cone $z^2 = x^2 + y^2$ is the one which cuts a perfect diagonal through the upper octants, and so we can bound the azimuthal angle ϕ by 0 and $\pi/4$. As we look out from the origin, the "roof" of this region is always the sphere, and so we have $0 \leq \rho \leq 1$. Because our region goes all the way around radially, we have $0 \leq \theta \leq 2\pi$. Together,

$$0 \leq \rho \leq 1 \quad ; \quad 0 \leq \theta \leq 2\pi \quad ; \quad 0 \leq \phi \leq \frac{\pi}{4}$$

We did play a little loose in coming up with the bounds on ϕ here. Usually, for the intersection of a cone and a hemisphere, we have to do algebra to discover what value of ϕ describes the intersection. ■

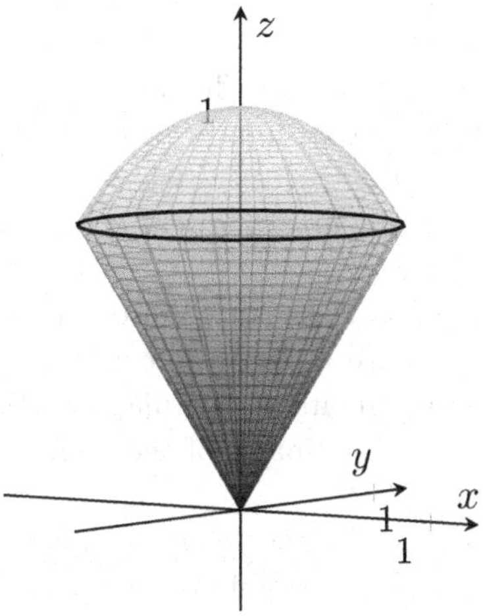

Fig. 15.30 Above $z^2 = x^2 + y^2$, inside $\rho = 1$.

You Try It

(9) Choose whether spherical or cylindrical coordinates might be best for each region, and delineate the region in the coordinate system you choose.

(a) The region bounded by the planes $z = 0$ and $z = x + y + 5$, and the cylinder $x^2 + y^2 = 9$.

(b) The region outside the sphere $x^2 + y^2 + z^2 = 1$ and inside the sphere $x^2 + y^2 + z^2 = 9$.

(c) The region in the first octant that's outside the cylinder $x^2 + y^2 = 1$ but inside the sphere $x^2 + y^2 + z^2 = 4$.

Cylindrical and Spherical Coordinates — Problem List

Cylindrical and Spherical Coordinates — You Try It

These appeared above; solutions begin on the next page.

(1) Locate the point whose cylindrical coordinates are $(r, \theta, z) = (2, \pi/4, 1)$ and determine its rectangular coordinates.
(2) What are the cylindrical coordinates of the point whose rectangular coordinates are $(1, -1, 4)$?
(3) What surface is described in cylindrical coordinates by the equation $r = 3$?
(4) Convert the equation $z = x^2/3 + y^2/2$ into cylindrical coordinates.
(5) Locate the point whose spherical coordinates are $(\rho, \theta, \phi) = (1, \pi/6, \pi/6)$ and determine its rectangular coordinates.
(6) What are the spherical coordinates of the point whose rectangular coordinates are $(1, \sqrt{3}, 2\sqrt{3})$?
(7) What surface is described by the spherical equation $\rho \cos\phi = 2$?
(8) Convert the equation $z = x^2 + y^2$ into spherical coordinates.
(9) Choose whether spherical or cylindrical coordinates might be best for each region, and delineate the region in the coordinate system you choose.

   (a) The region bounded by the planes $z = 0$ and $z = x + y + 5$, and the cylinder $x^2 + y^2 = 9$.
   (b) The region outside the sphere $x^2 + y^2 + z^2 = 1$ and inside the sphere $x^2 + y^2 + z^2 = 9$.
   (c) The region in the first octant that's outside the cylinder $x^2+y^2 = 1$ but inside the sphere $x^2 + y^2 + z^2 = 4$.

Cylindrical and Spherical Coordinates — Practice Problems

Try these as you get the hang of the You Try It problems. Solutions to these problems are available in Sec. B.3.2.

(1) Locate the point whose cylindrical coordinates are $(1, 3\pi/2, 2)$ and give its rectangular coordinates.
(2) What are the cylindrical coordinates of the point whose rectangular coordinates are $(3, 3, -2)$?
(3) What surface is described in cylindrical coordinates by the equation $r^2 + z^2 = 25$?

(4) Convert the cylindrical equation $r^2 - 2z^2 = 4$ into rectangular coordinates, and identify the kind of surface it is.
(5) Locate the point whose spherical coordinates are $(\rho, \theta, \phi) = (5, \pi, \pi/2)$ and give its rectangular coordinates.
(6) What are the spherical coordinates of the point whose rectangular coordinates are $(0, \sqrt{3}, 1)$?
(7) What surface is described by the spherical equation $\rho = 3$?
(8) Convert the equation $x^2 + y^2 + z^2 = 2$ into spherical coordinates.
(9) What are the spherical coordinates of the point whose cylindrical coordinates are $(r, \theta, z) = (\sqrt{6}, \pi/4, \sqrt{2})$?
(10) What are the cylindrical coordinates of the point whose spherical coordinates are $(\rho, \theta, \phi) = (2\sqrt{2}, 3\pi/2, \pi/2)$?
(11) Convert the rectangular equation $y^2 + z^2 = 1$ into both cylindrical and spherical coordinates.

Cylindrical and Spherical Coordinates — Challenge Problems

Try these problems to test your skills with the ideas in this section. Solutions to these problems are available in Sec. C.3.2.

(1) What are the cylindrical coordinates of the point whose spherical coordinates are $(\rho, \theta, \phi) = (2, 3\pi/4, \pi/6)$?
(2) Suppose a three-dimensional region is described with the following bounds in rectangular coordinates:
    - The full 3D region is bounded below by the paraboloid $z = x^2 + y^2$ and above by the cone $z = \sqrt{x^2 + y^2}$
    - In the xy-plane, the region covers the right half of the unit circle, thus we have $0 \leq x \leq \sqrt{1 - y^2}$ and $-1 \leq y \leq 1$.

    Describe the bounds of this same region in cylindrical coordinates by giving the bounds of r, θ, and z.
(3) Suppose a three-dimensional region is described with the following bounds in rectangular coordinates:
    - The full 3D region is underneath the hemisphere $z = \sqrt{4 - x^2 - y^2}$ and above the xy-plane
    - In the xy-plane, the region covers the upper half of a circle of radius 2 centered at the origin, thus we have $-2 \leq x \leq 2$ and $0 \leq y \leq \sqrt{4 - x^2}$.

    Describe the bounds of this same region in spherical coordinates by giving the bounds of ρ, θ, and ϕ.

Cylindrical and Spherical — You Try It — Solved

(1) Locate the point whose cylindrical coordinates are $(r, \theta, z) = (2, \pi/4, 1)$ and determine its rectangular coordinates.

□ With cylindrical coordinates $(2, \pi/4, 1)$, this point will be in the first octant. Its rectangular coordinates are found using Eq. (15.2):

$$x = 2\cos\left(\frac{\pi}{4}\right) = \sqrt{2}$$
$$y = 2\sin\left(\frac{\pi}{4}\right) = \sqrt{2}$$
$$z = 1$$

so this is the rectangular point $(\sqrt{2}, \sqrt{2}, 1)$. ■

(2) What are the cylindrical coordinates of the point whose rectangular coordinates are $(1, -1, 4)$?

□ With rectangular coordinates $(1, -1, 4)$ we have cylindrical coordinates

$$r = \sqrt{1^2 + (-1)^2} = \sqrt{2}$$
$$\theta = \tan^{-1}\left(-\frac{1}{1}\right) = -\frac{\pi}{4} = \frac{7\pi}{4}$$
$$z = 4$$

so this is the cylindrical point $(\sqrt{2}, 7\pi/4, 4)$. ■

(3) What surface is described in cylindrical coordinates by the equation $r = 3$?

□ This is a cylindrical equation describing all points whose r coordinate (distance from the z-axis) is 3. So, the surface is a cylinder of radius 3 centered on the z-axis. ■

(4) Convert the equation $z = x^2/3 + y^2/2$ into cylindrical coordinates.

□ This is an elliptic paraboloid; since traces (cross sections) at constant levels of z are ellipses and not pure circles, we lose the benefit of cylindrical coordinates. At best, we can do this:

$$z = \frac{x^2}{3} + \frac{y^2}{2} = \frac{(r\cos\theta)^2}{3} + \frac{(r\sin\theta)^2}{2} = r^2\left(\frac{\cos^2\theta}{3} + \frac{\sin^2\theta}{2}\right)$$

We can clean up a bit by multiplying both sides by 6:

$$6z = r^2(2\cos^2\theta + 3\sin^2\theta \quad ■$$

(5) Locate the point whose spherical coordinates are $(\rho, \theta, \phi) = (1, \pi/6, \pi/6)$ and determine its rectangular coordinates.

☐ This point will be in the first octant. We can convert to rectangular coordinates using Eq. (15.4):

$$x = \rho \sin\phi \cos\theta = 1 \sin\left(\frac{\pi}{6}\right) \cos\left(\frac{\pi}{6}\right) = \frac{1}{2} \cdot \frac{\sqrt{3}}{2} = \frac{\sqrt{3}}{4}$$

$$y = \rho \sin\phi \sin\theta = 1 \sin\left(\frac{\pi}{6}\right) \sin\left(\frac{\pi}{6}\right) = \frac{1}{2} \cdot \frac{1}{2} = \frac{1}{4}$$

$$z = \rho \cos\phi = 1 \cos\left(\frac{\pi}{6}\right) = \frac{\sqrt{3}}{2}$$

So this is the rectangular point $(x, y, z) = (\sqrt{3}/4, 1/4, \sqrt{3}/2)$. ■

(6) What are the spherical coordinates of the point whose rectangular coordinates are $(1, \sqrt{3}, 2\sqrt{3})$?

☐ With rectangular coords $(x, y, z) = (1, \sqrt{3}, 2\sqrt{3})$ we can get spherical coordinates using Eq. (15.3):

$$\rho = \sqrt{x^2 + y^2 + z^2} = \sqrt{1 + 3 + 12} = 4$$

$$\cos\phi = \frac{z}{\rho} = \frac{2\sqrt{3}}{4} = \frac{\sqrt{3}}{2} \quad \rightarrow \quad \phi = \frac{\pi}{6}$$

$$\cos\theta = \frac{x}{\rho \sin\phi} = \frac{1}{4\sin(\pi/6)} = \frac{1}{4(1/2)} = \frac{1}{2} \quad \rightarrow \quad \theta = \frac{\pi}{3}$$

So this is the spherical point $(\rho, \theta, \phi) = (4, \pi/3, \pi/6)$. ■

(7) What surface is described by the spherical equation $\rho \cos\phi = 2$?

☐ This directly matches the conversion equation in (15.4) $z = \rho \cos\phi$, so the surface is also known as $z = 2$. ■

(8) Convert the equation $z = x^2 + y^2$ into spherical coordinates.

☐ The conversion equation $x^2 + y^2 + z^2 = \rho^2$ gives $x^2 + y^2 = \rho^2 - z^2$, so this equation is (temporarily)

$$z = \rho^2 - z^2$$

But z itself is not a spherical coordinate, so we need to bring in $z = \rho\cos\phi$ to write

$$\begin{aligned}\rho\cos\phi &= \rho^2 - \rho^2\cos^2\phi\\ \cos\phi &= \rho(1-\cos^2\phi)\\ \cos\phi &= \rho\sin^2\phi\end{aligned}$$

About the only thing we can do here is divide both sides by $\cos\phi$ and say that the equation is also known as $\rho\sin\phi\tan\phi = 1$. ■

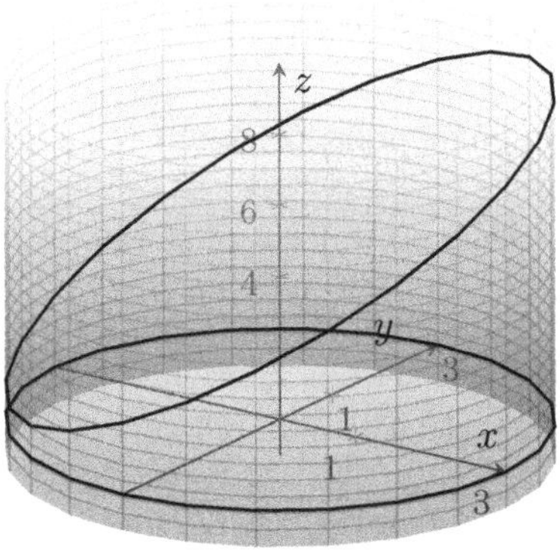

Fig. 15.31 $x^2 + y^2 = 9$ between $z = 0$, $z = x + y + 5$.

---

(9) Choose whether spherical or cylindrical coordinates might be best for each region, and delineate the region in the coordinate system you choose.

(a) The region bounded by the planes $z = 0$ and $z = x + y + 5$, and the cylinder $x^2 + y^2 = 9$.
(b) The region outside the sphere $x^2 + y^2 + z^2 = 1$ and inside the sphere $x^2 + y^2 + z^2 = 9$.
(c) The region in the first octant that's outside the cylinder $x^2+y^2 = 1$ but inside the sphere $x^2 + y^2 + z^2 = 4$.

□ (a) The region bounded by the planes $z = 0$ and $z = x+y+5$, and the cylinder $x^2 + y^2 = 9$, is shown in Fig. 15.31. The tilted circle is the upper boundary of the solid. Because the walls of this region come from a cylinder, let's use cylindrical coordinates; we'll have $0 \le r \le 3$ and $0 \le \theta \le 2\pi$. The lower bound for z is just $z = 0$. The "roof" of this solid is a slanted plane; we need to convert $z = 5 - x - y$ to cylindrical coordinates, and there's no nice way to do that except direct substitution. And so, we have

$$0 \le z \le 5 - r\cos\theta - r\sin\theta \quad ; \quad 0 \le r \le 3 \quad ; \quad 0 \le \theta \le 2\pi$$

Even though the upper bound on z is gross, it's worth it for the cylindrical cross sections.

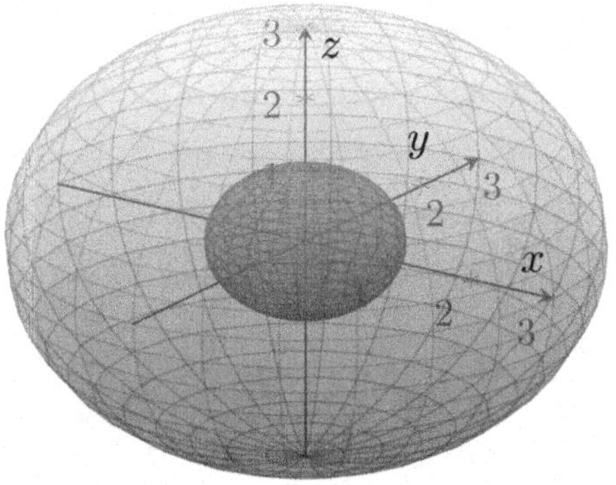

Fig. 15.32 Outside $\rho = 1$, inside $\rho = 3$.

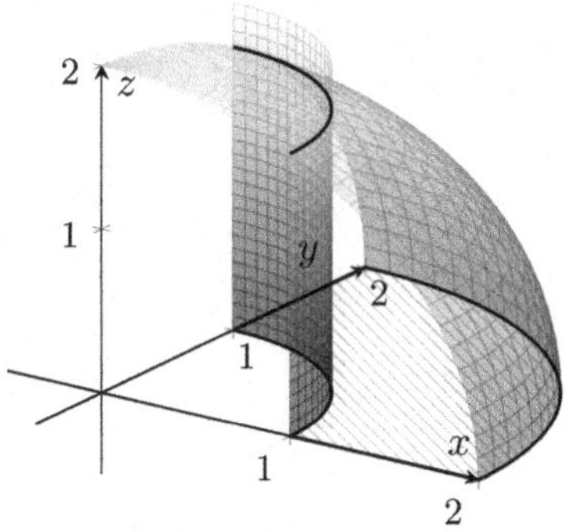

Fig. 15.33 Outside $r = 1$, inside $\rho = 2$.

(b) The region outside the sphere $x^2 + y^2 + z^2 = 1$ and inside the sphere $x^2 + y^2 + z^2 = 9$ is shown in Fig. 15.32. We are going all the way around in both angular directions; the radial coordinate begins at the radius of the inner sphere and ends at the radius of the outer sphere. Altogether,

$$1 \le \rho \le 3 \quad ; \quad 0 \le \theta \le 2\pi \quad ; \quad 0 \le \phi \le \pi$$

(c) The region in the first octant that's outside the cylinder $x^2+y^2 = 1$ but inside the sphere $x^2 + y^2 + z^2 = 4$ is shown in Fig. 15.33. There is no nice way to form the inner wall of this region (the cylinder) out of limits on ρ, so let's go for cylindrical coordinates. (Often, spherical coordinates have to defer to another set.) As we float around in this solid region, the "floor" is the lower hemisphere and the "roof" is the upper hemisphere. Therefore, we have $-\sqrt{4 - r^2} \le z \le \sqrt{4 - r^2}$. The lateral extent of the region has an inner edge from the cylinder (a circle of radius 1), and an outer edge from the sphere (a circle of radius 2). These give $1 \le r \le 2$. Since we only need to sweep out the first octant radially, we have $0 \le \theta \le \pi/2$. Altogether,

$$-\sqrt{4-r^2} \le z \le \sqrt{4-r^2} \quad ; \quad 1 \le r \le 2 \quad ; \quad 0 \le \theta \le \frac{\pi}{2} \quad \blacksquare$$

15.3 The Calc 3 Boss Fight: Triple Integrals in Cylindrical & Spherical Coordinates

Introduction

Your experience converting double integrals from rectangular to polar coordinates will serve you well in converting triple integrals from rectangular coordinates to cylindrical or spherical coordinates. Suppose you have a region of integration whose bounds in rectangular coordinates are just gross. Or suppose you have a triple integral containing a function you can't work with in rectangular coordinates. Or suppose you have both of these cases! Then you'll need to convert the problem from rectangular coordinates to either cylindrical or spherical coordinates, depending on which is more appropriate. To summarize what we know from before,

- Cylindrical coordinates may be useful when the region of integration is bound by a cylinder, or if the function being integrated contains $x^2 + y^2$. Also, if the region of integration is bound by a paraboloid, cylindrical coordinates may help, too, since $z = x^2 + y^2$ is easily written as $z = r^2$ in cylindrical coordinates.
- Spherical coordinates may be useful when the region of integration is bound by a sphere and/or a cone, or if the function being integrated contains $x^2 + y^2 + z^2$.

When we converted double integrals in rectangular coordinates to polar coordinates, we had to be careful to adjust the area element dA appropriately. In rectangular coordinates, dA could either be $dxdy$ or $dydx$. In polar coordinates, the area element is $dA = rdrd\theta$. In triple integrals, we will have to carefully convert the volume element dV. In rectangular coordinates, there are six possible configuration of dV ($dzdydz$, $dydzdx$, etc.). While the bad news is that both cylindrical and spherical coordinates have their own versions of dV to use, the good news is that we rarely deviate from one particular ordering in each:

Variations in dV

At this point, you can probably imagine the general partitioning process which leads to triple integrals in cylindrical and spherical coordinates. Diving in to the deep details would take us down some long geometric side roads that can distract us from our true goal, which is construction and evaluation of triple integrals. There are many resources you can consult if you want to see these details. For now, let's concentrate on construction and solution of triple integrals in non-rectangular coordinates.

The differential dV has six different forms rectangular coordinates (the possible rearrangements of $dxdydz$). In non-rectangular coordinates, the good news is that we usually only have one form for dV in each system. The bad news is that — especially for spherical coordinates — they are a bit messy.

Useful Fact 15.3. *In cylindrical coordinates, we write* $dV = r\,dzdrd\theta$*. In spherical coordinates, we write* $dV = \rho^2 \sin\phi\, d\rho d\phi d\theta$*.*

Again, remember that these are artifacts of the partitioning process. We may not like them (especially for spherical coordinates), but we have to accept them.

Triple Integrals in Cylindrical Coordinates

Given a triple integral $\iiint_E f\,dV$ that is awkward or impossible to do in rectangular coordinates, you may choose to do the integral in cylindrical coordinates instead. To do so, you may have to do one or more of the following:

- write the function in cylindrical form, using known conversion equations
$$r^2 = x^2 + y^2 \quad ; \quad \tan\theta = \frac{y}{x} \quad ; \quad x = r\cos\theta \quad ; \quad y = r\sin\theta$$
- write dV in cylindrical form, $dV = r\,dzdrd\theta$
- describe the solid E using cylindrical coordinates, creating proper limits of integration
- solve the integral

First, let's just practice "reading" triple integrals in cylindrical coordinates.

EX 1 Recall that we can find the volume of a general 3D region E by evaluating the triple integral $\iiint_E (1)dV$. What solid's volume is represented by the integral $\int_0^{\pi}\int_0^4\int_0^{r^2} r\,dzdrd\theta$?

Since $dV = rdzdrd\theta$ in cylindrical coordinates, this really is a volume integral. z is bound below by $z = 0$ and above by $z = r^2$. But $z = r^2$ is the paraboloid also known as $z = x^2 + y^2$. The variable r is bound by $r = 0$ and $r = 4$, and θ goes half-way around from 0 to π. Therefore, this integral computes the volume under the paraboloid $z = x^2 + y^2$ over the upper half of the circle $x^2 + y^2 = 16$ in the xy-plane. ■

You Try It

(1) What solid's volume is represented by the integral $\int_0^{\pi/2}\int_0^2\int_0^{9-r^2} r\,dzdrd\theta$?

Now let's try setting up and evaluating an integral in cylindrical coordinates. NOTE: In this section, we'll consider the most important part of any problem to be the correct determination of limits of integration and set-up of the triple integral. Therefore, once our integrals are set up, we can use tech to evaluate them.

EX 2 Evaluate $\iiint_E (x^3+xy^2)\,dV$, where E is the solid in the first octant beneath $z = 1 - x^2 - y^2$.

The surface $z = 1 - x^2 - y^2$ is the inverted paraboloid $z = 1 - r^2$, so cylindrical coordinates are likely most appropriate. The function $x^3 + xy^2$ can be rewritten as $x(x^2 + y^2)$ and so then also as $(r\cos\theta)(r^2)$, or $r^3\cos\theta$. Bounds on z are 0 and $1 - r^2$. Bounds on r and θ come from the region in the xy-plane used by our solid E. The inverted paraboloid intersects the xy-plane in the unit circle $x^2+y^2 = 1$ (we determine that by setting $z = 0$). Since we only want the portion of this circle in the first octant / quadrant, our limits on r and θ will be $0 \le r \le 1$ and $0 \le \theta \le \pi/2$. Remembering that the volume element in cylindrical coordinates is $dV = r\,dzdrd\theta$, the integral and its result are:

$$\iiint_E (x^3 + xy^2)\,dV = \int_0^{\pi/2}\int_0^1\int_0^{1-r^2} (r^3\cos\theta)r\,dzdrd\theta = \frac{2}{35}$$

Note that the result is NOT the volume of anything, since the integral is not of the form $\iiint_E (1)dV$. ■

EX 3 Find the volume of the solid inside the cylinder $x^2 + y^2 = 1$, above $z = 0$ and below the cone $z^2 = 4x^2 + 4y^2$.

This solid of integration is shown in Fig. 15.34. The cone forms the top of our region. In cylindrical coordinates, we can write the cone as $z^2 = 4r^2$, or $z = 2r$, so z is bound by 0 and $2r$. The cylinder forms the sides of the region, and helps us determine bounds on r and θ. The portion of the xy-plane used by this solid is the circle $x^2 + y^2 = 1$ (since the volume is inside the cylinder with the same equation), and so we have $0 \leq r \leq 1$ and $0 \leq \theta \leq 2\pi$. To compute a volume, we set up the integral $\iiint_E (1)dV$. Remembering that the volume element in cylindrical coordinates is $dV = r\,dzdrd\theta$, we have

$$V = \iiint_E (1)dV = \int_0^{2\pi} \int_0^1 \int_0^{2r} (1)rdzdrd\theta = \frac{4\pi}{3} \quad \blacksquare$$

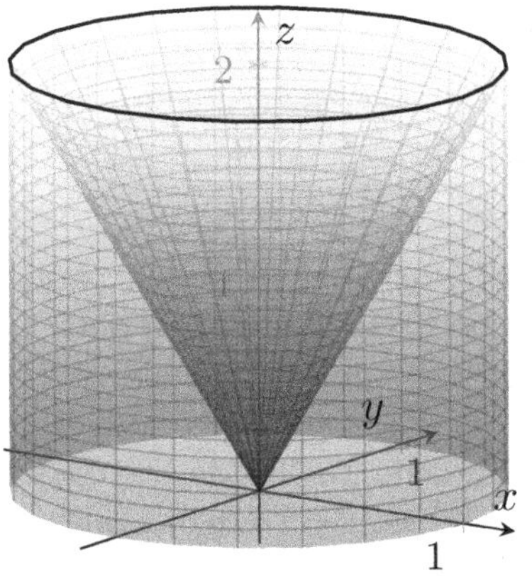

Fig. 15.34 Inside $r = 1$, below $z - 2r$, above $z = 0$.

You Try It

(2) Evaluate $\iiint_E e^z\, dV$, where E is the region between the paraboloid $z = 1 + x^2 + y^2$, the cylinder $x^2 + y^2 = 5$, and the xy-plane.

(3) Evaluate the following integral by first converting it to cylindrical coordinates:

$$\int_{-1}^{1} \int_{-\sqrt{1-x^2}}^{\sqrt{1-x^2}} \int_{x^2+y^2}^{2-x^2-y^2} (x^2 + y^2)^{3/2} dzdydx$$

Triple Integrals in Spherical Coordinates

Let's repeat the above shenanigans with spherical coordinates instead of cylindrical coordinates. Remember that the volume element in spherical coordinates is $dV = \rho^2 \sin\phi \, d\rho d\phi d\theta$. Perhaps a review of the conversion equations would be handy, too?

$$\rho^2 = x^2 + y^2 + z^2 \quad ; \quad \tan\theta = \frac{y}{x} \quad ; \quad \cos\phi = \frac{z}{\rho}$$

and

$$x = \rho\cos\theta\sin\phi \quad ; \quad y = \rho\sin\theta\sin\phi \quad ; \quad z = \rho\cos\phi$$

First, let's just practice "reading" triple integrals in spherical coordinates.

EX 4 What solid's volume is represented by this integral ?

$$\int_0^{\pi/2}\int_0^{\pi/6}\int_0^3 \rho^2 \sin\phi \, d\rho d\phi d\theta$$

Since $dV = \rho^2 \sin\phi d\rho d\phi d\theta$, this really is an integral of the form $\iiint_E (1)dV$, although it may not have looked like it at first. The limits of integration show that ρ is bound by $\rho = 0$ and $\rho = 3$, so our region E is at least part of the interior of the sphere $\rho = 3$, A.K.A. $x^2 + y^2 + z^2 = 9$. The variable ϕ is bound by $\phi = 0$ and $\phi = \pi/6$. Remembering that a spherical equation of the form $\phi = c$ is a cone, it looks like our region is also above the cone $\phi = \pi/6$. (We know it's above rather than below, since $\phi = 0$ point straight overhead.) Finally, θ only goes a quarter of the way around, from 0 to $\pi/2$. Therefore, this integral represents the volume above the cone $\phi = \pi/6$, inside the sphere $\rho = 3$, in the first octant. This solid is shown in Fig. 15.35. ■

You Try It

(4) What solid's volume is represented by this integral?

$$\int_0^{2\pi}\int_{\pi/2}^{\pi}\int_1^2 \rho^2 \sin\phi \, d\rho d\phi d\theta$$

Now let's try setting up and evaluating an integral in spherical coordinates.

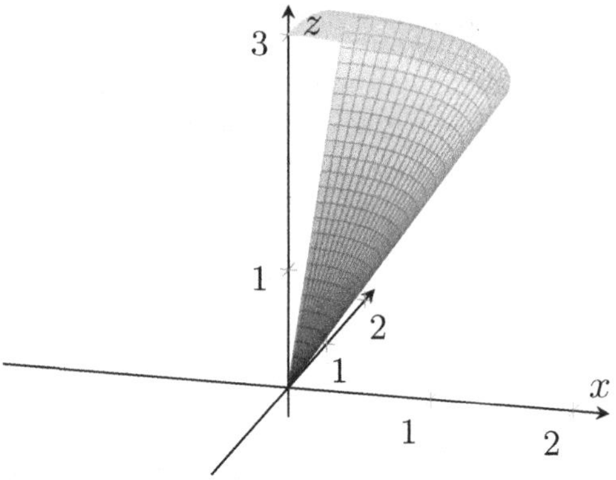

Fig. 15.35 Above $\phi = \pi/6$, below $\rho = 3$, inside first octant.

EX 5 Evaluate $\iiint_E (x^2+y^2+z^2)\, dV$, where E is the sphere $x^2+y^2+z^2 = 1$.

In spherical coordinates, the function becomes $x^2 + y^2 + z^2 = \rho^2$. The volume element is $dV = \rho^2 \sin\phi d\rho d\phi d\theta$. The solid E is just the unit sphere, and so we have

$$0 \leq \rho \leq 1 \quad ; \quad 0 \leq \phi \leq \pi \quad ; \quad 0 \leq \theta \leq 2\pi$$

The integral and its result are then:

$$\int_0^{2\pi} \int_0^{\pi} \int_0^{1} (\rho^2)\rho^2 \sin\phi d\rho d\phi d\theta = \int_0^{2\pi} \int_0^{\pi} \int_0^{1} \rho^4 \sin\phi d\rho d\phi d\theta = \frac{4\pi}{5}$$

If spherical coordinates are making you sweat, here's a reminder of why it's beneficial to have these alternate coordinate systems. As bad as that triple integral looks, the rectangular version would look even worse:

$$\int_{-1}^{1} \int_{-\sqrt{1-x^2}}^{\sqrt{1-x^2}} \int_{-\sqrt{1-x^2-y^2}}^{\sqrt{1-x^2-y^2}} x^2 + y^2 + z^2\, dV$$

So, be glad we have spherical coordinates! ■

You Try It

(5) Evaluate $\iiint_E z\,dV$, where E is the region between the spheres $x^2+y^2+z^2=1$ $x^2+y^2+z^2=4$ in the first octant.

(6) Evaluate the following integral by first converting the integral into spherical coordinates:

$$\int_{-3}^{3}\int_{-\sqrt{9-x^2}}^{\sqrt{9-x^2}}\int_{0}^{\sqrt{9-x^2-y^2}} z\sqrt{x^2+y^2+z^2}\,dzdydx$$

Choosing Between Cylindrical and Spherical Coordinates

In some problems, you won't be explicitly told which coordinate system to use, in which case the choice is yours. Just remember the guidelines stated at the beginning of this section,

- Cylindrical coordinates may be useful when the region of integration is bound by a cylinder or paraboloid, or if the function being integrated contains x^2+y^2.
- Spherical coordinates may be useful when the region of integration is bound by a sphere and/or a cone, or if the function being integrated contains $x^2+y^2+z^2$.

EX 6 Find the volume of the solid above the paraboloid $z=x^2+y^2$ and below the cone $z=\sqrt{x^2+y^2}$.

Here, we have both a cone and a paraboloid, so it's not immediately clear which coordinate system is best. However, if you attempt to write the equation of the paraboloid in spherical coordinates, you will quickly discover it's messy. On the other hand, the equation of a cone in cylindrical coordinates isn't that bad. So let's go with cylindrical coordinates. The equation of the paraboloid is $z=r^2$, and the equation of the cone is $z=r$. Those provide the lower and upper bounds of z, respectively. To find the bounds of r and θ, we must see what region of the xy-plane is spanned by this solid. The maximum horizontal extent of the solid comes from where these two surfaces intersect, i.e. where $x^2+y^2=\sqrt{x^2+y^2}$. This can only happen where $x^2+y^2=1$, and so the intersection of these two surfaces occurs when $x^2+y^2=1$; this gives r and θ bounds corresponding to the unit circle, and we've seen those before. Therefore, the volume is:

$$V = \iiint_E (1)dV = \int_0^{2\pi} \int_0^1 \int_{r^2}^{r} (1)r\, dzdrd\theta = \frac{\pi}{6}$$

(Reminder: Don't forget to use the correct dV!) ■

Triple Integrals in Cyl. & Spher. Coords — Problem List

Triple Ints in Cyl. and Spher. Coords — You Try It

These appeared above; solutions begin on the next page.

(1) What solid's volume is represented by the integral $\int_0^{\pi/2} \int_0^2 \int_0^{9-r^2} r\, dzdrd\theta$?
(2) Evalaute $\iiint_E e^z\, dV$, where E is the region between the paraboloid $z = 1 + x^2 + y^2$, the cylinder $x^2 + y^2 = 5$, and the xy-plane.
(3) Evaluate the following integral by first converting it to cylindrical coordinates:

$$\int_{-1}^{1} \int_{-\sqrt{1-x^2}}^{\sqrt{1-x^2}} \int_{x^2+y^2}^{2-x^2-y^2} (x^2+y^2)^{3/2}\, dzdydx$$

(4) What solid's volume is being represented by this integral?

$$\int_0^{2\pi} \int_{\pi/2}^{\pi} \int_1^2 \rho^2 \sin\phi\, d\rho d\phi d\theta$$

(5) Evaluate $\iiint_E z\, dV$, where E is the region between the spheres $x^2 + y^2 + z^2 = 1$ $x^2 + y^2 + z^2 = 4$ in the first octant.
(6) Evaluate the following integral by first converting it to spherical coordinates:

$$\int_{-3}^{3} \int_{-\sqrt{9-x^2}}^{\sqrt{9-x^2}} \int_0^{\sqrt{9-x^2-y^2}} z\sqrt{x^2+y^2+z^2}\, dzdydx$$

Triple Ints in Cyl. and Spher. Coords — Practice Problems

Try these as you get the hang of the You Try It problems. Solutions to these problems are available in Sec. B.3.3.

(1) Evaluate $\iiint_E x\,dV$, where E is the region bounded by the planes $z = 0$ and $z = x + y + 5$, and the cylinders $x^2 + y^2 = 4$ (i.e. $r^2 = 4$) and $x^2 + y^2 = 9$.

(2) Evaluate $\iiint_E e^{\sqrt{x^2+y^2+z^2}}\,dV$, where E is the region inside the sphere $x^2 + y^2 + z^2 = 9$ in the first octant.

(3) Evaluate the following integral by first converting it to the most appropriate coordinate system:

$$\int_0^1 \int_0^{\sqrt{1-y^2}} \int_{x^2+y^2}^{\sqrt{x^2+y^2}} (xyz)dzdxdy$$

(4) The region above the cone $z^2 = x^2 + y^2$ and below the sphere $x^2 + y^2 + z^2 = 5$ is shaped like an ice cream cone. What is the volume of this region?

(5) Find the volume of the region common to (inside) both the cylinder $x^2 + y^2 = 1$ and the sphere $x^2 + y^2 + z^2 = 4$.

(6) Evaluate $\iiint_E xyz\,dV$, where E is the region between the spheres $\rho = 2$ and $\rho = 4$ and above the cone $\phi = \pi/3$.

(7) Evaluate the following integral by first converting it to an appropriate coordinate system:

$$\int_0^3 \int_0^{\sqrt{9-y^2}} \int_{\sqrt{x^2+y^2}}^{\sqrt{18-x^2-y^2}} (x^2 + y^2 + z^2)\,dzdxdy$$

Triple Ints in Cyl. and Spher. Coords — Challenge Problems

Try these problems to test your skills with the ideas in this section. Solutions to these problems are available in Sec. C.3.3.

(1) Let's pretend the corn silo pictured in Fig. 15.36 is bounded on its sides by the cylinder $x^2 + y^2 = 225$ and above by the (inverted) cone $z = 50 - \sqrt{x^2 + y^2}/9$. Construct a triple integral in cylindrical coordinates that would give the volume of this silo, and compute it.

(2) Evaluate $\iiint_E xyz\,dV$, where E is the region between the spheres $\rho = 1$ and $\rho = 3$ and above the cone $\phi = 2\pi/3$.

(3) Evaluate the following integral by first converting it to an appropriate coordinate system; explain the geometric meaning of your final value:

$$\int_{-4}^4 \int_0^{\sqrt{16-y^2}} \int_{-\sqrt{16-x^2-y^2}}^{5} (1)\,dzdxdy$$

Fig. 15.36 Corn silo with cylinder and cone (with CP 1).

Triple Ints in Cyl. & Spher. Coords — You Try It — Solved

(1) What solid's volume is represented by the integral $\displaystyle\int_0^{\pi/2}\int_0^2\int_0^{9-r^2} r\,dz dr d\theta$?

☐ z is bound below by $= 0$ and above by $z = 9 - r^2$. But $z = 9 - r^2$ is the inverted paraboloid also known as $z = 9 - x^2 - y^2$. The variable r is bound by $r = 0$ and $r = 2$, and θ goes one-fourth of the way around from 0 to $\pi/2$. Therefore, this integral computes the volume under the inverted paraboloid $z = 9 - x^2 - y^2$ over the circle $x^2 + y^2 = 4$ in the first quadrant. ■

(2) Evaluate $\displaystyle\iiint_E e^z\, dV$, where E is the region between the paraboloid $z = 1 + x^2 + y^2$, the cylinder $x^2 + y^2 = 5$, and the xy-plane.

☐ Since a cylinder is involved, we should probably use cylindrical coordinates! The paraboloid $z = 1 + x^2 + y^2$ is also known as $z = 1 + r^2$ in cylindrical coordinates. The cylinder $x^2 + y^2 = 5$ becomes $r^2 = 5$ and the xy-plane is also known as $z = 0.$). Note that the cylinder encloses the paraboloid below their intersection, so this region is under the paraboloid and over a circle of radius $\sqrt{5}$ in the xy-plane. This 3D region is shown in Fig. 15.37. Thus, we have

$$1 \le z \le 1 + r^2 \quad ; \quad 0 \le r \le \sqrt{5} \quad ; \quad 0 \le \theta \le 2\pi$$

and the integral is (with the volume element $dV = r\,dzdrd\theta$ in cylindrical coordinates):

$$\iiint_E e^z dV = \int_0^{2\pi} \int_0^{\sqrt{5}} \int_1^{1+r^2} e^z r dzdrd\theta = \pi e(e^5 - 6) \quad \blacksquare$$

(3) Evaluate the following integral by first converting it to cylindrical coordinates:

$$\int_{-1}^{1} \int_{-\sqrt{1-x^2}}^{\sqrt{1-x^2}} \int_{x^2+y^2}^{2-x^2-y^2} (x^2+y^2)^{3/2}\, dzdydx$$

□ From the given limits of integration on x and y, we can recognize that in the xy-plane, our region of integration is the full unit circle. From the limits on z, we go from the paraboloid $z = x^2 + y^2$ to the upside down paraboloid $z = 2 - x^2 - y^2$. In cylindrical coordinates, this solid is bounded by $z = r^2$, $z = 2 - r^2$, and $r = 1$. This 3D region is shown in Fig. 15.38. Remember that for cylindrical coordinates, the volume element is $dV = r\,dzdrd\theta$. So we get

$$\begin{aligned}
&\int_{-1}^{1} \int_{-\sqrt{1-x^2}}^{\sqrt{1-x^2}} \int_{x^2+y^2}^{2-x^2-y^2} (x^2+y^2)^{3/2} dzdydx \\
&= \int_0^{2\pi} \int_0^1 \int_{r^2}^{2-r^2} (r^2)^{3/2} r dzdrd\theta \\
&= \int_0^{2\pi} \int_0^1 \int_{r^2}^{2-r^2} r^4 dzdrd\theta = \frac{8\pi}{35} \quad \blacksquare
\end{aligned}$$

(4) What solid's volume is represented by this integral?

$$\int_0^{2\pi} \int_{\pi/2}^{\pi} \int_1^2 \rho^2 \sin\phi \, d\rho d\phi d\theta$$

□ The limits of integration show that ρ is bound by $\rho = 1$ and $\rho = 2$, so our region E is at least part of the region between two concentric spheres of radius 1 and 2 (centered at the origin, of course). The variable ϕ is bound by $\phi = \pi/2$ and $\phi = \pi$. Remembering that $\phi = \pi/2$ is the xy-plane and $\phi = \pi$ looks straight down the negative z-axis, it looks like we want our region to be below the xy-plane. Finally, θ goes

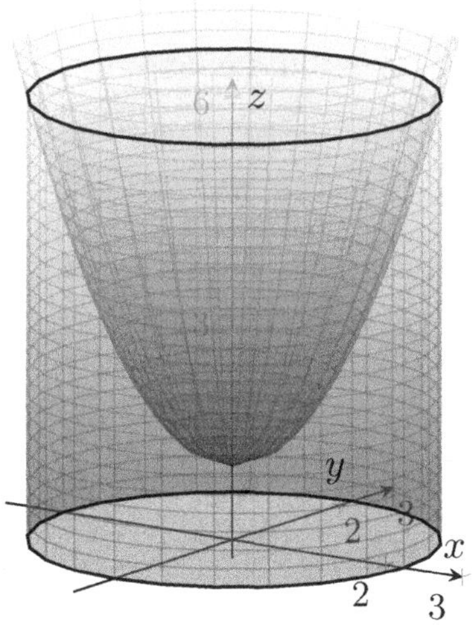

Fig. 15.37 Between $z = 1 + x^2 + y^2$, $x^2 + y^2 = 5$, and $z = 0$.

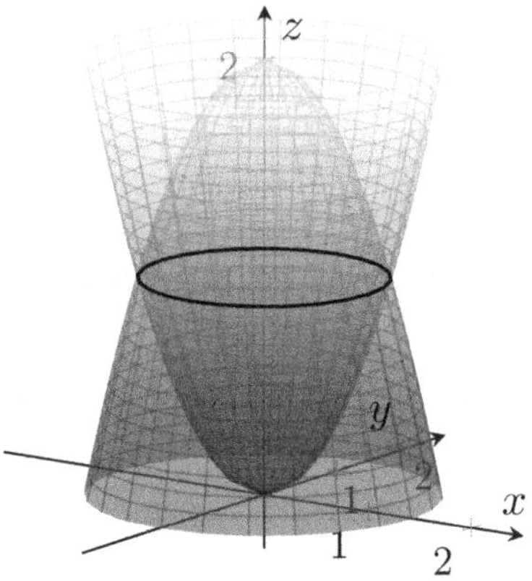

Fig. 15.38 Between $z = 2 - x^2 - y^2$ and $z = x^2 + y^2$.

all the way around, from 0 to 2π. Therefore, this integral computes the volume between the two spheres $x^2 + y^2 + z^2 = 1$ and $x^2 + y^2 + z^2 = 4$ below the xy-plane. This solid is shown in Fig. 15.39. ■

---

(5) Evaluate $\iiint_E z\, dV$, where E is the region between the spheres $x^2 + y^2 + z^2 = 1$ $x^2 + y^2 + z^2 = 4$ in the first octant.

□ Since spheres are involved, we should use spherical coordinates! The sphere $x^2 + y^2 + z^2 = 1$ is also known as $\rho = 1$ in spherical coordinates. The sphere $x^2 + y^2 + z^2 = 4$ is $\rho = 2$. Since we're in the first octant, we restrict θ to $0 \leq \theta \leq \pi/2$ and ϕ to $0 \leq \theta \leq \pi/2$. The volume element in spherical coordinates is $dV = \rho^2 \sin\phi d\rho d\phi d\theta$. So,

$$\iiint_E z\, dV = \int_0^{\pi/2} \int_0^{\pi/2} \int_1^2 (\rho\cos\phi)\rho^2 \sin\phi\, d\rho d\phi d\theta = \frac{15\pi}{16} \quad ■$$

---

(6) Evaluate the following integral by first converting the integral into spherical coordinates:

$$\int_{-3}^{3} \int_{-\sqrt{9-x^2}}^{\sqrt{9-x^2}} \int_0^{\sqrt{9-x^2-y^2}} z\sqrt{x^2 + y^2 + z^2}\, dzdydx$$

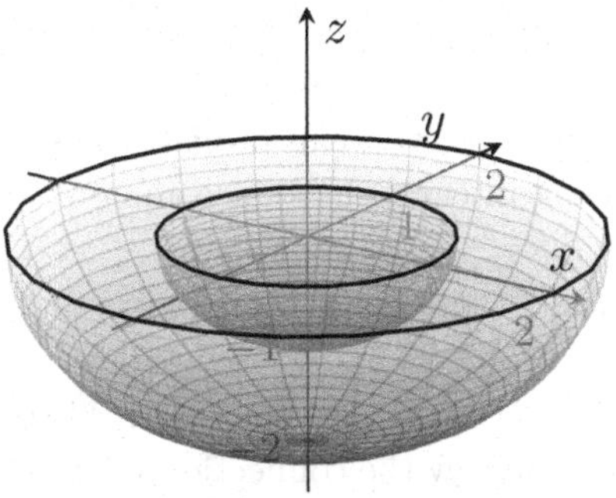

Fig. 15.39 Between $\rho = 1$ and $\rho = 2$, below $z = 0$.

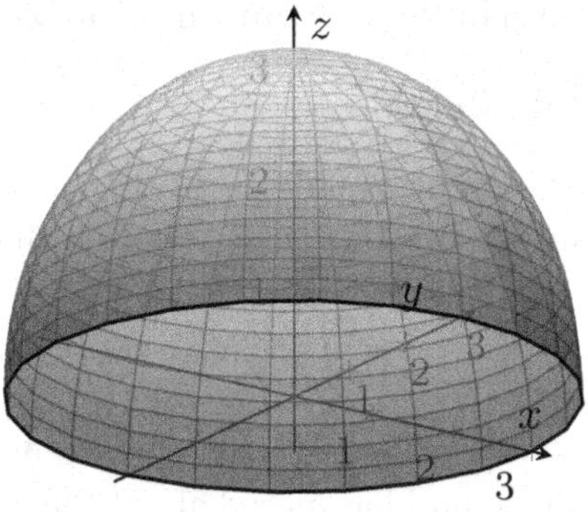

Fig. 15.40 Below $\rho = 3$, above $z = 0$.

☐ From the given limits of integration on x and y, we can recognize that in the xy-plane, our region of integration is a full circle of radius 3. From the limits on z, we go from $z = 0$ to $z = \sqrt{9 - x^2 - y^2}$, that is from $z = 0$ to $x^2 + y^2 + z^2 = 9$. So, we're inside the top half of a sphere of radius 3. This region is shown in Fig. 15.40. In spherical coordinates, we have restrict $0 \leq \phi \leq \pi/2$ for the upper half of the sphere, then $0 \leq \rho \leq 3$ and $0 \leq \theta \leq 2\pi$. Remember that for spherical coordinates, the volume element is $dV = \rho^2 \sin\phi d\rho d\phi d\theta$. So we get:

$$\begin{aligned}
\int_{-3}^{3} \int_{-\sqrt{9-x^2}}^{\sqrt{9-x^2}} \int_{0}^{\sqrt{9-x^2-y^2}} & z\sqrt{x^2 + y^2 + z^2}\, dzdydx \\
&= \int_0^{2\pi} \int_0^{\pi/2} \int_0^3 (\rho\cos\phi)(\rho)\rho^2 \sin\phi\, d\rho d\phi d\theta \\
&= \int_0^{2\pi} \int_0^{\pi/2} \int_0^3 \rho^4 \cos\phi \sin\phi\, d\rho d\phi d\theta \\
&= \frac{243\pi}{5} \quad \blacksquare
\end{aligned}$$

Interlude: Is Your Integral Zero?

Introduction

After your experience in single variable Calculus, you can probably easily answer this question: What do the following integrals have in common?

$$\int_{-1}^{1} x^3\,dx \qquad \int_{0}^{2\pi} \sin(x)\,dx \qquad \int_{-\pi}^{\pi} \cos(x)\,dx \tag{15.5}$$

If you say it's that all these integrals equal zero, you're correct! There was something special about the first time you saw the connection between the graph of $\sin(x)$ on $[0, 2\pi]$ and the value of the corresponding integral. The symmetry of the graph is pleasing, as was that brief rush of endorphins that came when you realized the positive / negative symmetry caused the value of $\int_0^{2\pi} \sin(x)\,dx$ to be zero. Be honest, you thought that was cool.

This partnership between the symmetry of a region of integration and the positive & negative values of the function being integrated there which leads to a result of zero for the integral continues into higher order integration. One way to practice visualizing the interplay between functions and regions of integration is to try to predict — without doing the full calculation in advance — whether given double and triple integrals (in any coordinate system) will yield a value of zero.

Is This Integral Zero?

In order to assess whether a given integral may have a value of zero, we have to consider a few things:

(1) Is the region of integration symmetric, such as around the origin, or an axis, or on both sides of a coordinate plane? If not, then making predictions about any results without calculations might be difficult. For example, the one-dimensional region of integration $[-\pi, \pi]$ is symmetric about the origin; $[0, 2\pi]$ is symmetric about $x = \pi$. So, if you understand how $\sin(x)$ behaves, it's easy to predict that both

$$\int_{-\pi}^{\pi} \sin(x)\,dx \qquad \text{and} \qquad \int_0^{2\pi} \sin(x)\,dx$$

are equal to zero. But recognizing that both

$$\int_{\pi/6}^{13\pi/6} \sin(x)\,dx = 0 \qquad \text{and} \qquad \int_{\pi/6}^{11\pi/6} \sin(x)\,dx = 0$$

while

$$\int_{5\pi/6}^{13\pi/6} \sin(x)\,dx \neq 0$$

is a bit harder.

(2) Given a "nicely shaped" region of integration for a function f (whether 1D, 2D, or 3D), can you assess that for every positive value of f within the region, there is a negative value of f to counteract it? For example, while $[-2, 2]$ is a region of integration nicely symmetric about the origin, we would have

$$\int_{-2}^{2} x^3\,dx = 0 \qquad \text{but} \qquad \int_{-2}^{2} \left(x - \frac{1}{4}\right)^3 dx \neq 0$$

and we should be able to make those determinations without actually using the Fundamental Theorem of Calculus to evaluate the integral.

Seeing a balance between the integrand and the region of integration can be helpful for integrals which cannot be evaluated by hand. For example, while the following integral cannot be solved using any techniques seen in standard Calculus courses, we can still determine that

$$\int_{-2}^{2} x\sin(x^2)\cos(x^4 + \pi)\,dx = 0$$

Now that we've opened up two and three-dimensional integrals in rectangular, polar, cylindrical, and spherical coordinate systems, we should still be able to consider the (more complicated) interplay between a region of integration and an integrand which generates both positive and negative values over that region, and make predictions about which integrals will yield a value of zero and which will not. And sure, this is lots of fun, but it's also good practice visualizing regions of integration.

EX 1 Without performing the full calculation, decide whether or not $\iint_D xy\,dA = 0$, where D is (a) the unit square centered at the origin, (b) the region between between the full circles $r = 1$ and $r = 2$, and (c) the region between between the circles $r = 1$ and $r = 2$ between the rays $\theta = \pi$ and $\theta = 3\pi/2$.

(a) The unit square centered at the origin uses $-1 \le x \le 1$ and $-1 \le y \le 1$; it's symmetric around the origin with equal areas (area of 1) in each Quadrant I to IV. In Quadrant I, both x and y are positive and in Quadrant III, both x and y are negative, but in both regions, the value of xy will be positive. But in Quadrants II and IV, the value of xy will be negative, because one of either x or y will be positive while the other is negative. Because of the symmetric nature of the overall region, the positive contributions of xy from QI and QIII are exactly balanced out by negative contributions from QII and QIV. So, overall, we will have $\iint_D xy\,dA = 0$.

(b) Given the circular geometry of this region, evaluation of the integral would be done in polar coordinates. But still, we can see — like in (a), that because the portions of the overall region of integration in each of the four quadrants are symmetrically shaped, and we would have positive contributions to the integral from QI and QIII balanced by negative contributions from QII and QIV, then we would have $\iint_D xy\,dA = 0$.

(c) The region between the rays $\theta = \pi$ and $\theta = 3\pi/2$ is Quadrant III, and so contributions from this region to the given integral will only be positive ($xy > 0$ when both $x < 0, y < 0$). Therefore we would have $\iint_D xy\,dA > 0$ here. ■

EX 2 Without performing the full calculation, decide whether or not $\iiint_D x^2yz\,dV = 0$, where E is (a) the unit cube centered at the origin, (b) the region between between the full spheres $\rho = 1$ and $\rho = 4$, and (c) the region between between the spheres $\rho = 1$ and $\rho = 4$ above the cone $\phi = \pi/3$.

The contribution to the integrand from x^2 is always positive, so it will not be a distinguishing factor in any of the three cases. We only need to consider y and z for contributions of negative values to balance positive values. In the perfect symmetry in all directions around the origin of cases (a) and (b), each contribution of y and z will be balanced by a negative contribution. In (c), the restriction to the region above $\theta = \pi/3$ results in having $z > 0$ for all points in the region, and so we only have to consider y. But even so, all positive y values have a corresponding negative value within the region. In all cases, net positive and negative contributions are equal but opposite, and so $\iiint_D x^2yz\,dV = 0$ for all cases. ■

There are plenty other varieties of integrals to be examined, but we have to wait until we see them in Chapter 18 — where they will be appended to problem sets in a couple of sections.

Is Your Integral Zero? — You Try It

These appeared above; solutions begin on the next page.

(1) We expect $\iint Df(x,y)\,dA = 0$ for which of the following combinations of function $f(x,y)$ and region of integration D?

I1) $f(x,y) = x^3\sqrt{y^2+1}$ and D is the rectangle $\{(x,y) : -1 \leq x \leq 1; -3 \leq y \leq 2\}$

I2) $f(x,y) = e^{-xy}$ and D is the upper half of the unit circle

I3) $f(x,y) = x/y^2$ and D is the region between the parabola $y = x^2+1$ and $y = 10$

(2) We expect $\iiint Ef(x,y,z)\,dV = 0$ for which of the following combinations of function $f(x,y,z)$ and region of integration E?

I4) $f(x,y,z) = xyz^2$ and E is the rectangle $\{(x,y,z) : -1 \leq x \leq 1; -2 \leq y \leq 2; -5 \leq z \leq 5\}$

I5) $f(x,y,z) = xyz$ and E is the region between two concentric spheres of radii 2 and 4

I6) $f(x,y,z) = x^2+y^2+z^2$ and E is the upper half of the unit sphere

(3) Which of the following integrals will yield 0?

I7) $\displaystyle\int_0^{2\pi}\int_0^4\int_0^{\sqrt{16-r^2}} (r\cos\theta)^2(r\sin\theta)z^2 \cdot r dz dr d\theta$

I8) $\displaystyle\int_0^{2\pi}\int_0^{\pi/2}\int_0^1 \rho^4 \sin\phi d\rho d\phi d\theta$

Is Your Integral Zero? — You Try It — Solved

(1) We expect $\iint Df(x,y)\,dA = 0$ for which of the following combinations of function $f(x,y)$ and region of integration D?

I1) $f(x,y) = x^3\sqrt{y^2+1}$ and D is the rectangle $\{(x,y) : -1 \le x \le 1; -3 \le y \le 2\}$

I2) $f(x,y) = e^{-xy}$ and D is the upper half of the unit circle

I3) $f(x,y) = x/y^2$ and D is the region between the parabola $y = x^2+1$ and $y = 10$

□ Integrals (I1) and (I3) will yield 0, while (I2) will not. The integrand in (I2) is always positive, so the integral will be positive. In (I1), consider a fixed value of y; for that value of y, every positive x has a negative counterpart, such as $1/3$ and $-1/3$. These contribute in equal but opposite fashions to the integrand via x^3. So the net integral will be zero. The result of (I3) arises similarly. I highly recommend drawing a diagram of each region of integration to visualize these results. ■

(2) We expect $\iiint Ef(x,y,z)\,dV = 0$ for which of the following combinations of function $f(x,y,z)$ and region of integration E?

I4) $f(x,y,z) = xyz^2$ and E is the rectangle $\{(x,y,z) : -1 \le x \le 1; -2 \le y \le 2; -5 \le z \le 5\}$

I5) $f(x,y,z) = xyz$ and E is the region between two concentric spheres of radii 2 and 4

I6) $f(x,y,z) = x^2+y^2+z^2$ and E is the upper half of the unit sphere

□ Integrals (I4) and (I5) will yield 0, while (I6) will not. In (I6), the integrand is always positive, so the net value of the integral will be positive. In (I4), consider a fixed value of z; for that fixed value, the cross section of the region of integration is a rectangle in the x, y directions that's symmetric in both directions around the origin. Each positive value of either x or y has a negative counterpart, and the two contributions to the integral balance to zero. In (I5), imagine a sphere for any fixed radius between 2 and 4; for every point on that sphere with coordinates (x,y,z), there is a negative counterpart $(-x,-y,-z)$. These contribute in equal but opposite fashions to the integrand xyz. Draw pictures! ■

(3) Which of the following integrals will yield 0?

$$\text{I7)} \quad \int_0^{2\pi}\int_0^4\int_0^{\sqrt{16-r^2}} (r\cos\theta)^2(r\sin\theta)z^2 \cdot r\,dzdrd\theta$$

$$\text{I8)} \quad \int_0^{2\pi}\int_0^{\pi/2}\int_0^1 \rho^4 \sin\phi\, d\rho d\phi d\theta$$

□ In (I7), the only part of the integrand which contributes both positive and negative values is the term $\sin\theta$. Since the integral takes place over $0 \leq \theta \leq 2\pi$, every value of $\sin\theta$ comes with an equal but opposite counterpart. So the net value of the integral will be zero. In (I8), the only term which *might* contribute negative values to the integral is $\sin\phi$; however, since the integral takes place over $0 \leq \phi \leq \pi/2$, we will see no negative values of $\sin\phi$ in the integral, and overall, the value of the integral must be positive. ■

FFT: We have not picked up this idea before, but it turns out that when double and triple integrals with constant limits have integrands that are "separable" according to the variables, i.e. $f(x)g(y)$, $f(\rho)g(\phi)h(\theta)$, and so on, then the double or triple integral can be separated as a product of single integrals for each variable. For example, if we have a double integral with an integrand that can be written as $F(x,y) = f(x)g(y)$, then

$$\int_c^d\int_a^b F(x,y)\,dydx = \int_c^d\int_a^b f(x)g(y)\,dydx = \int_a^b g(y)\,dy \cdot \int_c^d f(x)\,dx \tag{15.6}$$

Or, if an integrand $F(\rho,\phi,\theta)$ can be written as $f(\rho)g(\phi)h(\theta)$, then

$$\begin{aligned}\int_{\theta_1}^{\theta_2}\int_{\phi_1}^{\phi_2}\int_{\rho_1}^{\rho_2} F(\rho,\phi,\theta)\,d\rho d\phi d\theta &= \int_{\theta_1}^{\theta_2}\int_{\phi_1}^{\phi_2}\int_{\rho_1}^{\rho_2} f(\rho)g(\phi)h(\theta)\,d\rho d\phi d\theta \\ &= \int_{\theta_1}^{\theta_2} h(\theta)\,d\theta \cdot \int_{\phi_1}^{\phi_2} g(\phi)\,d\phi \cdot \int_{\rho_1}^{\rho_2} f(\rho)\,d\rho\end{aligned}$$

Does this help resolve which integrals from (I1) to (I8) would be zero?

Also, if you enjoy the constructive side of things that we occasionally encounter in "the Pit", can you establish relation (15.6) from the original definition of a double integral shown in Sec. 14.3?

Chapter 16

Player V Has Entered the Game

16.1 Vector Basics

Introduction

Most of you have encountered vectors in one form or another. They are the backbone of much of the mathematics which relates to the physical world. There's only so much we can do with scalars and scalar functions, and we can no longer avoid vectors in our development of tools. So let's review the basic concepts of, and simple operations we can do with, vectors. Then we'll start putting them to good use.

Climbing the Conceptual Ladder to Vectors

One of the fun things mathematicians do is invent multiple words that seem to mean the same thing, and it seems like the evil goal of this practice is to cause maximum confusion. However, what's really happening is you're learning that something you may have thought was special is just one specific case of a much broader concept. For example, those things you called "numbers" for a long time started being called "constants". At first, you just used numbers for counting, but then you started learning about functions which display *change*, yet those numbers can act as perfectly good functions, too — like $y = 5$ — and in that context, the function displays no change; in other words, it is *constant.*

Numbers (constants) are the building blocks for points, which determine locations in space. A single number, like 4, can be considered a point in one dimension. To get to a point in space, you need a map, and the *coordinates* of a point provide that map — with reference to some starting point, or *origin*. The coordinate to a point in one dimension comes from the number itself: to find the number (point) 4, we move four units to the right of the origin; to find the number (point) -4, we move four units to the left.

The one-dimensional coordinate system, i.e. the real number line, is denoted $\mathbb{R}$.

When we start pairing numbers (constants) together, we have the ability to describe two-dimensional space. A point in two dimensional space — which is denoted $\mathbb{R}^2$ — is described by two coordinates, and those coordinates locate the point relative to the origin of the coordinate system. To get from the origin to the point $(3, -2)$, we move 3 units to the right of the origin, then two units down. The coordinate axes align these directions; the x-axis designates "left" and "right", and the y-axis designates "up" and "down". (Based on your experience with polar coordinates, you know there are other ways to describe locations, but let's keep our focus on rectangular coordinates for now.)

In three dimensional space $\mathbb{R}^3$, we have more of the same. A point has three coordinates, all of which are needed to fix the position of a point relative to the origin of the 3D coordinate system. With the x-, y-, and z-axes in place, all perpendicular to each other, we can give road maps to points; the point $(3, 2, 1)$ is three units in the direction of the x-axis, and so on.

Once we move up to more dimensions, we can still describe points, such as $(1, 2, 3, 4)$, but now we can't directly visualize them.

Notice that even in this very basic conversation about things you already know, there is terminology that's undefined. What, exactly, is a "dimension"? What's do we mean by "space"? And there are other considerations: I'll bet when you draw $\mathbb{R}^2$ via x- and y-axes, you draw these parallel to the bottom and side edges of your paper. But what if you drew axes such that one pointed from, say, the lower left corner of your paper to the upper right corner, and the other one was perpendicular to it ... would these still be called x- and y-axes? Could we create unique locations of points in space with reference to axes that weren't perpendicular to each other? Are coordinates of a point unique? The answers to all of these questions go way beyond what we need here, but if these sound interesting, then you should take a class in Linear Algebra. It will blow your mind. We need to barely skim the surface of some concepts from Linear Algebra, and those concepts are all rooted in *vectors.*

Simple Vector Concepts and Properties

Where points in one, two, and three dimensions are tied to their positions in space, a *vector* is a mathematical creature which exists independent of position. A vector is defined by its size and direction, not its location. No matter whether we're in one, two, or three dimensions, it takes two points to define a single vector — an *initial point* and a *terminal point.* A vector is usually represented in a visual fashion by an arrow which points from the initial point to the terminal point. But the visual representation only gets us so far. Go ahead and draw an arrow on a piece of paper to represent a vector ... I'll wait ... OK, good. Now, if you had to tell someone over the phone how to draw exactly that same arrow, you can't say "look at my vector and draw the same thing". Rather, you'd probably describe the direction in which it points, and its length in that direction. You probably would not be so concerned about *where* on the paper the arrow is drawn, so position isn't an issue.

In geometry classes, you very likely named points by things like P and Q. Vectors also need names, but we have to do better than just assigning a letter, because we need to know when we're looking at a vector and not a constant or a point. And so, we usually pick one of two notations. If we want to name a vector with the letter v (because we're clever that way), we will either place a little arrow hat on it ($\mathbf{v}$), or put it in bold face ($\mathbf{v}$). The former is best when you're writing by hand, and the latter is best when there is mechanical typesetting, like here.

Vectors are built with their *components*, which describe the distance moved in each coordinate direction to get from the initial point to the terminal point of the vector. Components are to vectors what coordinates are to points; a vector has the same number of components as a point in the same space. But be alert to the difference: coordinates of a point give the distance in each coordinate direction from a fixed origin, and fix the point at one location in space. But the components of a vector simply locate the terminal point relative to the initial point, and therefore, the vector is independent of location. All of the vectors shown in Fig. 16.1 have components

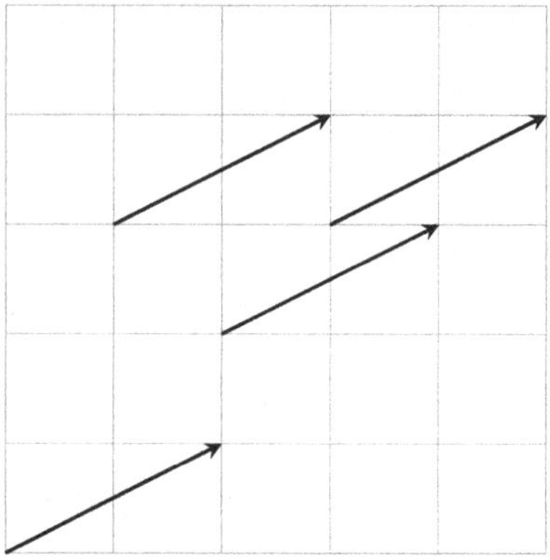

Fig. 16.1 Multiple representations of $\mathbf{v} = < 2, 1 >$.

2 and 1 because for each vector, we move two units horizontally and one unit vertically to get from the initial point to the terminal point.

To maintain distinction between points and vectors, we can't list out a vector's components by grouping them in parentheses; the creature $(1, 2)$ needs to be known distinctly as the point with coordinates $(1, 2)$. The standard notations for vector components are to group them either in pointy brackets, like $\mathbf{v} = \langle 2, 1 \rangle$, or in square brackets, like $\mathbf{v} = [2, 1]$. Here, we'll rely on the pointy brackets unless I accidentally slip and use square ones — but you'll understand either one, right?

Even though we want to keep points and vectors distinct conceptually, they are certainly related. The initial point and terminal points of a vector are *points*, after all, and so here is a first constructive step: the components of a vector can be found by subtracting the coordinates of the initial point from the coordinates of the terminal point. For example, consider the 2D vector $\langle 1, 2 \rangle$. If the initial point of this vector is at the origin, then this vector ends at the point $(1, 2)$. But if the initial point of this vector is, say, $(2, 2)$, then the terminal point of the vector is $(3, 4)$. As stated, it takes two points to define a single vector, and this is why.

EX 1 Find the components of the vector with initial point $(1, -2, 2)$ and terminal point $(0, 3, 1)$.

We need to take each pair of coordinates of the initial and terminal points and subtract them:

- The first pair: $0 - 1 = -1$.
- The second pair: $3 - (-2) = 5$.
- The third pair: $1 - 2 = -1$.

Therefore, the vector is identified as $\mathbf{v} = \langle -1, 5, -1 \rangle$. ■

You Try It

(1) Find the components of the vector from $A(2,3)$ to $B(-2,1)$.

Since a vector is determined by the straight line path from its initial point to its terminal point, you may think that its length would be one of its most important features, and you'd be right. Often, in the grand tradition of having more than one term available, we will refer to the length of a vector as its *magnitude*. There are a few reasons for this. First, saying "length" is fine in dimensions up to three since we have an intuitive idea of what it means, but in dimensions higher than 3, what does "length" mean anymore? So, we have a catch-all phrase that works for any dimension. Second, it turns out that what we're calling vectors are just one type of a huge category of mathematical objects, many of which have nothing to do with physical space — and so the term length would be meaningless.[1] Finally, saying "magnitude" is more impressive and makes you sound smarter, and the word is worth more points when playing Scrabble.

The length (or magnitude) of a vector $\mathbf{v}$ is given as $|\mathbf{v}|$. It's not an accident that we're using what we'd otherwise call absolute value bars. In fact, if you think about the fact that the size of both the one-dimensional vectors $\langle -5 \rangle$ and $\langle 5 \rangle$ are the same, and equal to 5, that notation should make a lot of sense! To compute the length, note that the horizontal and vertical lines that get you from the initial to terminal point of the vector are the two legs of a right triangle whose hypotenuse is the vector itself, and so we can invoke the Pythagorean Theorem.

Useful Fact 16.1. *The magnitude of a vector* $\mathbf{v}$ *is determined through its components in a recognizable way:*

- *In one dimension, the magnitude of* $\mathbf{v} = \langle v_1 \rangle$ *is* $|\mathbf{v}| = |v_1|$ *(literally the absolute value of its one component).*

[1]For example, it turns out that functions can be considered vectors, too .. for more information, take Linear Algebra!

- *In two dimensions, the magnitude of* $\mathbf{v} = \langle v_1, v_2 \rangle$ *is* $|\mathbf{v}| = \sqrt{v_1^2 + v_2^2}$ *(the Pythagorean Theorem applied to its two components).*
- *In three dimensions, the magnitude of* $\mathbf{v} = \langle v_1, v_2, v_3 \rangle$ *is* $|\mathbf{v}| = \sqrt{v_1^2 + v_2^2 + v_3^2}$ *(the 3D version of the Pythagorean Theorem).*

I'll bet you can now guess the pattern for $\mathbf{v} = \langle v_1, \ldots, v_4 \rangle$ and higher.

EX 2 Find the magnitude of the vector found in EX 1.

In EX 1, we found $\mathbf{v} = \langle -1, 5, -1 \rangle$. Therefore,

$$|\mathbf{v}| = \sqrt{(-1)^2 + (5)^2 + (-1)^2} = \sqrt{1 + 25 + 1} = \sqrt{27} \quad \blacksquare$$

You Try It

(2) Find the magnitude of the vector $\mathbf{v}$ that you found in YTI 1.

Combining Vectors With Vectors

Just like numbers (pardon me, *scalars*!) can be combined together to make new scalars (via addition, subtraction, and so forth), we can also combine vectors in various ways.

Operations on vectors are done using their components. The simplest operation on vectors is vector addition, and all we do is go through the components and add them pairwise. In 2D, we have

$$\mathbf{v} + \mathbf{w} = \langle v_1, v_2 \rangle + \langle w_1, w_2 \rangle = \langle v_1 + w_1, v_2 + w_2 \rangle$$

I'm sure you can imagine what this would look like in 3D or higher. To see what is happening geometrically, we (1) draw $\mathbf{v}$, (2) use the terminal point of $\mathbf{v}$ as the initial point of $\mathbf{w}$ and draw $\mathbf{w}$, and then (3) draw a new vector that completes the developing triangle, i.e., that connects the initial point of $\mathbf{v}$ to the terminal point of $\mathbf{w}$. Figure 16.2 shows $\mathbf{v} + \mathbf{w}$ for $\mathbf{v} = \langle 3, 1 \rangle$ and $\langle w \rangle = \langle -1, 3 \rangle$. Algebraically, we get $\mathbf{v} + \mathbf{w} = \langle 2, 4 \rangle$, and you can confirm that visually in the figure.

Subtraction should be the next obvious way to combine two vectors, but we'll defer that until a bit later.

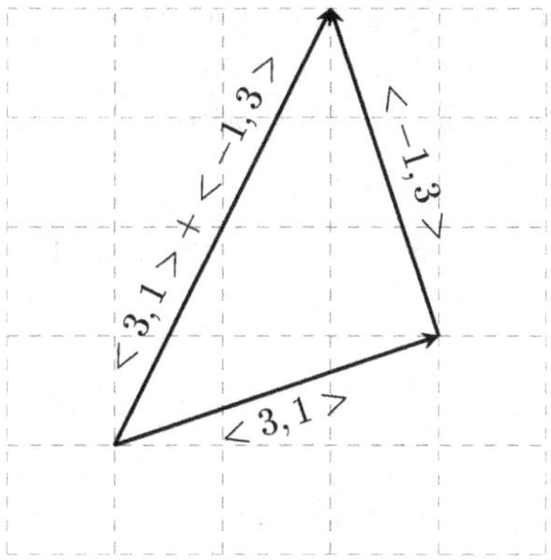

Fig. 16.2 Vector addition: $< 3, 1 > + < -1, 3 >$.

After addition and subtraction, you might think multiplication and division would get their turn. But there is no such thing as one vector "times" another, nor one vector divided by another. Rather, there are *two* ways to combine vectors that are akin to multiplication. These are called the *dot product* and *cross product*, respectively. In the dot product, we take two vectors, do something to them (via their components), and produce a scalar. In the cross product, we take two vectors, do something to them, and produce another vector. These operations are so much fun, they have their own section, coming up soon!

Combining Vectors With Scalars

There are things you can and cannot do with vectors and scalars. For example, if $\mathbf{v}$ is a vector and c is a scalar, the expression $\mathbf{v} + c$ makes no sense. You can't add apples to oranges. But there is an operation, called *scalar multiplication*, that does make sense, and it proceeds as follows in 2D:

$$c\mathbf{v} = c \cdot \langle v_1, v_2 \rangle = \langle cv_1, cv_2 \rangle$$

Again, I'm sure you can imagine what it would look like in 3D or higher. There are two potential effects of scalar multiplication:

- If $c > 0$, then the length of $c\mathbf{v}$ is c times the length of $\mathbf{v}$. If $0 < c < 1$, the new vector is shorter than the original; if $c > 1$, the new vector is longer. For example, if $\mathbf{v} = \langle 2, 1, 3 \rangle$, then $3\mathbf{v}$ is a new vector three times as long as $\mathbf{v}$ but having the same direction.

- If $c > 0$, the rescaling of the length is the same, but the direction of the vector gets reversed. The vector $-\mathbf{v}$ has the same length but opposite direction as $\mathbf{v}$. The vector $-(1/2)\mathbf{v}$ has half the length and the opposite direction as $\mathbf{v}$.

EX 3 If $\mathbf{v} = \langle -2, 0, 4\rangle$, then what are $-2\mathbf{v}$ and $(1/2)\mathbf{v}$?

$$-2\mathbf{v} = -2\langle -2,0,4\rangle = \langle 4,0,-8\rangle \quad ; \quad \frac{1}{2}\mathbf{v} = \frac{1}{2}\langle -2,0,4\rangle = \langle -1,0,2\rangle \quad \blacksquare$$

You Try It

(3) If $\mathbf{w} = \langle 5, 2\rangle$, then find $3\mathbf{w}$ and show using Useful Fact 16.1 that the length of $3\mathbf{w}$ is three times the length of $\mathbf{w}$.

With scalar multiplication in place, considering subtraction of vectors is now suitable, because rather than learning a bunch of new rules, we can just consider $\mathbf{v} - \mathbf{w}$ as $\mathbf{v} + (-\mathbf{w})$. If $\mathbf{v} = \langle 3,-1,3\rangle$ and $\mathbf{w} = \langle 1,1,1\rangle$, then algebraically we have $\mathbf{v} - \mathbf{w} = \langle 3-1, -1-1, 3-1\rangle = \langle 2,-2,2\rangle$. That's easy. Graphically, we attach $-\mathbf{w}$ to the terminal point of $\mathbf{v}$ and complete the resulting triangle. Figure 16.3 shows an example in $\mathbb{R}^2$, where we have $\mathbf{u} - \mathbf{w}$ for $\mathbf{u} = \langle 2,3\rangle$ and $\mathbf{v} = \langle 3,2\rangle$. Algebraically, we have $\mathbf{u} - \mathbf{w} = \langle -1,1\rangle$; is that consistent with what you see in the figure?

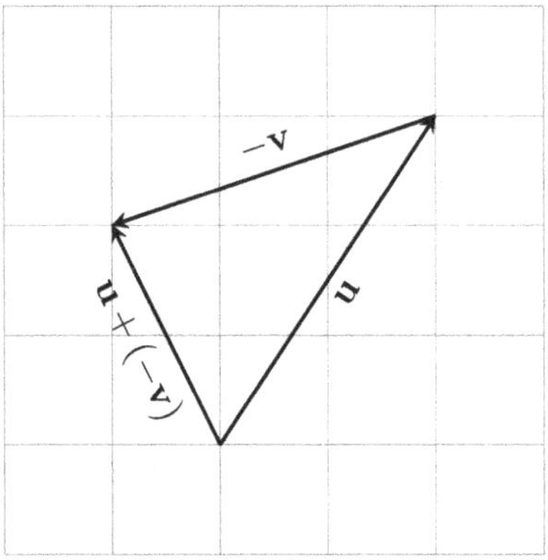

Fig. 16.3 $\mathbf{u} - \mathbf{v}$ for $\mathbf{u} = < 2, 3 >, \mathbf{v} = < 3, 1 >$.

EX 4 If $\mathbf{a} = \langle 1,-2,-2\rangle$ and $\mathbf{b} = \langle 3,2,0\rangle$, then predict whether each of the following will be a vector or a scalar, and compute them: (1) $\mathbf{a} - 2\mathbf{b}$, (2) $|\mathbf{a}+\mathbf{b}|$, (3) $\mathbf{b}/|\mathbf{b}|$, and (4) the length of $\mathbf{b}/|\mathbf{b}|$?

(1) is the difference of two vectors, and so will be a vector,

$$\mathbf{a} - 2\mathbf{b} = \langle 1, -2, -2\rangle - \langle 6, 4, 0\rangle = \langle -5, -6, -2\rangle$$

(2) is the length of a sum of vectors, and will be a scalar. Since $\mathbf{a} + \mathbf{b} = \langle 4, 0, -2\rangle$, then

$$|\mathbf{a} + \mathbf{b}| = \sqrt{4^2 + 0^2 + (-2)^2} = \sqrt{20} = 2\sqrt{5}$$

(3) is a vector divided by a scalar, or rather, a vector times the reciprocal of a scalar, so the result will be another vector. Since $|\mathbf{b}| = \sqrt{13}$, then

$$\frac{\mathbf{b}}{|\mathbf{b}|} = \frac{1}{|\mathbf{b}|}\mathbf{b} = \frac{1}{\sqrt{13}}\langle 3, 2, 0\rangle = \left\langle \frac{3}{\sqrt{13}}, \frac{2}{\sqrt{13}}, 0\right\rangle$$

(4) is asking for the length of the vector found in (3), and so we expect the result to be a scalar. We have

$$\left|\frac{\mathbf{b}}{|\mathbf{b}|}\right| = \sqrt{\left(\frac{3}{\sqrt{13}}\right)^2 + \left(\frac{2}{\sqrt{13}}\right)^2} = \sqrt{\frac{9}{13} + \frac{4}{13}} = \sqrt{\frac{13}{13}} = 1$$

Hey, I wonder if there's something to the fact that we got a length of 1 in (4)? ■

You Try It

(4) If $\mathbf{a} = \langle -4, 3\rangle$ and $\mathbf{b} = \langle 6, 2\rangle$, what are $|\mathbf{a}|$, $\mathbf{a} + \mathbf{b}$, $\mathbf{a} - \mathbf{b}$, $2\mathbf{a}$, and $3\mathbf{a} + 4\mathbf{b}$?

Unit Vectors

In the last part of EX 4 above, we saw a very important result. We took a vector, divided it (rescaled it) by its own length (sorry, *magnitude*!), then found the magnitude of the resulting vector. That new magnitude was 1, which is not an accident. Vectors with a magnitude of 1 are extremely special.

Definition 16.1. *A vector* $\mathbf{v}$ *for which* $|\mathbf{v}| = 1$ *is called a* ***unit vector****. For any vector* $\mathbf{v}$ *except* $\mathbf{0}$,[2] *we can construct a unit vector* $\mathbf{u}$ *in the same direction:*

$$\text{If } \mathbf{u} = \frac{\mathbf{v}}{|\mathbf{v}|} \quad \text{then} \quad |\mathbf{u}| = 1 \tag{16.1}$$

[2]The zero vector $\mathbf{0}$ is a vector whose components are all 0.

Unit vectors are very important. In fact, some unit vectors are so important they get their own names. In $\mathbb{R}^2$, the two most special unit vectors are

$$\mathbf{i} = \langle 1, 0 \rangle \qquad \text{and} \qquad \mathbf{j} = \langle 0, 1 \rangle$$

In $\mathbb{R}^3$, the three most special unit vectors are

$$\mathbf{i} = \langle 1, 0, 0 \rangle \quad \text{and} \quad \mathbf{j} = \langle 0, 1, 0 \rangle \quad \text{and} \quad \mathbf{k} = \langle 0, 0, 1 \rangle$$

Do you see why these are important vectors? Hopefully you are thinking along these lines:

- These vectors define the coordinate axes in their 2D or 3D plane. If we join their initial points, then that intersection can be considered the origin, and any point x on the x-axis is located by the terminal point of $x\mathbf{i}$, and and point y on the y-axis is located by the terminal point of $y\mathbf{j}$. Better yet,
- Any vector in 2D or 3D can be formed by a combination of these vectors; because $\langle x, y \rangle = x\mathbf{i} + y\mathbf{j}$. For example, the vector $\langle 17, 295 \rangle$ can be written as the vector sum $17\mathbf{i} + 295\mathbf{j}$.

The introduction of these unit vectors gives us a new way to denote vectors and their components. In $\mathbb{R}^3$, for example,

$$\langle v_1, v_2, v_3 \rangle = v_1\mathbf{i} + v_2\mathbf{j} + v_3\mathbf{k}$$

You need to be able to identify and handle vectors no matter which notation is used to presented them. It is common practice to give your answer in the same notation as the question was posed. Since many calculations are tidier when tracked using bracket notation, feel free to switch to it, but return your answer to unit vector notation if that's how the question was posed.

EX 5 If $\mathbf{a} = 3\mathbf{i} - \mathbf{j} + 2\mathbf{k}$, and $\mathbf{b} = \mathbf{i} + \mathbf{j} + 2\mathbf{k}$, what are $|\mathbf{a}|$ and $\mathbf{a} - 3\mathbf{b}$?

- The vector $\mathbf{a}$ is also known as $\langle 3, -1, 2 \rangle$; we have $|\mathbf{a}| = \sqrt{9 + 1 + 4} = \sqrt{14}$.
- Asking for $\mathbf{a} - 3\mathbf{b}$ is the same as asking for $\langle 3, -1, 2 \rangle - 3\langle 1, 1, 2 \rangle$, and so

$$\mathbf{a} - 3\mathbf{b} = \langle 3, -1, 2 \rangle - \langle 3, 3, 6 \rangle = \langle 0, -4, -4 \rangle \quad \blacksquare$$

You Try It

(5) Find a unit vector in the same direction as $\mathbf{v} = 9\mathbf{i} - 5\mathbf{j}$.

Vector Basics — Problem List

Vector Basics — You Try It

These appeared above; solutions begin on the next page.

(1) Find the components of the vector $\mathbf{v}$ from $A(2,3)$ to $B(-2,1)$.
(2) Find the magnitude of the vector $\mathbf{v}$ that you found in YTI 1.
(3) If $\mathbf{w} = \langle 5, 2\rangle$, then find $3\mathbf{w}$ and show using Useful Fact 16.1 that the length of $3\mathbf{w}$ is three times the length of $\mathbf{w}$.
(4) If $\mathbf{a} = \langle -4, 3\rangle$ and $\mathbf{b} = \langle 6, 2\rangle$, what are $|\mathbf{a}|$, $\mathbf{a}+\mathbf{b}$, $\mathbf{a}-\mathbf{b}$, $2\mathbf{a}$, and $3\mathbf{a}+4\mathbf{b}$?
(5) Find a unit vector in the same direction as $\mathbf{v} = 9\mathbf{i} - 5\mathbf{j}$.

Vector Basics — Practice Problems

Try these as you get the hang of the You Try It problems. Solutions to these problems are available in Sec. B.4.1.

(1) If $\mathbf{a} = 2\mathbf{i} - 3\mathbf{j}$ and $\mathbf{b} = \mathbf{i} + 5\mathbf{j}$, what are $|\mathbf{a}|$, $\mathbf{a}+\mathbf{b}$, $\mathbf{a}-\mathbf{b}$, $2\mathbf{a}$, and $3\mathbf{a}+4\mathbf{b}$?
(2) Find a unit vector in the same direction as $8\mathbf{i} - \mathbf{j} + 4\mathbf{k}$.
(3) Find a vector of length 6 in the same direction as $\mathbf{v} = \langle -2, 4, 2\rangle$.
(4) If $\mathbf{v} = \langle v_1, v_2\rangle$ and $c > 0$, prove that $|c\mathbf{v}| = c|\mathbf{v}|$, i.e. prove that the length of the vector $c\mathbf{v}$ is c times the length of $\mathbf{v}$. (You can use specific examples to test out the process, but in the end, you should prove the general rule for any vector $\mathbf{v} = \langle v_1, v_2\rangle$.)
(5) If $\mathbf{v} = \langle v_1, v_2\rangle$, prove that $|\mathbf{v}/|\mathbf{v}|| = 1$. (You can use specific examples to test out the process, but in the end, you should prove the general rule for any vector $\mathbf{v} = \langle v_1, v_2\rangle$.)

Vector Basics — Challenge Problems

Try these problems to test your skills with the ideas in this section. Solutions to these problems are available in Sec. C.4.1.

(1) Find a vector of length 3 in the opposite direction of $\mathbf{v} = \langle 1, -1, 2\rangle$.
(2) Find a unit vector that points in the direction of a minute hand on an analog 12-hour clock when it is exactly 10 minutes past the hour.
(3) If $\mathbf{v}$ is any vector $\langle v_1, v_2, v_3\rangle$ and $\mathbf{w} = 5\mathbf{v}$, use the definition of magnitude (length) to prove that the length of $\mathbf{w}$ is always 5 times the length of $\mathbf{v}$. (Sure, we can say "Well, duh, of course it is!", but can you prove it in the mathematical court of law?)

Vector Basics — You Try It — Solved

(1) Find the components of the vector **v** from $A(2,3)$ to $B(-2,1)$.

☐ Remember from your geometry that the symbol $\bar{AB}$ represents the line segment connecting A to B. We now have the vector from A to B: $\mathbf{AB} = \langle -2-2, 1-3\rangle = \langle -4, -2\rangle$. ■

(2) Find the magnitude of the vector **v** that you found in YTI 1.

☐ The vector $\mathbf{v} = \bar{AB}$ from YTI 1 was $\langle -4, -2\rangle$. The magnitude of this vector is

$$|\mathbf{v}| = \sqrt{(-4)^2 + (-2)^2} = \sqrt{16+4} = \sqrt{20} = 2\sqrt{5} \quad ■$$

(3) If $\mathbf{w} = \langle 5, 2\rangle$, then find $3\mathbf{w}$ and show using Useful Fact 16.1 that the length of $3\mathbf{w}$ is three times the length of **w**.

☐ Since $3\mathbf{w} = 3\langle 5,2\rangle = \langle 15, 6\rangle$, the length of this vector is

$$|3\mathbf{w}| = \sqrt{15^2 + 6^2} = \sqrt{261} = \sqrt{9\cdot 29} = 3\sqrt{29}$$

The original length of **w** is

$$|\mathbf{w}| = \sqrt{5^2 + 2^2} = \sqrt{29}$$

and therefore $|3\mathbf{w}| = 3|\mathbf{w}|$. ■

(4) If $\mathbf{a} = \langle -4, 3\rangle$ and $\mathbf{b} = \langle 6, 2\rangle$, what are $|\mathbf{a}|$, $\mathbf{a}+\mathbf{b}$, $\mathbf{a}-\mathbf{b}$, $2\mathbf{a}$ and $3\mathbf{a}+4\mathbf{b}$?

☐ The results are:

$$\begin{aligned}
|\mathbf{a}| &= \sqrt{16+9} = 5\\
\mathbf{a}+\mathbf{b} &= \langle 2, 5\rangle\\
\mathbf{a}-\mathbf{b} &= \langle -10, 1\rangle\\
2\mathbf{a} &= \langle -8, 6\rangle\\
3\mathbf{a}+4\mathbf{b} &= \langle -12+24, 9+8\rangle = \langle 12, 17\rangle \quad ■
\end{aligned}$$

(5) Find a unit vector in the same direction as $\mathbf{v} = 9\mathbf{i} - 5\mathbf{j}$.

☐ Since $|\mathbf{v}| = \sqrt{81+25} = \sqrt{106}$, then a unit vector in the same direction is

$$\frac{\mathbf{v}}{|\mathbf{v}|} = \frac{9}{\sqrt{106}}\mathbf{i} - \frac{5}{\sqrt{106}}\mathbf{j} \quad ■$$

Interlude: Determinants

Introduction

In order to proceed into the topic of cross products, you'll want to be familiar with *determinants.* If you already know how to set up and compute 3×3 determinants through the technique of expansion by minors, then you can skip this. If that sentence made no sense, then read on!

2-By-2 Determinants

For our purposes, all you need to know about determinants is that they are a computation performed on a square arrays of mathematical objects (most commonly numbers, but not necessarily). Suppose A represents a 2-by-2 array of numbers,

$$A = \begin{pmatrix} a_1 & a_2 \\ b_1 & b_2 \end{pmatrix}$$

(This is actually called a matrix, but we don't need to know that right now!) The determinant of this array of numbers is $\det(A)$ and is written using something like absolute value bars:

$$\det(A) = \begin{vmatrix} a_1 & a_2 \\ b_1 & b_2 \end{vmatrix}$$

This is evaluated by subtracting the products of the two diagonals, upper-left to lower-right and lower-left to upper-right:

$$\det(A) = \begin{vmatrix} a_1 & a_2 \\ b_1 & b_2 \end{vmatrix} = a_1b_2 - b_1a_2$$

For example,

$$\begin{vmatrix} 1 & 2 \\ 3 & 4 \end{vmatrix} = (1)(4) - (3)(2) = 4 - 6 = -2$$

As mentioned, the entries in the array do not need to be numbers. For example,

$$\begin{vmatrix} x^2 & -x \\ 1 & \frac{1}{x} \end{vmatrix} = (x^2)\left(\frac{1}{x}\right) - (1)(-x) = x + x = 2x$$

3-By-3 Determinants

Suppose A is a 3-by-3 array,

$$A = \begin{pmatrix} a_1 & a_2 & a_3 \\ b_1 & b_2 & b_3 \\ c_1 & c_2 & c_3 \end{pmatrix}$$

then the determinant is presented like this:

$$\det(A) = \begin{vmatrix} a_1 & a_2 & a_3 \\ b_1 & b_2 & b_3 \\ c_1 & c_2 & c_3 \end{vmatrix}$$

The value of this determinant is (are you ready?),

$$\det(A) = (a_1b_2c_3 + a_2b_3c_1 + a_3b_1c_2) - (c_1b_2a_3 + c_2b_3a_1 + c_3b_1a_2)$$

That's definitely a formula you don't want to have to memorize, but fortunately there is a very routine procedure to follow, which is called *expansion by minors*. Now don't get worried about that new terminology; a *minor determinant* is simply a determinant that's smaller than the one you already have. For a 3×3 array like our friend A here, there are "minor determinants" associated with A that are just 2×2 determinants made of some numbers pulled from A.

Specifically, a minor determinant is found like this:

- Select an entry from your given array. For example, let's say we select the entry a_2 from our 3-by-3 array A.
- Create the 2×2 array that's created when we strike out the row and column in which our selected entry sits. Note the position of the entry a_2 we've selected, so that we can yank out that row (Row 1) and column (Column 2) to see what's left:

$$\begin{pmatrix} a_1 & \textcircled{a_2} & a_3 \\ b_1 & b_2 & b_3 \\ c_1 & c_2 & c_3 \end{pmatrix} \quad \rightarrow \quad \begin{pmatrix} b_1 & b_3 \\ c_1 & c_3 \end{pmatrix}$$

- Find the determinant of the result. Since we struck out Row 1 and Column 2, this determinant is named the minor M_{12}, and we have

$$M_{12} = \begin{vmatrix} b_1 & b_3 \\ c_1 & c_3 \end{vmatrix} = b_1c_3 - c_1b_3$$

Now we can put all this together:

Useful Fact 16.2. *To evaluate a 3-by-3 determinant, we move along the top row, stop at each spot and form the resulting minor (so we find M_{11}, M_{12} and M_{13}) and then compute*

$$\det(A) = a_1 M_{11} - a_2 M_{12} + a_3 M_{13} \tag{16.2}$$

The three determinants given by the minors M_{11}, M_{12} and M_{13} are all 2-by-2 determinants, and are easy to compute. Pay very close attention to the fact that the second term in (16.2) is substracted; that's not a typo.[3]

Here's an example:

$$\begin{vmatrix} \mathbf{-1} & \mathbf{2} & \mathbf{1} \\ 2 & 1 & -1 \\ 0 & 1 & 2 \end{vmatrix} = (\mathbf{-1}) \begin{vmatrix} 1 & -1 \\ 1 & 2 \end{vmatrix} - (\mathbf{2}) \begin{vmatrix} 2 & -1 \\ 0 & 2 \end{vmatrix} + (\mathbf{1}) \begin{vmatrix} 2 & 1 \\ 0 & 1 \end{vmatrix}$$
$$= (-1)(2-(-1)) - (2)(4-0) + (1)(2-0) = -9$$

Here are a few for you to try. There are no separate solutions, just make sure you can produce the given result. In the third example, note that the top row has unit vectors in it rather than numbers, which has an influence on the form of the final result. If that third example works for you and makes sense, then congratulations, you have already learned how to compute a cross product of two vectors!

You Try It

(1) Show that $\det(A) = -38$ where

$$A = \begin{pmatrix} -2 & 3 & 1 \\ 0 & 1 & -4 \\ 2 & 1 & 2 \end{pmatrix}$$

You Try It

(2) Show that $\det(A) = -3x + 3x^2$ where

$$A = \begin{pmatrix} x & x^2 & x^3 \\ 1 & 1 & 2 \\ 2 & 2 & -1 \end{pmatrix}$$

---

[3] If you want to learn why the second term is negative, then ... here we go again ... take Linear Algebra!

You Try It

(3) Show that $\det(A) = -6\mathbf{i} + 5\mathbf{j} - 2\mathbf{k}$ where

$$A = \begin{pmatrix} \mathbf{i} & \mathbf{j} & \mathbf{k} \\ -1 & 0 & 3 \\ 2 & 2 & -1 \end{pmatrix}$$

and $\mathbf{i}$, $\mathbf{j}$, and $\mathbf{k}$ are the special 3D unit vectors.

16.2 Dot and Cross Products

Introduction

Although we can add and subtract vectors, there is no such thing for us as vector multiplication, nor vector division. Rather, there are *two* analogs to multiplication, and these are called the *dot product* and *cross product.* (At a higher level, these are tangled up in an operation that is more like what we might call vector multiplication, but we don't need to worry about that). In the dot product, we take two vectors, do something to them (via their components), and produce a scalar. In the cross product, we take two vectors, do something to them, and produce another vector. Both operations have a variety of applications.

The Dot Product

Definition 16.2. *If* $\mathbf{v}$ *and* $\mathbf{w}$ *are vectors in 3D, then their* **dot product** *is*

$$\mathbf{v} \cdot \mathbf{w} = v_1 w_1 + v_2 w_2 + v_3 w_3 \tag{16.3}$$

You can shrink or expand this definition to lower (2D) or higher dimensions.

EX 1 If $\mathbf{a} = \langle 2, -2, 1\rangle$ and $\mathbf{b} = \langle 0, -1, 1\rangle$, what is $\mathbf{a} \cdot \mathbf{b}$?

According to the above definition,

$$\mathbf{a} \cdot \mathbf{b} = (2)(0) + (-2)(-1) + (1)(1) = 0 + 2 + 1 = 3 \quad \blacksquare$$

You Try It

(1) If $\mathbf{a} = \langle 5, 0, 2\rangle$ and $\mathbf{b} = \langle 3, -1, 10\rangle$, what is $\mathbf{a} \cdot \mathbf{b}$?

Based on the definition of the dot product, which of the following properties do you think are true?

- $\mathbf{v} \cdot \mathbf{w} = \mathbf{w} \cdot \mathbf{v}$
- $\mathbf{v} \cdot (\mathbf{w} + \mathbf{u}) = \mathbf{v} \cdot \mathbf{w} + \mathbf{v} \cdot \mathbf{u}$
- If $\mathbf{v} \cdot \mathbf{w} = 0$ then either $\mathbf{v} = \mathbf{0}$ or $\mathbf{w} = \mathbf{0}$
- $\mathbf{v} \cdot \mathbf{0} = 0$ for all $\mathbf{v}$

The first and second are true; they follow pretty much immediately from the definition of the dot product (although proving the second one is a little messy — not hard, but messy). The third is NOT true. If $\mathbf{v} = \langle 1, 1\rangle$ and

$\mathbf{w} = \langle 1, -1 \rangle$ then $\mathbf{v} \cdot \mathbf{w} = 0$. So, there's an example of the fact that you can get a dot product of zero from two vectors which are not zero vectors. It turns out that having a dot product of zero tells you something important about your two vectors, and we'll see that below. The fourth one is true.

Does anything interesting happen when you take the dot product of a vector with itself? Let's see. If $\mathbf{v} = \langle v_1, v_2, v_3 \rangle$, then

$$\mathbf{v} \cdot \mathbf{v} = (v_1)(v_1) + (v_2)(v_2) + (v_3)(v_3) = v_1^2 + v_2^2 + v_3^2$$

Now that right hand side should look pretty familiar — it's the square of the magnitude of $\mathbf{v}$. So we've discovered an interesting relationship:

Useful Fact 16.3. *The magnitude of a vector is related to the dot product of the vector with itself, as*

$$\mathbf{v} \cdot \mathbf{v} = |\mathbf{v}|^2 \tag{16.4}$$

This doesn't really give us a new formula for the length of a vector, it just tidies up the ones we already have. The expression in Eq. (16.4) stays exactly the same regardless of whether the vector $\mathbf{v}$ lives in $\mathbb{R}^2$, $\mathbb{R}^3$, or higher. It's certainly much more efficient to write (16.4) than for all of the individual components to be on display.

The Angle Between Vectors

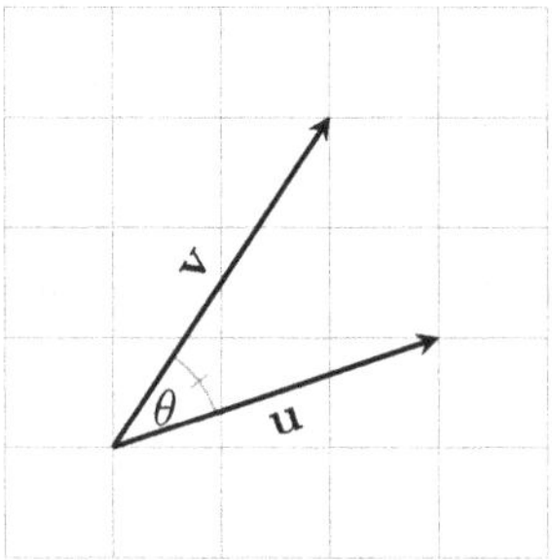

Fig. 16.4 $\mathbf{u} \cdot \mathbf{v} = |\mathbf{u}|\,|\mathbf{v}|\,\cos\theta$

There is geometry hiding behind the dot product.

Useful Fact 16.4. *The dot product between two vectors* $\mathbf{v}$ *and* $\mathbf{w}$ *is related to the angle* θ *between them (see Fig. 16.4):*

$$\mathbf{v} \cdot \mathbf{w} = |\mathbf{v}||\mathbf{w}| \cos\theta \tag{16.5}$$

Equation (16.5) is a consequence of the Law of Cosines, and I encourage you to do some digging to see how it's derived.[4] This equation is most commonly used to find the angle between two vectors.

EX 2 What is the angle between the vectors $\mathbf{a} = \langle 1, 1\rangle$ and $\mathbf{b} = \langle 0, -1\rangle$

Since $|\mathbf{a}| = \sqrt{2}$, $|\mathbf{b}| = 1$, and $\mathbf{a} \cdot \mathbf{b} = -1$, we have

$$\begin{aligned} \mathbf{a} \cdot \mathbf{b} &= |\mathbf{a}||\mathbf{b}| \cos\theta \\ -1 &= (\sqrt{2})(1)\cos\theta \\ \cos\theta &= -\frac{1}{\sqrt{2}} \\ \theta &= \frac{3\pi}{4} \quad \blacksquare \end{aligned}$$

You Try It

(2) What is the angle between the vectors $\mathbf{a} = \langle 1, 2, 3\rangle$ and $\mathbf{b} = \langle 4, 0, -1\rangle$?

But wait, Eq. (16.5) keeps on giving! Think about the angle between two vectors that are perpendicular: it's either $\theta = \pi/2$ or $\theta = -\pi/2$. But in either case, $\cos\theta = 0$. So we have an immediate way to tell whether two vectors are perpendicular.

Useful Fact 16.5. *If two vectors* $\mathbf{v}$ *and* $\mathbf{w}$ *are perpendicular, then* $\mathbf{v}\cdot\mathbf{w} = 0$.

If we can determine that two vectors are perpendicular with the dot product, can we determine whether two vectors are parallel? Sure! If two vectors are parallel and point in the same direction, the angle between them is $\theta = 0$. If they are parallel and point in the opposite direction, then the angle between them is $\theta = \pi$ (or $\theta = -\pi$, either one). And when $\theta = 0$ or $\theta = \pm\pi$, then $\cos\theta = \pm 1$.

Useful Fact 16.6. *If two vectors* $\mathbf{v}$ *and* $\mathbf{w}$ *are parallel, then* $|\mathbf{v}\cdot\mathbf{w}| = |\mathbf{v}||\mathbf{w}|$.

This particular result is not nearly as useful as the one that helps spot perpendicular vectors, though.

EX 3 Are the vectors $\mathbf{u} = \langle 1, -1, 2\rangle$ and $\mathbf{v} = \langle 2, -1, 1\rangle$ parallel, perpendicular, or neither?

[4] or take Linear Algebra!

We have $\mathbf{u} \cdot \mathbf{v} = 5$, $|\mathbf{u}| = \sqrt{6}$ and $|\mathbf{v}| = \sqrt{6}$ so that

$$\begin{aligned} \mathbf{u} \cdot \mathbf{v} &= |\mathbf{u}||\mathbf{v}| \cos\theta \\ 5 &= (\sqrt{6})(\sqrt{6}) \cos\theta \\ \cos\theta &= \frac{5}{6} \end{aligned}$$

This is enough to tell us that since $\mathbf{u} \cdot \mathbf{v} \neq 0$, the vectors are not perpendicular. Since $\theta \neq 0$ or π they are not parallel either. ■

You Try It

(3) Are the vectors $\mathbf{a} = \langle 4, 6 \rangle$ and $\mathbf{b} = \langle -3, 2 \rangle$ parallel, perpendicular, or neither?

(4) Are the vectors $\mathbf{a} = \langle -5, 3, 7 \rangle$ and $\mathbf{b} = \langle 6, -8, 2 \rangle$ parallel, perpendicular, or neither?

Here's another situation where a new word is going to be introduced, which might seem unnecessary at first:

Definition 16.3. *If* $\mathbf{v} \cdot \mathbf{w} = 0$, *then* $\mathbf{v}$ *and* $\mathbf{w}$ *are said to be* ***orthogonal***.

We really do need this new terminology. The concept of two vectors being *perpendicular* makes sense in 2D and 3D, when the angle between them can be physically measured as a right angle. But what about in 4D? If you have vectors with 4 components, it's still pretty special to have their dot product be zero, but the idea of them being perpendicular makes no sense anymore. So we use the word *orthogonal* to describe two vectors that have a dot product of zerp, and it just so happens that in 2D and 3D, orthogonal vectors are physically perpendicular. The word *perpendicular* is actually the less desirable term. If we all called things at right angles "orthogonal" all along, we wouldn't even need to have this conversation. So blame your geometry teachers, not me!

Components and Projections

A recurring theme here is that many things you learned along the way that seemed special really turn out to be just one case of a much broader concept. Here is another story like that. When we've considered the components of a vector so far, we've taken those components relative to the coordinate axes. The vector $\langle 1, 2 \rangle$ can be broken down into two pieces, each of which is parallel to a coordinate axis: we say that the components of $\mathbf{v}$ are 1 and 2, and we can write $\langle 1, 2\rangle = \langle 1, 0\rangle + 2\langle 0, 1\rangle = \mathbf{i} + 2\mathbf{j}$; those unit vectors $\mathbf{i}$ and $\mathbf{j}$ are parallel to the x- and y-axes, respectively, and their coefficients 1 and 2 are the components of the vector. To state this a bit more generally, we can take $\mathbf{j}$ out of the mix and state that we currently know how to decompose a vector into two pieces, one of which is parallel to $\mathbf{i}$ and the other is perpendicular to $\mathbf{i}$.

Well, it turns out that using the coordinate axes (or $\mathbf{i}$ specifically) to determine a vector's components is a totally arbitrary choice.

Definition 16.4. *Given two vectors* $\mathbf{v}$ *and* $\mathbf{w}$, *and also* $\mathbf{p}$ *which is perpendicular to* $\mathbf{w}$, *we can find two scalars* $comp_{\mathbf{w}}\mathbf{v}$ *and* $comp_{\mathbf{p}}\mathbf{v}$ *such that*

$$\mathbf{v} = (comp_{\mathbf{w}}\mathbf{v})\mathbf{w} + (comp_{\mathbf{p}}\mathbf{v})\mathbf{p}$$

The scalars $comp_{\mathbf{w}}\mathbf{v}$ *and* $comp_{\mathbf{p}}\mathbf{v}$ *are called the* ***components*** *of* $\mathbf{v}$ *along* $\mathbf{w}$ *and* $\mathbf{p}$.

What you've called the components of a vector all along have not been unique things; they were just the components of that vector along $\mathbf{i}$ and $\mathbf{j}$ specifically.

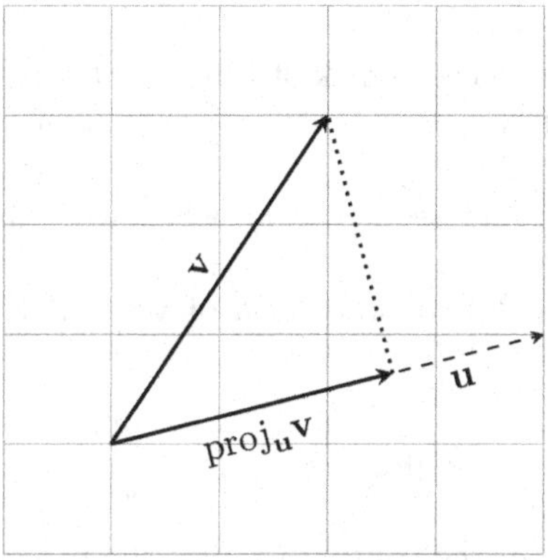

Fig. 16.5 The vector projection of $\mathbf{v}$ onto $\mathbf{u}$.

If we think about $\text{comp}_{\mathbf{w}}\mathbf{w}$ as the length of the shadow of $\mathbf{v}$ along $\mathbf{w}$, we should recognize that the shadow of $\mathbf{v}$ along $\mathbf{w}$ can be considered as a vector in itself. There are two quantities associated with this shadow. One is that component, which we've already defined. But also, when we take the shadow of $\mathbf{v}$ along $\mathbf{w}$ (which has length $\text{comp}_{\mathbf{w}}\mathbf{v}$) and make a vector out of that shadow, we've created the vector projection of $\mathbf{v}$ along $\mathbf{w}$. Let's put all this together and give means to find these quantities:

Definition 16.5. *The* ***component*** *of* $\mathbf{v}$ *along* $\mathbf{w}$ *is found by*

$$comp_{\mathbf{w}}\mathbf{v} = \frac{\mathbf{v}\cdot\mathbf{w}}{|\mathbf{w}|} \tag{16.6}$$

The ***vector projection*** *of* $\mathbf{v}$ *along* $\mathbf{w}$ *is the vector*

$$proj_{\mathbf{w}}\mathbf{v} = comp_{\mathbf{w}}\mathbf{v}\,\frac{\mathbf{w}}{|\mathbf{w}|} = \left(\frac{\mathbf{v}\cdot\mathbf{w}}{|\mathbf{w}|}\right)\frac{\mathbf{w}}{|\mathbf{w}|} \tag{16.7}$$

(Fig. 16.5 shows the projection of $\mathbf{v}$ along $\mathbf{w}$.) Be sure you see what's going on in Eq. (16.7): we simply take the length we need ($\text{comp}_{\mathbf{w}}\mathbf{v}$) times a unit vector in the direction of $\mathbf{w}$ to produce a vector of length $\text{comp}_{\mathbf{w}}\mathbf{v}$ in the direction of $\mathbf{w}$.

Here are two things to keep in mind when finding components and projections:

(1) When asked to compute components and projections, compute the component first, then use that result to compute the projection.

(2) It's important to keep track of which vector is being projected onto which, and place values in the expression carefully!

EX 4 If $\mathbf{v} = \langle 1, 0, 3\rangle$ and $\mathbf{w} = \langle 1, 1, 1\rangle$, what are $\text{comp}_{\mathbf{w}}\mathbf{v}$ and $\text{proj}_{\mathbf{w}}\mathbf{v}$?

We want the component and projection of $\mathbf{v}$ onto $\mathbf{w}$. To compute these, we need $\mathbf{v}\cdot\mathbf{w} = 4$ and $|\mathbf{w}| = \sqrt{3}$ so that

$$\text{comp}_{\mathbf{w}}\mathbf{v} = \frac{\mathbf{v}\cdot\mathbf{w}}{|\mathbf{w}|} = \frac{4}{\sqrt{3}}$$

We can use this value in the computation of the projection,

$$\text{proj}_{\mathbf{w}}\mathbf{v} = \text{comp}_{\mathbf{w}}\mathbf{v}\,\frac{\mathbf{w}}{|\mathbf{w}|} = \frac{4}{\sqrt{3}}\cdot\frac{\langle 1,1,1\rangle}{\sqrt{3}} = \frac{4}{3}\langle 1,1,1\rangle = \left\langle\frac{4}{3},\frac{4}{3},\frac{4}{3}\right\rangle \quad \blacksquare$$

You Try It

(5) If $\mathbf{a} = \langle 4, 2, 0\rangle$ and $\mathbf{b} = \langle 1, 1, 1\rangle$, what are $\text{comp}_{\mathbf{a}}\mathbf{b}$ and $\text{proj}_{\mathbf{a}}\mathbf{b}$?

The Cross Product

The cross product is an operation on two vectors, via their components, that results in a third vector.

Definition 16.6. *Given two vectors* $\mathbf{v} = \langle v_1, v_2, v_3 \rangle$ *and* $\mathbf{w} = \langle w_1, w_2, w_3 \rangle$, *their* ***cross product*** *is*

$$\mathbf{v} \times \mathbf{w} = \langle v_2 w_3 - v_3 w_2, v_3 w_1 - v_1 w_3, v_1 w_2 - v_2 w_1 \rangle$$

or, in a more convenient determinant form,

$$\mathbf{v} \times \mathbf{w} = \begin{vmatrix} \mathbf{i} & \mathbf{j} & \mathbf{k} \\ v_1 & v_2 & v_3 \\ w_1 & w_2 & w_3 \end{vmatrix}$$

You can see from the definition, particularly the determinant form, that the cross product is only defined for vectors in 3D (although you can also cheat and do the cross product of two vectors from $\mathbb{R}^2$ by assigning a third coordinate of 0 to both of them). The top row of the determinant contains the three special 3D unit vectors, **i**, **j**, and **k**.[5] Hopefully you can see that this determinant will result in the form $a\mathbf{i} + b\mathbf{j} + c\mathbf{k}$, which is another vector. It so happens that this vector resulting from the cross product will be perpendicular (orthogonal) to both of the original vectors.

EX 5 If $\mathbf{v} = \langle 1, 0, 3 \rangle$ and $\mathbf{w} = \langle 1, 1, 1 \rangle$, find $\mathbf{v} \times \mathbf{w}$ and demonstrate that it is orthogonal to both **v** and **w**.

Let's call the cross product **c**, so that

$$\mathbf{c} = \mathbf{v} \times \mathbf{w} = \begin{vmatrix} \mathbf{i} & \mathbf{j} & \mathbf{k} \\ 1 & 0 & 3 \\ 1 & 1 & 1 \end{vmatrix} = \begin{vmatrix} 0 & 3 \\ 1 & 1 \end{vmatrix} \mathbf{i} - \begin{vmatrix} 1 & 3 \\ 1 & 1 \end{vmatrix} \mathbf{j} + \begin{vmatrix} 1 & 0 \\ 1 & 1 \end{vmatrix}, \mathbf{k}$$

$$= (0-3)\mathbf{i} - (1-3)\mathbf{j} + (1-0)\mathbf{k} = -3\mathbf{i} + 2\mathbf{j} + \mathbf{k}$$

And then we have

$$\mathbf{c} \cdot \mathbf{v} = (-3)(1) + (2)(0) + (1)(3) = 0$$
$$\mathbf{c} \cdot \mathbf{w} = (-3)(1) + (2)(1) + (1)(1) = 0$$

so that $\mathbf{v} \times \mathbf{w}$ is orthogonal to both **v** and **w**. ■

[5]See Sec. 16.1 for a review on how to evaluate determinants.

You Try It

(6) If $\mathbf{a} = \langle 2, 1, -1 \rangle$ and $\mathbf{b} = \langle 0, 1, 2 \rangle$, demonstrate that $\mathbf{a} \times \mathbf{b}$ is orthogonal to $\mathbf{a}$ and $\mathbf{b}$.

(7) Find two unit vectors orthogonal to both $\mathbf{a} = \langle 1, -1, 1 \rangle$ and $\mathbf{b} = \langle 0, 4, 4 \rangle$.

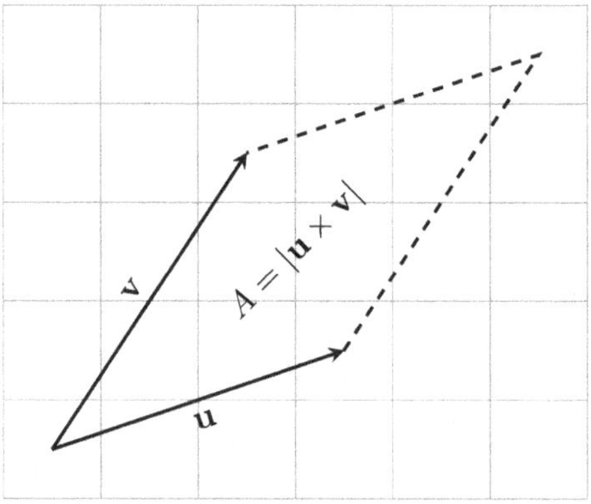

Fig. 16.6 Cross product gives area of a parallelogram.

The cross product is also useful for several other things, like finding the area of a parallelogram whose edges are defined by two vectors (see Fig. 16.6), or finding the volume of a parallelelpiped whose edges are defined by three vectors. Even these, then, have their own applications. For example, you can determine if four points (say A,B,C,D) are co-planar by finding the vectors **AB**, **AC**, and **AD** and then seeing if the volume of the resulting parallelepiped is zero.

Since we could spend several days going over geometrical applications of both the dot and cross products, we'll stick here to the aspects of them that that are of immediate interest in upcoming subjects. If you find that you like dealing with vectors and things like dot and cross products, then make sure to take Linear Algebra! In the meantime, here is the most concise summary of what you can get out of these examples:

- If you need to determine if two things are perpendicular or not, the dot product will likely be involved. You may have to create vectors to act as surrogates for whatever you're testing — for example, to find if two planes are perpendicular, we'll have to identify vectors that can be used to represent the plane in a calculation. That's coming up!

- If you need to find a vector that's ~~perpendicular~~, sorry, orthogonal to something, the cross product will probably get involved. Soon, we're going to need vectors that are orthogonal to curves, planes, and surfaces. In many cases, we'll need a *unit* vector that's orthogonal to something — but, perhaps you now realize that if we ever need a unit vector to do a job, all we have to do is find ANY vector that does the job, and then divide that vector by its own ~~length~~, sorry, magnitude to turn it into a unit vector.

The one last danger of learning new operations like the dot and cross product is losing track of what you can and cannot do, or losing track of proper notation. Vectors must be indicated as vectors, with an arrow or boldface, so that they are not confused with numbers. Operations should be understood by their context. Also, operations must be allowed to act on the types of things they are designed for. The dot and cross products require two vectors. So, for example, we all know that the expression $(a + b) + c$ makes perfect sense when a, b and c are numbers. But the expression $(\mathbf{a} \cdot \mathbf{b}) \cdot \mathbf{c}$ makes NO sense. The left parentheses $(\mathbf{a} \cdot \mathbf{b})$ results in a *scalar* and so trying to take another dot product of that value with $\mathbf{c}$ is silly. Or, make sure that you see $a \times b$ and $\mathbf{a} \times \mathbf{b}$ as two entirely different things. The first is multiplication of two scalars, the second is a cross product of two vectors. Both use the same symbol, $\times$. But we can tell them apart because we know how to interpret the symbol $\times$ based on how it's being used. So if you really mean $\mathbf{a} \times \mathbf{b}$, then don't get sloppy with your notation and write $a \times b$. (Remember, while I'm using $\mathbf{a} \times \mathbf{b}$ for typing purposes, you are likely using arrows in your writing: $\mathbf{a} \times \mathbf{b}$.)

Here's some practice identifying legitimate expressions.

You Try It

(8) Which of the following are legitimate expressions? (a) $\mathbf{a} \cdot (\mathbf{b} \times \mathbf{c})$, (b) $\mathbf{a} \times (\mathbf{b} \times \mathbf{c})$, and (c) $\mathbf{a} \cdot \mathbf{b} \times \mathbf{c} \cdot \mathbf{d}$.

Dot and Cross Products — Problem List

Dot and Cross Products — You Try It

These appeared above; solutions begin on the next page.

(1) If $\mathbf{a} = \langle 5, 0, 2\rangle$ and $\mathbf{b} = \langle 3, -1, 10\rangle$, what is $\mathbf{a} \cdot \mathbf{b}$?
(2) What is the angle between the vectors $\mathbf{a} = \langle 1, 2, 3\rangle$ and $\mathbf{b} = \langle 4, 0, -1\rangle$?
(3) Are the vectors $\mathbf{a} = \langle 4, 6\rangle$ and $\mathbf{b} = \langle -3, 2\rangle$ parallel, perpendicular, or neither?
(4) Are the vectors $\mathbf{a} = \langle -5, 3, 7\rangle$ and $\mathbf{b} = \langle 6, -8, 2\rangle$ parallel, perpendicular, or neither?
(5) If $\mathbf{a} = \langle 4, 2, 0\rangle$ and $\mathbf{b} = \langle 1, 1, 1\rangle$, what are $\text{comp}_{\mathbf{a}}\mathbf{b}$ and $\text{proj}_{\mathbf{a}}\mathbf{b}$?
(6) If $\mathbf{a} = \langle 2, 1, -1\rangle$ and $\mathbf{b} = \langle 0, 1, 2\rangle$, demonstrate that $\mathbf{a} \times \mathbf{b}$ is orthogonal to $\mathbf{a}$ and $\mathbf{b}$.
(7) Find two unit vectors orthogonal to both $\mathbf{a} = \langle 1, -1, 1\rangle$ and $\mathbf{b} = \langle 0, 4, 4\rangle$.
(8) Which of the following are legitimate expressions? (a) $\mathbf{a} \cdot (\mathbf{b} \times \mathbf{c})$, (b) $\mathbf{a} \times (\mathbf{b} \times \mathbf{c})$, and (c) $\mathbf{a} \cdot \mathbf{b} \times \mathbf{c} \cdot \mathbf{d}$.

Dot and Cross Products — Practice Problems

Try these as you get the hang of the You Try It problems. Solutions to these problems are available in Sec. B.4.2.

(1) What is the angle between $\mathbf{a} = \langle 6, -3, 2\rangle$ and $\mathbf{b} = \langle 2, 1, -2\rangle$?
(2) Are the vectors $\mathbf{a} = \langle -1, 2, 5\rangle$ and $\mathbf{b} = \langle 3, 4, -1\rangle$ parallel, perpendicular, or neither?
(3) Are the vectors $\mathbf{a} = \langle 2, 6, -4\rangle$ and $\mathbf{b} = \langle -3, -9, 6\rangle$ parallel, perpendicular, or neither?
(4) If $\mathbf{a} = \langle -1, -2, 2\rangle$ and $\mathbf{b} = \langle 3, 3, 4\rangle$, what are $\text{comp}_{\mathbf{a}}\mathbf{b}$ and $\text{proj}_{\mathbf{a}}\mathbf{b}$?
(5) If $\mathbf{a} = \langle 1, -1, 1\rangle$ and $\mathbf{b} = \langle 1, 1, 1\rangle$, demonstrate that $\mathbf{a} \times \mathbf{b}$ is perpendicular to both $\mathbf{a}$ and $\mathbf{b}$.
(6) Which of the following are legitimate expressions? (a) $\mathbf{a} \cdot (\mathbf{b} \cdot \mathbf{c})$, (b) $\mathbf{a} \times (\mathbf{b} \cdot \mathbf{c})$, and (c) $(\mathbf{a} \times \mathbf{b}) \cdot (\mathbf{c} \times \mathbf{d})$.
(7) If $\mathbf{a} = \langle 2, -3, 1\rangle$ and $\mathbf{b} = \langle 1, 6, -2\rangle$, what are $\text{comp}_{\mathbf{a}}\mathbf{b}$ and $\text{proj}_{\mathbf{a}}\mathbf{b}$?
(8) Find a unit vector orthogonal to both $\mathbf{i} + \mathbf{j}$ and $\mathbf{i} + \mathbf{k}$.
(9) We know that the dot product of a vector with itself is related to the length of the vector. What is special about the cross product of a vector with itself? Test this out for the general case $\mathbf{v} = \langle v_1, v_2, v_3\rangle$ and show all the details to support your conclusion.

Dot and Cross Products — Challenge Problems

Try these problems to test your skills with the ideas in this section. Solutions to these problems are available in Sec. C.4.2.

(1) If $\mathbf{a} = \langle 1, 6, -2\rangle$ and $\mathbf{b} = \langle 2, -3, 1\rangle$, what are $\text{comp}_{\mathbf{a}}\mathbf{b}$ and $\text{proj}_{\mathbf{a}}\mathbf{b}$?
(2) Find a unit vector orthogonal to both $\mathbf{i} - 2\mathbf{j}$ and $\mathbf{i} + \mathbf{k}$.

(3) Let $\mathbf{v} = \langle v_1, v_2, v_3\rangle$ be any vector in $\mathbf{R}^3$ and let $\mathbf{w}$ be any scalar multiple of $\mathbf{v}$, i.e. $w = c\mathbf{v}$. The cross product $\mathbf{v} \times \mathbf{w}$ will always be the same value; what is that value, and show how you know what it is.
(4) If $\mathbf{a} = \langle 2, -3, 1\rangle$ and $\mathbf{b} = \langle 1, 6, -2\rangle$, what are $\text{comp}_{\mathbf{a}}\mathbf{b}$ and $\text{proj}_{\mathbf{a}}\mathbf{b}$?
(5) Find a unit vector orthogonal to both $\mathbf{i} + \mathbf{j}$ and $\mathbf{i} + \mathbf{k}$.

(6) Let $\mathbf{v} = \langle p, q, r\rangle$ be any vector in $\mathbf{R}^3$ and let $\mathbf{w}$ be any scalar multiple of $\mathbf{v}$, i.e. $w = c\mathbf{v}$. The cross product $\mathbf{v} \times \mathbf{w}$ will always have the same result; find that result, and show how you know what it is.

Dot and Cross Products — You Try It — Solved

---

(1) If $\mathbf{a} = \langle 5, 0, 2\rangle$ and $\mathbf{b} = \langle 3, -1, 10\rangle$, what is $\mathbf{a} \cdot \mathbf{b}$?

□ $\mathbf{a} \cdot \mathbf{b} = (5)(3) + (0)(-1) + (2)(10) = 35$ ■

---

(2) What is the angle between the vectors $\mathbf{a} = \langle 1, 2, 3\rangle$ and $\mathbf{b} = \langle 4, 0, -1\rangle$?

□ We have $\mathbf{a} \cdot \mathbf{b} = 1$, $|\mathbf{a}| = \sqrt{14}$ and $|\mathbf{b}| = \sqrt{17}$ so that

$$\begin{aligned} \mathbf{a} \cdot \mathbf{b} &= |\mathbf{a}||\mathbf{b}| \cos\theta \\ 1 &= \sqrt{14}\sqrt{17} \cos\theta \\ \cos\theta &= \frac{1}{\sqrt{238}} \\ \theta &\approx 86^\circ \end{aligned}$$ ■

---

(3) Are the vectors $\mathbf{a} = \langle 4, 6\rangle$ and $\mathbf{b} = \langle -3, 2\rangle$ parallel, perpendicular, or neither?

□ Since $\mathbf{a} \cdot \mathbf{b} = 0$, the vectors are perpendicular. ■

---

(4) Are the vectors $\mathbf{a} = \langle -5, 3, 7\rangle$ and $\mathbf{b} = \langle 6, -8, 2\rangle$ parallel, perpendicular, or neither?

□ We have $\mathbf{a}\cdot\mathbf{b} = -40$, $|\mathbf{a}| = \sqrt{83}$ and $|\mathbf{b}| = \sqrt{104}$ so that

$$\begin{aligned} \mathbf{a}\cdot\mathbf{b} &= |\mathbf{a}||\mathbf{b}|\cos\theta \\ -40 &= \sqrt{83}\sqrt{104}\cos\theta \\ \cos\theta &= -\frac{40}{\sqrt{8362}} \end{aligned}$$

This is enough to tell us that since $\mathbf{a}\cdot\mathbf{b} \neq 0$, the vectors are not perpendicular. Since $\theta \neq 0, \pi$ they are not parallel either. ■

(5) If $\mathbf{a} = \langle 4, 2, 0\rangle$ and $\mathbf{b} = \langle 1, 1, 1\rangle$, what are $\text{comp}_{\mathbf{a}}\mathbf{b}$ and $\text{proj}_{\mathbf{a}}\mathbf{b}$?

□ We want the component and projection of $\mathbf{b}$ along $\mathbf{a}$. We need $\mathbf{a}\cdot\mathbf{b} = 6$ and $|\mathbf{a}| = \sqrt{20} = 2\sqrt{5}$ so that

$$\text{comp}_{\mathbf{a}}\mathbf{b} = \frac{\mathbf{a}\cdot\mathbf{b}}{|\mathbf{a}|} = \frac{3}{\sqrt{5}}$$

which we can then use in:

$$\begin{aligned} \text{proj}_{\mathbf{a}}\mathbf{b} = \text{comp}_{\mathbf{a}}\mathbf{b}\left(\frac{\mathbf{a}}{|\mathbf{a}|}\right) &= \left(\frac{3}{\sqrt{5}}\right)\left(\frac{1}{2\sqrt{5}}\right)\langle 4, 2, 0\rangle \\ &= \frac{3}{10}\langle 4, 2, 0\rangle = \left\langle \frac{6}{5}, \frac{3}{5}, 0\right\rangle \quad ■ \end{aligned}$$

(6) If $\mathbf{a} = \langle 2, 1, -1\rangle$ and $\mathbf{b} = \langle 0, 1, 2\rangle$, demonstrate that $\mathbf{a}\times\mathbf{b}$ is orthogonal to $\mathbf{a}$ and $\mathbf{b}$.

□ We have

$$\begin{aligned} \mathbf{c} = \mathbf{a}\times\mathbf{b} = \begin{vmatrix} ccc\mathbf{i} & \mathbf{j} & \mathbf{k} \\ 2 & 1 & -1 \\ 0 & 1 & 2 \end{vmatrix} &= \begin{vmatrix} 1 & -1 \\ 1 & 2 \end{vmatrix}\mathbf{i} - \begin{vmatrix} 2 & -1 \\ 0 & 2 \end{vmatrix}\mathbf{j} + \begin{vmatrix} 2 & 1 \\ 0 & 1 \end{vmatrix}\mathbf{k} \\ &= (2-(-1))\mathbf{i} - (4-0)\mathbf{j} + (2-0)\mathbf{k} = 3\mathbf{i} - 4\mathbf{j} + 2\mathbf{k} \end{aligned}$$

And then we have

$$\begin{aligned} \mathbf{c}\cdot\mathbf{a} &= (3)(2) + (-4)(1) + (2)(-1) = 0 \\ \mathbf{c}\cdot\mathbf{b} &= (3)(0) + (-4)(1) + (2)(2) = 0 \end{aligned}$$

so that $\mathbf{a}\times\mathbf{b}$ is orthogonal to both $\mathbf{a}$ and $\mathbf{b}$. ■

(7) Find two unit vectors orthogonal to both $\mathbf{a} = \langle 1, -1, 1\rangle$ and $\mathbf{b} = \langle 0, 4, 4\rangle$.

☐ The positive and negative cross product of $\mathbf{a} = \langle 1, -1, 1\rangle$ and $\mathbf{b} = \langle 0, 4, 4\rangle$ will both be orthogonal to both, and we can scale them to unit vectors:

$$\mathbf{c} = \mathbf{a} \times \mathbf{b} = \begin{vmatrix} \mathbf{i} & \mathbf{j} & \mathbf{k} \\ 1 & -1 & 1 \\ 0 & 4 & 4 \end{vmatrix} = \begin{vmatrix} -1 & 1 \\ 4 & 4 \end{vmatrix} \mathbf{i} - \begin{vmatrix} 1 & 1 \\ 0 & 4 \end{vmatrix} \mathbf{j} + \begin{vmatrix} 1 & -1 \\ 0 & 4 \end{vmatrix} \mathbf{k}$$

$$= (-4-4)\mathbf{i} - (4-0)\mathbf{j} + (4-0)\mathbf{k} = -8\mathbf{i} - 4\mathbf{j} + 4\mathbf{k}$$

We know that $\pm c/|\mathbf{c}|$ are unit vectors orthogonal to both of the given vectors. Since $|\mathbf{c}| = 4\sqrt{6}$, these are:

$$\pm\left\langle -\frac{2}{\sqrt{6}}, -\frac{1}{\sqrt{6}}, \frac{1}{\sqrt{6}} \right\rangle \quad \blacksquare$$

---

(8) Which of the following are legitimate expressions? (a) $\mathbf{a} \cdot (\mathbf{b} \times \mathbf{c})$, (b) $\mathbf{a} \times (\mathbf{b} \times \mathbf{c})$, and (c) $\mathbf{a} \cdot \mathbf{b} \times \mathbf{c} \cdot \mathbf{d}$.

☐ (a) is fine, it's the dot product of two vectors, $\mathbf{a}$ and $\mathbf{b} \times \mathbf{c}$.
(b) is fine, it is the cross product of two vectors, $\mathbf{a}$ and $\mathbf{b} \times \mathbf{c}$.
(c) is not valid, $\mathbf{a} \cdot \mathbf{b}$ and $\mathbf{c} \cdot \mathbf{d}$ are both scalars, so their cross product is not defined. ■

16.3 Vector Functions

Introduction

Think about how much fun we had with garden variety functions of the $y = f(x)$ variety, which had one input and one output. Now imagine how much more fun we can have with functions that have one input, but multiple outputs! A function of the $y = f(x)$ variety is called a *scalar function*, since it takes a scalar as input, and produces another scalar as output. That is, it does this: $f : \mathbb{R} \to \mathbb{R}$. Now we have the ability to create functions that take scalars as input, and create *vectors* as output. We will call these *vector-valued functions*, or "vector functions" for short. If such a function takes a scalar as input and uses it to produce a vector with, say, three components, we could write $f : \mathbb{R} \to \mathbb{R}^3$.

The idea of a vector function may seem strange, but there are many quantities you already know about that are natural vector functions. Velocity is one. Velocity is a vector; it has size (speed) and direction. If you drive your car through town, your velocity is a function of time; as t changes, so do both your speed and direction. In other words, different values of t give different velocity vectors. Voila, you have a vector function!

If you are worried about yet another thing to learn, the good news is that if you're comfortable with parametric equations, then you don't really need to learn anything new to handle vector functions. For example, we've seen that lines in three dimensions are described with parametric equations,

$$x(t) = a + bt \quad ; \quad y(t) = c + dt \quad ; \quad z(t) = m + nt \qquad \text{for } a \le t \le b$$

To make this a vector function, we just collect the three separate equations into vector brackets, so that we present three small component functions which each describe how the x, y, and z components of the vector function change with t. Like f is the usual generic name for a scalar function, we often use $\mathbf{r}(t)$ as the generic name for a vector function.[6] So a line segment is associated with a vector function

$$\mathbf{r}(t) = \langle x(t), y(t), z(t) \rangle = \langle a + bt, c + dt, m + nt \rangle \qquad \text{for } a \le t \le b$$

Similarly, do you already know what this vector function looks like in $\mathbb{R}^2$?

$$\mathbf{r}(t) = \langle x(t), y(t) \rangle = \langle \cos(t), \sin(t) \rangle \qquad \text{for } 0 \le t \le 2\pi$$

Just like scalar functions have algebra, limits, derivatives, and antiderivatives, so do vector functions.

[6] "r" for ... vector? Don't ask me!

Defining and Visualizing Vector Functions

The domain of a vector function is the set of values for t that are allowed (or specified) to be used in the component functions. For example, the domain of $\mathbf{r}(t) = \langle \sqrt{t}, (t-1)^{-1}, t^2 \rangle$ is all $t \geq 0$ except $t = 1$.

The range of a vector function is its collective output, and the collective output of a vector function is a curve in the appropriate $\mathbb{R}^n$. To visualize this, imagine rubber pencil with the non-writing end thumb-tacked at the origin, and the terminal (writing) end free to move; this pencil will be the vector function. As t varies, the pencil / vector moves around and stretches in and out, and the tip of the pencil draws a curve. This curve represents the evolution of the vector function as t increases. Figure 16.7 shows the evolution of a vector function's curve, as the curve is drawn out through successive values of t, indicated by t_1, t_2, t_3; these specific values are only there for display, we are using all values of t before, between, and after these values as well.

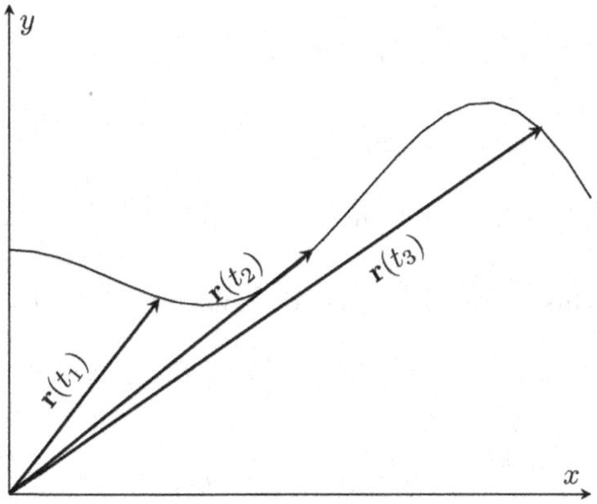

Fig. 16.7 Evolution of a vector function.

There is a match between the curve traced out by the vector function $\mathbf{v}(t) = \langle x(t), y(t), z(t) \rangle$ and the parametric curve with equations $\{x(t), y(t), z(t)\}$. The parametric curve is the collection of points generated by the equations $x(t), y(t)$, and $z(t)$; the vector function draws out the same set of points. So we can use what we know about parametric equations to decide what a vector function looks like.

EX 1 Identify the curve given by $\mathbf{r}(t) = \langle \sin t, \cos t, t \rangle$.

This is equivalent to the curve drawn out by the parametric equations $x(t) = \sin t$, $y(t) = \cos t$, $z = t$. There are no restrictions on t, so we want all possible points. In the horizontal cross section, this curve is circular, and going in the clockwise direction. But while the curve rotates circularly in the horizontal plane, the z-coordinate is increasing as t increases. Therefore, we get a helix. ■

You Try It

(1) Identify the curve given by $\mathbf{r}(t) = (1+t)\mathbf{i} + t^2\mathbf{j}$.

With vector functions, we can recast our descriptions of lines and line segments, and better determine the necessary ingredients for forming the vector equation of a line. In $\mathbb{R}^2$, we say that a slope and a point on a line are necessary ingredients to find the equation of a line; but in $\mathbb{R}^3$, how do we describe a slope? Let's follow the trail. We can start with the the parametric equations of a full line in $\mathbb{R}^3$:

$$x(t) = a + bt \quad ; \quad y(t) = c + dt \quad ; \quad z(t) = m + nt \qquad \text{for } -\infty \le t \le \infty$$

and collect them into a vector function

$$\mathbf{r}(t) = \langle x(t), y(t), z(t) \rangle = \langle a + bt, c + dt, m + nt \rangle \qquad \text{for } -\infty \le t \le \infty$$

Let's regroup this vector function to isolate the constants and the variable t:

$$\begin{aligned} \mathbf{r}(t) &= \langle a + bt, c + dt, m + nt \rangle \\ &= \langle a, c, m \rangle + \langle bt, dt, nt \rangle \\ &= \langle a, c, m \rangle + \langle b, d, n \rangle \cdot t \end{aligned}$$

Here we see that the vector function of a line is made of two constant vectors ($\langle a, c, m \rangle$ and $\langle b, d, n \rangle$) along with the variable t. These constant vectors play the role of a point on the line and its slope: one is a vector pointing to any point known to be on the line, and the other is a vector parallel to the direction of the line. Let's name them as $\mathbf{r}_0 = \langle a, c, m \rangle$ and $\mathbf{r}_1 = \langle b, d, n \rangle$. When we write $\mathbf{r}(t) = \mathbf{r}_0 + \mathbf{r}_1 t$, we see a structure similar to $y = b + mx$. However, normally, the t is placed before the vector $\mathbf{r}_1$:

Useful Fact 16.7. *The vector equation of a line is given by* $\mathbf{r}(t) = \mathbf{r}_0 + t \cdot \mathbf{r}_1$, *where*

- $\mathbf{r}_0$ *is a vector pointing to any point on the line*
- $\mathbf{r}_1$ *is a vector parallel to the direction of the line*

A full line requires $-\infty < t < \infty$, *whereas* $a \leq t \leq b$ *determines a line segment.*

A line segment in 3D from a point P to another point Q can be determined by a starting point / vector and a parallel vector. The parallel vector goes in the direction from P to Q, and can be found specifically as the vector $\mathbf{PQ}$. The act of drawing out the line segment consists of (1) starting at the initial vector $\mathbf{r}_0$, and then (2) following along the parallel vector $\mathbf{r}_1 = \mathbf{PQ}$. How do we follow along the parallel vector? Remember that any multiple of the parallel vector $\mathbf{r}_1$ goes in the same direction, but perhaps with different length. So when we start at $\mathbf{r}_0$ and then add on vectors $t \cdot \mathbf{r}_1$, we march our way from P to Q. Figure 16.8 shows a vector $\mathbf{r}_0$ pointing to one fixed point on the line, a vector $\mathbf{r}_1$ which is parallel to the line, some multiple $t\,\mathbf{r}_1$ (for $t\rangle 1$), and the resulting sum $\mathbf{r}_0 + t\mathbf{r}_1$, which lands us farther down the line. When this is done continuously for a suite of t values $a \leq t \leq b$, we will trace out a full line segment. There is no reference to specific points P or Q, because once we've used them to find the parallel vector $\mathbf{r}_1$, their job is done.

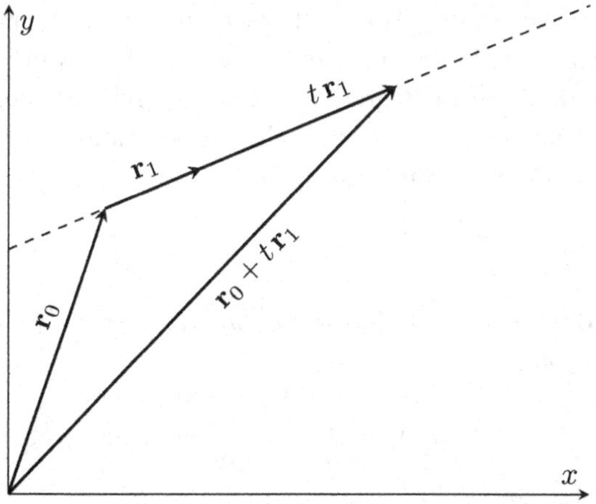

Fig. 16.8 The building blocks of the vector equation of a line.

A fact that really helps with this is that the vector pointing from the origin to a point $P(a, b, c)$ is the vector $\langle a, b, c \rangle$. So if we are given a point on a line, we also know the vector $\mathbf{r}_0$ right away.

EX 2 Find the vector equation of the line segment joining the points $P(-2, 4, 0)$ and $Q(6, -1, 2)$.

The initial point is $(-2, 4, 0)$, and so the initial vector is $\mathbf{r}_0 = \langle -2, 4, 0 \rangle$. A vector parallel to our line segment is $\mathbf{r}_1 = \mathbf{PQ} = \langle 8, -5, 2 \rangle$. Therefore, we can form the equation(s) of the line through P and Q as

$$\mathbf{r}_0 + t\mathbf{r}_1 = \langle -2, 4, 0 \rangle + t\langle 8, -5, 2 \rangle = \langle -2 + 8t, 4 - 5t, 2t \rangle$$

To restrict this to the line segment that starts at P and ends at Q, we need $0 \leq t \leq 1$. If we want this back in parametric form, then we have

$$x = -2 + 8t \quad ; \quad y = 4 - 5t \quad ; \quad z = 2t \quad ; \quad 0 \leq t \leq 1 \quad \blacksquare$$

You Try It

(2) Give both the vector function and parametric equations for the line segment joining the points $P(0, 0, 0)$ and $Q(1, 2, 3)$.

Limits, Derivatives, and Antiderivatives of Vector Functions

There are really no surprises in the computation of limits, derivatives, and antiderivatives of vector functions. They're just more work, because we have to deal with individual component functions. For example, the derivative of a vector function in $\mathbb{R}^3$ requires three individual "regular" derivatives. We are going to speed through these operations quickly, with little to no background information.

Limits

Useful Fact 16.8. *The limit of a vector function* $\mathbf{r}(t) - \langle f(t), g(t), h(t) \rangle$ *in* $\mathbb{R}^3$ *is defined as follows:*

$$\lim_{t \to t_0} \mathbf{r}(t) = \lim_{t \to t_0} \langle f(t), g(t), h(t) \rangle = \left\langle \lim_{t \to t_0} f(t), \lim_{t \to t_0} g(t), \lim_{t \to t_0} h(t) \right\rangle$$

In other words, just find the limits of the individual component functions, then slap them back together again as a the vector.

EX 3 Evaluate $\lim_{t\to 0}\left\langle \frac{e^t-1}{t}, \frac{\sqrt{1+t}-1}{t}, \frac{3}{1+t}\right\rangle$.

The limits of the first and second components will require L-Hopital's rule, while the third is straightforward:

$$\lim_{t\to 0}\frac{e^t-1}{t} = \lim_{t\to 0}\frac{e^t}{1} = 1$$

$$\lim_{t\to 0}\frac{\sqrt{1+t}-1}{t} = \lim_{t\to 0}\frac{1/(2\sqrt{1+t})}{1} = \frac{1}{2}$$

$$\lim_{t\to 0}\frac{3}{1+t} = 3$$

So

$$\lim_{t\to 0}\left\langle \frac{e^t-1}{t}, \frac{\sqrt{1+t}-1}{t}, \frac{3}{1+t}\right\rangle = \left\langle 1, \frac{1}{2}, 3\right\rangle \quad \blacksquare$$

You Try It

(3) Evaluate $\lim_{t\to 0^+}\langle \cos t, \sin t, t\ln t\rangle$.

Derivatives

The official definition of the derivative of a vector function $\mathbf{r}(t)$ at a location fixed by $t = t_0$ looks very much like other definitions of derivatives:

$$\frac{d\mathbf{r}(t)}{dt} = \mathbf{r}'(t_0) = \lim_{h\to 0}\frac{\mathbf{r}(t_0+h)-\mathbf{r}(t_0)}{h}$$

The geometry hiding in this limit process determines that $\mathbf{r}'(t_0)$ gives a vector tangent to the curve of $\mathbf{r}(t)$ at $t = t_0$. (I encourage you to draw some pictures and convince yourself of this.)

In practice, we find derivatives of vector functions this way:

Useful Fact 16.9. *For a vector function* $\mathbf{r}(t) - \langle f(t), g(t), h(t)\rangle$ *in* $\mathbb{R}^3$,

$$\frac{d\mathbf{r}(t)}{dt} = \mathbf{r}'(t) = \left\langle \frac{d}{dt}f(t), \frac{d}{dt}g(t), \frac{d}{dt}h(t)\right\rangle$$

In other words, we just operate on each component function one at a time.

EX 4 If $\mathbf{r}(t) = \sin^{-1}(t)\,\mathbf{i} + \sqrt{1-t^2}\,\mathbf{j} + \mathbf{k}$, what is $\mathbf{r}'(t)$?

Hey, we switched to unit vector notation! That should not be a cause for concern. The derivatives of the component functions are:

$$\frac{d}{dt}\sin^{-1}(t) = \frac{1}{\sqrt{1-t^2}}$$
$$\frac{d}{dt}\sqrt{1-t^2} = \frac{t}{\sqrt{1-t^2}}$$
$$\frac{d}{dt}1 = 0$$

and putting those together,

$$\mathbf{r}'(t) = \frac{1}{\sqrt{1-t^2}}\,\mathbf{i} - \frac{t}{\sqrt{1-t^2}}\,\mathbf{j} + 0\,\mathbf{k} \quad \blacksquare$$

You Try It

(4) If $\mathbf{r}(t) = e^{t^2}\,\mathbf{i} - \mathbf{j} + \ln(1+3t)\,\mathbf{k}$, what is $\mathbf{r}'(t)$?

EX 5 Find a vector tangent to the curve $\mathbf{p}(t) = \langle t^5, t^4, t^3\rangle$ at $t = 2$.

Since $\mathbf{p}'(t) = \langle 5t^4, 4t^3, 3t^2\rangle$, then a vector tangent to the curve at $t = 2$ is $\mathbf{p}'(2) = \langle 80, 32, 12\rangle$. (We say *a* vector tangent to the curve, not *the* vector, since any scalar multiple of this vector will also be tangent to the curve.) Just to summarize the information we have now, at $t = 2$ the vector $\mathbf{p}(2) = \langle 32, 16, 8\rangle$ points from the origin to the curve, and the vector $\langle 80, 32, 12\rangle$ is tangent to the curve at that location. $\blacksquare$

You Try It

(5) Find a vector tangent to the curve $\mathbf{r}t = (1+t)\,\mathbf{i} + t^2\,\mathbf{j}$ at $t = 1$.

Now let's try putting some concepts together:

EX 6 Find the equations of the line tangent to $\mathbf{r}(t) = \langle \ln t, 2\sqrt{t}, t^2\rangle$ at $\langle 0, 2, 1\rangle$.

Note that we're looking for a tangent *line* now, not a tangent vector. But to make the vector equation of a line, we need a vector parallel to that line; this vector will be the tangent vector we can find using the derivative! Now, we need to know the value of t which leads us to where all the action is;

comparing the components of $\mathbf{r}(t)$ to the location $\langle 0, 2, 1\rangle$ we can see that the vector $\langle 0, 2, 1\rangle$ is given by $t = 1$. Now, the derivative vector function is

$$\mathbf{r}'(t) = \left\langle \frac{1}{t}, \frac{1}{\sqrt{t}}, 2t \right\rangle$$

and so at $t = 1$, we have $\mathbf{r}'(1) = \langle 1, 1, 2\rangle$.

Now we have enough information for our tangent line. Its initial vector is the vector given as the location for our tangent line, $\mathbf{r}_0 = \langle 0, 2, 1\rangle$. A vector parallel to the tangent line is the tangent vector $\mathbf{r}'(1) = \langle 1, 1, 2\rangle$. Therefore, the parametric equations of the line are built by

$$\mathbf{r}_0 + t\mathbf{r}'(1) = \langle 0, 2, 1\rangle + t\langle 1, 1, 2\rangle$$

giving the vector equation of the line as $\langle t, 2+t, 1+2t\rangle$; in parametric form, we have

$$x = t \quad ; \quad y = 2 + t \quad ; \quad z = 1 + 2t \quad \blacksquare$$

You Try It

(6) Find the vector equation of the line tangent to $\mathbf{r}(t) = \langle t^5, t^4, t^3\rangle$ at $\langle 1, 1, 1\rangle$.

In single variable calculus, the introduction of derivatives was followed by a cascade of further derivative techniques, like the product rule, chain rule, and quotient rule. Each of these rules is still effective when acting on individual components of vector functions, but there can be further extensions to new structures. For example, here is a product-rule type of derivative relationship for the dot product of two vector functions. The derivation is below, in *the Pit*!

Useful Fact 16.10. *Suppose* $\mathbf{p}(t)$ *and* $\mathbf{r}(t)$ *are vector functions with differentiable component functions. Then*

$$\frac{d}{dt}\mathbf{p}(t) \cdot \mathbf{r}(t) = \mathbf{p}'(t) \cdot \mathbf{r}(t) + \mathbf{p}(t) \cdot \mathbf{r}'(t) \tag{16.8}$$

EX 7 Let $\mathbf{p}(t)$ and $\mathbf{r}(t)$ be the vector functions from EX 5 and EX 6, respectively. Confirm Useful Fact 16.10 by (a) computing the dot product and then finding the derivative, and (b) applying the relation in Eq. (16.8).

With $\mathbf{p}(t) = \langle t^5, t^4, t^3\rangle$ and $\mathbf{r}(t) = \langle \ln t, 2\sqrt{t}, t^2\rangle$, we have

$$\mathbf{p}(t)\cdot\mathbf{r}(t) = t^5(\ln t) + t^4(2\sqrt{t}) + t^3(t^2) = t^5\ln t + 2t^{9/2} + t^5$$

so that with a "regular" product rule on the first term,

$$\frac{d}{dt}\mathbf{p}(t)\cdot\mathbf{r}(t) = (5t^4\ln t + t^4) + 9t^{7/2} + 5t^4 = 5t^4\ln t + 6t^4 + 9t^{7/2}$$

On the other hand, with individual derivatives $\mathbf{p}'(t) = \langle 5t^4, 4t^3, 3t^2\rangle$ and $\mathbf{r}'(t) = \langle 1/t, 1/\sqrt{t}, 2t\rangle$, we get

$$\begin{aligned}\mathbf{p}'(t)\cdot\mathbf{r}(t) &= \langle 5t^4, 4t^3, 3t^2\rangle\cdot\langle \ln t, 2\sqrt{t}, t^2\rangle\\ &= 5t^4(\ln t) + 4t^3(2\sqrt{t}) + 3t^2(t^2) = t^4(5\ln t + 3) + 8t^{7/2}\\ \mathbf{p}(t)\cdot\mathbf{r}'(t) &= \langle t^5, t^4, t^3\rangle\cdot\langle\frac{1}{t}, \frac{1}{\sqrt{t}}, 2t\rangle\\ &= t^5\left(\frac{1}{t}\right) + t^4\left(\frac{1}{\sqrt{t}}\right) + t^3(2t) = 3t^4 + t^{7/2}\end{aligned}$$

So that by Useful Fact 16.10,

$$\begin{aligned}\frac{d}{dt}\mathbf{p}(t)\cdot\mathbf{r}(t) &= \mathbf{p}'(t)\cdot\mathbf{r}(t) + \mathbf{p}(t)\cdot\mathbf{r}'(t)\\ &= t^4(5\ln t + 3) + 8t^{7/2} + (3t^4 + t^{7/2})\\ &= 5t^4\ln t + 6t^4 + 9t^{7/2}\end{aligned}$$

and the results of the two methods match. Note that while Eq. (16.8) gives a relation that is analogous to the product rule, it still seems easier to compute the derivative of $\mathbf{p}(t)\cdot\mathbf{r}(t)$ by finding the dot product first, and then performing the derivative — rather than finding the individual derivatives and employing Eq. (16.8). ■

You Try It

(7) Let $\mathbf{p}(t) = \langle \sin t, \cos t, e^{-t}\rangle$ and $\mathbf{r}(t) = \langle 2\sin t, 2\cos t, e^t\rangle$. Confirm Useful Fact 16.10 by (a) computing the dot product and then finding the derivative, and (b) applying the relation in Eq. (16.8).

Antiderivatives

Having done derivatives of vector functions, you can probably imagine how antiderivatives are going to go. There's one extra wrinkle, though: what used to be a single arbitrary constant is now going to be multiple constants ... which get bundled into an arbitrary *vector*. Just like we find derivatives of vector functions component by component, we also find antiderivatives component by component. The general structure for the indefinite integral of $\mathbf{r}(t) = \langle f(t), g(t), h(t)\rangle$ is this:

$$\begin{aligned}\int \mathbf{r}(t)\,dt &= \int \langle f(t), g(t), h(t)\rangle\,dt = \left\langle \int f(t)dt, \int g(t)dt, \int h(t)dt \right\rangle \\ &= \langle F(t) + C_1, G(t) + C_2, H(t) + C_3\rangle \\ &= \langle F(t), G(t), H(t)\rangle + \langle C_1, C_2, C_3\rangle\end{aligned}$$

The antiderivative of each component function comes with an arbitrary constant; those constants are then collected into an arbitrary *vector.* Definite integrals are structured similarly, although no arbitrary constants are generated. Here's a summary:

Useful Fact 16.11. *For the vector function* $\mathbf{r}(t) = \langle f(t), g(t), h(t)\rangle$, *we have*

$$\int \mathbf{r}(t)\,dt = \left\langle \int f(t)\,dt, \int g(t)\,dt, \int h(t)\,dt \right\rangle + \mathbf{C}$$

and

$$\int_a^b \mathbf{r}(t)dt = \left\langle \int_a^b f(t)dt, \int_a^b g(t)dt, \int_a^b h(t)dt \right\rangle$$

EX 7 Find $\int \mathbf{r}(t)\,dt$ for $\mathbf{r}(t) = \langle e^t, 2\cos t, 1/(1+t)\rangle$.

According to Useful Fact 16.11,

$$\int \mathbf{r}(t)dt = \left\langle \int e^t\,dt, \int 2\cos t\,dt, \int \frac{1}{1+t}\,dt \right\rangle$$

Taking the components one at a time,

$$\int e^t\,dt = e^t + c_1$$

$$\int 2\cos t\,dt = 2\sin t + c_2$$

$$\int \frac{1}{1+t}\,dt = \ln|1+t| + c_3$$

Putting these together and separating the constants,

$$\begin{aligned}\int \mathbf{r}(t)dt &= \langle e^t + c_1, 2\sin t + c_2, \ln|1+t| + c_3\rangle \\ &= \langle e^t, 2\sin t, \ln|1+t|\rangle + \langle c_1, c_2, c_3\rangle \\ &= \langle e^t, 2\sin t, \ln|1+t|\rangle + \mathbf{C}\end{aligned}$$

Reminder: the arbitrary vector should be reported properly as $+\mathbf{C}$, not just $+C$. C represents one constant, $\mathbf{C}$ represents many. You don't want graders taking off one point for each constant missed! ■

Here's an example of a definite integral; there are no surprises here:

EX 8 Evaluate $\int_1^4 \left(\sqrt{t}\,\mathbf{i} + te^{-t}\,\mathbf{j} + \frac{1}{t^2}\,\mathbf{k}\right) dt$.

$$\begin{aligned}\int_1^4 \left(\sqrt{t}\mathbf{i} + te^{-t}\mathbf{j} + \frac{1}{t^2}\mathbf{k}\right) dt &= \left(\frac{2}{3}t^{3/2}\mathbf{i} - (t+1)e^{-t}\mathbf{j} - \frac{1}{t}\mathbf{k}\right)\Bigg|_1^4 \\ &= \left(\frac{16}{3}\mathbf{i} - 5e^{-4}\mathbf{j} - \frac{1}{4}\mathbf{k}\right) - \left(\frac{2}{3}\mathbf{i} - 2e^{-1}\mathbf{j} - \mathbf{k}\right) \\ &= \frac{14}{3}\mathbf{i} + \left(\frac{2}{e} - \frac{5}{e^4}\right)\mathbf{j} + \frac{3}{4}\mathbf{k}\end{aligned}$$

(The integral in the second component required integration by parts, which was not shown here as it should be old news.) ■

You Try It

(8) Evaluate $\int_0^1 (16t^3\,\mathbf{i} - 9t^2\,\mathbf{j} + 25t^4\,\mathbf{k})\,dt$.

Arc Length

We can put definite integrals of vector functions to good use right away.

Useful Fact 16.12. *The arc length L of the curve given by the vector function $\mathbf{r}(t) = \langle f(t), g(t), h(t)\rangle$ for $a \le t \le b$ is:*

$$\begin{aligned}L &= \int_a^b |\mathbf{r}'(t)|dt \\ &= \int_a^b \sqrt{[f'(t)]^2 + [g'(t)]^2 + [h'(t)]^2}dt\end{aligned}$$

The two expressions shown are equivalent, you can use whichever one you like better.

EX 9 Find the arc length of $\mathbf{r}(t) = \langle 2t, 1-3t, 5+4t\rangle$ from $(0,1,5)$ to $(4,-5,13)$.

First, note that $(0,1,5)$ is given by $t=0$ and $(4,-5,13)$ is given by $t=2$. With

$$f(t)=2t \quad ; \quad g(t)=1-3t \quad ; \quad h(t)=5+4t$$

we have

$$f'(t)=2 \quad ; \quad g'(t)=-3 \quad ; \quad h'(t)=4$$

so that

$$\begin{aligned} L &= \int_a^b \sqrt{[f'(t)]^2+[h'(t)]^2+[h'(t)]^2}dt = \int_0^2 \sqrt{2^2+(-3)^2+4^2}\,dt \\ &= \int_0^2 \sqrt{29}\,dt = \sqrt{29}\,t\Big|_0^2 = 2\sqrt{29} \end{aligned}$$

Note that the vector function describes a line, and so this arc length is just the straight-line distance between the two points, $\sqrt{4^2+(-6)^2+8^2} = \sqrt{116} = 2\sqrt{29}$. ■

You Try It

(9) Find the arc length of $\mathbf{r}(t)=\langle\sqrt{2}t, e^t, e^{-t}\rangle$ for $0\le t\le 1$.

Into the Pit!!

A Vector Function Product Rule

We are going to prove Useful Fact 16.10 for vector functions in $\mathbb{R}^3$ that have differentiable scalar component functions. Let $\mathbf{p}(t)=\langle p_1(t),p_2(t),p_3(t)\rangle$ and $\mathbf{r}(t)=\langle r_1(t),r_2(t),r_3(t)\rangle$, where each function $p_i(t)$ and $r_i(t)$ are differentiable. We can compute the derivative of $\mathbf{p}(t)\cdot\mathbf{r}(t)$ by building the dot product first, and then seeking the derivative. Since

$$\mathbf{p}(t)\cdot\mathbf{r}(t)=p_1(t)r_1(t)+p_2(t)r_2(t)+p_3(t)r_3(t)$$

then with several individual scalar product rules,

$$\begin{aligned} \frac{d}{dt}\mathbf{p}(t)\cdot\mathbf{r}(t) &= p_1'(t)r_1(t)+p_1(t)r_1'(t) \\ &\quad + p_2'(t)r_2(t)+p_2(t)r_2'(t)+p_3'(t)r_3(t)+p_3(t)r_3'(t) \qquad (16.9) \end{aligned}$$

Now we can assemble the right side of Eq. (16.8) for comparison. With

$$\begin{aligned}\mathbf{p}'(t)\cdot\mathbf{r}(t) &= \langle p_1'(t), p_2'(t), p_3'(t)\rangle \cdot \langle r_1(t), r_2(t), r_3(t)\rangle \\ &= p_1'(t)r_1(t) + p_2'(t)r_2(t) + p_3'(t)r_3(t) \\ \mathbf{p}(t)\cdot\mathbf{r}'(t) &= \langle p_1(t), p_2(t), p_3(t)\rangle \cdot \langle r_1'(t), r_2'(t), r_3'(t)\rangle \\ &= p_1(t)r_1'(t) + p_2(t)r_2'(t) + p_3(t)r_3'(t)\end{aligned}$$

we get

$$\begin{aligned}\mathbf{p}'(t)\cdot\mathbf{r}(t) + \mathbf{p}(t)\cdot\mathbf{r}'(t) &= [p_1'(t)r_1(t) + p_2'(t)r_2(t) + p_3'(t)r_3(t)] \\ &\quad + [p_1(t)r_1'(t) + p_2(t)r_2'(t) + p_3(t)r_3'(t)]\end{aligned}$$

or, rearranged,

$$\begin{aligned}\mathbf{p}'(t)\cdot\mathbf{r}(t) + \mathbf{p}(t)\cdot\mathbf{r}'(t) &= p_1'(t)r_1(t) + p_1(t)r_1'(t) \\ &\quad + p_2'(t)r_2(t) + p_2(t)r_2'(t) + p_3'(t)r_3(t)) + p_3(t)r_3'(t))\end{aligned} \tag{16.10}$$

Comparing (16.9) and (16.10) confirms that for vector functions in $\mathbb{R}^3$,

$$\frac{d}{dt}\mathbf{p}(t)\cdot\mathbf{r}(t) = \mathbf{p}'(t)\cdot\mathbf{r}(t) + \mathbf{p}(t)\cdot\mathbf{r}'(t)$$

The proof would be similar for vector functions with other numbers of components, and could even be expanded to consider all cases at once.

Vector Functions — Problem List

Vector Functions — You Try It

These appeared above; solutions begin on the next page.

(1) Identify the curve given by $\mathbf{r}(t) = (1+t)\mathbf{i} + t^2\mathbf{j}$.
(2) Give both the vector function and parametric equations for the line segment joining the points $P(0,0,0)$ and $Q(1,2,3)$.

(3) Evaluate $\lim_{t\to 0^+} \langle \cos t, \sin t, t\ln t\rangle$.
(4) If $\mathbf{r}(t) = e^{t^2}\mathbf{i} - \mathbf{j} + \ln(1+3t)\mathbf{k}$, what is $\mathbf{r}'(t)$?
(5) Find a vector tangent to the curve $\mathbf{r}t = (1+t)\mathbf{i} + t^2\mathbf{j}$ at $t = 1$.
(6) Find the vector equation of the line tangent to $\mathbf{r}t = \langle t^5, t^4, t^3\rangle$ at $\langle 1,1,1\rangle$.
(7) Let $\mathbf{p}(t) = \langle \sin t, \cos t, e^{-t}\rangle$ and $\mathbf{r}(t) = \langle 2\sin t, 2\cos t, e^t\rangle$. Confirm Useful Fact 16.10 by (a) computing the dot product and then finding the derivative, and (b) applying the relation in Eq. (16.8).
(8) Evaluate $\int_0^1 (16t^3\,\mathbf{i} - 9t^2\,\mathbf{j} + 25t^4\,\mathbf{k})\,dt$.
(9) Find the arc length of $\mathbf{r}(t) = \langle \sqrt{2}t, e^t, e^{-t}\rangle$ for $0 \le t \le 1$.

Vector Functions — Practice Problems

Try these as you get the hang of the You Try It problems. Solutions to these problems are available in Sec. B.4.3.

(1) Find the vector and parametric equations of the line segment joining the points $P(1,0,1)$ and $Q(2,3,1)$.

(2) Evaluate $\lim_{t\to 1}\left(\sqrt{t+3}\,\mathbf{i}+\frac{t-1}{t^2-1}\mathbf{j}+\frac{\tan t}{t}\mathbf{k}\right)$.

(3) If $\mathbf{r}(t)=(at\cos 3t)\,\mathbf{i}+(b\sin^3 t)\,\mathbf{j}+(c\cos^3 t)\,\mathbf{k}$, what is $\mathbf{r}'(t)$?

(4) Identify the curve given by $\mathbf{r}(t)=e^t\,\mathbf{i}+e^{-t}\,\mathbf{j}$ and find a vector tangent to the curve at $t=0$.

(5) Find the vector equation of the line tangent to $\mathbf{r}(t)=\langle t^2-1,t^2+1,t+1\rangle$ at $\langle -1,1,1\rangle$.

(6) Find the vector equation of the line tangent to $\mathbf{r}(t)=\langle \ln t, 2\sqrt{t}, t^2\rangle$ at $\langle 0,2,1\rangle$.

(7) Based on Useful Fact 16.10, you may be suspicious that there could be an expression like this:

$$\frac{d}{dt}\mathbf{p}(t)\times\mathbf{r}(t)=\mathbf{p}'(t)\times\mathbf{r}(t)+\mathbf{p}(t)\times\mathbf{r}'(t)$$

Test this relation on the vector functions $\mathbf{p}(t)=\langle 1,t,t^2\rangle$ and $\mathbf{r}(t)=\langle t^2,t,1\rangle$ by constructing both sides separately and comparing the results.

(8) Evaluate $\int_0^1\left(\frac{4}{1+t^2}\mathbf{j}+\frac{2t}{1+t^2}\mathbf{k}\right)dt$.

(9) Write the integral that gives the arc length of the vector function from Practice Problem 5 from the point $(0,2,2)$ to the point $(15,17,5)$.

Vector Functions — Challenge Problems

Try these problems to test your skills with the ideas in this section. Solutions to these problems are available in Sec. C.4.3.

(1) Suppose object A is moving along the path $\mathbf{r}_1(t)=\langle t,t^2,t^3\rangle$ and object B is moving along the path $\mathbf{r}_2(s)=\langle 1+2s,1+6s,1+14s\rangle$.

 (a) Find a vector that points from object A to object B at $t=s=2$.
 (b) How far apart are the objects at that instant?
 (c) Are there any points that the two paths share? If so, find them. If not, say how you know.

(2) Consider the vector curve $\mathbf{r}(t) = \langle 2\cos t, 2t/\pi, 2\sin t\rangle$.

(a) Find the equation(s) of the line tangent to the vector curve at the location given by $\langle 0, 1, 2\rangle$.

(b) Write the integral that would give the total arc length of the curve from $\langle 2, 0, 0\rangle$ to $\langle 0, 1, 2\rangle$.

(3) If tangent lines to a vector function $\mathbf{r}(t)$ can be found as

$$\left\langle \frac{1}{1+t^2}, \frac{1}{2\sqrt{t}}, e^{-t} \right\rangle$$

and the vector function passes through the point $(\pi/4, 1, 1-1/e)$, what is the vector function $\mathbf{r}(t)$?

(4) (Bonus Time in *the Pit*!) Can you apply the technique used to prove Useful Fact 16.10 (in *the Pit*!) to prove the expression which Practice Problem 8 suggests might be true?

Vector Functions — You Try It — Solved

---

(1) Identify the curve given by $\mathbf{r}(t) = (1+t)\mathbf{i} + t^2\mathbf{j}$.

□ By eliminating t from $x(t) = 1+t, y(t) = t^2$, we get $t = x-1$, so that $y = t^2$ becomes $y = (x-1)^2$. This is a parabola with vertex $(1, 0)$. ■

---

(2) Give both the vector function and parametric equations for the line segment joining the points $P(0, 0, 0)$ and $Q(1, 2, 3)$.

□ The vector function has initial vector $\mathbf{r_0} = \langle 0, 0, 0\rangle$ and parallel vector $\mathbf{r}_1 = \mathbf{PQ} = \langle 1, 2, 3\rangle$ so that its equation is

$$\mathbf{r_0} + t\mathbf{r}_1 = \langle 0, 0, 0\rangle + t\langle 1, 2, 3\rangle = \langle t, 2t, 3t\rangle$$

To restrict this to the line segment that starts at P and ends at Q, we need $0 \le t \le 1$. So we get

$$x = t \quad ; \quad y = 2t \quad ; \quad z = 3t \quad ; \quad 0 \le t \le 1 \quad \blacksquare$$

---

(3) Evaluate $\lim\limits_{t\to 0+} \langle \cos t, \sin t, t\ln t\rangle$.

□ For the limit of $\langle \cos t, \sin t, t\ln t\rangle$ as $t \to 0^+$, note that the third component's limit is indeterminate (of the form $0\cdot\infty$) and so L-Hopital's

Rule is needed:

$$\begin{aligned}\lim_{t\to 0^+}\langle \cos t, \sin t, t\ln t\rangle &= \lim_{t\to 0^+}\left\langle \cos t, \sin t, \frac{\ln t}{1/t}\right\rangle \\ &= \lim_{t\to 0^+}\left\langle \cos t, \sin t, \frac{1/t}{-1/t^2}\right\rangle \\ &= \lim_{t\to 0^+}\langle \cos t, \sin t, (-t)\rangle = \langle 1,0,0\rangle \quad \blacksquare\end{aligned}$$

---

(4) If $\mathbf{r}(t) = e^{t^2}\,\mathbf{i} - \mathbf{j} + \ln(1+3t)\mathbf{k}$, what is $\mathbf{r}'(t)$?

☐ By direct evaluation, we have

$$\mathbf{r}'(t) = 2te^{t^2}\mathbf{i} + \frac{3}{1+3t}\mathbf{k} \quad \blacksquare$$

---

(5) Find a vector tangent to the curve $\mathbf{r}(t) = (1+t)\,\mathbf{i} + t^2\,\mathbf{j}$ at $t = 1$.

☐ Since $\mathbf{r}'(t) = \mathbf{i} + 2t\,\mathbf{j}$, then at $t = 1$, a tangent vector is $\mathbf{r}'(1) = \mathbf{i} + 2\mathbf{j}$. ■

---

(6) Find the vector equation of the line tangent to $\mathbf{r}(t) = \langle t^5, t^4, t^3\rangle$ at $\langle 1,1,1\rangle$.

☐ Note that since we want the tangent line at $\langle 1,1,1\rangle$, we're going to use $t = 1$. We need:

$$\begin{aligned}\mathbf{r}'(t) &= \langle 5t^4, 4t^3, 3t^2\rangle \\ \mathbf{r}'(1) &= \langle 5,4,3\rangle\end{aligned}$$

So we have an initial vector $\mathbf{r_0} = \langle 1,1,1\rangle$ and a parallel vector $\mathbf{r}'(1)$, and the equation of the line is

$$\mathbf{r}(t) = \mathbf{r}_0 + t\mathbf{r}'(1) = \langle 1,1,1\rangle + t\langle 5,4,3\rangle = \langle 1+5t, 1+4t, 1+3t\rangle \quad \blacksquare$$

---

(7) Let $\mathbf{p}(t) = \langle \sin t, \cos t, e^{-t}\rangle$ and $\mathbf{r}(t) = \langle 2\sin t, 2\cos t, e^t\rangle$. Confirm Useful Fact 16.10 by (a) computing the dot product and then finding the derivative, and (b) applying the relation in Eq. (16.8).

☐ Since $\mathbf{p}(t)\cdot\mathbf{r}(t) = 2\sin^2 t + 2\cos^2 t + 1 = 3$, then

$$\frac{d}{dt}\,\mathbf{p}(t)\cdot\mathbf{r}(t) = \frac{d}{dt}(3) = 0$$

Separately, we have

$$\mathbf{p}'(t) = \langle \cos t, -\sin t, -e^{-t}\rangle \quad ; \quad \mathbf{r}'(t) = \langle 2\cos t, -2\sin t, e^t\rangle$$

so that

$$\begin{aligned}\mathbf{p}'(t)\cdot\mathbf{r}(t) &= \langle\cos t, -\sin t, -e^{-t}\rangle\cdot\langle 2\sin t, 2\cos t, e^{t}\rangle\\ &= 2\sin t\cos t - 2\sin t\cos t - 1 = -1\\ \mathbf{p}(t)\cdot\mathbf{r}'(t) &= \langle\sin t, \cos t, e^{-t}\rangle\cdot\langle 2\cos t, -2\sin t, e^{t}\rangle\\ &= 2\sin t\ cost - 2\sin t\cos t + 1 = 1\end{aligned}$$

So that by Useful Fact 16.10,

$$\frac{d}{dt}\mathbf{p}(t)\cdot\mathbf{r}(t) = \mathbf{p}'(t)\cdot\mathbf{r}(t) + \mathbf{p}(t)\cdot\mathbf{r}'(t) = -1+1=0$$

and the resulting derivative expressions match. ■

---

(8) Evaluate $\int_0^1 (16t^3\,\mathbf{i} - 9t^2\,\mathbf{j} + 25t^4\,\mathbf{k})\,dt$.

☐ It really is this simple:

$$\int_0^1 (16t^3\mathbf{i} - 9t^2\mathbf{j} + 25t^4\mathbf{k})dt = (4t^4\mathbf{i} - 3t^3\mathbf{j} + 5t^5\mathbf{k})\Big|_0^1 = 4\mathbf{i} - 3\mathbf{j} + 5\mathbf{k} \quad ■$$

---

(9) Find the arc length of $\mathbf{r}(t) = \langle\sqrt{2}t, e^t, e^{-t}\rangle$ for $0 \le t \le 1$.

☐ We're going to use Useful Fact 16.12; let's build the integrand first. Since $\mathbf{r}'(t) = \langle\sqrt{2}, e^t, -e^{-t}$, then the integrand will be

$$\sqrt{(\sqrt{2})^2 + (e^t)^2 + (-e^{-t})^2} = \sqrt{e^{2t} + 2 + e^{-2t}} = \sqrt{(e^t + e^{-t})^2} = e^t + e^{-t}$$

and so the arc length is

$$L = \int_0^1 e^t + e^{-t}\,dt = e^t - e^{-t}\Big|_0^1 = e - \frac{1}{e} \quad ■$$

16.4 Vector Fields and the Gradient

Introduction

Ask yourself if you have ever seen a *vector field* before. If your self answers, "No", then your self is misinformed. Everyone has seen vector fields, even if that specific term was not used. If you've seen a picture with a bunch of arrows on it, indicating that the magnitude and direction of some quantity varies with location, then you have seen a vector field. A wind speed map on a weather report might be presented as a vector field. The magnetic field around the north and south poles of a magnet might be displayed as a vector field. Figure 16.9 shows a vector field representation of the solar magnetic field. Figure 16.10 shows a vector field representation of wind speed in the upper part of North America.[7]

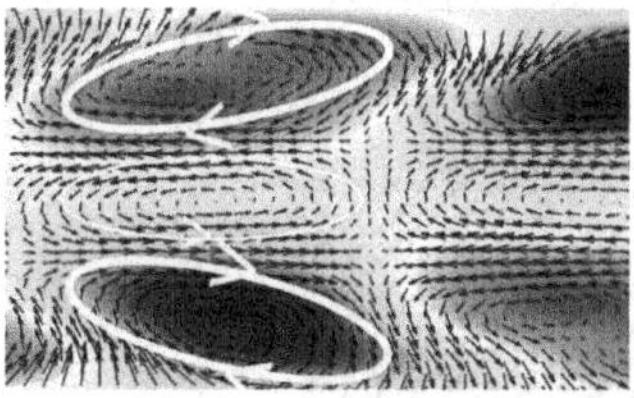

Fig. 16.9 The solar magnetic field (NCAR / HAO).†

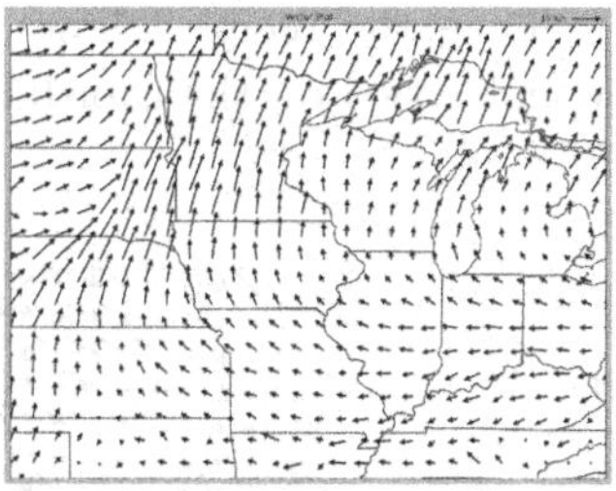

Fig. 16.10 A vector field representation of wind speed.‡

Vector fields should not be confused with vector functions. Vector functions use vectors to *select* points in space, and we often collect those points together into a curve. On the other hand, vector fields assign a single vector to every point in space, and then we can visualize that vector field by drawing a bunch of those vectors.

Vector Fields

Vector fields associate a vector with each point in space. While vector fields can exist in all dimensions, useful visualizations only happen in $\mathbb{R}^2$ and $\mathbb{R}^3$. A vector field comes with some recipe that takes the coordinates of a point

[7]†`https://www2.hao.ucar.edu/news/2019-aug/what-causes-seasons-space-weather`;
‡Courtesy of K. Goebbert, Dept. of Geography and Meteorology, Valparaiso University.

and produces a vector out of them. So to "see" a vector field, we go to a point, use its coordinates to come up with a vector using the given recipe, and draw the vector right there. We end up with a plot that has arrows all over it. The following are several examples of vector fields:

(1) $\mathbf{F}_1(x, y) = \langle x, -y \rangle$

(2) $\mathbf{F}_2(x, y) = \left\langle -\dfrac{y}{\sqrt{x^2+y^2}}, \dfrac{x}{\sqrt{x^2+y^2}} \right\rangle$

(3) $\mathbf{F}_3(x, y) = \langle y, \sin x \rangle$

(4) $\mathbf{F}_4(x, y, z) = \langle y, -x, z \rangle$

(5) $\mathbf{F}_5(x, y, z) = \left\langle \dfrac{1}{y}, \dfrac{1}{x+z}, \sin(xyz) \right\rangle$

A capital **F** is often the generic name for a vector field, and the arguments (i.e. the information passed to the recipe for **F** are the components of any point. Then the recipe follows, and you can see how to construct each component based on the coordinates of the given point.

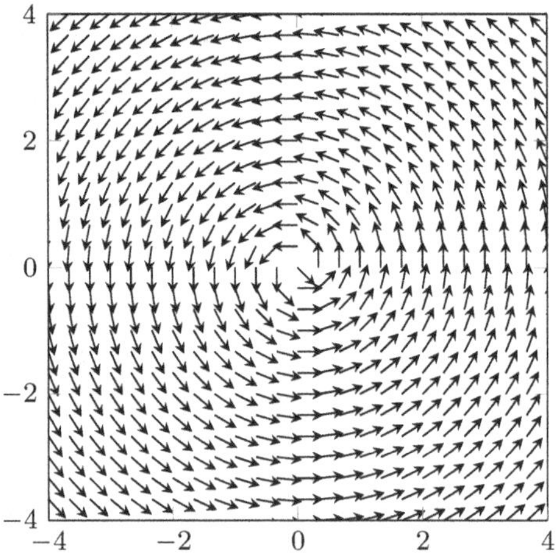

Fig. 16.11 Vortex vector field.

Figure 16.11 shows one of the first two vector fields in that list. The vector field in the figure represents a *vortex*; if you dropped a small feather into a bathtub that had water continually swirling around the drain, these arrows show the path the vector might follow. Can you decide which of those first two vector fields this figure shows? Here are some hints:

- What would happen to the size of the vectors in $\mathbf{F}_1$ as we moved farther and farther away from the origin? Do you see that trend in the figure?
- Examine the signs of the components in $\mathbf{F}_1$. Do the vectors displayed in the figure reflect that pattern?

The figure shows the second vector field, $\mathbf{F}_2$. The denominator $\sqrt{x^2+y^2}$ normalizes the lengths of the vectors so that their sizes do not get out of hand.

EX 1 Which vector field in shown in Fig. 16.12?

$$\mathbf{F}_1(x,y) = \left\langle \frac{x}{\sqrt{x^2+y^2}}, \frac{y}{\sqrt{x^2+y^2}} \right\rangle \qquad \text{or}$$

$$\mathbf{F}_2(x,y) = \left\langle \sqrt{x^2+y^2}, -\sqrt{x^2+y^2} \right\rangle$$

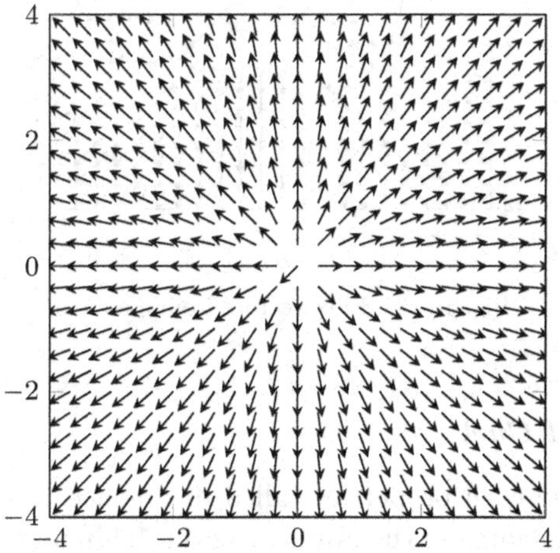

Fig. 16.12 Sample vector field, with EX 1.

The figure shows the vector field $\mathbf{F}_1$. The magnitude of the vectors in $\mathbf{F}_2$ will get larger with increased distance from the origin; that's not happening in the figure. Also, all the vectors in $\mathbf{F}_2$ would point downwards, because the second component is always negative.

It turns out that $\mathbf{F}_1$ is the vector field of a *source* of a flow field — such as fluid flow or electric charge. Something in the middle of the image is generating flow or current. The vector field of a *sink* would look similar, except the vectors would point inwards towards the center. ■

You Try It

(1) Which vector field in shown in Fig. 16.14?

$\mathbf{F}_1(x,y) = \langle x^2\cos(y), 2x\sin(y)\rangle$; $\mathbf{F}_2(x,y) = \langle y^2\cos(x), 2y\sin(x)\rangle$

(The figure is displayed at the end of this section.)

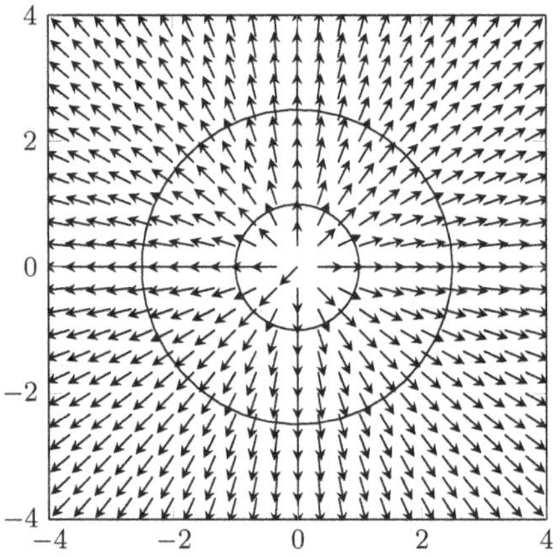

Fig. 16.13 Flux vectors from a source.

Divergence and Curl

Vector fields come with properties called divergence and curl, which each have significant meaning in the physical world. Thinking about fluid flow is a good way to help visualize vector fields and their properties, we'll do that a lot. To use this setting for our vector fields, it helps to understand that the vectors in the vector field, seen as the arrows, represent flux at a given point — which is a measure of the volume of flow moving past that point per unit area per unit time. Figure 16.13 reproduces Fig. 16.11 except with two circles around the source of flow (we'll use them soon). At any point, imagine the page is one unit deep, then each arrow shows the magnitude and direction of flux; that is, the amount of flow past that point through one square unit of area perpendicular to the page. A bigger arrow means there is more flux, a smaller arrow represents less.

It may be hard to tell on the scale of the figure, but the arrows are getting smaller as we move away from the source in the center. This is because the flow is spreading out. If we measure the total flow around the perimeter of each circle in the figure (by integrating), it should be the same. The total flow entering the field at the center per unit time must be the same as the total flow per unit time across each circle — but since the outer circle is larger, there is more area to work with, and so we have smaller amounts of flow at each point per unit area.

Now, suppose we measured the total amount of flux across the outer circle in one unit of time and discovered that it was larger than the total amount moving across the inner circle in that same unit of time. This means the flow field is exhibiting **divergence**. In other words, divergence measures the ability of a vector field to "spread" and make room for new flow. But that spread is not the macro scale and natural spreading out of flow that's exhibited in the figure. Rather, it would mean additional flow merging in on the micro scale from the medium itself even in the absence of other sources. The vector field in Fig. 16.13 has no divergence. A flow field that displays no divergence is called **incompressible**; the flow cannot be "compressed" at the micro scale to absorb more flow and — in the case of Fig. 16.13 — make the total flow across the outer circle larger than the flow across the inner circle.

The **curl** of a vector field measures how much rotation there is in the vector field. A vector field with no curl is called **irrotational**. Like divergence, this is a measure of what happens on the micro scale. Imagine a leaf floating in the current of a stream. As the stream itself bends, the leaf's path will curve too. This is not curl or rotation, it's just a macro-scale change in direction. But if the leaf is spinning in circles while it's floating down the stream, then we have curl / rotation in the flow field.

That's a lot of words, now let's see some math! To show the formulations of divergence and curl, we need a new symbol, which is a "differential operator".[8] The "del operator" is a vector built out of derivative operators:

[8]Don't freak out. An operator is just a mathematical object that has a job to do.

Definition 16.7. *The differential operator called "del" or "nabla" is a vector quantity:*

$$\nabla = \left\langle \frac{\partial}{\partial x}, \frac{\partial}{\partial y}, \frac{\partial}{\partial z} \right\rangle$$

The term nabla comes from the Greek word for harp, since the the symbol apparently looked like a harp to one of those old Greek math guys. In fact, while typing this content, I have to call the command `\nabla` to make the symbol ∇ appear.

The first thing to notice about the del operator is that, like a square root symbol, it is totally meaningless unless it's attached to something else. Del is a vector, and we can put it into dot and cross products. That may seem strange, but you'll know what to do in any situation by tracking how the derivative operators get attached to the other things. For example, let's have a sneak preview of a full definition coming next: the divergence of a vector field $\mathbf{F} = \langle P, Q, R \rangle$ will be

$$\text{div}(\mathbf{F}) = \frac{\partial P}{\partial x} + \frac{\partial Q}{\partial y} + \frac{\partial R}{\partial z}$$

That structure looks an awful lot like a dot product. If we pull it apart, we can see that

$$\text{div}(\mathbf{F}) = \left\langle \frac{\partial}{\partial x}, \frac{\partial}{\partial y}, \frac{\partial}{\partial z} \right\rangle \cdot \langle P.Q, R \rangle$$

And there is our differential operator in action. More formally,

Definition 16.8. *Given a 3D vector field* $\mathbf{F}(x, y, z)$, *the* ***divergence*** *of* $\mathbf{F}$ *is*

$$div(\mathbf{F}) = \nabla \cdot \mathbf{F}$$

And guess what the cross product with del does:

Definition 16.9. *The* ***curl*** *of a 3D vector field* $\mathbf{F}(x, y, z)$ *is*

$$curl(\mathbf{F}) = \nabla \times \mathbf{F}$$

Now, here is how that cross product plays out:

$$\text{curl}(\mathbf{F}) = \begin{vmatrix} \mathbf{i} & \mathbf{j} & \mathbf{k} \\ \frac{\partial}{\partial x} & \frac{\partial}{\partial y} & \frac{\partial}{\partial z} \\ P & Q & R \end{vmatrix}$$

(remember that P, Q, R are the component functions of $\mathbf{F}$). Here are a couple of important things to note about these quantities:

- The divergence of a vector field is a *scalar function.*
- The curl of a vector field is another *vector field.*

Don't be worried about the strange determinant we use for curl. You'll know what to do with those derivative operators, the computations sort of build themselves. The important thing is to maintain notation; in this determinant scheme, we are worried about *operation* not multiplication. So in the process of evaluating one of these determinants, don't write something like

$$P \cdot \frac{\partial}{\partial y}$$

That makes no sense. You wouldn't write "$16\sqrt{\ }$" for the square root of 16, so don't write derivative operations out of order either. Multiplication can be scrambled up, these operations cannot.

With divergence and curl, we can now do better with some other terms mentioned above:

Definition 16.10. *Given a 3D vector field* $\mathbf{F}(x, y, z)$,

- $\mathbf{F}$ *is* ***irrotational*** *if* $\nabla \times \mathbf{F} = \mathbf{0}$.
- $\mathbf{F}$ *is* ***incompressible*** *if* $\nabla \cdot \mathbf{F} = 0$.

EX 2 Find the curl and divergence of $\mathbf{F}(x, y, z) = \langle xz, xyz, xy \rangle$. Is this vector field irrotational or incompressible?

The curl of $\mathbf{F}$ is

$$\begin{vmatrix} \mathbf{i} & \mathbf{j} & \mathbf{k} \\ \frac{\partial}{\partial x} & \frac{\partial}{\partial y} & \frac{\partial}{\partial z} \\ xz & xyz & xy \end{vmatrix} = \left(\frac{\partial}{\partial y}(xy) - \frac{\partial}{\partial z}(xyz)\right)\mathbf{i} - \left(\frac{\partial}{\partial x}(xy) - \frac{\partial}{\partial z}(xz)\right)\mathbf{j}$$

$$+ \left(\frac{\partial}{\partial x}xyz - \frac{\partial}{\partial y}(xz)\right)\mathbf{k}$$

$$= (x - xy)\,\mathbf{i} + (x - y)\,\mathbf{j} - yz\,\mathbf{k}$$

Since $\nabla \times \mathbf{F} \neq \mathbf{0}$, the vector field is not irrotational. I guess it's "rotational"? The divergence of $\mathbf{F}$ is

$$\frac{\partial}{\partial x}(xz) + \frac{\partial}{\partial y}(xyz) + \frac{\partial}{\partial z}(xy) = z + xz + 0 = z(1 + x)$$

Since $\nabla \cdot \mathbf{F} \neq 0$, the vector field is not incompressible. ■

You Try It

(2) Find the curl and divergence of $\mathbf{r}(x, y, z) = \langle x, y, z \rangle$. Is this vector field irrotational or incompressible?

(3) Find the curl and divergence of $\mathbf{F}(x, y, z) = \langle 1, x + yz, xy - \sqrt{z} \rangle$. Is this vector field irrotational or incompressible?

We will use divergence and curl a lot in upcoming sections. First, we need to encounter one more application of ∇.

The Gradient

In divergence and curl calculations, the differential operator ∇ acted on a vector field. But ∇ is a very versatile symbol; it can also interact with scalar functions — and this operation creates a quantity called a *gradient.* The gradient of a scalar function is the single most important vector field you'll ever see.

Definition 16.11. *Suppose $f(x, y, z)$ is a scalar function. Then the* ***gradient*** *of f is given by*

$$\nabla f = \left\langle \frac{\partial}{\partial x}, \frac{\partial}{\partial y}, \frac{\partial}{\partial z} \right\rangle (f) = \left\langle \frac{\partial f}{\partial x}, \frac{\partial f}{\partial y}, \frac{\partial f}{\partial z} \right\rangle$$

In the gradient operation, we take a scalar function and created a vector field out of derivatives of its components. This vector field contains a lot of good information which we will see in upcoming sections. But for now, we just practice.

EX 3 Find the gradient of $f(x, y, z) = xy^2z^3$.

We have

$$\nabla f = \left\langle \frac{\partial f}{\partial x}, \frac{\partial f}{\partial y}, \frac{\partial f}{\partial z} \right\rangle = \langle y^2z^3, 2xyz^3, 3xy^2z^2 \rangle \quad ■$$

So you see, gradients are not hard to compute; they are only as hard as any of the individual derivatives of f. If we want the gradient of a function of only two variables, then we just use the x and y derivatives. Also, note that we can use subscript notation to write $\nabla f = \langle f_x, f_y, f_z \rangle$. Use whichever derivative notation you like better.

You Try It

(4) Find the gradient of $f(x,y) = 5xy^2 - 4x^3y$, and determine $\nabla f(1,2)$.
(5) Find the gradient of $g(x,y,z) = x^2 + 2y^2 + 3z^2$.

Here are a couple of the interesting questions we'll soon ask about gradients:

- We can always take a scalar function f and find its gradient, which is a vector field. But, if we pick a vector field at random, does there have to be a scalar function f of which our vector field is the gradient?
- In "real life", you probably recognize the word *gradient* to mean something to do with slope — like on a hill. In the context of scalar functions and their gradients, do you think the word "gradient" was picked by accident?
- Will knowing how to find gradients make our mathematics lives easier?

The answer to two of these questions is YES! But which two?

Math Symbol Soup

In this section, we've learned about three new operations; each requires certain input and produces a certain kind of output, and you have to keep them straight:

- The gradient takes in a scalar function and spits out a vector field.
- The divegence takes in a vector field and spits out a scalar function.
- The curl takes in a vector field and spits out another vector field.

All of these are based on the differential operator ∇, however, the combinations of this operator with other quantities is not unlimited. Some combinations work and some don't. You have to become adept at knowing how to create and recognize valid combinations of operations related to ∇. Can we find the divergence of a divergence? Can we find the curl of a curl? Can we find the divergence of a curl? And so on...

EX 4 Let f be a scalar function and $\mathbf{F}$ a vector field. Describe why each operation is defined or undefined: (a) $\nabla \times f$; (b) $\nabla \mathbf{F}$; (c) $\nabla \cdot \mathbf{F}$; (d) $\nabla \times (\nabla f)$.

(a) $\nabla \times f$ is undefined since f is a scalar function and we find the curl of vector functions.
(b) $\nabla \mathbf{F}$ is undefined; with no dot or cross product symbol, we have to assume this is an attempt at a gradient; but since $\mathbf{F}$ is a vector field, it does not have a gradient
(c) $\nabla \cdot \mathbf{F}$ is OK since $\mathbf{F}$ is a vector function and we find the divergence of vector functions.
(d) $\nabla \times (\nabla f)$ is OK since ∇f is a vector function and we find the curl of vector functions. ■

You Try It

(6) Let f be a scalar function and $\mathbf{F}$ a vector field. Describe why each operation is defined or undefined: (a) $\nabla \mathbf{F}$; (b) $\nabla(\nabla \cdot \mathbf{F})$; (c) $\nabla \cdot (\nabla f)$; (d) $\nabla(\nabla \cdot f)$.

In Volume 1, the chapter in which definite integrals were introduced had the title, "The Best Mathematics Symbol There Is ... So Far." And sure, it's fun to draw the integral symbol. It looks cool to the Calculus novice. It's both impressive and scary to people who are not familiar with it. But ∇ is even better. It's symmetric. It's very flexible in how it gets used. I mean, just look at all the expressions there in You Try It 6. It doesn't get better than this. ∇ gets my vote for the best mathematical symbol there is. Sorry, integration.

Vector Fields and the Gradient — Problem List

Vector Fields and the Gradient — You Try It

These appeared above; solutions begin on the next page.

(1) Which vector field in shown in Fig. 16.14?

$$\mathbf{F}_1(x,y) = \langle x^2\cos(y), 2x\sin(y)\rangle \text{ or } \mathbf{F}_2(x,y) = \langle y^2\cos(x), 2y\sin(x)\rangle$$

(2) Find the curl and divergence of $\mathbf{r}(x,y,z) = \langle x,y,z\rangle$. Is this vector field irrotational or incompressible?
(3) Find the curl and divergence of $\mathbf{F}(x,y,z) = \langle 1, x+yz, xy-\sqrt{z}\rangle$. Is this vector field irrotational or incompressible?
(4) Find the gradient of $f(x,y) = 5xy^2 - 4x^3y$, and determine $\nabla f(1,2)$.
(5) Find the gradient of $g(x,y,z) = x^2 + 2y^2 + 3z^2$.
(6) Let f be a scalar function and $\mathbf{F}$ a vector field. Describe why each operation is defined or undefined: (a) $\nabla\mathbf{F}$; (b) $\nabla(\nabla\cdot\mathbf{F})$; (c) $\nabla\cdot(\nabla f)$; (d) $\nabla(\nabla\cdot f)$.

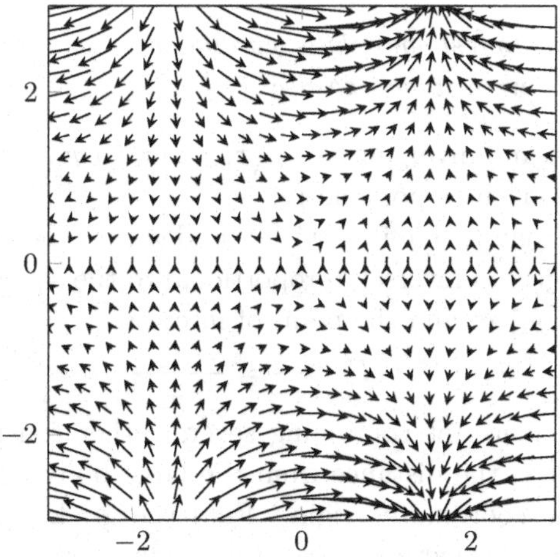

Fig. 16.14 Sample vector field, with YTI 1.

Vector Fields and the Gradient — Practice Problems

Try these as you get the hang of the You Try It problems. Solutions to these problems are available in Sec. B.4.4.

(1) Find the curl and divergence of $\mathbf{F}(x, y, z) = \langle 0, \cos(xz), -\sin(xy)\rangle$. Is this vector field irrotational or incompressible?
(2) Is this vector field irrotational or incompressible?
$$\mathbf{F}(x, y, z) = \left\langle \frac{x}{x^2 + y^2 + z^2}, \frac{y}{x^2 + y^2 + z^2}, \frac{z}{x^2 + y^2 + z^2} \right\rangle$$
(3) Find the gradient of $f(x, y) = y \ln x$, and determine $|\nabla f(1, -3)|$.
(4) Assuming s and t are still rectangular coordinates, find the gradient of $g(s, t) = e^{-s} \sin t$.
(5) Let f be a scalar function and $\mathbf{F}$ a vector field. Describe why each operation is defined or undefined: (a) $\nabla \times (\nabla \times \mathbf{F})$; (b) $\nabla \cdot (\nabla \cdot \mathbf{F})$; (c) $\nabla f \times \nabla \cdot \mathbf{F}$; (d) $\nabla \cdot (\nabla \times (\nabla f))$.
(6) Use the partial derivatives found in PP 9 of Sec. 13.5 to construct gradient vectors for the function $z = 3x^2 - 2y^2$ at the points $(1, 0)$, $(1, 1)$, and $(0, 1)$. Draw the gradient vectors at these locations as carefully as you can into a new version of Fig. B.3. Do you notice anything interesting?

Vector Fields and the Gradient — Challenge Problems

Try these problems to test your skills with the ideas in this section. Solutions to these problems are available in Sec. C.4.4.

(1) Find the curl and divergence of $\mathbf{F}(x, y, z) = \langle xyz, x^2y^2z^2, y^2z^3\rangle$. (Simplify each as much as possible.)
(2) Some of you may be familiar with the idea that the gravitational force due to an object is inversely proportional to the square of the distance between the object and the point of interest,
$$F = \frac{c}{r^2} = \frac{c}{x^2 + y^2}$$
where c is a constant containing several other constants mushed together, and r is the distance from the large body (presuming the object is at the coordinate origin. Find the gradient of this function. (Optional: In your expression for the gradient, try to introduce r anywhere you see an equivalent expression in x and y.)
(3) This is a problem for those of you who like puzzlers. Make up a couple of simple scalar functions $f(x, y, z)$. Find the gradient of each function, then find the curl of each gradient. What do you get? Make a conjecture as to what you'll *always* get for $\nabla \times (\nabla f)$, the curl of a gradient. Demonstrate why your conjecture will be true no matter what scalar function $f(x, y, z)$ you start with.

(4) *(Bonus! This problem has been developing in CP 4 of Secs. 13.3 and 13.5.)* Use the partial derivatives found in CP 4 of Sec. 13.5 to construct gradient vectors for the function $w = 2x^2 + y^2 + z$ at the point $(1, 1, 1)$. Draw the projection of this gradient vector into a new version of Fig. C.3. Do you notice anything interesting?

Vector Fields and the Gradient — You Try It — Solved

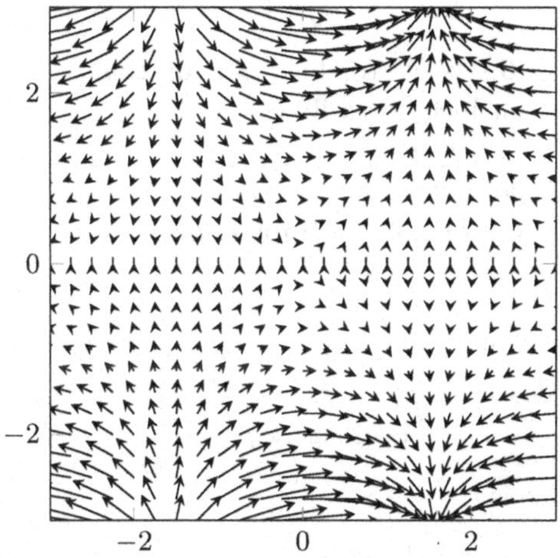

Fig. 16.15 Sample vector field, with YTI 1.

---

(1) Which vector field in shown in Fig. 16.15?

$\mathbf{F}_1(x, y) = \langle x^2 \cos(y), 2x \sin(y)\rangle$ or $\mathbf{F}_2(x, y) = \langle y^2 \cos(x), 2y \sin(x)\rangle$

□ The image goes with $\mathbf{F}_2$. The vectors are pretty much constant at any y, apart from a slight oscillation that varies in the x-direction. It can't be $\mathbf{F}_1$, since oscillations in that vector field would be along the y-direction. ■

---

(2) Find the curl and divergence of $\mathbf{r}(x, y, z) = \langle x, y, z\rangle$. Is this vector field irrotational or incompressible?

☐ The curl is

$$\begin{aligned}\nabla \times \mathbf{r} &= \begin{vmatrix} \mathbf{i} & \mathbf{j} & \mathbf{k} \\ \frac{\partial}{\partial x} & \frac{\partial}{\partial y} & \frac{\partial}{\partial z} \\ x & y & z \end{vmatrix} \\ &= \left(\frac{\partial}{\partial y}(z) - \frac{\partial}{\partial z}(y)\right)\mathbf{i} - \left(\frac{\partial}{\partial x}(z) - \frac{\partial}{\partial z}(x)\right)\mathbf{j} + \left(\frac{\partial}{\partial x}(y) - \frac{\partial}{\partial y}(x)\right)\mathbf{k} \\ &= \mathbf{0}\end{aligned}$$

The divergence is

$$\nabla \cdot \mathbf{r} = \frac{\partial}{\partial x}(x) + \frac{\partial}{\partial y}(y) + \frac{\partial}{\partial z}(z) = 1 + 1 + 1 = 3$$

Since $\nabla \times \mathbf{r} = \mathbf{0}$, the vector field is irrotational. Since $\nabla \cdot \mathbf{r} \neq 0$, the vector field is not incompressible. ■

---

(3) Find the curl and divergence of $\mathbf{F}(x, y, z) = \langle 1, x + yz, xy - \sqrt{z} \rangle$. Is this vector field irrotational or incompressible?

☐ The curl is:

$$\nabla \times \mathbf{F} = \begin{vmatrix} \mathbf{i} & \mathbf{j} & \mathbf{k} \\ \frac{\partial}{\partial x} & \frac{\partial}{\partial y} & \frac{\partial}{\partial z} \\ 1 & x + yz & xy - \sqrt{z} \end{vmatrix} = \langle x - y, -y, 1 \rangle$$

The divergence is

$$\begin{aligned}\nabla \cdot \mathbf{F} &= \left\langle \frac{\partial}{\partial x}, \frac{\partial}{\partial y}, \frac{\partial}{\partial z} \right\rangle \cdot \langle 1, x + yz, xy - \sqrt{z} \rangle \\ &= \frac{\partial}{\partial x}(1) + \frac{\partial}{\partial y}(x + yz) + \frac{\partial}{\partial z}(xy - \sqrt{z}) = z - \frac{1}{2\sqrt{z}}\end{aligned}$$

This vector field is neither irrotational nor incompressible. ■

---

(4) Find the gradient of $f(x, y) = 5xy^2 - 4x^3y$, and determine $\nabla f(1, 2)$.

☐ This is only a 2D example, so we just don't worry about the third component. Let's find the vector field ∇f and then plug in the point $(1, 2)$.

$$\begin{aligned}\nabla f(x, y) &= \langle f_x, f_y \rangle = \langle 5y^2 - 12x^2y, 10xy - 4x^3 \rangle \\ \nabla f(1, 2) &= \langle -4, 16 \rangle \quad ■\end{aligned}$$

(5) Find the gradient of $g(x, y, z) = x^2 + 2y^2 + 3z^2$.

□ No problem:

$$\nabla g(x, y, z) = \langle g_x, g_y, g_z \rangle = \langle 2x, 4y, 6z \rangle \quad \blacksquare$$

(6) Let f be a scalar function and $\mathbf{F}$ a vector field. Describe why each operation is defined or undefined: (a) $\nabla\mathbf{F}$; (b) $\nabla(\nabla \cdot \mathbf{F})$; (c) $\nabla \cdot (\nabla f)$; (d) $\nabla(\nabla \cdot f)$.

□ (a) $\nabla\mathbf{F}$ is undefined since $\mathbf{F}$ is a vector function and we find the gradient of scalar functions.
(b) $\nabla(\nabla \cdot \mathbf{F})$ is OK since $\nabla \cdot \mathbf{F}$ is a scalar function and we find the gradient of scalar functions.
(c) $\nabla \cdot (\nabla f)$ is OK since ∇f is a vector function and we find the divergence of vector functions.
(d) $\nabla(\nabla \cdot f)$ is undefined since f is a scalar function and we find the divergence of vector functions. $\blacksquare$

16.5 Planes and Tangent Planes

Introduction

A long time ago, we noticed that the since the equation of a line in 2D is of the standard form $ax + by + c = 0$, it might be reasonable to guess that a line in 3D would have the equation $ax + by + cz + d = 0$. We then learned, though, that this is actually the equation of a plane! But we never pursued that in any detail.

And, you may have noticed that while we've been dealing with (partial) derivatives quite a bit, we never actually got around to talking about tangents (except in the context of vector functions). But if you think about it, the problem of tangency is a bit more complicated for a surface $f(x, y)$. If we go to a point on the surface described by $f(x, y)$, how many lines are tangent to the surface at any one point? For example, how many tangent lines can you imagine touching the top of a balloon / sphere? An infinite number of them! (Better yet, think about a propeller beanie cap like the one shown in Fig. 16.16; the propeller is attached at a point on the surface of the cap — OK, you have to use your imagination a bit here — and as the propeller spins around, it is moving through all of the infinitely many lines tangent to the surface of the cap there.) There is only one type of object that can be uniquely tangent to a surface at a point, and that's a plane. So rather than finding tangent lines on a surface, we're looking for tangent *planes*.

It's now time to fill in both of these earlier omissions. We had to wait until now because we'll be using vectors and gradients in these -discussions.

How to Build a Plane

Being lazy creatures, we always want to get by with the minimum effort, right? When describing a line, we know that the minimum set of data requires two items — either two points, or a point and a slope. We already know that it takes three points to define a plane, so it makes sense that the minimum pieces of data needed to describe a plane in 3D is three. But, I'll argue that in order to fully describe a plane, we still need only two things. Do you believe me?

Fig. 16.16 A high tech visualization.

First, we need a point on the plane, that's kind of obvious. But one point alone isn't enough.

For the second piece of information we need, imagine a pencil stuck through the center of a flat piece of paper. As you move the pencil around, the paper moves around, too. In fact, the direction of the pencil determines the direction the paper is facing, right? Well, pretend the pencil is a vector. The direction of this "steering vector" determines the direction of the plane. And how is the pencil oriented to the paper? It's perpendicular! So if we have one point on a plane, and we also know a vector which is perpendicular to a plane, we have uniquely positioned the plane in space and determined its orientation.

And these two things are enough. With knowledge of (1) one point on a plane, and (2) a vector perpendicular to the plane, we cannot confuse that plane with any other, it is uniquely determined. Now how do we turn this information into the equation of a plane? First, let's name these two things. The point on the plane will be (x_0, y_0, z_0). The vector perpendicular to the plane will be called $\mathbf{n} = \langle a, b, c \rangle$, and it is called the *normal* vector. (There we go again, using a new word for something we can already describe!) Now consider,

- From our dealings with vector functions, we know there is a direct association between the point (x_0, y_0, z_0) and the vector $\mathbf{r}_0 = \langle x_0, y_0, z_0 \rangle$ pointing from the origin to our point, since this vector ends at our point.

- If the vector $\mathbf{r} = \langle x, y, z \rangle$ points from the origin to any other point on the plane, then the vector $\mathbf{r} - \mathbf{r}_0$ lies in the plane itself.
- If $\mathbf{r} - \mathbf{r}_0$ is in the plane, and $\mathbf{n}$ is normal to the plane, then $\mathbf{n} \cdot (\mathbf{r} - \mathbf{r}_0) = 0$.
- These vectors are illustrated in Fig. 16.17.

And voila, we have the ingredients for the equation of our plane:

$$\begin{aligned} \mathbf{n} \cdot (\mathbf{r} - \mathbf{r}_0) &= 0 \\ \langle a, b, c \rangle \cdot \langle x - x_0, y - y_0, z - z_0 \rangle &= 0 \\ a(x - x_0) + b(y - y_0) + c(z - z_0) &= 0 \end{aligned}$$

Definition 16.12. *A plane in* $\mathbb{R}^3$ *is uniquely determined by one point on the plane* (x_0, y_0, z_0) *and a vector normal (perpendicular) to the plane,* $\mathbf{n} = \langle a, b, c \rangle$*. The equation of the plane is, in standard form,*

$$a(x - x_0) + b(y - y_0) + c(z - z_0) = 0$$

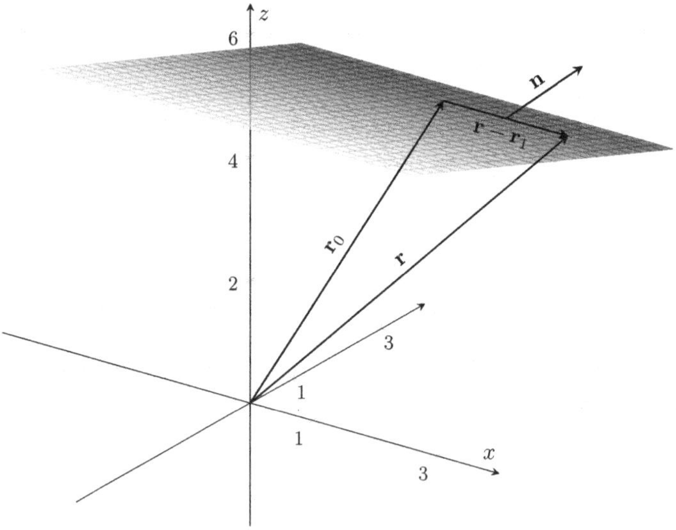

Fig. 16.17 Determining geometry of a plane.

Note that if we multiply out all the terms in this equation, we'll get:

$$ax + by + cz - (ax_0 + by_0 + cz_0) = 0$$

and the terms grouped in the parentheses are all constants, so we can set $d = -(ax_0 + by_0 + cz_0)$ and have another form of the equation of a plane,

$$ax + by + cz + d = 0$$

Having two forms of an equation is not new to you. After all, you know both the standard form of the equation of a line, $y - y_0 = m(x - x_0)$, as well as the slope-intercept form, $y = mx + b$.

The standard form is always preferable to the multiplied-out form. Here's why. Given a plane in that second form, we can immediately see what normal vector the plane has, but that's it. The plane $2x - 3y + 7z = -13$ is seen immediately to have a normal vector $\langle 2, -3, 7\rangle$, but it would take a bit more work to determine a point on the plane from that equation. On the other hand, given that plane's equation in standard form,

$$2(x-2) - 3(y-1) + 7(z+2) = 0$$

we can immediately see *both* the normal vector $\langle 2, -3, 7\rangle$ and a point on the plane, $(2, 1, -2)$.

EX 1 Find the equation of a plane containing the point $(-2, 2, 3)$ and with normal vector $\langle -1, 1, 2\rangle$.

This is just a plug-n-chug operation, since we are explicitly given all the information we need. We already know $(x_0, y_0, z_0) = (-2, 2, 3)$ and $\langle a, b, c\rangle = \langle -1, 1, 2\rangle$ so all that's left is to do this:

$$\begin{aligned} a(x - x_0) + b(y - y_0) + c(z - z_0) &= 0 \\ -1(x - (-2)) + 1(y - 2) + 2(z - 3) &= 0 \\ -(x + 2) + (y - 2) + 2(z - 3) &= 0 \end{aligned}$$

We can leave the equation of the plane in standard form, or collect the constants together and write the equation as $-x + y + 2z = 10$. ■

Problems in which you're given all the information you need right away are no fun, it's better to have to figure something out for yourself. There are lots of ways we can be given information about a plane, and it can be interesting to see how to distill a normal vector from that information.

EX 2 Find the equation of the plane containing the points $P(3, -1, 2)$, $Q(8, 2, 4)$ and $R(-1, -2, -3)$.

We're certainly OK in the area of needing one point on the plane, we have three. But we don't yet know a vector perpendicular to this plane. However, if we form the two vectors **PQ** and **PR**, then both of those vectors will be in the plane, and therefore their cross product will be perpendicular to the plane.

$$\begin{aligned}\mathbf{PQ} &= \langle 5, 3, 2\rangle \\ \mathbf{PR} &= \langle -4, -1, -5\rangle \\ \mathbf{PQ} \times \mathbf{PR} &= \langle -13, 17, 7\rangle\end{aligned}$$

So the equation of the plane is (using P as the point) $-13(x-3)+(17)(y+1)+(7)(z-2)=0$, or $-13x+17y+7z=-42$. ■

You Try It

(1) Find the equation of the plane containing the point $(1, -1, 1)$ and with normal vector $\langle 1, 1, -1\rangle$.
(2) Find the equation of the plane containing the points $P(0, 1, 1)$, $Q(1, 0, 1)$, and $R(1, 1, 0)$.

We can use the normal vectors of planes to assess how planes are positioned to each other. If two planes' normal vectors are parallel, then the planes are parallel. If their normal vectors are perpendicular, then the planes themselves are perpendicular.

EX 3 Are the planes $2x - y - z = 1$ and $4x - 2y - 2z = 0$ parallel, perpendicular, or neither?

The normal vector of the first plane is $\mathbf{n}_1 = \langle 2, -1, -1\rangle$ and the normal vector of the second plane is $\mathbf{n}_2 = \langle 4, -2, -2\rangle$. Since $\mathbf{n}_2 = 2\mathbf{n}_1$, these normal vectors are parallel, and so the planes are parallel, too. There's also the possiblity that these equations both describe the *same* plane; but the second plane contains the point $(0, 0, 0)$ and the first one does not (make sure you know how we can tell!) — so they can't be the same plane. ■

You Try It

(3) Are the planes $x + 4y - 3z = 1$ and $-3x + 6y + 7z = 0$ parallel, perpendicular, or neither?

Tangent Planes

Hopefully you remember that the value of a derivative $f'(x_0)$ gives the slope of a line tangent to a function $f(x)$ at x_0, and the equation of that line could be written as:

$$y - y_0 = f'(x_0)(x - x_0)$$

It should then not be much of a surprise that the equation of the *plane* tangent to the surface $z = f(x, y)$ at (x_0, y_0, z_0) is

$$z - z_0 = f_x(x_0, y_0)(x - x_0) + f_y(x_0, y_0)(y - y_0) \tag{16.11}$$

And even if it is a surprise, it should (soon) make sense. In order to see why, I need you to believe one geometric fact: Suppose you know an infinite number of lines are contained in a single plane (that is, all the lines are *coplanar*), and you have a candidate plane that may or may not be the right one that contains them. If your candidate plane contains only *one* of the lines you know must be in the target plane, there are no guarantees that your plane contains more of the lines. But if your candidate plane contains is positioned properly to capture *two* of the lines, it has to capture *all* of them.

So, how does this tie to tangent planes? Well, we noted before that there are infinite number of lines tangent to $z = f(x, y)$ at any point (x_0, y_0). The tangent plane must contain *all* of these lines. We can show that the tangent plane given by the above surface contains at least two of them, and therefore contains all of them.

Consider the intersection of the plane $y = y_0$ with the surface $z = f(x, y)$. The candidate equation of the tangent plane (16.11) reduces there to

$$z - z_0 = f_x(x_0, y_0)(x - x_0)$$

and this is precisely the equation of the *line* tangent to the surface in the plane $y = y_0$, with slope $f_x(x_0, y_0)$. If we take the intersection of the plane $x = x_0$ with the surface, the candidate equation of the tangent plane (16.11) reduces there to

$$z - z_0 = f_y(x_0, y_0)(y - y_0)$$

which is precisely the equation of the line tangent to the surface in the plane $x = x_0$, with slope $f_y(x_0, y_0)$.

Therefore, the plane described by (16.11) contains two of the lines the actual tangent plane must contain, and so it must contain all of them. Let's make it official:

Useful Fact 16.13. *Given a surface $z = f(x, y)$ and a point on the surface (x_0, y_0, z_0) such that the first derivatives of $f(x, y)$ are continuous at (x_0, y_0, z_0), then the equation of the plane tangent to $z = f(x, y)$ at that point is:*

$$z - z_0 = f_x(x_0, y_0)(x - x_0) + f_y(x_0, y_0)(y - y_0)$$

EX 4 Find the equation of the plane tangent to $z = 4x^2 - y^2 + 2y$ where $(x, y) = (-1, 2)$.

We know $x_0 = -1$ and $y_0 = 2$, so we're still going to need z_0 and the values of f_x and f_y at this point. So,

$$\begin{aligned} z_0 = f(x_0, y_0) &= 4(-1)^2 - (2)^2 + 2(2) = 4 \\ f_x(x, y) &= 8x \to f_x(-1, 2) = -8 \\ f_y(x, y) &= -2y + 2 \to f_y(-1, 2) = -2 \end{aligned}$$

so the equation of the tangent plane is

$$\begin{aligned} z - z_0 &= f_x(x_0, y_0)(x - x_0) + f_y(x_0, y_0)(y - y_0) \\ z - 4 &= (-8)(x - (-1)) + (-2)(y - 2) \\ z - 4 &= -8x - 2y - 4 \end{aligned}$$

That last step reduces to the equation of our tangent plane, $8x+2y+z = 0$. Figure 16.18 shows the surface and point of tangency, and then Fig. 16.19 shows a small patch of the tangent plane at the point of tangency. ■

You Try It

(4) Find the the equation of the plane tangent to $z = \sqrt{4 - x^2 - 2y^2}$ at $(1, -1, 1)$.

Here is a fun fact about the equation of a tangent plane given in Useful Fact 16.13, and I'll let you stew over it for a while. We can rearrange

$$z - z_0 = f_x(x_0, y_0)(x - x_0) + f_y(x_0, y_0)(y - y_0)$$

to

$$f_x(x_0, y_0)(x - x_0) + f_y(x_0, y_0)(y - y_0) - (z - z_0) = 0$$

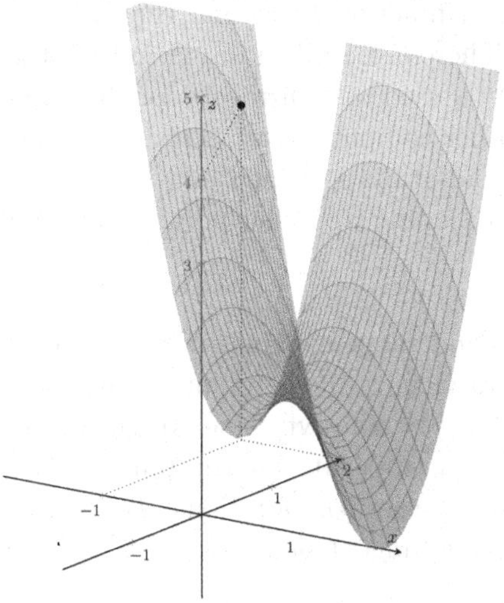

Fig. 16.18 The surface $z = 4x^2 - y^2 + 2y$ and point $(-1, 2, 4)$, with EX 4.

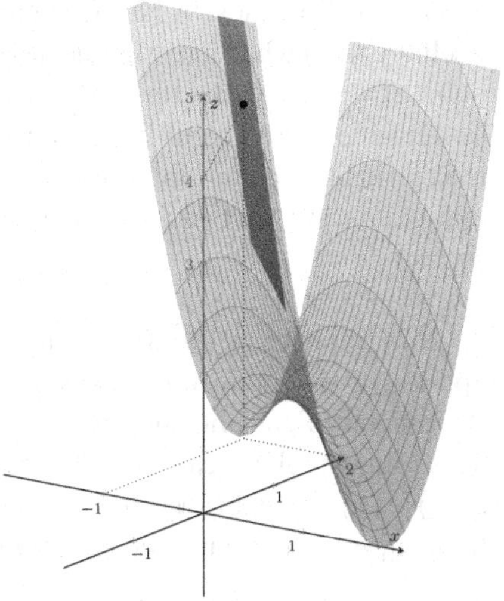

Fig. 16.19 A tangent plane on $z = 4x^2 - y^2 + 2y$, with EX 4.

The left side of this new version looks like a dot product! In fact, we can write the equation as:

$$\langle f_x(x_0, y_0), f_y(x_0, y_0), -1\rangle \cdot \langle x - x_0, y - y_0, z - z_0\rangle = 0$$

The vector on the right, $\langle x - x_0, y - y_0, z - z_0\rangle$ is a vector that's in the plane, connecting any (x, y, z) in the plane to the fixed point (x_0, y_0, z_0) we use to define the plane. Can you see anything interesting about the vector $\langle f_x(x_0, y_0), f_y(x_0, y_0), -1\rangle$ on the left? It looks sort of like a gradient. But the gradient of what? The gradient of $f(x, y)$ itself would only have *two* components ... But regardless of all that, we need to flag this bit of information for use elsewhere, because it's a Useful Fact!

Useful Fact 16.14. *Assuming the partial derivatives exist, the vectors* $\langle f_x(x_0, y_0), f_y(x_0, y_0), -1\rangle$, *and so also* $\langle -f_x(x_0, y_0), -f_y(x_0, y_0), 1\rangle$, *will be perpendicular to the surface* $z = f(x, y)$ *at the point* (x_0, y_0, z_0).

Total Differentials

Here's a quick application of tangent planes. You should recall from single variable Calculus that we can have a process called "local linear approximation". This is based on the idea that if you zoom in close enough to a point of tangency, a line tangent to a function there and the function itself are somewhat indistinguishable. Therefore, in the local vicinity of the point of tangency, the tangent line can be used as a surrogate for the function itself — often resulting in much easier computations. (This is the idea that's then exploited later for local quadratic approximation, local cubic approximation, and so forth, ultimately leading to Taylor polynomials and series.)

Now that we're discussing surfaces and tangent planes, the same idea applies: if we zoom in on the very local vicinity of the point of tangency, the tangent plane and the surface itself will be somewhat indistinguishable. The tangent plane can then be used as a surrogate for the full function $z = f(x, y)$ — again resulting, perhaps, in easier computations. To see this happen, we begin with the equation of the tangent plane,

$$z - z_0 = f_x(x_0, y_0)(x - x_0) + f_y(x_0, y_0)(y - y_0)$$

and rename $x - x_0$ to dx, $y - y_0$ to dy and $z - z_0$ to dz, to get this:

Definition 16.13. *Given a function $z = f(x, y)$ and a point (x_0, y_0) where the first derivatives of $f(x, y)$ are continuous, then the* ***total differential*** *of z at that point is defined as:*

$$dz = f_x(x_0, y_0)\, dx + f_y(x_0, y_0)\, dy$$

The total differential quantifies the idea that the *total* change of $f(x, y)$ at a given location is a combination of the change in f due to a change in x (i.e. f_x) and the change in f due to a change in y (i.e. f_y). It's no accident that we call these "partial" derivatives, since they each give a *part* of the total chance in $f(x, y)$. Total differentials have applications in error estimation and numerical computation.

We can use general expressions for f_x and f_y to build a general expression for dz before zooming in to a specific point.

EX 5 Find the total differential of $z = \ln\sqrt{x^2 + y^2}$. What is the total change dz found when (x, y) changes from $(1/\sqrt{2}, 1/\sqrt{2})$ to $(1/2, 1/2)$?

Let's rewrite the function to make the derivatives easier, it's

$$z = \ln\sqrt{x^2 + y^2} = \frac{1}{2}\ln(x^2 + y^2)$$

Then the partial derivatives are:

$$f_x(x,y) = \frac{x}{x^2+y^2} \qquad \text{and} \qquad f_y(x,y) = \frac{y}{x^2+y^2}$$

so that the total differential dz is

$$dz = \frac{x}{x^2+y^2}\,dx + \frac{y}{x^2+y^2}\,dy$$

We can now use this to investigate total change at any location. Specifically, if (x, y) changes from $(1/\sqrt{2}, 1/\sqrt{2})$ to $(1/2, 1/2)$, then dx and dy are both

$$dx = dy = \frac{1}{2} - \frac{1}{\sqrt{2}}$$

When measuring change, we generally take the first point as the anchor point, so

$$\begin{aligned} dz &= \frac{1/\sqrt{2}}{(1/\sqrt{2})^2+(1/\sqrt{2})^2}\left(\frac{1}{2}-\frac{1}{\sqrt{2}}\right) + \frac{1/\sqrt{2}}{(1/\sqrt{2})^2+(1/\sqrt{2})^2}\left(\frac{1}{2}-\frac{1}{\sqrt{2}}\right) \\ &= \frac{1}{\sqrt{2}}\left(\frac{1}{2}-\frac{1}{\sqrt{2}}\right) + \frac{1}{\sqrt{2}}\left(\frac{1}{2}-\frac{1}{\sqrt{2}}\right) \\ &= \frac{2}{\sqrt{2}}\left(\frac{1}{2}-\frac{1}{\sqrt{2}}\right) = \frac{1}{\sqrt{2}} - 1 \end{aligned}$$

The total change in z is $1/\sqrt{2} - 1$. ■

You Try It

(5) Find the total differential of $z = x^3 \ln(y^2)$. What is the total change dz found when (x, y) changes from $(1, 1)$ to $(1, \sqrt{e})$?

Planes and Tangent Planes — Problem List

Planes and Tangent Planes — You Try It

These appeared above; solutions begin on the next page.

(1) Find the equation of the plane containing the point $(1,-1,1)$ and with normal vector $\langle 1,1,-1\rangle$.
(2) Find the equation of the plane containing the points $P(0,1,1)$, $Q(1,0,1)$ and $R(1,1,0)$.
(3) Are the planes $x+4y-3z=1$ and $-3x+6y+7z=0$ parallel, perpendicular, or neither?
(4) Find the the equation of the plane tangent to $z=\sqrt{4-x^2-2y^2}$ at $(1,-1,1)$.
(5) Find the total differential of $z=x^3\ln(y^2)$. What is the total change dz found when (x,y) changes from $(1,1)$ to $(0,e)$?

Planes and Tangent Planes — Practice Problems

Try these as you get the hang of the You Try It problems. Solutions to these problems are available in Sec. B.4.5.

(1) Find the equation of the plane containing the point $(4,0,3)$ and with normal vector $\langle 0,1,2\rangle$.
(2) Find the equation of the plane containing the points $P(0,0,0)$, $Q(2,-4,6)$, and $R(5,1,3)$.
(3) Are the planes $2z=4y-x$ and $3x-12y+6z=1$ parallel, perpendicular, or neither?
(4) Find the equation of the plane tangent to $z=y\ln x$ at $(1,4,0)$.
(5) If $z=5x^2+y^2$ and (x,y) changes from $(1,2)$ to $(1.05,2.1)$, then what is the resulting dz?
(6) Find the equation of the plane containing the point $(-2,8,10)$ and perpendicular to the line $x=1+t, y=2t, z=4-3t$.
(7) Find the equation of the plane tangent to $z=e^{x^2-y^2}$ where $(x,y)=(1,-1)$.
(8) The length and width of a rectangle are measured as 30cm and 24cm respectively. There is a possible error in measurement of 0.1cm in each direction. Estimate the maximum possible error in the calculated area of the rectangle. (Hint: This is a total differential problem.)

Planes and Tangent Planes — Challenge Problems

Try these problems to test your skills with the ideas in this section. Solutions to these problems are available in Sec. C.4.5.

(1) Find the equation of the plane containing the point $(3, -5, 4)$ and perpendicular to the line $x = 1 + 2t, y = 1 - t, z = 4 - 3t$.
(2) Find the equation of the plane tangent to $z = \sqrt{x^2 - y^2}$ where $(x, y) = (2, -1)$.
(3) The radius and height of a cylinder are measured as 30cm and 24cm respectively. There is a possible error in measurement of 0.1cm in each direction. Estimate the maximum possible error in the calculated volume of the cylinder.

Planes and Tangent Planes — You Try It — Solved

(1) Find the equation of the plane containing the point $(1, -1, 1)$ and with normal vector $\langle 1, 1, -1 \rangle$.

□ We have all the information we need: a point $(x_0, y_0, z_0) = (1, -1, 1)$ and a normal vector $\mathbf{n} = \langle a, b, c \rangle = \langle 1, 1, -1 \rangle$ so using the standard form of a plane,

$$\begin{aligned} a(x - x_0) + b(y - y_0) + c(z - z_0) &= 0 \\ 1(x - 1) + 1(y + 1) + (-1)(z - 1) &= 0 \\ x + y - z &= -1 \quad \blacksquare \end{aligned}$$

(2) Find the equation of the plane containing the points $P(0, 1, 1)$, $Q(1, 0, 1)$, and $R(1, 1, 0)$.

□ We have our choice of three points on the plane, but we still need a vector perpendicular to this plane. But $\mathbf{PQ} \times \mathbf{PR}$ is such a vector:

$$\begin{aligned} \mathbf{PQ} &= \langle 1, 0 \rangle \\ \mathbf{PR} &= \langle 1, 0, -1 \rangle \\ \mathbf{PQ} \times \mathbf{PR} &= \langle 1, 1, 1 \rangle \end{aligned}$$

So the equation of the plane is (using P as the point),

$$\begin{aligned} a(x - x_0) + b(y - y_0) + c(z - z_0) &= 0 \\ 1(x - 0) + (1)(y - 1) + (1)(z - 1) &= 0 \\ x + y + z &= 2 \quad \blacksquare \end{aligned}$$

(3) Are the planes $x + 4y - 3z = 1$ and $-3x + 6y + 7z = 0$ parallel, perpendicular, or neither?

☐ From the equation of the plane $x + 4y - 3z = 1$, we know its perpendicular vector is $\mathbf{n}_1 = \langle 1, 4, -3 \rangle$. From the equation of the plane $-3x + 6y + 7z = 0$, we know its perpendicular vector is $\mathbf{n}_2 = \langle -3, 6, 7 \rangle$. Since $\mathbf{n}_1 \cdot \mathbf{n}_2 = 0$, these vectors, and so also the planes, are perpendicular to each other. ■

(4) Find the the equation of the plane tangent to $z = \sqrt{4 - x^2 - 2y^2}$ at $(1, -1, 1)$.

☐ We need

$$f_x = -\frac{x}{\sqrt{4 - x^2 - 2y^2}} \to f_x(1, -1) = -\frac{1}{\sqrt{4 - 1 - 2}} = -1$$

$$f_y = -\frac{2y}{\sqrt{4 - x^2 - 2y^2}} \to f_y(1, -1) = \frac{2}{\sqrt{4 - 1 - 2}} = 2$$

so that the plane is:

$$z - z_0 = f_x(x_0, y_0)(x - x_0) + f_y(x_0, y_0)(y - y_0)$$

$$z - 1 = -1(x - 1) + 2(y + 1)$$

$$x - 2y + z = 4 \quad ■$$

(5) Find the total differential of $z = x^3 \ln(y^2)$. What is the total change dz found when (x, y) changes from $(1, 1)$ to $(0, e)$?

☐ The total differential is:

$$dz = f_x(x, y)dx + f_y(x, y)dy = 3x^2 \ln(y^2)\, dx + \frac{2x^3}{y}\, dy$$

When we move from $(1, 1)$ to $(0, e)$, then we have $dx = -1$ and $dy = e - 1$. Using $(x, y) = (1, 1)$ as the reference point,

$$dz = 3(1)^2 \ln(1)^2 (-1) + \frac{2(1)^3}{1} (e - 1) = 0 + 2(e - 1) = 2(e - 1) \quad ■$$

16.6 Directional Derivatives

Introduction

Did you think you were done with derivatives? Hardly! You were only done with derivatives that didn't require the use of vectors. Regular partial derivatives are fun, but somewhat inflexible in terms of the information they give by themselves. The partial derivative of $f(x, y)$ with respect to x quantifies how $f(x, y)$ changes with x assuming y is constant; in other words, it looks at the rate of change of $f(x, y)$ only in a direction parallel to the x-axis. Similarly, f_y examines the rate of change of $f(x, y)$ only in a direction parallel to the y-axis. This is like going on a hike and having a map that only shows you the terrain looking due east and due north. What if you wanted to walk northeast?

Directional Derivatives

The definitions of the partial derivative of $f(x, y)$ with respect to x and y were:

$$\lim_{h\to 0} \frac{f(x+h, y) - f(x, y)}{h} \quad \text{and} \quad \lim_{h\to 0} \frac{f(x, y+h) - f(x, y)}{h} \tag{16.12}$$

(Do you remember how they work?) Now, here is a limit structure that allows us to observe the same trends, but not just in the coordinate directions:

$$\lim_{h\to 0} \frac{f(x_0 + ha, y_0 + hb) - f(x_0, y_0)}{h} \tag{16.13}$$

Exploiting the equivalence between vectors and points that we have when we assume the initial point of all our vectors is the origin, and also abusing our functional notation a bit, we can rewrite this as

$$\lim_{h\to 0} \frac{f(\langle x_0, y_0\rangle + h\langle a, b\rangle) - f(\langle x_0, y_0\rangle)}{h} \tag{16.14}$$

Take a close look at the argument in the left hand term of the numerator. The vector $\langle x_0, y_0\rangle + h\langle a, b\rangle$ is a vector offset from $\langle x_0, y_0\rangle$ in a direction parallel to some fixed vector $\langle a, b\rangle$ by a distance scaled with h — see Fig. 16.20. In partial derivatives, we have either $\langle a, b\rangle = \langle 1, 0\rangle$ or $\langle a, b\rangle = \langle 0, 1\rangle$, whereas now we allow $\langle a, b\rangle$ to be any vector. And as h goes to 0, the vector $\langle x_0, y_0\rangle + h\langle a, b\rangle$ creeps back towards $\langle x_0, y_0\rangle$ itself. What this new and improved difference quotient does, with the limit, is looks at the rate of change in f at (x_0, y_0) in the direction of the vector $\langle a, b\rangle$.

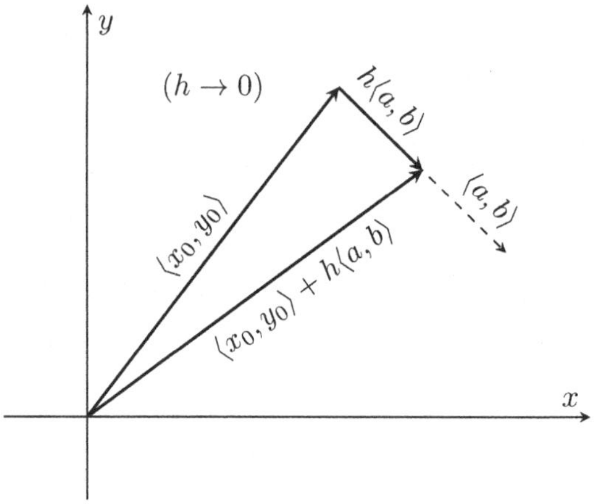

Fig. 16.20 Building blocks of a directional derivative.

It's important to keep in mind that all this action regarding location and direction is taking place in the xy-plane; as we stand at (x_0, y_0) in the xy-plane and then face in the direction of the vector $\langle a, b\rangle$, we then look up (or down) to the surface $f(x, y)$ to see how "steep" the surface is in that direction. This improvement in the difference quotient frees up our prior derivative definitions from being restricted to directions parallel to the x- or y-axes. While the "vectorized" version of this limit process, shown as (16.14), has been useful for visualizing what's happening, the formalizing of this process requires us to go back to (16.13):

Definition 16.14. *The **directional derivative** of $f(x, y)$ at (x_0, y_0) in the direction of the vector $\mathbf{u} = \langle a, b\rangle$ is denoted $D_{\mathbf{u}}f(x_0, y_0)$, and is defined as*

$$D_{\mathbf{u}}f(x_0, y_0) = \lim_{h\to 0} \frac{f(x_0 + ha, y_0 + hb) - f(x_0, y_0)}{h} \tag{16.15}$$

NOTE: In order for the scaling by h in the difference quotient to work properly, the vector $\mathbf{u}$ must be a unit vector.

We need to convert Eq. (16.15) into a usable expression. After all, we don't want to actually compute our directional derivatives with this formula, any more than we wanted to compute partial derivatives by their technical definitions in (16.12). The details of the derivation of a better expression for $D_{\mathbf{u}}$ are pushed to the end of this section so that we can get to work right away. But I definitely recommend spending some time with that derivation because your ability to hang in there and follow it is a good check on how much you've retained from several prior concepts. Here is the payoff:

Useful Fact 16.15. *The directional derivative of $f(x, y)$ at (x_0, y_0) in the direction of the unit vector $\mathbf{u} = \langle a, b\rangle$ can be computed as*

$$D_{\mathbf{u}} f(x_0, y_0) = \nabla f(x_0, y_0) \cdot \langle a, b\rangle \tag{16.16}$$

(assuming that the first derivatives of f exist).

Useful Fact 16.15 tells us that in order to determine the directional derivative of a function f at some point, we must

- Find a unit vector that describes the direction of interest.
- Compute the gradient of the function at the point of interest.
- Compute the dot product of the gradient with the (unit) direction vector.

EX 1 Find the derivative of $f(x, y) = x^2y^2 - xy$ at $(x, y) = (1, 1)$ in the direction given by $\langle 1, 1\rangle$.

First, note that the direction vector is not a unit vector, so let's fix that to get

$$\mathbf{u} = \frac{\langle 1, 1\rangle}{|\langle 1, 1\rangle|} = \frac{\langle 1, 1\rangle}{\sqrt{2}} = \left\langle \frac{1}{\sqrt{2}}, \frac{1}{\sqrt{2}} \right\rangle$$

Then, let's find the gradient of f at the point $(1, 1)$:

$$\begin{aligned}\nabla f(x, y) &= \langle f_x, f_y\rangle \\ &= \langle 2xy^2 - y, 2x^2y - x\rangle \\ \nabla f(1, 1) &= \langle 2 - 1, 2 - 1\rangle = \langle 1, 1\rangle\end{aligned}$$

and therefore

$$D_{\mathbf{u}} f(1, 1) = \nabla f(1, 1) \cdot \mathbf{u} = \langle 1, 1\rangle \cdot \left\langle \frac{1}{\sqrt{2}}, \frac{1}{\sqrt{2}} \right\rangle = \frac{2}{\sqrt{2}} = \sqrt{2} \quad \blacksquare$$

You Try It

(1) Find the directional derivative of $f(x,y) = 5xy^2 - 4x^3y$ at (1,2) in the direction of $\mathbf{u} = \langle 5/13, 12/13 \rangle$.
(2) Find the directional derivative of $g(s,t) = s^2e^t$ at (2,0) in the direction of $\mathbf{v} = \mathbf{i} + \mathbf{j} = \langle 1, 1 \rangle$.

Sometimes the direction of interest can be given more cryptically, by specification of a polar angle θ. But of course you know that a unit vector $(r = 1)$ in the direction of a polar angle θ would be $\mathbf{u} = \langle \cos\theta, \sin\theta \rangle$.

EX 2 Find the derivative of $f(x,y) = 1 + 2x\sqrt{y}$ at $(3,4)$ in the direction given by $\theta = \pi/3$.

A unit vector pointing in the direction of the angle $\theta = \pi/3$ is

$$\mathbf{u} = \left\langle \cos\frac{\pi}{3}, \sin\frac{\pi}{3} \right\rangle = \left\langle \frac{1}{2}, \frac{\sqrt{3}}{2} \right\rangle$$

Since the gradient of $f(x,y)$ is $\nabla f = \langle 2/\sqrt{y}, x/\sqrt{y} \rangle$, then $\nabla f(3,4) = \langle 4, 3/2 \rangle$ and

$$D_{\mathbf{u}}f(3,4) = \nabla f(3,4) \cdot \mathbf{u} = \left\langle 4, \frac{3}{2} \right\rangle \cdot \left\langle \frac{1}{2}, \frac{\sqrt{3}}{2} \right\rangle = 2 + \frac{3\sqrt{3}}{4} \quad \blacksquare$$

You Try It

(3) Find the directional derivative of $f(x,y) = \sqrt{5x - 4y}$ at (4,1) in the direction given by $\theta = -\pi/6$.

The Direction of Steepest Ascent

The question in hand so far has been: given a function $f(x,y)$, a location (x_0, y_0) and a predetermined direction $\mathbf{u}$, what is the rate of change of f at that point in the given direction? We could turn this question around and ask something a bit different, and often more important: If we're at a location given by (x_0, y_0) on a surface $f(x,y)$, in what direction should we look to see the *largest* rate of change possible from that point? This is called the *direction of steepest ascent.* When you're hiking up a mountainside, you probably do not want to walk in the direction of steepest ascent, so you may want to find and avoid the direction of steepest ascent from your location. The *direction of steepest descent* is exactly the opposite of the direction of steepest ascent. If we are standing at (x_0, y_0) and set a ball on the ground, it will roll in the direction of steepest descent. Finding these directions is important in many physical problems. Fluid or air flows from high pressure to low, following the direction of steepest descent within a pressure field. The following is stated without proof, you'll just have to trust me!

Useful Fact 16.16. *The derivatives of $f(x, y)$ behave themselves at a point (x_0, y_0), then...*

- *The maximum value of the directional derivative $D_{\mathbf{u}}f(x, y)$ at $(x_0, y_0$ is $|\nabla f(x_0, y_0)|$ and it occurs in the direction of $\nabla f(x_0, y_0)$.*
- *The minimum value of the directional derivative $D_{\mathbf{u}}f(x, y)$ at $(x_0, y_0$ is $-|\nabla f(x_0, y_0)|$ and it occurs in the direction $-\nabla f(x_0, y_0)$.*

In other words, the direction of steepest ascent at any location is the direction indicated by the gradient vector; the direction of steepest descent is therefore opposite the gradient. The size of the maximum and minimum rates of change in these directions are the positive and negative magnitudes of the gradient vector.

In Useful Fact 16.16, don't let the phrase "minimum" fool you. Often this makes us think of a number that's smallest, but positive. But it really means "farthest to the left on the number line". The minimum of the numbers $2, 6, -5, 7, 11, -3.7$ is -5. So we usually expect a minimum rate of change (i.e. a minimum value of a directional derivative) to be negative. The only time a minimum directional derivative would not be negative is if it's zero — such as at the vertex of an upwards paraboloid.

EX 3 Find the minimum rate of change of $f(x, y) = x^2y - y^2x$ at the point $(1, -1)$, and the direction in which this rate of change occurs.

We are looking for the direction of steepest descent (since we want the minimum rate of change) and the magnitude of that steepest descent. Planning to use Useful Fact 16.16, we build up some data. First, we have $f_x = 2xy - y^2$ and $f_y = x^2 - 2xy$, so that $f_x(1, -1) = -3$ and $f_y(1, -1) = 3$. And then,

$$\nabla f(1, -1) = \langle -3, 3 \rangle \quad ; \quad |\nabla f(-3, 3)| = \sqrt{18} = 3\sqrt{2}$$

So, the minimum rate of change of f at $(1, -1)$ is then $-|\nabla f(1, -1)| = -3\sqrt{2}$, and this minimum rate occurs in the direction opposite the gradient, i.e. $-\nabla f(1, -1) = \langle 3, -3 \rangle$. ■

You Try It

(4) Find the maximum rate of change of $f(x, y) = y^2/x$ at the point $(2, 4)$, and the direction in which this rate of change occurs.

All the examples provided here have used a function of two variables, $f(x, y)$. However, all the concepts can extend to $f(x, y, z)$ or even more variables. You just do the same computations, but with more components.

Into the Pit!!

Into the Pit! The Details Behind Equation (16.16)

Let's dig into Eq. (16.15),

$$D_{\mathbf{u}}f(x_0, y_0) = \lim_{h\to 0} \frac{f(x_0 + ha, y_0 + hb) - f(x_0, y_0)}{h}$$

and see how it turns into Equation (16.16).

We can consider the expression $f(x_0 + ha, y_0 + hb)$ (in the numerator) as a function of h: different values of h lead to different results right? So let's name it $g(h) = f(x_0 + ha, y_0 + hb)$. As an example of this naming, we have $g(0) = f(x_0, y_0)$. Having installed that name, we can tidy the limit expression in the definition of $D_{\mathbf{u}}f(x_0, y_0)$ as

$$D_{\mathbf{u}}f(x_0, y_0) = \lim_{h\to 0} \frac{g(h) - g(0)}{h}$$

You should recognize the right side of this expression as the limit definition of the plain old derivative $g'(0)$. So in one sense, our directional derivative is equivalent to $g'(0)$:

$$g'(0) = D_{\mathbf{u}}f(x_0, y_0) \tag{16.17}$$

Hold that thought.

Now step back and reconsider the new naming situation we have going on. We've created a function $g(h)$ for which $g(h) = f(x, y)$ for $x = x_0 + ha$ and $y = y_0 + hb$ — that is, g uses x and y as intermediate variables: g depends on x and y, but then x and y depend on h. Does this overall scenario sound familiar? This is a *chain rule* situation! And so if we ask what $g'(0)$ looks like from that perspective, we get:

$$g'(h) = \frac{\partial f}{\partial x}\frac{\partial x}{\partial h} + \frac{\partial f}{\partial y}\frac{\partial y}{\partial h} = f_x(x, y)a + f_y(x, y)b$$

and so when $h = 0$,

$$g'(0) = f_x(x_0, y_0)a + f_y(x_0, y_0)b \tag{16.18}$$

Now we have two versions of $g'(0)$, in (16.17) and (16.18):

$$g'(0) = D_{\mathbf{u}}f(x_0, y_0)$$
$$g'(0) = f_x(x_0, y_0)a + f_y(x_0, y_0)b$$

Since both expressions give $g'(0)$, they must be equal:

$$D_{\mathbf{u}}f(x_0, y_0) = f_x(x_0, y_0)a + f_y(x_0, y_0)b$$

Do you notice anything interesting about that right hand side? Like the fact that it looks an awful lot like a dot product? Let's write it like this:

$$D_{\mathbf{u}}f(x_0, y_0) = f_x(x_0, y_0)a + f_y(x_0, y_0)b = \langle f_x(x_0, y_0), f_y(x_0, y_0)\rangle \cdot \langle a, b\rangle \tag{16.19}$$

We're almost done! Remember that $\langle f_x(x_0, y_0), f_y(x_0, y_0)\rangle$ is just the gradient of $f(x, y)$ at $(x_0, y_0$, and $\langle a, b\rangle$ is the *unit* direction vector $\mathbf{u}$. So now, we can package up (16.19) as

$$D_{\mathbf{u}}f(x_0, y_0) = \nabla f(x_0, y_0) \cdot \langle a, b\rangle$$

and this is (16.16).

Directional Derivatives — Problem List

Directional Derivatives — You Try It

These appeared above; solutions begin on the next page.

(1) Find the directional derivative of $f(x, y) = 5xy^2 - 4x^3y$ at (1,2) in the direction of $\mathbf{u} = \langle 5/13, 12/13 \rangle$.
(2) Find the directional derivative of $g(s, t) = s^2e^t$ at (2,0) in the direction of $\mathbf{v} = \mathbf{i} + \mathbf{j} = \langle 1, 1 \rangle$.
(3) Find the directional derivative of $f(x, y) = \sqrt{5x - 4y}$ at (4,1) in the direction given by $\theta = -\pi/6$.
(4) Find the maximum rate of change of $f(x, y) = y^2/x$ at the point $(2, 4)$, and the direction in which this rate of change occurs.

Directional Derivatives — Practice Problems

Try these as you get the hang of the You Try It problems. Solutions to these problems are available in Sec. B.4.6.

(1) Find the directional derivative of $f(x, y) = y \ln x$ at (1,-3) in the direction of $\mathbf{u} = \langle -4/5, 3/5 \rangle$.
(2) Find the directional derivative of $g(x, y) = e^{-x} \sin y$ at $(0, \pi/3)$ in the direction of $\mathbf{v} = \langle 3, -2 \rangle$.
(3) Find the the directional derivative of $f(x, y) = x^2y^3 - y^4$ at (2,1) in the direction given by $\theta = \pi/4$.
(4) Find the maximum rate of change of $f(p, q) = qe^{-p} + pe^{-q}$ at (0,0), and the direction in which this rate of change occurs. (Assume p, q are renamed rectangular coordinates.)
(5) Find the directional derivative of $f(x, y, z) = \sqrt{x + yz}$ at (1,3,1) in the direction of $\mathbf{u} = \langle 2/7, 3/7, 6/7 \rangle$.
(6) Find the directional derivative of $f(x, y, z) = \dfrac{x}{y + z}$ at (4,1,1) in the direction of $\mathbf{v} = \langle 1, 2, 3 \rangle$.
(7) Given the 3D temperature function $T(x, y, z) = 20e^{-x^2-y^2-2z^2}$, what is the maximum rate of change of the temperature at the point $(2, -1, 2)$, and in what direction does this rate of change occur?

Directional Derivatives — Challenge Problems

Try these problems to test your skills with the ideas in this section. Solutions to these problems are available in Sec. C.4.6.

(1) Find the directional derivative of $f(x, y, z) = \sqrt{xy + z}$ at (1,3,1) in the direction of $\mathbf{w} = \langle 3, 2, 6 \rangle$.
(2) Find the directional derivative of $f(x, y, z) = y/(x + z)$ at the point $P = (1, 4, 1)$ in the direction of $\mathbf{v} = \langle 1, 2, 1 \rangle$. If S is the surface represented by the graph of f, and a bug standing at P started walking on S in the direction of $\mathbf{v}$, would the bug be walking uphill or downhill?
(3) Given the 3D temperature function $T(x, y, z) = 5e^{-x^2-y^2-2z^2}$, what is the (simplified) maximum rate of change of the temperature at the point $(1/\sqrt{2}, 0, 1/\sqrt{2})$, and in what direction does this rate of change occur?

Directional Derivatives — You Try It — Solved

---

(1) Find the directional derivative of $f(x, y) = 5xy^2 - 4x^3y$ at (1,2) in the direction of $\mathbf{u} = \langle 5/13, 12/13 \rangle$.

□ The direction vector $\mathbf{u}$ is already a unit vector, so computing the gradient and then bringing in the direction vector,

$$\nabla f(x, y) = \langle f_x, f_y \rangle = \langle 5y^2 - 12x^2y, 10xy - 4x^3 \rangle$$

$$\nabla f(1, 2) = \langle -4, 16 \rangle$$

$$D_u f(1, 2) = \nabla f(1, 2) \cdot \mathbf{u} = \langle -4, 16 \rangle \cdot \left\langle \frac{5}{13}, \frac{12}{13} \right\rangle = \frac{172}{13} \quad \blacksquare$$

---

(2) Find the directional derivative of $g(s, t) = s^2e^t$ at (2,0) in the direction of $\mathbf{v} = \mathbf{i} + \mathbf{j} = \langle 1, 1 \rangle$.

□ The given direction vector is not a unit vector, so let's fix that. A unit vector $\mathbf{u}$ in the direction of $\mathbf{v}$ can be found this way:

$$\mathbf{u} = \frac{\mathbf{v}}{|\mathbf{v}|} = \left\langle \frac{1}{\sqrt{2}}, \frac{1}{\sqrt{2}} \right\rangle$$

And then, computing the gradient and using the direction vector:

$$\nabla g(s,t) = \langle g_s, g_t \rangle = \langle 2se^t, s^2e^t \rangle$$
$$\nabla g(2,0) = \langle 4,4 \rangle$$
$$D_u g(2,0) = \nabla g(2,0) \cdot \mathbf{u} = \langle 4,4 \rangle \cdot \left\langle \frac{1}{\sqrt{2}}, \frac{1}{\sqrt{2}} \right\rangle = 4\sqrt{2} \quad \blacksquare$$

---

(3) Find the directional derivative of $f(x,y) = \sqrt{5x-4y}$ at (4,1) in the direction given by $\theta = -\pi/6$.

☐ We need a unit vector in the given direction; this is

$$\mathbf{u} = \left\langle \cos\left(-\frac{\pi}{6}\right), \sin\left(-\frac{\pi}{6}\right) \right\rangle = \left\langle \frac{\sqrt{3}}{2}, -\frac{1}{2} \right\rangle$$

So, computing the gradient and then bringing in the direction vector,

$$\nabla f(x,y) = \langle f_x, f_y \rangle = \left\langle \frac{5}{2\sqrt{5x-4y}}, -\frac{2}{\sqrt{5x-4y}} \right\rangle$$
$$\nabla f(4,1) = \left\langle \frac{5}{8}, -\frac{1}{2} \right\rangle$$
$$D_u f(4,1) = \nabla f(4,1) \cdot \mathbf{u} = \left\langle \frac{5}{8}, -\frac{1}{2} \right\rangle \cdot \left\langle \frac{\sqrt{3}}{2}, -\frac{1}{2} \right\rangle$$
$$= \frac{5\sqrt{3}}{16} + \frac{1}{4} \quad \blacksquare$$

---

(4) Find the maximum rate of change of $f(x,y) = y^2/x$ at the point $(2,4)$, and the direction in which this rate of change occurs.

☐ We can easily get $f_x = -y^2/x^2$ and $f_y = 2y/x$, so that $f_x(2,4) = -4$ and $f_y(2,4) = 4$. Then $\nabla f(2,4) = \langle -4, 4 \rangle$ and $|\nabla f(2,4)| = 4\sqrt{2}$.

So, the maximum rate of change of f at (2,4) is then $|\nabla f(2,4)| = 4\sqrt{2}$ and this maximum rate occurs in the direction of the gradient, $\langle -4, 4 \rangle$ — which is also the direction of $\langle -1, 1 \rangle$. ■

Chapter 17

We've Been Framed!

17.1 Arc Length Parameterization

Introduction

Vector functions are useful for conceptualizing many physical processes, such as tracking the path of a moving object. We usually visualize vector functions as curves, or contours, in two or three-dimensional space. Finding the arc length of (a portion of) a curve presented as a vector function is a natural calculation to do. Fortunately, we've already laid the groundwork for this calculation.

Arc Length

In (Volume 1) Sec. 12.2, we saw that the arc length of a parametric curve with equations $x = x(t)$, $y = y(t)$, and $z = z(t)$ on an interval $a \leq t \leq b$ is given by

$$L = \int_a^b \sqrt{[x'(t)]^2 + [y'(t)]^2 + [z'(t)]^2} dt \tag{17.1}$$

Now that we've become smarter about such quantities, we can recognize the integrand as the magnitude of $\mathbf{r}'(t)$ for the vector function $\mathbf{r}(t) = \langle x(t), y(t), z(t) \rangle$; so, a new and improved definition is:

Definition 17.1. *The length of a piecewise continuous vector function* $\mathbf{r}(t)$ *on* $a \leq t \leq b$ *is given by*

$$L = \int_a^b |\mathbf{r}'(t)| dt \tag{17.2}$$

We can put this to use right away.

EX 1 Find the arc length of $\mathbf{r}(t) = \langle 2t, 1 - 3t, 5 + 4t \rangle$ from $(0, 1, 5)$ to $(4, -5, 13)$.

First, note that $(0, 1, 5)$ is given by $t = 0$ and $(4, -5, 13)$ is given by $t = 2$. To use (17.2), we must build the magnitude of $\mathbf{r}'(t)$ — but that's easy in this case. From $\mathbf{r}(t) = \langle 2t, 1 - 3t, 5 + 4t \rangle$ we get $\mathbf{r}'(t) = \langle 2, -3, 4 \rangle$, so that

$$|\mathbf{r}'(t)| = \sqrt{2^2 + (-3)^2 + 4^2} = \sqrt{29}$$

By (17.2),

$$L = \int_a^b |\mathbf{r}'(t)|\, dt = \int_0^2 \sqrt{29}\, dt = \sqrt{29}\, t \Big|_0^2 = 2\sqrt{29}$$

If you happened to notice that $\mathbf{r}(t)$ is just a simple line segment, then it's no surprise that this arc length is just the straight-line distance between the two given points,

$$\sqrt{(4-0)^2 + (-5-1)^2 + (13-5)^2} = \sqrt{116} = 2\sqrt{29} \quad \blacksquare$$

EX 2 Find the arc length of $\mathbf{r}(t) = \langle 2\sin t, 2\cos t, t + 1 \rangle$ from $t = 0$ to $t = 2\pi$.

For this vector function, $\mathbf{r}'(t) = \langle 2\cos t, -2\sin t, 1 \rangle$, so

$$|\mathbf{r}'(t)| = \sqrt{4\cos^2 t + 4\sin^2 t + 1} = \sqrt{5}$$

And

$$L = \int_a^b |\mathbf{r}'(t)|\, dt = \int_0^{2\pi} \sqrt{5}\, dt = 2\pi\sqrt{5}$$

Keep tabs on this EX 2 as we continue below, because it will be referenced as an on-going example. $\blacksquare$

You Try It

(1) Find the arc length of of $\mathbf{r}(t) = \langle t^2/2, t^3/3 \rangle$ for $0 \le t \le 4$.
(2) Find the arc length of $\mathbf{r}(t) = \langle \sin(2t), 2t, \cos(2t) \rangle$ for $0 \le t \le \pi$.

Arc Length Parameterization

Since $\mathbf{r}(t)$ commonly represents the path of a moving object, then if t represents time, the parameter has an intuitive meaning. But this parameter t (time?) we've been using in vector functions is not the only one that is available. In fact, as we know from our experience crafting lines in three dimensions, there is an unlimited supply of parameters and corresponding ranges that can produce the same function or curve.

In EX 2 above, we saw a vector function $\mathbf{r}(t) = \langle 2\sin t, 2\cos t, t+1\rangle$ and learned that at the location corresponding to $t = 2\pi$, the total arc length covered from $t = 0$ was $2\pi\sqrt{5}$. If we repeated similar calculations, we could have learned that at $t = 6$ we would have achieved a total arc length of $6\sqrt{5}$; at $t = 100$, we would have an arc length of $100\sqrt{5}$. In general, for any interval $[0, t]$, we would have an arc length of $t\sqrt{5}$. That is, we have a relation between the parameter t and the arc length. We will now introduce s to represent arc length (don't ask why s means arc length, just go with it). For this particular vector function, we have determined that the total arc length on some interval $[0, t]$ can be found by $s = t\sqrt{5}$.

Now, it's great to be able to calculate arc length for any given value of t, but wouldn't it also be neat to specify an arc length and be handed the point at which we reach that particular arc length? That is the job of *arc length parameterization*; if we correctly swap the original parameter t for a new "arc length parameter" s, the new version of the vector function will allow us to do just that.

In our continuing example at hand, since we have $s = t\sqrt{5}$, then we also have $t = s/\sqrt{5}$; if we substitute that into $\mathbf{r}(t) = \langle 2\sin t, 2\cos t, t+1\rangle$, we get a newly parameterized version of the same vector function:

$$\mathbf{r}(s) = \left\langle 2\sin\frac{s}{\sqrt{5}}, 2\cos\frac{s}{\sqrt{5}}, \frac{s}{\sqrt{5}} + 1\right\rangle$$

Now with this new version of the vector function, let's calculate the total arc length achieved at the point marked by $s = 2\pi\sqrt{5}$. First, we need to see the derivative of the vector function:

$$\mathbf{r}'(s) = \left\langle \frac{2}{\sqrt{5}}\cos\frac{s}{\sqrt{5}}, -\frac{2}{\sqrt{5}}\sin\frac{s}{\sqrt{5}}, \frac{1}{\sqrt{5}}\right\rangle$$

Then, we compute the magnitude of this new (derivative) vector function,

$$|\mathbf{r}'(s)| = \sqrt{\frac{4}{5}\cos^2\frac{s}{\sqrt{5}} + \frac{4}{5}\sin^2\frac{s}{\sqrt{5}} + \frac{1}{5}} = 1 \tag{17.3}$$

On the interval $[0, 2\pi\sqrt{5}]$ for s, then,

$$L = \int_a^b |\mathbf{r}'(s)|\, ds = \int_0^{2\pi\sqrt{5}} (1)\, ds = 2\pi\sqrt{5} \tag{17.4}$$

So at the point located by the arc length parameter $s = 2\pi\sqrt{5}$, we have achieved an arc length of $2\pi\sqrt{5}$. If we repeat a similar calculation for $s = 1$, we can find that the total arc length achieved at the point located by $s = 1$ is

$$L = \int_a^b |\mathbf{r}'(s)|\, ds = \int_0^1 (1)\, ds = 1$$

The pattern here (presuming we start at $s = 0$) is that when we move along the curve to a point marked by a particular value of the parameter s, the arc length traveled to get to that point is also s. This is why s is called the **arc length parameter**.

The natural question is, then, if we start with a vector curve designed with a generic parameter t, can we "reparameterize" the function and write it in terms of the arc length parameter s? And the answer is ... sometimes! The relation between t and s may or may not be helpful for doing so. But here's how we can at least try to do it: We go to the original integral formula for arc length (17.2) and say, Hey formula, what is the arc length achieved over *any* parameter interval $[0, t]$? The way to do that is prepare this integral:

$$s = \int_0^t |\mathbf{r}'(\tau)|\, d\tau \tag{17.5}$$

Note, (1) if we install t as the upper endpoint of integration, it cannot also remain the variable of integration, and so the latter is changed to τ. Also (2), just as a flashback, this puts us in Fundamental Theorem of Calculus territory, where we have a function defined as an integral, such that the "live" variable in the overall function is the upper endpoint of integration — that is, Eq. (17.5) presents a function $s(t)$.

For the ongoing example spawned from EX 2, we'd have

$$s = \int_0^t \sqrt{5}\, d\tau = t\sqrt{5}$$

which is exactly the relation between s and t that we discovered in a more informal way already.

In general, to introduce an arc length parameter to a vector function, we do the following;

(1) Find the relation between the original parameter t and the new arc length parameter s; that is, find $s = s(t)$ via

$$s = \int_0^t |\mathbf{r}'(\tau)|\, d\tau$$

(2) Having developed a relationship $s = s(t)$, turn it inside-out to find t written in terms of s; that is, find $t = t(s)$.
(3) Substitute $t = t(s)$ into the vector function to change $\mathbf{r}(t)$ into $\mathbf{r}(s)$, and also convert the endpoints of any specific interval given from bounding values of t to bounding values for s.

EX 3 Revisit the vector valued function from EX 1 and find its arc length parameterized representation.

EX 1 presented $\mathbf{r}(t) = \langle 2t, 1 - 3t, 5 + 4t \rangle$; the arc length on an interval $[0, t]$ is given by

$$s = \int_0^t \sqrt{(2)^2 + (-3)^2 + (4)^2}\, d\tau = \sqrt{29} \int_0^t d\tau = t\sqrt{29}$$

We solve $s = t/\sqrt{29}$ for t as $t = s/\sqrt{29}$, and substitute this back into the original vector function:

$$\begin{aligned} \mathbf{r}(s) &= \langle 2 \cdot \frac{s}{\sqrt{29}}, 1 - 3 \cdot \frac{s}{\sqrt{29}}, 5 + 4 \cdot \frac{s}{\sqrt{29}} \rangle \\ &= \langle \frac{2s}{\sqrt{29}}, 1 - \frac{3s}{\sqrt{29}}, 5 + \frac{4s}{\sqrt{29}} \rangle \quad \blacksquare \end{aligned}$$

EX 4 Find an arc length parameterization of $\mathbf{r}(t) = \langle \cos t, \sin t, 2t^{3/2}/3 \rangle$ for $t \geq 0$, and determine the location at which the curve accumulates a total arc length of $s = 14/3$.

For this vector function, we have

$$\mathbf{r}'(t) = \left\langle -\sin t, \cos t, \sqrt{t} \right\rangle$$
$$|\mathbf{r}'(t)| = \sqrt{\sin^2 t + \cos^2 t + t} = \sqrt{1+t}$$

On the interval $[0, t]$, then, we find the general relation between s and t via

$$s = \int_0^t |\mathbf{r}'(\tau)|\, d\tau = \int_0^t \sqrt{1+\tau}\, d\tau = \frac{2}{3}(1+\tau)^{3/2}\Big|_0^t = \frac{2}{3}\left((1+t)^{3/2} - 1\right)$$

Taking

$$s = \frac{2}{3}\left((1+t)^{3/2} - 1\right)$$

and solving for t,

$$(1+t)^{3/2} = \frac{3}{2}s + 1 \quad \Rightarrow \quad t = \left(\frac{3}{2}s + 1\right)^{2/3} - 1$$

Substituting this expression for t back into the original vector function $\mathbf{r}(t) = \langle \cos t, \sin t, 2t^{3/2}/3 \rangle$, we get a very large expression of the form $\mathbf{r}(s) = \langle x(s), y(s), z(s) \rangle$, where

$$x(s) = \cos\left(\left(\frac{3}{2}s + 1\right)^{2/3} - 1\right) \tag{17.6}$$

$$y(s) = \sin\left(\left(\frac{3}{2}s + 1\right)^{2/3} - 1\right) \tag{17.7}$$

$$z(s) = \frac{2}{3}\left[\left(\frac{3}{2}s + 1\right)^{2/3} - 1\right]^{3/2} \tag{17.8}$$

The appearance of $\mathbf{r}(s)$ is terrible! Sadly, the suite of vector functions which have pleasant arc length parameter forms is quite limited. In fact, the suite of vector functions for which it's even *possible* to extract a closed form arc length parameterization is relatively small.

The utility of the arc length parameter version is, again, that in order to find where we reach an arc length of, say, $s = 14/3$, we just plug in $s = 14/3$ to the component functions of (17.6) to get $\mathbf{r}(14/3) = \langle x(14/3), y(14/3), z(14/3)\rangle$, where

$$x\left(\frac{14}{3}\right) = \cos\left(\left(\frac{3}{2}\cdot\frac{14}{3}+1\right)^{2/3} - 1\right) = \cos\left((8)^{2/3} - 1\right) = \cos(3)$$

$$y\left(\frac{14}{3}\right) = \sin\left(\left(\frac{3}{2}\cdot\frac{14}{3}+1\right)^{2/3} - 1\right) = \sin\left((8)^{2/3} - 1\right) = \sin(3)$$

$$\begin{aligned} z\left(\frac{14}{3}\right) &= \frac{2}{3}\left[\left(\frac{3}{2}\cdot\frac{14}{3}+1\right)^{2/3} - 1\right]^{3/2} \\ &= \frac{2}{3}\left[(8)^{2/3} - 1\right]^{3/2} = \frac{2}{3}(3)^{3/2} = 2\sqrt{3} \end{aligned}$$

So, assuming we started measuring arc length from $s = 0$ (i.e. the point (1,0,0)), the curve achieves an arc length of $s = 14/3$ at the terminal point of the vector $\langle\cos(3), \sin(3), 2\sqrt{3}\rangle$, i.e. the point $(\cos(3), \sin(3), 2\sqrt{3})$. ■

You Try It

(3) Find an arc length parameterization of the vector function in You Try It 2, for $t \geq 0$, , and determine the location at which the curve accumulates a total arc length of $s = 1$.

Here is a messy, but nifty, fact that also makes arc length parameterization special. An evolving story that began in EX 2 has shown us that the vector function $\mathbf{r}(t) = \langle 2\sin t, 2\cos t, t+1\rangle$ can be reparameterized according to an arc length parameter as

$$\mathbf{r}(s) = \left\langle 2\sin\frac{s}{\sqrt{5}}, 2\cos\frac{s}{\sqrt{5}}, \frac{s}{\sqrt{5}} + 1\right\rangle$$

Back in Eq. (17.3), we happened to find that for this function,

$$|\mathbf{r}'(s)| = 1$$

That was awfully convenient for the subsequent calculation in (17.4). I wonder if that was just a happy accident, or if there is more to it? Let's investigate by finding $|\mathbf{r}'(s)|$ for the vector valued function that appeared first in

EX 1, and was converted to its arc length parameterization in EX 3. That arc length parameterized version of this function was:

$$\mathbf{r}(s) = \left\langle \frac{2s}{\sqrt{29}}, 1 - \frac{3s}{\sqrt{29}}t, 5 + \frac{4s}{\sqrt{29}} \right\rangle \tag{17.9}$$

Now, here's a PRO TIP: if you're ultimately looking for $|\mathbf{r}'(s)|$, you should assemble $|\mathbf{r}'(s)|^2$ first; this way, you're not dragging around a big messy square root term. Then, having built $|\mathbf{r}'(s)|^2$, you can then reduce it to $|\mathbf{r}'(s)|$. For the function in (17.9), we have:

$$\mathbf{r}'(s) = \left\langle \frac{2}{\sqrt{29}}, -\frac{3}{\sqrt{29}}, \frac{4}{\sqrt{29}} \right\rangle$$

so that

$$|\mathbf{r}'(s)|^2 = \left(\frac{2}{\sqrt{29}}\right)^2 + \left(\frac{-3}{\sqrt{29}}\right)^2 + \left(\frac{4}{\sqrt{29}}\right)^2 = 1$$

Then $|\mathbf{r}'(s)|^2 = 1$ means $|\mathbf{r}'(s)| = 1$. Gosh. Are you ready to put some money on a wager that vector functions parameterized by arc length always generate $|\mathbf{r}'(s)| = 1$? Maybe we should try a few more just to build confidence.

EX 5 Revisit the vector valued function from EX 4 and find a general expression for $|\mathbf{r}'(s)|$.

This arc length parameterized version of this vector function is really nasty. I debated whether to be mean and give this one to you as a You Try It, but decided to be nice and do it myself. Here we go. Let's take the arc length parameterized vector function in Eq. (17.6) and name each component function as $r_1(s), r_2(s), r_3(s)$ respectively. Then, with a little simplification done behind the scenes,

$$r_1(s) = \cos\left(\left(\frac{3}{2}s + 1\right)^{2/3} - 1\right)$$

$$r_1'(s) = -\left(\frac{3}{2}s + 1\right)^{-1/3} \sin\left(\left(\frac{3}{2}s + 1\right)^{2/3} - 1\right)$$

$$(r_1'(s))^2 = \left(\frac{3}{2}s + 1\right)^{-2/3} \sin^2\left(\left(\frac{3}{2}s + 1\right)^{2/3} - 1\right)$$

and

$$\begin{aligned}
r_2(s) &= \sin\left(\left(\frac{3}{2}s+1\right)^{2/3}-1\right)\\
r_2'(s) &= \left(\frac{3}{2}s+1\right)^{-1/3}\cos\left(\left(\frac{3}{2}s+1\right)^{2/3}-1\right)\\
(r_2'(s))^2 &= \left(\frac{3}{2}s+1\right)^{-2/3}\cos^2\left(\left(\frac{3}{2}s+1\right)^{2/3}-1\right)
\end{aligned}$$

and

$$\begin{aligned}
r_3(s) &= \frac{2}{3}\left[\left(\frac{3}{2}s+1\right)^{2/3}-1\right]^{3/2}\\
r_1'(s) &= \left(\frac{3}{2}s+1\right)^{-1/3}\left[\left(\frac{3}{2}s+1\right)^{2/3}-1\right]^{1/2}\\
(r_1'(s))^2 &= \left(\frac{3}{2}s+1\right)^{-2/3}\left[\left(\frac{3}{2}s+1\right)^{2/3}-1\right] = 1-\left(\frac{3}{2}s+1\right)^{-2/3}
\end{aligned}$$

so that $|\mathbf{r}'(s)|^2 = r_1^2(s) + r_2^2(s) + r_3^2(s)$ expands as:

$$\begin{aligned}
&\left(\frac{3}{2}s+1\right)^{-2/3}\sin^2\left(\left(\frac{3}{2}s+1\right)^{2/3}-1\right)\\
&\quad+\left(\frac{3}{2}s+1\right)^{-2/3}\cos^2\left(\left(\frac{3}{2}s+1\right)^{2/3}-1\right)+1-\left(\frac{3}{2}s+1\right)^{-2/3}\\
&\quad=\left(\frac{3}{2}s+1\right)^{-2/3}+1-\left(\frac{3}{2}s+1\right)^{-2/3}=1
\end{aligned}$$

And since $|\mathbf{r}'(s)|^2 = 1$, then of course, $|\mathbf{r}'(s)| = 1$. Hey, this is fun! Without formal proof, let's pose the fact that our evidence has been leading towards:

Useful Fact 17.1. *If $\mathbf{r}(s)$ is the arc length parameterized representation of a vector valued function, then it will satisfy $|\mathbf{r}'(s)| = 1$. Conversely, if you determine that $|\mathbf{r}'(s)| = 1$ for a particular representation of a vector function, then you know the parameter involved is the arc length parameter. Further, at any location on the curve, the vector $\mathbf{r}''(s)$ is perpendicular to the vector $\mathbf{r}'(s)$.*

We have not conclusively proven that $|\mathbf{r}'(s)| = 1$, but we can use that relation to develop further information. Since $|\mathbf{r}'(s)|^2 = \mathbf{r}'(s) \cdot \mathbf{r}'(s)$, then $|\mathbf{r}'(s)| = 1$ means $\mathbf{r}'(s) \cdot \mathbf{r}'(s) = 1$. Finding the derivative of both sides with respect to s (using Useful Fact 16.10) gives:

$$\begin{aligned} \frac{d}{ds}\left(\mathbf{r}'(s) \cdot \mathbf{r}'(s)\right) &= \frac{d}{ds}(1) \\ \mathbf{r}'(s) \cdot \mathbf{r}''(s) + \mathbf{r}''(s) \cdot \mathbf{r}'(s) &= 0 \\ 2\mathbf{r}'(s) \cdot \mathbf{r}''(s) &= 0 \\ \mathbf{r}'(s) \cdot \mathbf{r}''(s) &= 0 \end{aligned}$$

But $\mathbf{r}'(s) \cdot \mathbf{r}''(s) = 0$ means that $\mathbf{r}''(s)$ is perpendicular to $\mathbf{r}'(s)$.

You Try It

(4) Revisit the vector function from You Try It 2 and You Try It 3, and confirm that $|\mathbf{r}'(s)| = 1$.

Arc Length and Parameterization — Problem List

Arc Length and Parameterization — You Try It

These appeared above; solutions begin on the next page.

(1) Find the arc length of of $\mathbf{r}(t) = \langle t^2/2, t^3/3 \rangle$ for $0 \leq t \leq 4$.
(2) Find the arc length of $\mathbf{r}(t) = \langle \sin(2t), 2t, \cos(2t) \rangle$ for $0 \leq t \leq \pi$.
(3) Find an arc length parameterization of the vector function in You Try It 2, for $t \geq 0$, and determine the location at which the curve accumulates a total arc length of $s = 1$.
(4) Revisit the vector function from You Try It 2 and You Try It 3, and confirm that $|\mathbf{r}'(s)| = 1$.

Arc Length and Parameterization — Practice Problems

Try these as you get the hang of the You Try It problems. Solutions to these problems are available in Sec. B.5.1.

(1) Find the arc length of $\mathbf{r}(t) = \langle \sqrt{2}t, e^t, e^{-t} \rangle$ for $0 \leq t \leq 1$. (Hint: When you create $|\mathbf{r}'(t)| = \sqrt{g(t)}$, the function $g(t)$ will be a perfect square! Really!)
(2) Find the arc length of t $\mathbf{r}(t) = \langle t^2, 2t, \ln t \rangle$ for $1 \leq t \leq e$. (Hint: When you create $|\mathbf{r}'(t)| = \sqrt{g(t)}$, the function $g(t)$ will be a perfect square! Really!)
(3) Find an arc length parameterization of the vector function in You Try It 1. for $t \geq 0$, and determine the location at which the curve accumulates a total arc length of $s = 2$.
(4) Revisit the vector function from You Try It 1 and Practice Problem 3, and confirm that $|\mathbf{r}'(s)| = 1$.

Arc Length and Parameterization — Challenge Problems

Try these problems to test your skills with the ideas in this section. Solutions to these problems are available in Sec. C.5.1.

(1) Find the arc length of of $\mathbf{r}(t) = \langle t^2, \sin t - t\cos t, \cos t + t\sin t \rangle$ for $0 \leq t \leq \pi$.
(2) Find the arc length (use computational aid to estimate it, if needed) of the segment of $y = x^3$ from $(-1, 1)$ to $(1, 1)$. (Hint: Can you parameterize this curve?)

(3) Find an arc length parameterization of $\mathbf{r}(t) = \langle \sin 2t, 2t^{3/2}/3, \cos 2t \rangle$ for $t \geq 0$, and determine the location at which the curve accumulates a total arc length of $s = 1$.

Arc Length Parameterization — You Try It — Solved

(1) Find the arc length of of $\mathbf{r}(t) = \langle t^2/2, t^3/3 \rangle$ for $0 \leq t \leq 4$.

☐ First, build:

$$\mathbf{r}'(t) = \langle t, t^2 \rangle$$
$$|\mathbf{r}'(t)| = \sqrt{t^2 + t^4} = t\sqrt{1 + t^2}$$

So that

$$L = \int_0^4 |\mathbf{r}'(t)|\, dt = \int_0^4 t\sqrt{1 + t^2}\, dt = \frac{1}{3}(1+t^2)^{3/2}\Big|_0^4 = \frac{1}{3}(17^{3/2} - 1) \quad \blacksquare$$

(2) Find the arc length of $\mathbf{r}(t) = \langle \sin(2t), 2t, \cos(2t) \rangle$ for $0 \leq t \leq \pi$.

☐ First, build:

$$\mathbf{r}'(t) = < 2\cos 2t, 2, -2\sin 2t >$$
$$|\mathbf{r}'(t)| = \sqrt{4\cos^2 2t + 4 + 4\sin^2 2t} = \sqrt{8} = 2\sqrt{2}$$

So that

$$L = \int_0^\pi |\mathbf{r}'(t)|\, dt = \int_0^\pi 2\sqrt{2}\, dt = 2\pi\sqrt{2} \quad \blacksquare$$

(3) Find an arc length parameterization of the vector function in You Try It 2, for $t \geq 0$, and determine the location at which the curve accumulates a total arc length of $s = 1$.

☐ Picking up You Try It 2, the arc length of $\mathbf{r}(t) = \langle \sin(2t), 2t, \cos(2t) \rangle$ at any t is

$$s = \int_0^t |\mathbf{r}'(\tau)|\, d\tau = \int_0^t 2\sqrt{2}\, d\tau = 2\sqrt{2}\, t$$

Then since $s = 2\sqrt{2}\, t$, we have $t = s/(2\sqrt{2})$, and so

$$\mathbf{r}(s) = \left\langle \sin \frac{s}{\sqrt{2}}, \frac{s}{\sqrt{2}}, \cos \frac{s}{\sqrt{2}} \right\rangle$$

This is the arc length parameterized version of the vector function. To now find where we accumulate an arc length of 1, we look for $\mathbf{r}(1)$:

$$\mathbf{r}(1) = \left\langle \sin \frac{1}{\sqrt{2}}, \frac{1}{\sqrt{2}}, \cos \frac{1}{\sqrt{2}} \right\rangle$$

Well, those aren't attractive coordinates for our location, but what are you gonna do? ■

(4) Revisit the vector function from You Try It 2 and You Try It 3, and confirm that $|\mathbf{r}'(s)| = 1$.

□ The arc length parameterized version of this vector function is

$$\mathbf{r}(s) = \left\langle \sin\frac{s}{\sqrt{2}}, \frac{s}{\sqrt{2}}, \cos\frac{s}{\sqrt{2}} \right\rangle$$

from which we get

$$\mathbf{r}'(s) = \left\langle \frac{1}{\sqrt{2}}\cos\frac{s}{\sqrt{2}}, \frac{1}{\sqrt{2}}, -\frac{1}{\sqrt{2}}\sin\frac{s}{\sqrt{2}} \right\rangle$$

and then

$$|\mathbf{r}'(s)|^2 = \frac{1}{2}\cos^2\frac{s}{\sqrt{2}} + \frac{1}{2} + \frac{1}{2}\sin^2\frac{s}{\sqrt{2}} = \frac{1}{2} + \frac{1}{2} = 1$$

Since $|\mathbf{r}'(s)|^2 = 1$, then $|\mathbf{r}'(s)| = 1$. It works! ■

17.2 Contours, Orientation, and Pointers

Introduction

With the ability to draw curves through two and three-dimensional space comes the need to develop tools to analyze those curves. This involves some new terminology, development of vectors that give a framework to the curve (a la tangent lines for functions), and a way to describe positional orientation. If you have taken a physics class and solved problems that involved falling objects, you are familiar with the need to assign ahead of time the direction in which acceleration will be considered positive. We have to make similar decisions when moving along curves. With assignment of positive and negative *orientation* several "navigation" tools that help us analyze vector curves.

Contours and Orientation

We generally assume that the initial point of a vector function $\mathbf{r}(t)$ is held fixed at the origin while the terminal point is free to move in space as the parameter t sweeps through the values in some domain, $a \leq t \leq b$. Whether in two or three dimensions, we can think of that terminal point as a pen that draws out a curve, or *contour*, in space. Contours have several descriptors associated with them.

A contour can be **open** or **closed**. A closed contour is one that starts and ends at the same point in space, such as a circle or an ellipse. A contour is open if it's not closed.[1] For example, the vector function $\mathbf{r}(t) = \langle \cos t, \sin t \rangle$ for $0 \leq t \leq \pi$ is an open contour (remember, it's the upper half of the unit circle).

A contour can be formed with multiple segments, each of which has its own vector function. Imagine drawing the triangle connecting the three points $(0, 0)$, $(2, 0)$, and $(0, 2)$. This is a single closed contour, but we would have to specify three different vector functions to collectively describe the three sides of the triangle, and the parametric equations that could create this particular contour are certainly not unique.

[1]Why make definitions more complicated than they need to be?

A contour is **piecewise smooth** if it does not have breaks, jumps, etc. The triangle just mentioned is piecewise smooth. There are corners at each junction between sides, but each individual "piece" is smooth. There is a deeper definition for smoothness that involves derivatives of the vector functions, but the loose definition will do for now.

A contour is **simple** if it does not intersect itself. A circle is simple, a figure-eight is not.

When dealing with vector functions and contours, we must make a distinction between **direction** and **orientation**. From our previous experience, we already know vector functions have direction associated with them. The direction of a contour is inherited from its parametric equation and domain (such as $a \leq t \leq b$) specified for the parameter. The direction of the contour arises from the order in which we encounter points as we move from the location marked by $t = a$ to the location marked by $t = b$. For example, the vector function $\mathbf{r}(t) = \langle \cos t, \sin t \rangle$ with $0 \leq t \leq 2\pi$ establishes that we begin at $(1, 0)$ (where $t = 0$), go *counterclockwise* around the circle, and end again back at $(1, 0)$ (where $t = 2\pi$).

On the other hand, the **orientation** of a contour is a property that is *assigned* as either positive or negative; this assignment is often only given for closed contours, or contours that could be closed if followed in a predictable trajectory. The usual convention is that traversal of the contour in a counterclockwise direction corresponds to *positive* orientation.

If we strengthen the description of the triangular contour given above to be, "a triangle connecting the three points $(0, 0)$, $(2, 0)$, and $(0, 2)$, *with positive orientation*" then we now know we are supposed to move around the triangle in a counterclockwise direction, and this helps us design proper parametric equations for each segment of the contour.

If the idea of visualizing directions as clockwise or counterclockwise isn't appealing, we can define positive orientation in a different way, at least for *simple* closed contours. A simple closed contour $\mathbf{r}(t)$ will surround a 2D region; if the captured region is always to the left of a pencil that is drawing the contour in the direction specified by increasing t, the contour is positively oriented.

EX 1 Determine parametric equations and bounds for t ($a \leq t < b$) that will give us the unit circle, starting and ending at $(0, 1)$, oriented negatively.

Since we're making a circle, let's set the domain for t as $0 \leq t < 2\pi$. Therefore we must be at $(0, 1)$ when $t = 0$, go around the circle *clockwise*, and be back at $(0, 1)$ when $t = 2\pi$ (technically we do not mark the point at $t = 2\pi$). The equations

$$x(t) = \sin(t)$$
$$y(t) = \cos(t)$$

will put us at $(0, 1)$ for both $t = 0$ and $t = 2\pi$. And, when $t = \pi/2$, we are at the point $(\sin(\pi/2), \cos(\pi/2)) = (1, 0)$. If we continue through other "easy" markers set by t, we see that we traverse the curve from $(0, 1)$, then through $(1, 0)$, $(0, -1)$, $(-1, 0)$, and back to $(0, 1)$. This is clockwise. Or equivalently, if we are looking forward from a pencil drawing this curve, the interior of the circle is to the *right*. Either way, we have determined negative orientation. The vector function that gives the prescribed contour and direction is indeed $\mathbf{r}(t) = \langle \sin t, \cos t \rangle$ for $0 \leq t \leq 2\pi$. ■

You Try It

(1) Determine parametric equations and bounds for t ($a \leq t < b$) that will give us the unit circle, starting and ending at $(-1, 0)$, oriented negatively.
(2) Determine parametric equations and bounds for t ($a \leq t < b$) that will give us the unit circle, starting and ending at $(1, 0)$, oriented negatively.

Note that the domains for t given in YTI 1 and YTI 2 here are requested as $a \leq t < b$; since the contours are closed circles, we are trying not to mark the same (starting and ending) point twice. However, there is no harm in marking that point twice, and so specifying these domains as $a \leq t \leq b$.

Pointers: T *and* N

Contours can represent the path of a moving object, or flowlines of fluid, heat, or electrostatic charge. It is often useful to have vectors acting as "navigational aids" along a contour, which point us in directions tangent to (i.e. forwards along) and perpendicular (i.e. normal) to the contour.

But what's better than just any old vectors tangent or normal to a contour? Why, *unit* vectors tangent or perpendicular to the contour!

If we are lucky enough to have a contour $\mathbf{r}(s)$ parameterized by arc length, then we know from Sec. 17.1 that $\mathbf{r}'(s)$ is always tangent to the contour at any point, and $\mathbf{r}''(s)$ is always perpendicular to the contour. But the gold standard here is to have *unit* vectors tangent and normal to the contour We already happen to know that $|\mathbf{r}'(s)| = 1$, so if we designate by $\mathbf{T}(s)$ a vector function which produces a *unit* vector tangent to $\mathbf{r}(s)$ at any point, then $\mathbf{T}(s) = \mathbf{r}'(s)$.

Here's the tricky bit: since $\mathbf{r}'(s)$ is already a unit vector, is $\mathbf{r}''(s)$ also necessarily a unit vector? The answer is *no*. The vector function $\mathbf{r}''(s)$ is not guaranteed to produce unit vectors; in fact, we use $|\mathbf{r}''(s)|$ to define a property of contours called *curvature*, and we'll explore that more in the next section. For now, at least, we can pose the following result:

Definition 17.2. *If* $\mathbf{r}(s)$ *is an arc length parameterized vector function with differentiable components, then unit tangent and unit normal vector functions,* $\mathbf{T}(s)$ *and* $\mathbf{N}(s)$, *are defined by*

$$\mathbf{T}(s) = \mathbf{r}'(s) \quad \textit{and} \quad \mathbf{N}(s) = \frac{\mathbf{r}''(s)}{|\mathbf{r}''(s)|} \tag{17.10}$$

As we know by now, our vector functions are often parameterized by t (time?) rather than arc length, and determining the corresponding arc length version is not always feasible or recommended. So, given a generic $\mathbf{r}(t)$, let's designate as $\mathbf{T}(t)$ and $\mathbf{N}(t)$ the corresponding unit tangent and unit normal vector functions. You might imagine that these are a bit messier, and you'd be right. At any point on $\mathbf{r}(t)$, we know $\mathbf{r}'(t)$ is tangent to the contour, pointing in the direction of motion. Therefore a *unit* tangent vector function is given by:

$$\boxed{\mathbf{T}(t) = \frac{\mathbf{r}'(t)}{|\mathbf{r}'(t)|} = \frac{\mathbf{r}'(t)}{ds}} \tag{17.11}$$

EX 2 Find a unit vector tangent for $\mathbf{r}(t) = \langle 6t^5, 4t^3, 2t\rangle$ at $t = 1$.

We have

$$\begin{aligned} \mathbf{r}'(t) &= \langle 30t^4, 12t^2, 2\rangle \\ \mathbf{r}'(1) &= \langle 30, 12, 2\rangle \\ |\mathbf{r}'(1)| &= \sqrt{30^2 + 12^2 + 4} = \sqrt{1048} = 2\sqrt{262} \end{aligned}$$

so that

$$\mathbf{T}(1) = \frac{\mathbf{r}'(1)}{|\mathbf{r}'(1)|} = \left\langle \frac{15}{\sqrt{262}}, \frac{6}{\sqrt{262}}, \frac{1}{\sqrt{262}} \right\rangle \quad \blacksquare$$

An alternate way to proceed through this calculation would be to build a general expression for $\mathbf{T}(t)$ and wait until the last step to plug in $t = 1$. That calculation goes like this:

$$\mathbf{r}'(t) = \langle 30t^4, 12t^2, 2\rangle$$
$$|\mathbf{r}'(t)| = \sqrt{900t^8 + 144t^4 + 4}$$

So that

$$\mathbf{T}(t) = \frac{\mathbf{r}'(t)}{|\mathbf{r}'(t)|} = \frac{\langle 30t^4, 12t^2, 2\rangle}{\sqrt{900t^8 + 144t^4 + 4}}$$

and specifically,

$$\mathbf{T}(1) = \frac{\langle 30(1)^4, 12(1)^2, 2\rangle}{\sqrt{900(1)^8 + 144(1)^4 + 4}} = \left\langle \frac{15}{\sqrt{262}}, \frac{6}{\sqrt{262}}, \frac{1}{\sqrt{262}} \right\rangle$$

I think you'll agree the first method is preferable. The story here is to be very aware of when you can stop and plug in a specific value for t rather than carrying on with a general formula. This theme will be repeated in other examples below. ■

You Try It

(3) Find the unit tangent vector $\mathbf{T}$ for $\mathbf{r}(t) = (4t^{3/2}/3)\mathbf{i} + t^2\mathbf{j} + t\mathbf{k}$ at $t = 1$.

(4) Find the unit tangent vector $\mathbf{T}$ for $\mathbf{r}(t) = 2\sin t\mathbf{i} + 2\cos t\mathbf{j} + (\sin t - \cos t)\mathbf{k}$ at $t = 3\pi/4$.

For the unit normal vector, we'll start with the 2D case, since that can be handled in a relatively painless way. Given $\mathbf{r}(t) = \langle x(t), y(t)\rangle$, where $x(t)$ and $y(t)$ are both differentiable, the vector $\langle x'(t), y'(t)\rangle$ is tangent to the contour at any point. Our job is now to find a vector *perpendicular* to $\langle x'(t), y'(t)\rangle$. That is, we need to find a new vector $\langle A, B\rangle$ such that

$$\langle A, B\rangle \cdot \langle x'(t), y'(t)\rangle = Ax'(t) + By'(t) = 0$$

Now, we could go through all sorts of geometric and algebraic shenanigans to find A and B ... or, we can just be clever. What if $A = y'(t)$ and $B = -x'(t)$? Then we have

$$\begin{aligned}\langle A, B\rangle \cdot \langle x'(t), y'(t)\rangle &= \langle y'(t), -x'(t)\rangle \cdot \langle x'(t), y'(t)\rangle \\ &= y'(t)x'(t) - x'(t)y'(t) = 0\end{aligned}$$

And there we go! At a given point on a 2D contour, $\langle x'(t), y'(t)\rangle$ is tangent to the contour and $\langle y'(t), -x'(t)\rangle$ is perpendicular to the contour. (Note that neither of these are necessarily unit vectors.) While $\langle x'(t), y'(t)\rangle$ is already known as $\mathbf{r}'(t)$, let's name $\langle y'(t), -x'(t)\rangle$ as $\mathbf{r}'(t)^{\perp}$ — where the superscript $\perp$ is there to remind us that this vector is perpendicular to $\mathbf{r}'(t)$ (and so is also perpendicular to $\mathbf{T}(t)$). A fun fact about $\mathbf{r}'(t)^{\perp}$ is that it can be tied to the arc length parameter as follows:

$$|\mathbf{r}'(t)^{\perp}| = \sqrt{(y'(t))^2 + (-x'(t))^2} = \sqrt{(x'(t))^2 + (y'(t))^2} = ds$$

Now, let's ask again: what's even better than a generic vector perpendicular to a contour? We find this as:

$$\boxed{\mathbf{N}(t) = \frac{\mathbf{r}'(t)^{\perp}}{|\mathbf{r}'(t)^{\perp}|} = \frac{\mathbf{r}'(t)^{\perp}}{ds} \quad (2D \text{ only})} \tag{17.12}$$

EX 3 Find $\mathbf{T}$ and $\mathbf{N}$ for the contour $\mathbf{r}(t) = \langle 2t^2, t\rangle$ at $(2, 1)$.

Note that the point $(2, 1)$ corresponds to $t = 1$. In general we have $\mathbf{r}'(t) = \langle x'(t), y'(t)\rangle$ and $\mathbf{r}'(t)^{\perp} = \langle y'(t), -x'(t)\rangle$, so for this function specifically, we have $\mathbf{r}'(t) = \langle 4t, 1\rangle$ and $\mathbf{r}'(t)^{\perp} = \langle 1, -4t\rangle$. At $t = 1$, then,

$$\begin{aligned} \mathbf{r}'(1) &= \langle x'(1), y'(1)\rangle = \langle 4, 1\rangle \\ \mathbf{r}'(1)^{\perp} &= \langle y'(1), -x'(1)\rangle = \langle 1, -4\rangle \\ ds &= \sqrt{x'(1)^2 + y'(1)^2} = \sqrt{4^2 + 1^2} = \sqrt{17} \end{aligned}$$

and so, using (17.11) and (17.12),

$$\begin{aligned} \mathbf{T}(1) &= \frac{\mathbf{r}'(t)}{ds} = \frac{\langle 4, 1\rangle}{\sqrt{17}} = \frac{1}{\sqrt{17}}\langle 4, 1\rangle \\ \mathbf{N}(1) &= \frac{\mathbf{r}'(t)^{\perp}}{ds} = \frac{\langle 1, -4\rangle}{\sqrt{17}} = \frac{1}{\sqrt{17}}\langle 1, -4\rangle \end{aligned}$$

As a sort of "quality assurance" check, we can confirm that $|\mathbf{T}(1)| = 1$, $|\mathbf{N}(1)| = 1$, and $\mathbf{T}(1) \cdot \mathbf{N}(1) = 0$. ■

You Try It

(5) Find $\mathbf{T}$ and $\mathbf{N}$ for the contour $\mathbf{r}(t) = \langle \cos t, \sin t\rangle$ at $t = \pi/4$.

To express $\mathbf{N}(t)$ in the general case (whether 2D or 3D), consider this: from the relation between dot product and length, we know that

$$\mathbf{T}(t) \cdot \mathbf{T}(t) = |\mathbf{T}(t)|^2$$

But since $\mathbf{T}(t)$ is always a unit vector, then $|\mathbf{T}(t)| = 1$ and so we have

$$\mathbf{T}(t) \cdot \mathbf{T}(t) = 1$$

If we find the derivative of both sides of this expression with respect to t using the vector function dot product from the end of Sec. 16.3,

$$\begin{aligned} \frac{d}{dt}\mathbf{T}(t) \cdot \mathbf{T}(t) &= \frac{d}{dt}1 \\ \mathbf{T}'(t) \cdot \mathbf{T}(t) + \mathbf{T}(t) \cdot \mathbf{T}'(t) &= 0 \\ \mathbf{T}(t) \cdot \mathbf{T}'(t) &= 0 \end{aligned}$$

and the result $\mathbf{T}(t) \cdot \mathbf{T}'(t) = 0$ means that $\mathbf{T}'(t)$ is perpendicular to $\mathbf{T}(t)$, and so normal to $\mathbf{r}(t)$. This is the vector we're after. While $\mathbf{T}'(t)$ itself is not a *unit* vector, we know what to do to form the unit normal vector:

$$\boxed{\mathbf{N}(t) = \frac{\mathbf{T}'(t)}{|\mathbf{T}'(t)|}} \tag{17.13}$$

It bears mentioning that there is a subtle implication in this expression for $\mathbf{N}(t)$. It's easy to presume that because $\mathbf{T}(t)$ represents a unit vector, then $\mathbf{T}'(t)$ will be a unit vector too. But that's not true — $\mathbf{T}'(t)$ is not necessarily a unit vector, and we have to account for that when finding $\mathbf{N}(t)$.

EX 4 Determine $\mathbf{T}$ and $\mathbf{N}$ for the helix $\mathbf{r}(t) = \langle 3\cos t, 3\sin t, 4t\rangle$ at the point $(0, 3, 2\pi)$.

Note that the point $(0, 3, 2\pi)$ corresponds to $t = \pi/2$. In problems like this, it pays to be alert to when it's either helpful or dangerous to plug in the specific value $t = \pi/2$. Because we have to carry this calculation all the way to a specific value of $\mathbf{N}(t)$, we are going to need a general expression for $\mathbf{T}'(t)$. That hurts. But even so, we can be strategic.

To discover $\mathbf{T}(\pi/2)$, we'll first need this cascade of items:

$$\begin{aligned} \mathbf{r}'(t) &= \langle -3\sin t, 3\cos t, 4\rangle \\ |\mathbf{r}'(t)| &= \sqrt{9\sin^2 t + 9\cos^2 t + 16} = \sqrt{25} = 5 \\ \mathbf{T}(t) &= \frac{\mathbf{r}'(t)}{|\mathbf{r}'(t)|} = \frac{1}{5}\langle -3\sin t, 3\cos t, 4\rangle \end{aligned}$$

With this general expression for $\mathbf{T}(t)$, we can stop and find specifically that

$$\mathbf{T}\left(\frac{\pi}{2}\right) = \frac{1}{5}\langle 3, 0, 4\rangle = \left\langle \frac{3}{5}, 0, \frac{4}{5}\right\rangle$$

(As a quick check, we should confirm that this vector is indeed a unit vector. Is it?)

Allow me to pause for a quick strategy session: had this example asked us to calculate *only* $\mathbf{T}(\pi/2)$, we would not have needed the full expression for $\mathbf{T}(t)$; rather, we could get away with finding the full expression for $\mathbf{r}'(t)$, so that we could compute the specific vector $\mathbf{r}'(\pi/2)$ and then also $|\mathbf{r}'(\pi/2)|$, which together form $\mathbf{T}(\pi/2)$. But since this example asks us to continue and find $\mathbf{N}(\pi/2)$, then we have to have a general expression for $\mathbf{T}(t)$ and then also $\mathbf{T}'(t)$.

Carrying on, then, we have the general expression

$$\mathbf{T}(t) = \frac{1}{5}\langle -3\sin t, 3\cos t, 4\rangle$$

from which we determine

$$\mathbf{T}'(t) = \frac{1}{5}\langle -3\cos t, -3\sin t, 0\rangle$$

Now it's safe to deal with $t = \pi/2$ specifically. We have:

$$\mathbf{T}'\left(\frac{\pi}{2}\right) = \frac{1}{5}\left\langle -3\cos\frac{\pi}{2}, -3\sin\frac{\pi}{2}, 0\right\rangle = \frac{1}{5}\langle 0, -3, 0\rangle = \left\langle 0, -\frac{3}{5}, 0\right\rangle$$

so that

$$\left|\mathbf{T}'\left(\frac{\pi}{2}\right)\right| = \frac{3}{5}$$

and together,

$$\mathbf{N}\left(\frac{\pi}{2}\right) = \frac{\mathbf{T}'(\pi/2)}{|\mathbf{T}'(\pi/2)|} = \frac{\langle 0, -3/5, 0\rangle}{3/5} = \langle 0, -1, 0\rangle$$

In summary, we have:

$$\mathbf{T}\left(\frac{\pi}{2}\right) = \left\langle \frac{3}{5}, 0, \frac{4}{5}\right\rangle$$
$$\mathbf{N}\left(\frac{\pi}{2}\right) = \langle 0, -1, 0\rangle$$

For quality assurance purposes, we can confirm that each is a unit vector, and they are all mutually perpendicular, as they should be. ■

Now, it just so happens that there is a third vector which, along with **T** and **N**, completes a nice triplet of unit reference vectors which are all mutually perpendicular in three dimensions ... a lot like the coordinate axes! This third vector is called the binormal vector, **B**, and we'll see it in the next section.

> **You Try It**
>
> (4) Determine **T** and **N** for $\mathbf{r}(t) = \langle e^t, e^t \sin t, e^t \cos t\rangle$ at the point (1,0,1). Can you demonstrate some quality assurance about your results?

Contours, Orientation, and Pointers — Problem List

Contours, Orientation, and Pointers — You Try It

These appeared above; solutions begin on the next page.

(1) Determine parametric equations and bounds for t $(a \le t < b)$ that will give us the unit circle, starting and ending at $(0,1)$, oriented negatively.
(2) Determine parametric equations and bounds for t $(a \le t < b)$ that will give us the unit circle, starting and ending at $(-1,0)$, oriented negatively.
(3) Find the unit tangent vector **T** for $\mathbf{r}(t) = \frac{4}{3}t^{3/2}\mathbf{i} + t^2\mathbf{j} + t\mathbf{k}$ at $t = 1$.
(4) Find the unit tangent vector **T** for $\mathbf{r}(t) = 2\sin t\mathbf{i} + 2\cos t\mathbf{j} + (\sin t - \cos t)\mathbf{k}$ at $t = 3\pi/4$.
(5) Find **T** and **N** for the contour $\mathbf{r}(t) = \langle \cos t, \sin t\rangle$ at $t = \pi/4$.
(6) Determine **T** and **N** for $\mathbf{r}(t) = \langle e^t, e^t \sin t, e^t \cos t\rangle$ at the point (1,0,1). Can you demonstrate some quality assurance about your results?

Contours, Orientation, and Pointers — Practice Problems

Try these as you get the hang of the You Try It problems. Solutions to these problems are available in Sec. B.5.2.

(1) Determine parametric equations and bounds for t $(a \le t < b)$ that will give us an ellipse containing points $(2,0)$ and $(0,3)$, oriented positively.
(2) Determine parametric equations and bounds for t $(a \le t < b)$ that will give us an ellipse containing points $(4,0)$ and $(0,3)$, oriented negatively.
(3) Find **T** and **N** for the contour $\mathbf{r}(t) = \langle 2t^2, t\rangle$ at $(2,1)$.

(4) Find the unit tangent vector $\mathbf{T}$ and unit normal vector $\mathbf{N}$ for $\mathbf{r}(t) = \langle 2e^{-t}, e^{-2t}, t\rangle$ at $t = 0$. Can you demonstrate some quality assurance about your results?

(5) Find the unit normal vector $\mathbf{N}$ for $\mathbf{r}(t) = (4t^{3/2}/3)\mathbf{i} + t^2\mathbf{j} + t\mathbf{k}$ at $t = 1$. (Hint: Has $\mathbf{T}$ been calculated elsewhere?)

(6) Find the unit tangent vector $\mathbf{T}$ and unit normal vector $\mathbf{N}$ for $\mathbf{r}(t) = \langle \cos t + t\sin t, \sin t - t\cos t, 1\rangle$ at the point $(1, 0, 1)$.

Contours, Orientation, and Pointers — Challenge Problems

Try these problems to test your skills with the ideas in this section. Solutions to these problems are available in Sec. C.5.2.

(1) The orbital path of a comet follows the vector function $\mathbf{r}(t) = \langle \cosh t, \sinh t\rangle$. Two aliens riding on the comet decide to jump off before the comet gets close to Earth, because right now, Earth is a pretty dumb place to be. They both leap at $t = \ln 2$; one leaps off in the direction of $\mathbf{T}$, and one leaps off in the direction of $\mathbf{N}$. If they fly away in straight lines, find the vector equations of those lines. (This identity might be useful: $\sinh^2 t + \cosh^2 t = \cosh(2t)$.)

(2) Determine $\mathbf{T}$ and $\mathbf{N}$ for $\mathbf{r}(t) = \langle t^2, 2t^3/3, t\rangle$ at the point $(1, 2/3, 1)$. Can you demonstrate some quality assurance about your results?

(3) A track where horses race is in the shape of the ellipse $\mathbf{r}(t) = \langle 3\cos t, 2\sin t\rangle$ (for $0 \le t \le 2\pi$). The coordinate system is centered at the beer tent in the center of the lawn inside the racetrack. One particularly moody horse decides enough is enough, and when he reaches the spot on the track marked by $t = \pi/4$, he breaks off the track and runs away in a direction perfectly tangent to his original path on the track. So:

    (a) What are the coordinates at which he crosses the y-axis and escapes to freedom?

    (b) If a gust of wind adds an acceleration vector of $\langle 0, 10\rangle$ to aid the horse, what is the component of that acceleration vector in the (tangent) direction of escape?

Contours, Orientation, and Pointers — You Try It — Solved

---

(1) Determine parametric equations and bounds for t ($a \leq t < b$) that will give us the unit circle, starting and ending at $(0, 1)$, oriented negatively.

□ The component functions will be a combination of $\sin t$ and $\cos t$. For negative orientation, we must go around the circle in a clockwise direction. So after starting at $(0, 1)$, we need to have x increase from 0 and y decrease from 1. Setting $\mathbf{r}(t) = \langle \sin t, \cos t \rangle$ for $0 \leq t < 2\pi$ accomplishes this, and we go around the unit circle as follows:

$$(0,1)_{t=0} \to (1,0)_{t=\pi/2} \to (0,-1)_{t=\pi} \to (-1,0)_{t=3\pi 2} \to (0,1)_{t=2\pi}$$

FFT: This answer is not unique. Can you think of another combination of parametric equations with bounds on t which do the job? ■

---

(2) Determine parametric equations and bounds for t ($a \leq t < b$) that will give us the unit circle, starting and ending at $(-1, 0)$, oriented negatively.

□ The component functions will be a combination of $\sin t$ and $\cos t$. For negative orientation, we must go around the circle in a clockwise direction. So after starting at $(-1, 0)$, we need to have x increase from -1 and y increase from 0. Setting $\mathbf{r}(t) = \langle \sin t, \cos t \rangle$ for $-\pi/2 \leq t \leq 3\pi/2$ accomplishes this, and we go around the unit circle as follows:

$$(-1,0)_{t=-\pi/2} \to (0,1)_{t=0} \to (1,0)_{t=\pi/2} \to (0,-1)_{t=\pi} \to\to (-1,0)_{t=3\pi/2}$$

FFT: This answer is not unique. Can you think of another combination of parametric equations with bounds on t which works as well? ■

---

(3) Find $\mathbf{T}$ and $\mathbf{N}$ for the contour $\mathbf{r}(t) = \langle \cos t, \sin t \rangle$ at $t = \pi/4$.

□ Note that this is just the unit circle at the halfway mark through the first quadrant; so, we already know that a tangent vector here will be $\langle -1, 1 \rangle$ and so the *unit* tangent vector is

$$\mathbf{T}\left(\frac{\pi}{4}\right) = \langle -\frac{1}{\sqrt{2}}, \frac{1}{\sqrt{2}} \rangle$$

Similarly, a normal vector will be $\langle 1, 1 \rangle$, and the *unit* normal vector is

$$\mathbf{N}\left(\frac{\pi}{4}\right) = \langle \frac{1}{\sqrt{2}}, \frac{1}{\sqrt{2}} \rangle$$

But, we should confirm this through the general expressions (17.11) and (17.12).

Since $\mathbf{r}'(t) = \langle x'(t), y'(t)\rangle = \langle -\sin t, \cos t\rangle$ then

$$ds = \sqrt{(-\sin t)^2 + (\cos t)^2} = 1$$

and so

$$\mathbf{T}(t) = \frac{\mathbf{r}'(t)}{ds} = \frac{\langle -\sin t, \cos t\rangle}{1} = \langle -\sin t, \cos t\rangle$$

$$\mathbf{T}\left(\frac{\pi}{4}\right) = \langle -\sin\frac{\pi}{4}, \cos\frac{\pi}{4}\rangle = \langle -\frac{1}{\sqrt{2}}, \frac{1}{\sqrt{2}}\rangle$$

Since $\mathbf{r}'(t)^{\perp} = \langle y'(t), -x'(t)\rangle = \langle \cos t, \sin t\rangle$ and $ds = 1$, we have

$$\mathbf{N}(t) = \frac{\mathbf{r}'(t)^{\perp}}{ds} = \frac{\langle \cos t, \sin t\rangle}{1} = \langle \cos t, \sin t\rangle$$

$$\mathbf{N}\left(\frac{\pi}{4}\right) = \langle \cos\frac{\pi}{4}, \sin\frac{\pi}{4}\rangle = \langle \frac{1}{\sqrt{2}}, \frac{1}{\sqrt{2}}\rangle$$

And our suspicions were confirmed. ■

---

(4) Find the unit tangent vector $\mathbf{T}(t)$ for $\mathbf{r}(t) = (4t^{3/2}/3)\mathbf{i} + t^2\mathbf{j} + t\mathbf{k}$ at $t = 1$.

☐ The building blocks of $\mathbf{T}(1)$ are:

$$\mathbf{r}'(t) = 2\sqrt{t}\mathbf{i} + 2t\mathbf{j} + \mathbf{k}$$

$$\mathbf{r}'(1) = 2\mathbf{i} + 2\mathbf{j} + \mathbf{k}$$

$$|\mathbf{r}'(1)| = \sqrt{4+4+1} = 3$$

So that

$$\mathbf{T}(1) = \frac{\mathbf{r}'(1)}{|\mathbf{r}'(1)|} = \frac{2}{3}\mathbf{i} + \frac{2}{3}\mathbf{j} + \frac{1}{3}\mathbf{k} \quad ■$$

---

(5) Find the unit tangent vector $\mathbf{T}$ for $\mathbf{r}(t) = 2\sin t\mathbf{i} + 2\cos t\mathbf{j} + (\sin t - \cos t)\mathbf{k}$ at $t = 3\pi/4$.

☐ The building blocks of $\mathbf{T}(3\pi/4)$ are:

$$\mathbf{r}'(t) = 2\cos t\mathbf{i} - 2\sin t\mathbf{j} + (\cos t - \sin t)\mathbf{k}$$

$$\mathbf{r}'\left(\frac{3\pi}{4}\right) = -\sqrt{2}\mathbf{i} - \sqrt{2}\mathbf{j} + 0\mathbf{k}$$

$$\left|\mathbf{r}'\left(\frac{3\pi}{4}\right)\right| = \sqrt{2+2+0} = 2$$

So that

$$\mathbf{T}\left(\frac{3\pi}{4}\right) = \frac{\mathbf{r}'(3\pi/4)}{|\mathbf{r}'(3\pi/4)|} = -\frac{\sqrt{2}}{2}\mathbf{i} - \frac{\sqrt{2}}{2}\mathbf{j} + 0\mathbf{k} \quad \blacksquare$$

---

(6) Determine $\mathbf{T}$ and $\mathbf{N}$ for $\mathbf{r}(t) = \langle e^t, e^t \sin t, e^t \cos t\rangle$ at the point (1,0,1). Can you demonstrate some quality assurance about your results?

□ Note that the given point corresponds to $t = 0$. The computation of $\mathbf{T}$, $\mathbf{N}$, and $\mathbf{B}$ proceeds through the following cascade of results. Pay close attention to when we need a full blown expression and when it's OK to stop and plug in constants:

$$\mathbf{r}'(t) = \langle e^t, e^t(\cos t + \sin t), e^t(\cos t - \sin t)\rangle$$

$$|\mathbf{r}'(t)| = \sqrt{3}e^t \text{ (tedious details went on behind the scenes)}$$

$$\mathbf{T}(t) = \frac{\mathbf{r}'(t)}{|\mathbf{r}'(t)|} = \frac{1}{\sqrt{3}}\langle 1, \cos t + \sin t, \cos t - \sin t\rangle$$

$$\rightarrow \mathbf{T}(0) = \frac{1}{\sqrt{3}}\langle 1, 1, 1\rangle$$

$$\mathbf{T}'(t) = \frac{1}{\sqrt{3}}\langle 0, -\sin t + \cos t, -\sin t - \cos t\rangle$$

$$\mathbf{T}'(0) = \frac{1}{\sqrt{3}}\langle 0, 1, -1\rangle$$

$$|\mathbf{T}'(0)| = \sqrt{\frac{2}{3}}$$

$$\rightarrow \mathbf{N}(0) = \frac{\mathbf{T}'(0)}{|\mathbf{T}'(0)|} = \frac{1}{\sqrt{2}}\langle 0, 1, -1\rangle$$

Collecting our results, we have

$$\mathbf{T}(0) = \left\langle \frac{1}{\sqrt{3}}, \frac{1}{\sqrt{3}}, \frac{1}{\sqrt{3}} \right\rangle \quad \text{and} \quad \mathbf{N}(0) = \left\langle 0, \frac{1}{\sqrt{2}}, -\frac{1}{\sqrt{3}} \right\rangle$$

As a double-check, we can confirm that $|\mathbf{T}(0)| = 1$, $|\mathbf{N}(0)| = 1$, and $\mathbf{T}(0) \cdot \mathbf{N}(0) = 0$. ■

17.3 The Fresnet–Serret Frame

The unit vectors **i**, **j**, and **k** are super awesome vectors. They all have length 1, and they are all perpendicular to each other. They are so awesome that they are the vectors we use to construct every other vector in $\mathbb{R}^3$ via

$$\langle a, b, c\rangle = a\langle 1,0,0\rangle + b\langle 0,1,0\rangle + c\langle 0,0,1\rangle = a\mathbf{i} + b\mathbf{j} + c\mathbf{k}$$

(this is called a linear combination of **i**, **j**, and **k**, for those of you en route to taking Linear Algebra). This triplet of vectors forms our usual 3D coordinate system: the vector **i** defines the x-axis, and **j** and **k** define the y- and z-axes. Life without **i**, **j**, and **k** would be incredibly dull.

Having said that, well ... these three vectors aren't so special. There is an infinite number of sets of three vectors which are (a) all unit vectors, and (b) perpendicular to each other. And in many applications, the coordinate system specified by **i**, **j**, **k** might not be as useful to us as one of the other triplets. Imagine you are in a car is going around a curve — like an exit ramp — at a high (but safe!) rate of speed; you know that you are acted on by forces as you go around the curve, this is why you end up leaning to the side. At this instant, you are being "pushed" in a direction perpendicular to your path of travel. So perhaps being able to examine the components of your acceleration vector in this direction is more useful than being able to examine the components of your acceleration vector relative to some arbitrary coordinate axes which might have their origin who knows where ... Cleveland?

In Sec. 17.2, we unlocked the method to resolve this. We learned how to develop unit vectors tangent and normal to a path of motion. If we can come up with one more vector, then we have a mini coordinate system (i.e. a set of mutually perpendicular unit vectors) which travels with a moving object along the path of motion. The missing vector is called the "binormal" vector, and the resulting set of three mutually orthogonal unit vectors is called the TNB-frame, or the Fresnet–Serret frame.

The TNB Frame (Fresnet–Serret Frame)

Given a vector function $\mathbf{r}(t)$, we would like to find a trio of mutually perpendicular unit vectors at any point along the contour. In Sec. 17.2, we discovered how to form two of these:

$$\mathbf{T}(t) = \frac{\mathbf{r}'(t)}{|\mathbf{r}'(t)|} \quad , \quad \mathbf{N}(t) = \frac{\mathbf{T}'(t)}{|\mathbf{T}'(t)|}$$

where $\mathbf{T}$ is a unit tangent vector and $\mathbf{N}$ is a unit normal vector. The third vector which completes the set is simple to find. We need a third vector which is perpendicular to the first two, and we already know that the cross product of two vectors is perpendicular to both. And so, we introduce the "binormal" vector,

$$\mathbf{B}(t) = \mathbf{T}(t) \times \mathbf{N}(t) \tag{17.14}$$

FFT: You should be a bit skeptical about this ... sure, this vector is going to be perpendicular to both $\mathbf{T}(t)$ and $\mathbf{N}(t)$, but are we sure it's a unit vector? The answer is yes, it is for sure a unit vector. You should challenge yourself to figure out why, on your own.

Note that the computation of $\mathbf{T}(t)$, $\mathbf{N}(t)$ and $\mathbf{B}(t)$ must proceed in order: we need $\mathbf{T}(t)$ to get $\mathbf{N}(t)$, and we need both $\mathbf{T}(t)$ and $\mathbf{N}(t)$ to get $\mathbf{B}(t)$. You might also notice that these calculations have the potential to get very ugly! As when finding just $\mathbf{T}$ and $\mathbf{N}$ in Sec. 17.2, be alert to when you can jump out of computation of a general formula for any one of these quantities, and instead consider only the needed values at a single point. Here is a general guide:

(1) Calculate $\mathbf{r}'(t)$ and $|\mathbf{r}'(t)|$ in their full glory.
(2) Build the general expression for the unit tangent vector, $\mathbf{T}(t) = \mathbf{r}'(t)/|\mathbf{r}'(t)|$.
(3) Compute the derivative $\mathbf{T}'(t)$. This will be the worst step by far. (See the Useful Fact below about a product-rule type of process.)
(4) But at this point, we only need numerical computations. For the specific value of t you are interested in, say $t = a$:

   (a) From the general formulas, find $\mathbf{r}'(a)$ and $|\mathbf{r}'(a)|$, and use them to compute

$$\boxed{\mathbf{T}(a) = \frac{\mathbf{r}'(a)}{|\mathbf{r}'(a)|}}$$

   (b) From the general formula, find $\mathbf{T}'(a)$.
   (c) From the previous step, then calculate $|\mathbf{T}'(a)|$.

(d) Use the above two to build

$$\boxed{\mathbf{N}(a) = \frac{\mathbf{T}'(a)}{|\mathbf{T}'(a)|}}$$

(e) Having found $\mathbf{T}(a)$ and $\mathbf{N}(a)$, build

$$\boxed{\mathbf{B}(a) = \mathbf{T}(a) \times \mathbf{N}(a)}$$

In these first examples, a lot of the vectors we need have been found already in other examples or problems.

EX 1 Determine the TNB-frame for the helix $\mathbf{r}(t) = \langle 3\cos t, 3\sin t, 4t\rangle$ at the point $(0, 3, 2\pi)$.

Note that the point $(0, 3, 2\pi)$ corresponds to $t = \pi/2$. We found $\mathbf{T}$ and $\mathbf{N}$ at this point in EX 4 of Sec. 17.2:

$$\mathbf{T}\left(\frac{\pi}{2}\right) = \left\langle \frac{3}{5}, 0, \frac{4}{5} \right\rangle \quad , \quad \mathbf{N}\left(\frac{\pi}{2}\right) = \langle 0, -1, 0\rangle$$

And so, the only thing left to do is find $\mathbf{B}$ with (17.14):

$$\mathbf{B}\left(\frac{\pi}{2}\right) = \mathbf{T}\left(\frac{\pi}{2}\right) \times \mathbf{N}\left(\frac{\pi}{2}\right) = \left\langle \frac{3}{5}, 0, \frac{4}{5} \right\rangle \times \langle 0, -1, 0\rangle = \left\langle \frac{4}{5}, 0, -\frac{3}{5} \right\rangle$$

Quality assurance: we can confirm that $\mathbf{T}$, $\mathbf{N}$, and $\mathbf{B}$ are all unit vectors, and they are all mutually perpendicular. ■

EX 2 Determine the TNB-frame for $\mathbf{r}(t) = \langle t^2, 2t^3/3, t\rangle$ at the point $(1, 2/3, 1)$.

Note that the given point corresponds to $t = 1$. We found $\mathbf{T}$ and $\mathbf{N}$ at this point in CP 2 of Sec. 17.2:

$$\mathbf{T}(1) = \left\langle \frac{2}{3}, \frac{2}{3}, \frac{1}{3} \right\rangle \quad \text{and} \quad \mathbf{N}(1) = \left\langle -\frac{1}{3}, \frac{2}{3}, -\frac{2}{3} \right\rangle$$

And so, the only thing left to do is find $\mathbf{B}(1)$ with (17.14):

$$\mathbf{B}(1) = \mathbf{T}(1) \times \mathbf{N}(1) = \left\langle \frac{2}{3}, \frac{2}{3}, \frac{1}{3} \right\rangle \times \left\langle \frac{-1}{3}, \frac{2}{3}, -\frac{2}{3} \right\rangle = \left\langle -\frac{2}{3}, \frac{1}{3}, \frac{2}{3} \right\rangle$$

Quality assurance: we can confirm that $\mathbf{T}$, $\mathbf{N}$, and $\mathbf{B}$ are all unit vectors, and they are all mutually perpendicular. ■

You Try It

(1) In YTI 3 and PP 5 of Sec. 17.2, we found the unit tangent and normal vectors $\mathbf{T}$ and $\mathbf{N}$ for $\mathbf{r}(t) = (4t^{3/2}/3)\mathbf{i} + t^2\mathbf{j} + t\mathbf{k}$ at $t = 1$. Find the associated unit binormal vector, $\mathbf{B}$.

(2) In YTI 5 of Sec. 17.2, we found the unit tangent vector $\mathbf{T}$ for $\mathbf{r}(t) = 2\sin t\mathbf{i} + 2\cos t\mathbf{j} + (\sin t - \cos t)\mathbf{k}$ at $t = 3\pi/4$. Only one of the following vectors could be the unit normal there; choose it, and then calculate the unit binormal vector $\mathbf{B}$.

$$(a)\ \left\langle \frac{3}{\sqrt{22}}, \frac{3}{\sqrt{22}}, \frac{2}{\sqrt{22}} \right\rangle \qquad (b)\ \left\langle -\frac{3}{\sqrt{22}}, \frac{3}{\sqrt{22}}, -\frac{2}{\sqrt{22}} \right\rangle$$

$$(c)\ \left\langle \frac{3}{\sqrt{22}}, \frac{3}{\sqrt{22}}, -\frac{2}{\sqrt{22}} \right\rangle \qquad (d)\ \left\langle -\frac{3}{\sqrt{22}}, \frac{3}{\sqrt{22}}, \frac{2}{\sqrt{22}} \right\rangle$$

When finding a Fresnet–Serret frame from scratch, the worst step is finding the derivative $\mathbf{T}'(t)$, where $\mathbf{T}(t) = \mathbf{r}'(t)/|\mathbf{r}'(t)|$. Since we can write $\mathbf{r}(t) = \langle x(t), y(t), z(t)\rangle$ and $\mathbf{r}'(t) = \langle x'(t), y'(t), z'(t)\rangle$, then we can think of $\mathbf{T}(t)$ as having either of these forms:

$$\mathbf{T}(t) = \frac{1}{|\mathbf{r}'(t)|}\langle x'(t), y'(t), z'(t)\rangle \quad \text{or} \quad \mathbf{T}(t) = \left\langle \frac{x'(t)}{|\mathbf{r}'(t)|}, \frac{y'(t)}{|\mathbf{r}'(t)|}, \frac{z'(t)}{|\mathbf{r}'(t)|} \right\rangle$$

In the latter form, finding $\mathbf{T}'(t)$ requires three quotient rules, one for each component. What sane person would want to do that? If we consider the former version, finding $\mathbf{T}'(t)$ is a bit easier *if* there is a product-rule type of operation for the product of a scalar function with a vector function. Luckily, there is!

Useful Fact 17.2. *The derivative of a scalar function multiple of a vector function can be found using a product rule construction; if* $\beta(t)$ *is the scalar multiple and* $\mathbf{r}(t) = \langle f(t), g(t), h(t)\rangle$ *is the vector function, then*

$$\frac{d}{dt}(\beta(t)\mathbf{r}(t)) = \beta'(t)\mathbf{r}(t) + \beta(t)\mathbf{r}'(t) \tag{17.15}$$

or in more detail,

$$\frac{d}{dt}\beta(t)\langle f(t), g(t), h(t)\rangle = \beta'(t)\langle f(t), g(t), h(t)\rangle + \beta(t)\langle f'(t), g'(t), h'(t)\rangle$$

So, the derivative of

$$\frac{1}{|\mathbf{r}'(t)|}\langle x'(t), y'(t), z'(t)\rangle$$

can be found using this Useful Fact by assigning $\beta(t) = 1/|\mathbf{r}'(t)|$.

The relation (17.15) in Useful Fact 17.2 is another example of a familiar scalar function derivative rule extended to include vector functions — for comparison, see Useful Fact 16.10 in Sec. 16.3, and also Challenge Problem 4 in that same section. See Challenge Problem 4 in this section for your chance to test your derivation skills on this new one.

You Try It

(3) Determine the complete TNB-frame for $\mathbf{r}(t) = \langle e^{-t}, e^{-2t}, 2t\rangle$ at $t = 0$.

Simple Velocity and Acceleration Problems

Suppose $\mathbf{r}(t)$ represents the position of a particle moving through space, as a function of time t. Then you can guess that vector functions for the velocity $\mathbf{v}(t)$ and acceleration $\mathbf{a}(t)$ of the moving particle are:

$$\begin{aligned}\mathbf{v}(t) &= \mathbf{r}'(t)\\ \mathbf{a}(t) &= \mathbf{v}'(t) = \mathbf{r}''(t)\end{aligned}$$

Thus, given a position function, it is easy to discover the velocity and acceleration functions.

EX 3 Find the velocity and acceleration functions for the particle moving according to $\mathbf{r}(t) = \langle t^2+1, t^3, t^2-1\rangle$. Does the particle move at a constant speed?

For the position function $\mathbf{r}(t) = \langle t^2+1, t^3, t^2-1\rangle$, the velocity function is

$$\mathbf{v}(t) = \mathbf{r}'(t) = \langle 2t, 3t^2, 2t\rangle$$

and acceleration is

$$\mathbf{a}(t) = \mathbf{v}'(t) = \langle 2, 6t, 2\rangle$$

Since speed is the magnitude of velocity, we have

$$s = |\mathbf{v}(t)| = \sqrt{9t^4 + 8t^2} = |t|\sqrt{9t^2+8}$$

and so this particle does not move at constant speed. ■

You Try It

(4) Find the velocity and acceleration functions for the particle moving according to $\mathbf{r}(t) = \langle 2\cos t, 3t, 2\sin t\rangle$. Does the particle move at a constant speed?

We can also take a given acceleration function and determine the associated velocity and position functions — but only if we are given extra data about the velocity and position of the particle. To go from acceleration to velocity to position requires *antiderivatives*, which come along with arbitrary constants of integration; thus, we need data points or *initial conditions* to pin down specific velocity and acceleration functions.

EX 4 Find the velocity and positions functions for a particle moving with acceleration $\mathbf{a}(t) = \langle 0, 0, 1\rangle$, if we know that $\mathbf{v}(0) = \langle 1, -1, 0\rangle$ and $\mathbf{r}(0) = \langle 0, 0, 0\rangle$.

We have $\mathbf{a}(t) = \mathbf{k} = \langle 0, 0, 1\rangle$ with initial conditions $\mathbf{v}(0) = \mathbf{i} - \mathbf{j} = \langle 1, -1, 0\rangle$ and $\mathbf{r}(0) = \mathbf{0} = \langle 0, 0, 0\rangle$. The velocity function can be found from the acceleration function,

$$\mathbf{v}(t) = \int \mathbf{a}(t)dt = \int \langle 0, 0, 1\rangle\, dt = \langle c_1, c_2, t + c_3\rangle$$

Since we know that $\mathbf{v}(0) = \langle 1, -1, 0\rangle$, we can get

$$\mathbf{v}(0) = \langle c_1, c_2, c_3\rangle = \langle 1, -1, 0\rangle$$

so that $c_1 = 1, c_2 = -1, c_3 = 0$ and the specific velocity function is then $\mathbf{v}(t) = \langle 1, -1, t\rangle$. Then, from this velocity function we can get the position function,

$$\mathbf{r}(t) = \int \mathbf{v}(t)dt = \left\langle t + c_1, -t + c_2, \frac{1}{2}t^2 + c_3\right\rangle$$

The initial condition for position lets us find these new constants of integration:

$$\mathbf{r}(0) = \langle c_1, c_2, c_3\rangle = \langle 0, 0, 0\rangle$$

so that $c_1 = c_2 = c_3 = 0$ and the specific position function is then

$$\mathbf{r}(t) = \left\langle t, -t, \frac{1}{2}t^2\right\rangle \quad \blacksquare$$

You Try It

(5) Find the velocity and positions functions for a particle moving with acceleration $\mathbf{a}(t) = -10\mathbf{k}$, if we know that $\mathbf{v}(0) = \mathbf{i} + \mathbf{j} - \mathbf{k}$ and $\mathbf{r}(0) = 2\mathbf{i} + 3\mathbf{j}$.

Tangential and Normal Components of Acceleration

When you take an exit ramp off a highway, the components of acceleration that you feel as you go around the sharp curve are certainly not oriented with respect to some arbitrary x- and y-axes set at the universal center of coordinates. Rather, those forces (which are related to components of your acceleration) are directed along and perpendicular to your direction of motion. So, acceleration problems are often better solved by finding the components of acceleration along directions that are tangential and normal to the direction of motion. That is, we would like to find the components of acceleration of a moving object along the vectors $\mathbf{T}(t)$ and $\mathbf{N}(t)$. These components are called a_T and a_N, and the full acceleration function can then be written

$$\mathbf{a}(t) = a_T\mathbf{T}(t) + a_N\mathbf{N}(t)$$

Let's find a_T first by recognizing that a_T is simply the scalar component (projection) of $\mathbf{a}(t)$ onto $\mathbf{T}(t)$:

$$a_T = \text{comp}_{\mathbf{T}}\mathbf{a} = \frac{\mathbf{a(t)} \cdot \mathbf{T(t)}}{|\mathbf{T(t)}|}$$

But remember that $\mathbf{T}(t)$ is a unit vector, and so $|\mathbf{T(t)}| = 1$. Also remember that $\mathbf{a}(t) = \mathbf{r}''(t)$ and

$$\mathbf{T}(t) = \frac{\mathbf{r}'(t)}{|\mathbf{r}'(t)|}$$

so

$$a_T = \mathbf{a(t)} \cdot \mathbf{T(t)} = \mathbf{r}''(t) \cdot \frac{\mathbf{r}'(t)}{|\mathbf{r}'(t)|}$$

or in another order,

$$a_T = \frac{\mathbf{r}'(t) \cdot \mathbf{r}''(t)}{|\mathbf{r}'(t)|}$$

To find a_N, we simply must recognize that a_T and a_N are components of $\mathbf{a}(t)$ in two perpendicular directions, and so they are the lengths of two legs of a right triangle in which the vector $\mathbf{a}(t)$ itself is the hypotenuse:

The Pythagorean Theorem tells us that

$$a_N = \sqrt{|\mathbf{a}(t)|^2 - a_T^2}$$

EX 5 Find the tangential and normal components of acceleration for $\mathbf{r}(t) = (3t - t^3)\mathbf{i} + 3t^2\mathbf{j}$, and interpret the results.

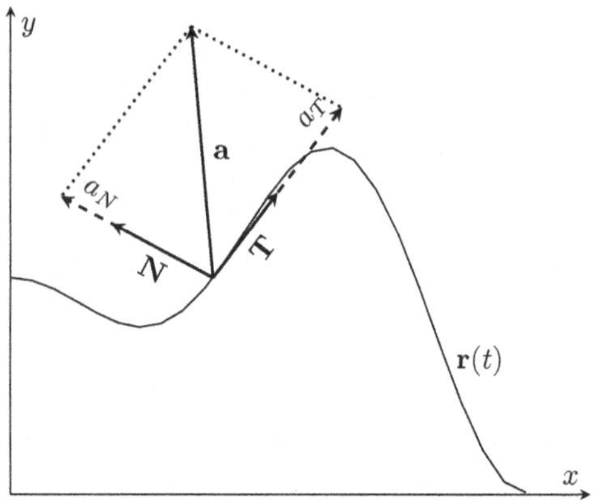

Fig. 17.1 Tangential and normal components of an acceleration vector.

To find the tangential component of acceleration for $\mathbf{r}(t) = (3t - t^3)\mathbf{i} + 3t^2\mathbf{j}$, we need:

$$\begin{aligned}
\mathbf{r}'(t) &= (3 - 3t^2)\mathbf{i} + 6t\mathbf{j} \\
|\mathbf{r}'(t)| &= \sqrt{(9 - 18t^2 + 9t^4) + 36t^2} \\
&= \sqrt{9t^4 + 18t^2 + 9} = \sqrt{(3t^2 + 3)^2} = 3(t^2 + 1) \\
\mathbf{r}''(t) &= -6t\mathbf{i} + 6\mathbf{j} \\
\mathbf{r}'(t) \cdot \mathbf{r}''(t) &= -6t(3 - 3t^2) + 6(6t) = 18t^3 + 18t = 18t(t^2 + 1)
\end{aligned}$$

so then

$$a_T = \frac{\mathbf{r}'(t) \cdot \mathbf{r}''(t)}{|\mathbf{r}'(t)|} = \frac{18t(t^2 + 1)}{3(t^2 + 1)} = 6t$$

To find the normal component of acceleration, we need $|\mathbf{a}(t)|^2$. Note that $\mathbf{a}(t) = \mathbf{r}''(t)$, which is found above; so,

$$\begin{aligned}
\mathbf{a}(t) &= -6t\mathbf{i} + 6\mathbf{j} \\
|\mathbf{a}(t)|^2 &= 36t^2 + 36
\end{aligned}$$

so that

$$a_N = \sqrt{|\mathbf{a}(t)|^2 - a_T^2} = \sqrt{(36t^2 + 36) - (6t)^2} = 6$$

Together, then, we can write the acceleration vector as $\mathbf{a}(t) = 6t\mathbf{T}(t) + 6\mathbf{N}(t)$. This means that the component of acceleration normal (perpendicular) to the direction of motion is constant, while the component tangent to the direction of motion varies. ■

You Try It

(6) Find the tangential and normal components of acceleration for $\mathbf{r}(t) = (1+t)\mathbf{i} + (t^2 - 2t)\mathbf{j}$, and interpret the results.

The Straight Story on Curvature

With the TNB-frame (Fresnet–Serret frame) unlocked, we can start to take a deeper look at characteristics of curves in space. Two of the primary properties of space curves are *curvature* and *torsion*. The curvature of curve at a point tells us how sharply the curve bends in the plane of motion there. The curvature at any point on a line is 0, while the curvature of a circle is constant all around the circle, and is related to the radius. FFT: Would you like to predict if a larger radius gives a larger or smaller curvature? When you are riding the Tilt-A-Whirl at an amusement park, you are very aware that your personal $\mathbf{T}$ changes a LOT as your absolute position changes only slightly; you are experiencing a large curvature on your path of motion. Torsion measures (somewhat) how much a curve is bending *out of* the plane of motion; while curvature can be examined in 2D or 3D, torsion has meaning only in 3D.

The good news is that these properties can have relatively straightforward, although cumbersome, computational formulas. The bad news is that they are simpler to examine when you have written your space curves with arc length s as the parameter. (This is "bad news" because as we saw in Sec. 17.1, we often have curves delivered with a parameter that represents time t, for which generating an explicit arc length parameterization isn't always simple or even possible.)

What we can do, though, is start by assuming we have curves parameterized by arc length, develop information about curvature and torsion in such cases, and then expand that information so that we can apply it to any curve. The most fundamental definition of curvature is this:

Definition 17.3. *Let $\mathbf{r}(s)$ be a vector function parameterized by arc length. The curvature κ of $\mathbf{r}$ at any point is given by $\kappa(s) = |\mathbf{r}''(s)|$.*

This definition won't be the one we normally use to compute curvature, but it's the one that sets the stage intuitively. As we know, $\mathbf{r}''(s)$ is the rate of change of $\mathbf{r}'(s)$. Now $\mathbf{r}'(s)$ is a vector quantity, and so any change must be quantified by change in magnitude or direction. But, we know that the magnitude of $\mathbf{r}'(s)$ is $|\mathbf{r}'(s)| = 1$, which does not change. So any "change" measured in $\mathbf{r}'(s)$ must be a change in direction; therefore, $|\mathbf{r}''(s)|$ measures how much the direction of tangent lines to $\mathbf{r}(s)$ are changing at a given point. So in effect, we are indeed measuring how much the curve is bending.

This gets even better: we know from Sec. 17.1 that $\mathbf{r}''(s)$ is perpendicular to the contour at any point. So since our unit normal vectors $\mathbf{N}(s)$ are perpendicular to the contour at a point, and so give the direction of $\mathbf{r}''(s)$, and we have now established that $\kappa(s)$ measures the *magnitude* of $\mathbf{r}''(s)$, then together,

$$\mathbf{r}''(s) = \kappa(s)\mathbf{N}(s) \tag{17.16}$$

Further, since our unit tangent vectors $\mathbf{T}(s)$ are given by $\mathbf{T}(s) = \mathbf{r}'(s)$, then $\mathbf{T}'(s) = \mathbf{r}''(s)$, and so by (17.16),

$$\mathbf{T}'(s) = \kappa(s)\mathbf{N}(s) \tag{17.17}$$

It seems that curvature is tangled up with both our unit tangent and normal vectors.

Note that Eq. (17.17) emphasizes that $\mathbf{T}'(s)$ is parallel to $\mathbf{N}(s)$. Similarly, it's true that $\mathbf{B}'(s)$ is also parallel to $\mathbf{N}(s)$; the proof of this is given below, in *the Pit*! But we can use this fact to define torsion. Since $\mathbf{B}'(s)$ is parallel to $\mathbf{N}(s)$, then there is some function $\tau(s)$ for which we can write

$$\mathbf{B}'(s) = \tau(s)\mathbf{N}(s) \tag{17.18}$$

which looks a lot like (17.17) in structure, and in fact, it helps define the torsion $\tau(s)$ of our curve. So we can say that curvature is established via $\mathbf{T}'(s) = \kappa(s)\mathbf{N}(s)$ and torsion is established via $\mathbf{B}'(s) = \tau(s)\mathbf{N}(s)$. If you're really on your toes, you've seen Eqs. (17.17) and (17.18) and are asking, "Well, if there are tidy expressions involving curvature and torsion for $\mathbf{T}'(s)$ and $\mathbf{B}'(s)$, what about $\mathbf{N}'(s)$?" That's a great question! We can indeed generate such an expression, but it's not as nice as the other two. Because of their orientations relative to each other (as a right-hand-rule triad), we have $\mathbf{B} = \mathbf{T} \times \mathbf{N}$, $\mathbf{T} = \mathbf{N} \times \mathbf{B}$, and $\mathbf{N} = \mathbf{B} \times \mathbf{T}$. Starting with the latter, we can look for $\mathbf{N}'(s)$:

$$\begin{aligned}\frac{d}{ds}\mathbf{N} &= \frac{d}{ds}\mathbf{B}\times\mathbf{T}\\ &= \mathbf{B}'\times\mathbf{T}+\mathbf{B}\times\mathbf{T}'\\ &= (\tau\mathbf{N})\times\mathbf{T}+\mathbf{B}\times(\kappa\mathbf{N})\\ &= \tau(\mathbf{N}\times\mathbf{T})+\kappa(\mathbf{B}\times\mathbf{N})\\ &= \tau(-\mathbf{B})+\kappa(-\mathbf{T})\end{aligned}$$

or, finally,

$$\mathbf{N}'(s) = -\kappa(s)\mathbf{T}(s) - \tau(s)\mathbf{B}(s) \tag{17.19}$$

Equations (17.17), (17.18), and (17.19) together are often called the Fresnet–Serret formulas,

$$\begin{aligned}\mathbf{T}'(s) &= \kappa(s)\mathbf{N}(s)\\ \mathbf{N}'(s) &= -\kappa(s)\mathbf{T}(s) - \tau(s)\mathbf{B}(s)\\ \mathbf{B}'(s) &= \tau(s)\mathbf{N}(s)\end{aligned}$$

If you find all this development intriguing, then once you are done with Multivariable Calculus and a good proof-based Linear Algebra course, head for your nearest course in Differential Geometry.

At this point, though, you may be saying, "Come on, man, how are we going to compute these things?" To get more useful formulas, we have to switch from using arc length parameter s to the more generic t. The defining expression $\kappa(s) = \mathbf{r}''(s)$ is equivalent to either of

$$\kappa(t) = \frac{|\mathbf{T}'(t)|}{|\mathbf{r}'(t)|} \qquad \text{or} \qquad \kappa(t) = \frac{|\mathbf{r}'(t)\times\mathbf{r}''(t)|}{|\mathbf{r}'(t)|^3} \tag{17.20}$$

If we have already computed $\mathbf{T}(t)$ for a particular curve, then the first formula in (17.20) is probably the one to go with. If we are starting the curvature computation from scratch, then the second might be the most efficient. FFT: Can you convert the arc length version $\kappa(s)$ to either of the more generic versions $\kappa(t)$?

Torsion is defined via

$$\tau(t) = \frac{(\mathbf{r}'(t)\times\mathbf{r}''(t))\cdot\mathbf{r}'''(t)}{|\mathbf{r}'(t)\times\mathbf{r}''(t)|^2} \tag{17.21}$$

There is not universal agreement on the signage of torsion, so in some sources, the right side of (17.21) actually defines $-\tau(t)$.

EX 6 Find the full curvature function $\kappa(t)$ and torsion function $\tau(t)$ of $\mathbf{r}(t) = \langle t^2, 1, t \rangle$.

For the curvature of $\mathbf{r}(t) = \langle t^2, 1, t \rangle$ we need:

$$\begin{aligned} \mathbf{r}'(t) &= \langle 2t, 0, 1 \rangle \\ |\mathbf{r}'(t)| &= \sqrt{4t^2+1} \\ \mathbf{r}''(t) &= \langle 2, 0, 0 \rangle \\ \mathbf{r}'(t) \times \mathbf{r}''(t) &= \langle 0, -2, 0 \rangle \\ |\mathbf{r}'(t) \times \mathbf{r}''(t)| &= 2 \end{aligned}$$

so that

$$\kappa(t) = \frac{|\mathbf{r}'(t) \times \mathbf{r}''(t)|}{|\mathbf{r}'(t)|^3} = \frac{2}{(4t^2+1)^{3/2}}$$

To use Eq. (17.21) to compute torsion, we need the same functions $\mathbf{r}'(t)$ and $\mathbf{r}''(t)$, as well as $\mathbf{r}'''$; but since $\mathbf{r}''(s) = \langle 2, 0, 0 \rangle$, we have $\mathbf{r}''' = \langle 0, 0, 0 \rangle$. The result of (17.21) is $\tau(t) = 0$. This makes sense, since torsion is inherently a 3D quantity, and this curve — because of it's constant second component $y(t) = 1$, only has motion in a two-dimensional plane. ■

You Try It

(7) Find the curvature function $\kappa(t)$ of $\mathbf{r}(t) = \langle t, t, 1 + t^2 \rangle$, and the specific torsion $\tau(2)$.

Into the Pit!!

Into the Pit! $\mathbf{B}'(s)$ *is Parallel to* $\mathbf{N}(s)$

Proving that $\mathbf{B}'(s)$ is parallel to $\mathbf{N}(s)$ is equivalent to proving that $\mathbf{B}'(s)$ is perpendicular to $\mathbf{T}(s)$. Let's start at the beginning, where $\mathbf{B}(s) = \mathbf{T}(s) \times \mathbf{N}(s)$. Then using the result of CP 4 in Sec. 16.3 (you did do that problem, didn't you?), we get

$$\mathbf{B}'(s) = \mathbf{T}'(s) \times \mathbf{N}(s) + \mathbf{T}(s) \times \mathbf{N}'(s)$$

We know that $\mathbf{T}'(s)$ is parallel to $\mathbf{N}(s)$, as established in (17.17). So, $\mathbf{T}'(s) \times \mathbf{N}(s) = \mathbf{0}$. Therefore, we now have simply,

$$\mathbf{B}'(s) = \mathbf{T}(s) \times \mathbf{N}'(s)$$

Since $\mathbf{B}'(s)$ is the result of the cross product between $\mathbf{T}(s)$ and another vector, $\mathbf{B}'(s)$ is perpendicular to $\mathbf{T}(s)$. But by definition, $\mathbf{N}(s)$ is also perpendicular to $\mathbf{T}(s)$. So $\mathbf{B}'(s)$ is parallel to $\mathbf{N}(s)$.

Gosh, that wasn't so bad, was it?

Fresnet–Serret Frame — Problem List

Fresnet–Serret Frame — You Try It

These appeared above; solutions begin on the next page.

(1) In YTI 4 and PP 5 of Sec. 17.2, we found the unit tangent and normal vectors $\mathbf{T}$ and $\mathbf{N}$ for $\mathbf{r}(t) = (4t^{3/2}/3)\mathbf{i} + t^2\mathbf{j} + t\mathbf{k}$ at $t = 1$. Find the associated unit binormal vector, $\mathbf{B}$.

(2) In YTI 5 of Sec. 17.2, we found the unit tangent vector $\mathbf{T}$ for $\mathbf{r}(t) = 2\sin t\mathbf{i} + 2\cos t\mathbf{j} + (\sin t - \cos t)\mathbf{k}$ at $t = 3\pi/4$. Only one of the following vectors could be the unit normal there; choose it, and then calculate the unit binormal vector $\mathbf{B}$.

$$(a)\ \left\langle \frac{3}{\sqrt{22}}, \frac{3}{\sqrt{22}}, \frac{2}{\sqrt{22}} \right\rangle \qquad (b)\ \left\langle -\frac{3}{\sqrt{22}}, \frac{3}{\sqrt{22}}, -\frac{2}{\sqrt{22}} \right\rangle$$

$$(c)\ \left\langle \frac{3}{\sqrt{22}}, \frac{3}{\sqrt{22}}, -\frac{2}{\sqrt{22}} \right\rangle \qquad (d)\ \left\langle -\frac{3}{\sqrt{22}}, \frac{3}{\sqrt{22}}, \frac{2}{\sqrt{22}} \right\rangle$$

(3) Determine the complete TNB-frame for $\mathbf{r}(t) = \langle e^{-t}, 2e^t, 2t \rangle$ at $t = 0$.

(4) Find the velocity and acceleration functions for the particle moving according to $\mathbf{r}(t) = \langle 2\cos t, 3t, 2\sin t \rangle$. Does the particle move at a constant speed?

(5) Find the velocity and positions functions for a particle moving with acceleration $\mathbf{a}(t) = -10\mathbf{k}$, if we know that $\mathbf{v}(0) = \mathbf{i} + \mathbf{j} - \mathbf{k}$ and $\mathbf{r}(0) = 2\mathbf{i} + 3\mathbf{j}$.

(6) Find the tangential and normal components of acceleration for $\mathbf{r}(t) = (1 + t)\mathbf{i} + (t^2 - 2t)\mathbf{j}$, and interpret the results.

(7) Find the curvature function $\kappa(t)$ of $\mathbf{r}(t) = \langle t, t, 1 + t^2 \rangle$, and the specific torsion $\tau(2)$.

Fresnet–Serret Frame — Practice Problems

Try these as you get the hang of the You Try It problems. Solutions to these problems are available in Sec. B.5.3.

(1) Find the unit binormal vector $\mathbf{B}$ for $\mathbf{r}(t) = \langle e^t, e^t \sin t, e^t \cos t\rangle$ at the point (1,0,1). (Hint: Have $\mathbf{T}$ and $\mathbf{N}$ been calculated elsewhere?)
(2) Find the unit binormal vector $\mathbf{B}$ for $\mathbf{r}(t) = \langle \cos t + t \sin t, \sin t - t \cos t, 1\rangle$ at the point $(1, 0, 1)$. (Hint: Have $\mathbf{T}$ and $\mathbf{N}$ been calculated elsewhere?)
(3) Determine the complete TNB-frame for $\mathbf{r}(t) = \langle e^t, e^t \sin t, e^t \cos t\rangle$ at the point $(e^\pi, 0, e^\pi)$. (Hint: A previously solved problem will be very useful.)
(4) Find the velocity and acceleration functions for the particle moving according to $\mathbf{r}(t) = \langle t^2, \ln t, t\rangle$. Does the particle move at a constant speed?
(5) Find the velocity and positions functions for a particle moving with acceleration $\mathbf{a}(t) = t\mathbf{i} + t^2\mathbf{j} + \cos 2t\mathbf{k}$, if we know that $\mathbf{v}(0) = \mathbf{i} + \mathbf{k}$ and $\mathbf{r}(0) = \mathbf{j}$.
(6) Find the tangential and normal components of acceleration for $\mathbf{r}(t) = t\mathbf{i} + t^2\mathbf{j} + 3t\mathbf{k}$.
(7) Find the curvature and torsion of $\mathbf{r}(t) = \langle e^t \cos t, e^t \sin t, t\rangle$ at the point (1,0,0).

Fresnet–Serret Frame — Challenge Problems

Try these problems to test your skills with the ideas in this section. Solutions to these problems are available in Sec. C.5.3.

(1) The shuttlecraft *Galileo* is lifting off from the planet Taurus 2 along the path given by $\mathbf{r}(t) = \langle t^5/5, 2t^3/3, t\rangle$ in the galactic coordinate system. When at the point marked by $t = 1$, the shuttle launches a probe in the direction of the binormal vector $\mathbf{B}(1)$. If the probe travels in a straight line, it will cross through two of three galactic coordinate planes (xy, xz, or yz). Which two planes will it cross, and at which galactic coordinates?
(2) Find the tangential and normal components of acceleration for $\mathbf{r}(t) = \langle e^{-t} \cos t, e^{-t} \sin t\rangle$ at $t = \pi/4$. (Hint: How can you rig up a cross product involving vectors that are only two-dimensional?)
(3) Show that the curvature of a circle of radius a is $\kappa = 1/a$. What is the torsion anywhere on this circle?
(4) (Bonus Time in *the Pit*!) Can you derive the relation (17.15) in Useful Fact 17.2?

The Fresnet–Serret Frame — You Try It — Solved

(1) In YTI 4 and PP 5 of Sec. 17.2, we found the unit tangent and normal vectors $\mathbf{T}$ and $\mathbf{N}$ for $\mathbf{r}(t) = (4t^{3/2}/3)\mathbf{i} + t^2\mathbf{j} + t\mathbf{k}$ at $t = 1$. Find the associated unit binormal vector, $\mathbf{B}$.

☐ In YTI 4 of the previous section, we found

$$\mathbf{T}(1) = \frac{2}{3}\mathbf{i} + \frac{2}{3}\mathbf{j} + \frac{1}{3}\mathbf{k}$$

In PP 5, we found:

$$\mathbf{N}(1) = -\frac{1}{3}\mathbf{i} + \frac{2}{3}\mathbf{j} - \frac{2}{3}\mathbf{k}$$

The binormal vector follows immediately as:

$$\begin{aligned}\mathbf{B}(1) = \mathbf{T}(1) \times \mathbf{N}(1) &= \left(\frac{2}{3}\mathbf{i} + \frac{2}{3}\mathbf{j} + \frac{1}{3}\mathbf{k}\right) \times \left(-\frac{1}{3}\mathbf{i} + \frac{2}{3}\mathbf{j} - \frac{2}{3}\mathbf{k}\right)\\ &= -\frac{2}{3}\mathbf{i} + \frac{1}{3}\mathbf{j} + \frac{2}{3}\mathbf{k}\end{aligned}$$

In summary, the entire TNB-frame for $\mathbf{r}(t)$ at $t = 1$ is

$$\begin{aligned}\mathbf{T}(1) &= \frac{2}{3}\mathbf{i} + \frac{2}{3}\mathbf{j} + \frac{1}{3}\mathbf{k}\\ \mathbf{N}(1) &= -\frac{1}{3}\mathbf{i} + \frac{2}{3}\mathbf{j} - \frac{2}{3}\mathbf{k}\\ \mathbf{B}(1) &= -\frac{2}{3}\mathbf{i} + \frac{1}{3}\mathbf{j} + \frac{2}{3}\mathbf{k} \quad \blacksquare\end{aligned}$$

(2) In YTI 5 of Sec. 17.2, we found the unit tangent vector $\mathbf{T}$ for $\mathbf{r}(t) = 2\sin t\mathbf{i} + 2\cos t\mathbf{j} + (\sin t - \cos t)\mathbf{k}$ at $t = 3\pi/4$. Only one of the following vectors could be the unit normal there; choose it, and then calculate the unit binormal vector $\mathbf{B}$.

$(a)\ \left\langle \frac{3}{\sqrt{22}}, \frac{3}{\sqrt{22}}, \frac{2}{\sqrt{22}} \right\rangle \qquad (b)\ \left\langle -\frac{3}{\sqrt{22}}, \frac{3}{\sqrt{22}}, -\frac{2}{\sqrt{22}} \right\rangle$

$(c)\ \left\langle \frac{3}{\sqrt{22}}, \frac{3}{\sqrt{22}}, -\frac{2}{\sqrt{22}} \right\rangle \qquad (d)\ \left\langle -\frac{3}{\sqrt{22}}, \frac{3}{\sqrt{22}}, \frac{2}{\sqrt{22}} \right\rangle$

☐ From YTI 5 in Sec. 17.2, we found

$$\mathbf{T}\left(\frac{3\pi}{4}\right) = -\frac{\sqrt{2}}{2}\mathbf{i} - \frac{\sqrt{2}}{2}\mathbf{j} + 0\mathbf{k}$$

We know that $\mathbf{N}(3\pi/4)$ must be a unit vector, and must also be perpendicular to $\mathbf{T}(3\pi/4)$. All of the candidate vectors in (a)–(d) are unit vectors, so that condition is no help at all. However, only the candidate in (b) satisfies $\mathbf{T}(3\pi/4) \cdot \mathbf{N}(3\pi/4) = 0$. Therefore,

$$\mathbf{N}\left(\frac{3\pi}{4}\right) = -\frac{3}{\sqrt{22}}\mathbf{i} + \frac{3}{\sqrt{22}}\mathbf{j} - \frac{2}{\sqrt{22}}\mathbf{k}$$

So the completion of the TNB-frame for this function at the given point is

$$\begin{aligned}
\mathbf{B}\left(\frac{3\pi}{4}\right) &= \mathbf{T}\left(\frac{3\pi}{4}\right) \times \mathbf{N}\left(3\frac{\pi}{4}\right) \\
&= \left(-\frac{\sqrt{2}}{2}\mathbf{i} - \frac{\sqrt{2}}{2}\mathbf{j} + 0\mathbf{k}\right) \times \left(-\frac{3}{\sqrt{22}}\mathbf{i} + \frac{3}{\sqrt{22}}\mathbf{j} - \frac{2}{\sqrt{22}}\mathbf{k}\right) \\
&= -\frac{1}{\sqrt{11}}\mathbf{i} - \frac{1}{\sqrt{11}}\mathbf{j} - \frac{3}{\sqrt{11}}\mathbf{k}
\end{aligned}$$

(The details of the cross product took place behind the scenes.) A bit of quality assurance: **B** is a unit vector and is perpendicular to both **T** and **N**. ■

(3) Determine the complete TNB-frame for $\mathbf{r}(t) = \langle e^{-t}, 2e^t, 2t\rangle$ at $t = 0$.

□ Since $\mathbf{r}'(t) = \langle -e^{-t}, 2e^t, 2\rangle$, then

$$|\mathbf{r}'(t)| = \sqrt{e^{-2t} + 4e^{2t} + 4} = \sqrt{(2e^t + e^{-t})^2} = 2e^t + e^{-t}$$

and our first result at $t = 0$ is:

$$\begin{aligned}
\mathbf{r}'(0) &= \langle -1, 2, 2\rangle \\
|\mathbf{r}'(0)| &= 3 \\
\rightarrow \mathbf{T}(0) &= \left\langle -\frac{1}{3}, \frac{2}{3}, \frac{2}{3}\right\rangle
\end{aligned}$$

We can also build the general expression

$$\mathbf{T}(t) = \frac{1}{2e^t + e^{-t}}\langle -e^{-t}, 2e^t, 2\rangle$$

from which we get

$$\mathbf{T}'(t) = \frac{-(2e^t - e^{-t})}{(2e^t + e^{-t})^2}\langle -e^{-t}, 2e^t, 2\rangle + \frac{1}{2e^t + e^{-t}}\langle e^{-t}, 2e^t, 0\rangle$$

This is the final general expression we need, and now everything else can be built specifically for $t = 0$:

$$\begin{aligned}\mathbf{T}'(0) &= -\frac{1}{9}\langle -1,2,2\rangle + \frac{1}{3}\langle 1,2,0\rangle \\ &= \left\langle \frac{4}{9}, \frac{4}{9}, -\frac{2}{9}\right\rangle \\ |\mathbf{T}'(0)| &= \sqrt{\frac{36}{81}} = \frac{2}{3} \\ \rightarrow \mathbf{N}(0) &= \frac{\mathbf{T}'(0)}{|\mathbf{T}'(0)|} = \left\langle \frac{2}{3}, \frac{2}{3}, -\frac{1}{3}\right\rangle\end{aligned}$$

Finally,

$$\mathbf{B}(0) = \mathbf{T}(0) \times \mathbf{N}(0) \quad = \left\langle -\frac{1}{3}, \frac{2}{3}, \frac{2}{3}\right\rangle \times \left\langle \frac{2}{3}, \frac{2}{3}, -\frac{1}{3}\right\rangle = \left\langle -\frac{2}{3}, \frac{1}{3}, -\frac{2}{3}\right\rangle$$

Summarizing, the complete TNB-frame is:

$$\begin{aligned}\mathbf{T}(0) &= \left\langle -\frac{1}{3}, \frac{2}{3}, \frac{2}{3}\right\rangle \\ \mathbf{N}(0) &= \left\langle \frac{2}{3}, \frac{2}{3}, -\frac{1}{3}\right\rangle \\ \mathbf{B}(0) &= \left\langle -\frac{2}{3}, \frac{1}{3}, -\frac{2}{3}\right\rangle\end{aligned}$$

Quality assurance: each of $\mathbf{T}(0), \mathbf{N}(0), \mathbf{B}(0)$ is a unit vector, and all three are mutually perpendicular. ■

---

(4) Find the velocity and acceleration functions for the particle moving according to $\mathbf{r}(t) = \langle 2\cos t, 3t, 2\sin t\rangle$. Does the particle move at a constant speed?

□ For the position function $\mathbf{r}(t) = \langle 2\cos t, 3t, 2\sin t\rangle$, the velocity function is

$$\mathbf{v}(t) = \mathbf{r}'(t) = \langle -2\sin t, 3, 2\cos t\rangle$$

and acceleration is

$$\mathbf{a}(t) = \mathbf{v}'(t) = \langle -2\cos t, 0, -2\sin t\rangle$$

Since speed is the magnitude of velocity,

$$s = |\mathbf{v}(t)| = \sqrt{4\sin^2 t + 9 + 4\cos^2 t} = \sqrt{13}$$

and this particle does move at a constant speed. ■

(5) Find the velocity and positions functions for a particle moving with acceleration $\mathbf{a}(t) = -10\mathbf{k}$, if we know that $\mathbf{v}(0) = \mathbf{i} + \mathbf{j} - \mathbf{k}$ and $\mathbf{r}(0) = 2\mathbf{i} + 3\mathbf{j}$.

□ We have $\mathbf{a}(t) = -10\mathbf{k} = \langle 0, 0, -10 \rangle$ with initial conditions $\mathbf{v}(0) = \mathbf{i} + \mathbf{j} - \mathbf{k} = \langle 1, 1, -1 \rangle$ and $\mathbf{r}(0) = 2\mathbf{i} + 3\mathbf{j} = \langle 2, 3, 0 \rangle$. The velocity and acceleration functions then come from antidifferentiation and resolution of the constants of integration:

$$\begin{aligned}
\mathbf{v}(t) &= \int \mathbf{a}(t) dt = \langle c_1, c_2, -10t + c_3 \rangle \\
\mathbf{v}(0) &= \langle c_1, c_2, c_3 \rangle = \langle 1, 1, -1 \rangle \\
&\to \mathbf{v}(t) = \langle 1, 1, -10t - 1 \rangle \\
\mathbf{r}(t) &= \int \mathbf{v}(t) dt = \langle t + c_1, t + c_2, -5t^2 - t + c_3 \rangle \\
\mathbf{r}(0) &= \langle c_1, c_2, c_3 \rangle = \langle 2, 3, 0 \rangle \\
&\to \mathbf{r}(t) = \langle t + 2, t + 3, -5t^2 - t \rangle
\end{aligned}$$

In original notation,

$$\mathbf{v}(t) = \mathbf{i} + \mathbf{j} - (10t + 1)\mathbf{k} \quad ; \quad \mathbf{r}(t) = (t + 2)\mathbf{i} + (t + 3)\mathbf{j} - (5t^2 + t)\mathbf{k} \quad \blacksquare$$

(6) Find the tangential and normal components of acceleration for $\mathbf{r}(t) = (1 + t)\mathbf{i} + (t^2 - 2t)\mathbf{j}$, and interpret the results.

□ To find the tangential component of acceleration for $\mathbf{r}(t) = (1 + t)\mathbf{i} + (t^2 - 2t)\mathbf{j}$, we need:

$$\begin{aligned}
\mathbf{r}'(t) &= \mathbf{i} + 2(t - 1)\mathbf{j} \\
|\mathbf{r}'(t)| &= \sqrt{1 + 4(t^2 - 2t + 1)} = \sqrt{4t^2 - 8t + 5} \\
\mathbf{r}''(t) &= 2\mathbf{j} \\
\mathbf{r}'(t) \cdot \mathbf{r}''(t) &= 4(t - 1)
\end{aligned}$$

so then

$$a_T = \frac{\mathbf{r}'(t) \cdot \mathbf{r}''(t)}{|\mathbf{r}'(t)|} = \frac{4(t - 1)}{\sqrt{4t^2 - 8t + 5}}$$

To find the normal component of acceleration, we need $|\mathbf{a}(t)|^2$. Note that $\mathbf{a}(t) = \mathbf{r}''(t)$, which is found above; so,

$$\begin{aligned}
\mathbf{a}(t) &= 2\mathbf{j} \\
|\mathbf{a}(t)|^2 &= 4
\end{aligned}$$

so that

$$\begin{aligned}
a_N &= \sqrt{|\mathbf{a}(t)|^2 - a_T^2} = \sqrt{4 - \left(\frac{4(t-1)}{\sqrt{4t^2-8t+5}}\right)^2} \\
&= \sqrt{4 - \frac{16(t-1)^2}{4t^2-8t+5}} = \sqrt{\frac{16t^2-32t+20}{4t^2-8t+5} - \frac{16t^2-32t+16}{4t^2-8t+5}} \\
&= \sqrt{\frac{4}{4t^2-8t+5}} = \frac{2}{\sqrt{4t^2-8t+5}}
\end{aligned}$$

Neither component is constant. So in this case, finding a_T and a_N may not have been particularly useful or illuminating. ■

---

(7) Find the curvature function $\kappa(t)$ of $\mathbf{r}(t) = \langle t, t, 1+t^2 \rangle$, and the specific torsion $\tau(2)$.

□ For the curvature of $\mathbf{r}(t) = \langle t, t, 1+t^2 \rangle$ we need:

$$\begin{aligned}
\mathbf{r}'(t) &= \langle 1, 1, 2t \rangle \\
|\mathbf{r}'(t)| &= \sqrt{4t^2+2} = \sqrt{2}\sqrt{2t^2+1} \\
\mathbf{r}''(t) &= \langle 0, 0, 2 \rangle \\
\mathbf{r}'(t) \times \mathbf{r}''(t) &= \langle 2, 2, 0 \rangle \\
|\mathbf{r}'(t) \times \mathbf{r}''(t)| &= 2\sqrt{2}
\end{aligned}$$

so that

$$\kappa(t) = \frac{|\mathbf{r}'(t) \times \mathbf{r}''(t)|}{|\mathbf{r}'(t)|^3} = \frac{2\sqrt{2}}{(\sqrt{2}\sqrt{2t^2+1})^3} = \frac{1}{(2t^2+1)}$$

For the torsion, we also need $\mathbf{r}'''$. Since $\mathbf{r}''(t) = \langle 0, 0, 2 \rangle$, then $\mathbf{r}'''(t) = \langle 0, 0, 0 \rangle$. Therefore, By (17.21), we will then have $\tau(2) = 0$. FFT: Since the computation of torsion requires a dot product involving the third derivative of $\mathbf{r}(t)$, if the component functions of $\mathbf{r}(t)$ are polynomials, what degree polynomial do we have to see in at least one component of $\mathbf{r}(t)$ for the torsion to have a chance of not being zero? ■

17.4 Lagrange Multipliers

In Sec. 5.3 (Volume 1), we learned how to find extreme values of single variable functions $f(x)$. This process was generally straightforward, although it might not have seemed like it at the time. We computed first and second derivatives of $f(x)$, applied appropriate tests, and boom, we were done.

In Sec. 14.2, we learned how to find local extremes and absolute extremes of a function $f(x, y)$. The search for local extremes began by identifying critical points, where both partial derivatives f_x and f_y were equal to zero. Then from those critical points, we selected those which were revealed to be local maximums or minimums by the (multivariable) second derivative test. This hunt for local extremes could take place from a given restricted region, or over the entire domain of the function.

The search for absolute extremes of a function $f(x, y)$ took place on a defined region, and it was "simply" a matter of determining locations at which the function attained the maximum and minimum values possible; sometimes these locations were critical points, and sometimes they were points on the boundary of the specified region. The examination of the boundary for potential extremes required that we trace the edges of the boundary to look for Calculus-I (single variable) type critical points, and that we check any corners that the boundary might have. In all, the variety of regions we could examine for absolute extremes were pretty limited; you may recall, they were quite often rectangular or triangular regions.

Now we are ready to expand our ability to find extreme values of a given function. As before, this search will require identification of an objective function and a constraint. Unlike before, our new technique will be applicable to functions of more than just two independent variables.

Constrained Optimization

As before in Sec. 14.2, we will be seeking the maximum and minimum values a function takes on according to some rule that must be obeyed by the independent variables. This is called *constrained optimization*, and we apply some terminology: the function we are trying to maximize or minimize, $f(x, y)$, is called the **objective function**. The rule that the variables need to obey is called the **constraint**, and it usually comes in a form such as $g(x, y) = c$ or $g(x, y) \leq c$. When we say, "Find the absolute maximum and minimum values of $f(x.y) = x^2 - y^2$ over the region $x^2 + y^2 \leq 4$," we establish the objective function $f(x, y) = x^2 - y^2$ and the constraint $x^2 + y^2 \leq 4$.

EX 1 Identify the objective function and the constraint in the following problem statements:

- Find the extreme values of the function $2xy$ subject to the rule $2x^2 + 3y^2 = 8$.
- The temperature of a metal plate is given by $T(x, y) = x^2 + 2xy + y^2$; the plate is elliptical, and the equation describing the plate is $x^2 + 4y^2 \leq 24$. Find the maximum temperature on the plate.

In the first problem statement, the objective function is $f(x, y) = 2xy$ and the constraint is $2x^2 + 3y^2 = 8$. In the second problem statement, the objective function is $T(x, y) = x^2 + 2xy + y^2$ and the constraint is $x^2 + 4y^2 \leq 24$. ■

You Try It

(1) Identify the objective function and the constraint in the following problem statements:

- The height of a rocket t seconds after launch is given by $f(t, u) = \frac{1}{2}(u - 32)t^2$ feet, where u is a measure of thrust in ft^2/s. The fuel usage is limited by $u^2 t = 10,000$. Find the value of u that maximizes the height that the rocket reaches.
- Find the points on the sort-of-sphere $x^4 + y^4 + z^4 = 1^2$ that are closest to the origin.

These examples aside, we are going to constrain our initial conversation about constrained optimization by using only constraints which are equalities, not inequalities. That is, our constraints will be of the form $g(x, y) = c$. Figure 17.2 shows a graphical display of such a constrained optimization scenario: we are seeking absolute extremes of an objective function $f(x, y)$ subject to a constraint $g(x, y) = c$. Some selected level curves of $f(x, y)$ are shown, as is the level curve of $g(x, y)$ corresponding to the constraint $g(x, y) = c$. The constraint function $g(x, y)$ certainly also has level curves other than the one shown, but for the optimization problem at hand, we are

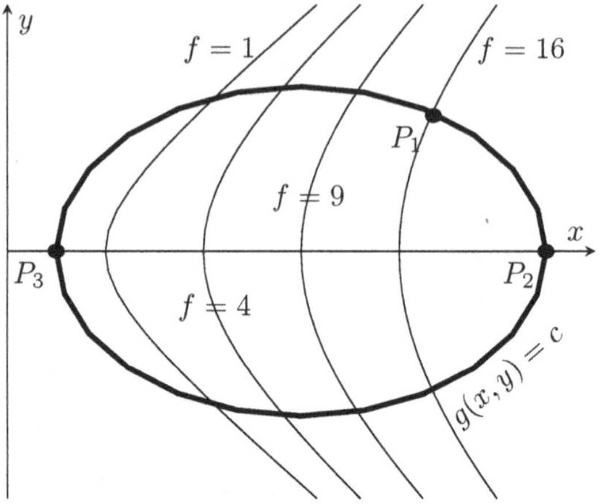

Fig. 17.2 Level curves of a function $f(x, y)$ and a constraint $g(x, y) = c$.

only interested in the one specific level curve that goes with the constraint $g(x, y) = c$. Similarly, there are level curves of the objective function $f(x, y)$ other than the ones shown in the figure. In fact, every point in the plane is associated with some level curve of $f(x, y)$, we just happen to be seeing the level curves corresponding to $f(x, y) = 1$, 4, 9, and 16.

The constrained optimization problem could solved by a smart Calculus Ant. Imagine the ant walking around the constraint curve $g(x, y) = c$, noting the values of $f(x, y)$ generated by each point (x, y) along the way. When it comes to a point on the curve which gives the largest value of $f(x, y)$ that will be seen anywhere along the constraint curve, our intrepid ant starts waving its little forelegs around to get our attention. Where might this point be in Fig. 17.2? Since it appears that values of $f(x, y)$ tend to increase as we move farther to the right, it's likely that the point on $g(x, y) = c$ that gives the maximum value of $f(x, y)$ is point P_2. Similarly, it's likely that the point on $g(x, y) = c$ that gives the minimum value of $f(x, y)$ is point P_3.

Since our Calculus Ant is nowhere to be found, though, we have to learn how the ant would know it has found an extreme of $f(x, y)$; this requires that you believe a few things about the relation between $g(x, y)$ and $f(x, y)$. The first is this:

- At a point on the constraint curve $g(x, y) = c$ which locates an absolute extreme of $f(x, y)$, the constraint curve will touch some unknown level curve of $f(x, y)$.

This should be fairly self evident. Every point in the plane belongs to some level curve of $f(x, y)$, so if we are trapped on the curve $g(x, y) = c$, when we do come across an extreme of $f(x, y)$, we will be on the level curve of $f(x, y)$ corresponding to that extreme value. Let's say we're looking for the absolute maximum of $f(x, y)$, and this value is f_{max}. In Fig. 17.2, this extreme won't be at one of the four level curves of $f(x, y)$ that are shown, though. Surely $f_{max} > 16$. But whatever this value is, the point where we find it will be on two level curves: $g(x, y) = c$ and $f(x, y) = f_{max}$.

A more specific thing I need you to believe, then, is this:

- At the location of the extreme point of $f(x, y)$, the constraint curve $g(x, y) = c$ will be *tangent to* the (unknown) level curve of $f(x, y) = f_{max}$.

This seems like it would be awfully complicated to prove, but it's actually fairly simple. There are only two ways the level curve $g(x, y) = c$ could interact with the level curve of the extreme, $f(x, y) = f_{max}$: either they cross, or they are tangent to each other. If the level curve $g(x, y) = c$ crosses over the level curve $f(x, y) = f_{max}$, then the value f_{max} can't be the maximum value after all — because all we'd have to do is keep moving along $g(x, y) = c$ in the direction of increasing f. This is what's happening at point P_1 in Fig. 17.2; we know $f(x, y) = 16$ cannot represent the maximum value of f along $g(x, y) = c$, because we can just move along $g(x, y) = c$ to the right of $f(x, y) = 16$ to encounter larger values of f. It's necessary for $g(x, y) = c$ and $f(x, y) = f_{max}$ to be tangent to one another.

Just so we can play with these two ideas again, Fig. 17.3 shows a different optimization scenario involving a constraint curve $g(x, y) = c$ along with several level curves of a function $f(x, y)$. A couple of things are evident already — we will not find a maximum of $f(x, y)$ subject to $g(x, y) = c$ (at least according to the visible trends), and the minimum value f_{min} of $f(x, y)$ subject to $g(x, y) = c$ is between 10 and 15, that is, $10 < f_{min} < 15$. At whatever point we find both $g(x, y) = c$ and $f(x, y) = f_{min}$, the level curve $g(x, y) = c$ will be tangent to the level curve $f(x, y) = f_{min}$. Based on what we see, that absolute minimum is likely located at P_1.

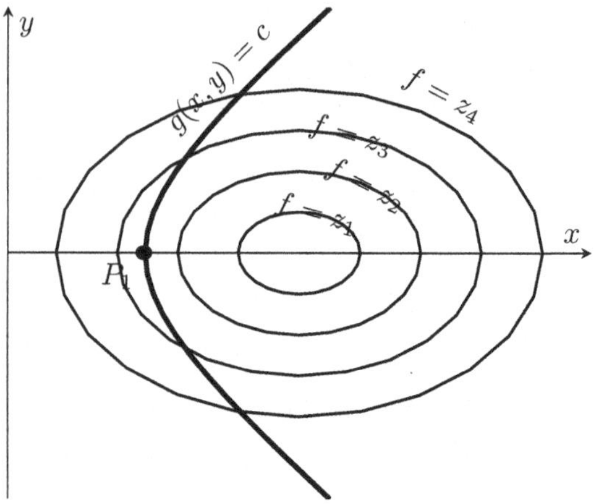

Fig. 17.3 Level curves of a function $f(x, y)$ and a constraint $g(x, y) = c$.

Finally, I need to remind you of the following fact, which has already been developed in previous sections:

- Gradient vectors of a function $f(x, y)$ are normal to level curves of $f(x, y)$.

Figure 17.4 shows several level curves of a function $f(x, y)$ along with some gradient vectors at selected points; note that each gradient vector is perpendicular to the level curve that goes through the point where the gradient originates.

So let's put all these ideas together. Consider the location (x, y) where we have an extreme point of $f(x, y)$ subject to the constraint $g(x, y) = c$. Let's say the extreme is a maximum. At this magic point (x, y), the gradient of f is normal to the level curve $f(x, y) = f_{max}$, and the gradient of g is normal to the level curve $g(x, y) = c$. But since this point marks an extreme of $f(x, y)$ subject to $g(x, y) = c$, we know that the level curves $f(x, y) = f_{max}$ and $g(x, y) = c$ are tangent to each other. And this means, then, that the gradient vectors $\nabla f(x, y)$ and $\nabla g(x, y)$ must be parallel to each other! This sets up our ability to locate the extreme of $f(x, y)$.

In general, if a vector $\mathbf{v}$ is parallel to another vector $\mathbf{w}$, then they are related by a scalar multiple: $\mathbf{v} = \lambda \mathbf{w}$. So at a location where $\nabla f(x, y)$ and

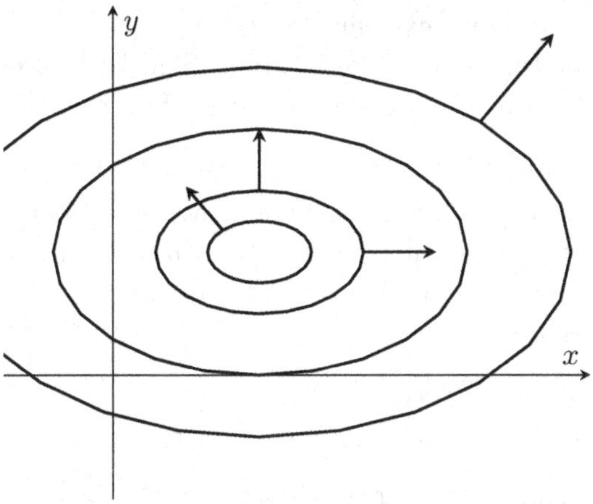

Fig. 17.4 Level curves of a function $f(x, y)$ and gradient vectors.

$\nabla g(x, y)$ must be parallel to each other, we must have $\nabla f(x, y) = \lambda \nabla g(x, y)$ for some scalar value λ. And now you have been introduced to the Lagrange Multiplier λ.

Useful Fact 17.3. *At the location of an absolute extreme value of a function $f(x, y)$ subject to a constraint $g(x, y) = c$, we must have $\nabla f(x, y) = \lambda \nabla g(x, y)$ for some scalar value λ. This vector equation, along with the constraint $g(x, y) = c$, presents a system of equations in unknowns x, y, λ that can be solved for the location (x, y) of the extreme.*

This Useful Fact provides an offhand mention of a "system of equations" which sounds harmless. But note the absence of the word "linear". The system of equations provided by the relation $\nabla f(x, y) = \lambda \nabla g(x, y)$ and the constraint $g(x, y) = c$ will very possibly not be linear. Also, while the build-up to this Useful Fact centered on functions of two variables, we can immediately generalize it to any number of variables. And so, this harmless "system of equations" can actually become quite intimidating. We should expect to use tech to solve these systems. The purists among us may be offended, but if it's a trade-off between doing fewer Lagrange Multiplier problems because we spend so much time solving the systems by hand, versus saving time on solving the systems so that we can do more Lagrange Multiplier problems, I'll take the latter. Let's give it a shot.

EX 2 Find the absolute extremes of $f(x, y) = x^2 - y^2$ subject to the constraint $16(x-3)^2 + 25y^2 = 100$. (This is the scenario shown in Fig. 17.2, although there are no axis labels to prove it.)

The objective function is $f(x, y) = x^2 - y^2$, and the constraint function is $g(x, y) = 16(x-3)^2 + 25y^2$. (Remember, the value 100 is not part of the named constraint *function*, it is simply the single level curve of the constraint function chosen to provide the full constraint itself.) The gradients of these functions are:

$$\nabla f(x, y) = \langle 2x, 2y \rangle$$
$$\nabla g(x, y) = \langle 32(x-3), 50y \rangle$$

By matching the components of these gradients by the relation $\nabla f(x, y) = \lambda \nabla g(x, y)$, we get two equations:

$$2x = \lambda(32(x-3))$$
$$2y = \lambda(50y)$$

or, tidied up,

$$x = 16\lambda(x-3)$$
$$y = 25\lambda y$$

Note that this system has two equations but three unknowns (x, y, λ). There are infinitely many solutions. This just means, so far, that there are lots of places where we'd find $\nabla f(x, y) = \lambda \nabla g(x, y)$, and that's fine. We are looking, specifically, for the place where that relationship holds along with the constraint! So when we toss our constraint into this system, we get the full set of three equations in three unknowns:

$$x = 16\lambda(x-3)$$
$$y = 25\lambda y$$
$$16(x-3)^2 + 25y^2 = 100$$

We could track down the solutions by hand by weeding through all the possible cases, such as, "What if $y = 0$? What if $y \neq 0$, What if $x = 0$?" and so on. Or, we can use tech. I'm not sure what your favorite platform is, but Wolfram Alpha is pretty convenient; if we go there and type in (using L for λ):

```
solve {x = 16*L*(x-3); y = 25*L*y; 16(x-3)^2 + 25 y^2 = 100}
```

then we will be provided with four solutions; two of them show imaginary numbers for y, and we can ignore those. The other two are:

$$x = \frac{1}{2}, y = 0, \lambda = -\frac{1}{80}$$

$$x = \frac{11}{2}, y = 0, \lambda = \frac{11}{80}$$

The values of λ are not needed, as λ has now served its purpose as a scalar multiple relating gradients. The values of x and y provide two key points, $(x, y) = (1/2, 0)$ and $(x, y) = (11/2, 0)$. These are locations where we satisfy $\nabla f(x, y) = \lambda \nabla g(x, y)$ as well as $16(x-3)^2 + 25y^2 = 100$. That is, these are the locations of our extremes. To find which one locates the absolute maximum and which one locates the absolute minimum, we have to calculate $f(x, y)$ at each of these points. Since $f(1/2, 0) = 1/4$ and $f(11/2, 0) = 121/4$, the former is the absolute minimum and the latter is the absolute maximum. Hey, I'll bet these locations correspond to the points P_2 and P_3 from Fig. 17.2. Figure 17.5 shows a new version of Fig. 17.2, now with axis labels included and level curves corresponding to $f(x, y) = 1/4$ and $f(x, y) = 121/4$ displayed. This figure shows everything coming together: the level curve $f(x, y) = 1/4$ goes through point $P_3 = (1/2, 0)$ (the absolute minimum) and level curve $f(x, y) = 121/4$ goes through point $P_2 = (11/2, 0)$ (the absolute maximum). ■

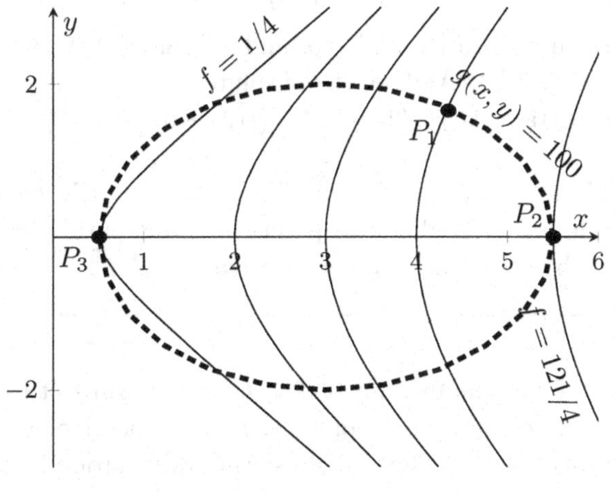

Fig. 17.5 Level curves of a function $f(x, y)$ and a constraint $g(x, y) = 100$.

EX 3 Now let's solve the problem displayed in Fig. 17.4, which flips around functions from EX 1: Find the absolute extremes of $f(x,y) = 16(x-3)^2 + 25y^2$ subject to the constraint $x^2 - y^2 = 2$ for $x \geq 0$.

Using the relation established in Useful Fact 17.3, $\nabla f(x,y) = \lambda \nabla g(x,y)$, we have

$$\langle 32(x-3), 50y \rangle = \lambda \langle 2x, 2y \rangle$$

Matching components and cleaning up those relationships, and also bringing in the constraint leads to the system

$$16(x-3) = \lambda x$$
$$25y = \lambda y$$
$$x^2 - y^2 = 2$$

(It's tempting to simplify the middle equation to just $\lambda = 25$, but when doing so, we must set aside the case $y = 0$ for separate examination.) Now, remember our prediction from Fig. 17.4 that in this scenario, we'd find no absolute maximum, and one absolute minimum, at point P_1. By entering the following into Wolfram Alpha,

```
solve {16*(x-3) = L*x; 25*y = L*y; x^2 - y^2 = 2}
```

we receive four locations where all three equations hold:

$$\left(-\frac{16}{3}, -\frac{\sqrt{238}}{3}\right) \quad , \quad \left(-\frac{16}{3}, \frac{\sqrt{238}}{3}\right) \quad , \quad \left(-\sqrt{2}, 0\right) \quad , \quad \left(\sqrt{2}, 0\right)$$

Only one of them also satisfies $x \geq 0$, and that is $(\sqrt{2}, 0)$, which happens to be P_1 from Fig. 17.3. And so, the absolute minimum of $f(x,y)$ subject to the given constraint is $f(\sqrt{2}, 0) = 16(\sqrt{(2)} - 3)^2 \approx 40.2$. ■

FFT: The solution set from Wolfram Alpha in EX 2 contains three points for which $x < 0$. Might these points have any particular significance relative to the objective function and constraint?

You Try It

(2) Find the maximum value of $f(x,y) = 4x^2y$ subject to $x^2 + y^2 = 3$.
(3) The temperature at a point (x,y) on a metal plate is $T(x,y) = 4x^2 - 4xy + y^2$. The Calculus Ant walks around the plate on a circle of radius 5 centered at the origin. What are the highest and lowest temperatures encountered by the ant?

Now that we have the scheme for solving constrained optimization problems with Lagrange Multipliers, we aren't restricted to functions of just two independent variables. Anything that was true about level curves for some $f(x, y)$ and $g(x, y)$ remains true for level surfaces of $f(x, y, z)$ and $g(x, y, z)$ and so on.

EX 4 Find the point on the unit sphere (centered at the origin) that provides the minimum value of $f(x, y, z) = x^4 - 2y^2 + z^4$.

Our objective is to minimize the function $f(x, y, z) = x - 2y^2 - 2z$ subject to the constraint $x^2 + y^2 + z^2 = 1$. The relation $\nabla f(x, y, z) = \lambda \nabla g(x, y, z)$ sets up the following vector equation:

$$\langle 1, -4y, -2 \rangle = \lambda \langle 2x, 2y, 2x \rangle$$

Matching components on both sides of this vector equation and tossing in the constraint gives the system of equations

$$\begin{aligned} 1 &= \lambda(2x) \\ -4y &= \lambda(2y) \\ -2 &= \lambda(2z) \\ x^2 + y^2 + z^2 &= 1 \end{aligned}$$

Wolfram Alpha returns four solution sets for this system, $(x, y, z) = \ldots$

$$\ldots \left(-\frac{1}{4}, -\frac{\sqrt{11}}{4}, \frac{1}{2}\right), \left(-\frac{1}{4}, \frac{\sqrt{11}}{4}, \frac{1}{2}\right), \left(-\frac{1}{\sqrt{5}}, 0, \frac{2}{\sqrt{5}}\right), \left(\frac{1}{\sqrt{5}}, 0, -\frac{2}{\sqrt{5}}\right)$$

At these locations, we have the following function values:

$$\begin{aligned} f\left(-\frac{1}{4}, -\frac{\sqrt{11}}{4}, \frac{1}{2}\right) &= -\frac{21}{8} \approx -2.62 \\ f\left(-\frac{1}{4}, \frac{\sqrt{11}}{4}, \frac{1}{2}\right) &= -\frac{21}{8} \approx -2.62 \\ f\left(-\frac{1}{\sqrt{5}}, 0, \frac{2}{\sqrt{5}}\right) &= -\sqrt{5} \approx -2.23 \\ f\left(\frac{1}{\sqrt{5}}, 0, -\frac{2}{\sqrt{5}}\right) &= \sqrt{5} \approx 2.23 \end{aligned}$$

And so, we find the minimum value of $f(x, y, z)$ occurring at two locations:

$$f(x, y, z) = -\frac{21}{8} \text{ at } (x, y, z) = \left(-\frac{1}{4}, -\frac{\sqrt{11}}{4}, \frac{1}{2}\right), \left(-\frac{1}{4}, \frac{\sqrt{11}}{4}, \frac{1}{2}\right)$$

A reminder! In solving the system of equations, it's tempting to simplify the second equation from $-4y = \lambda(2y)$ to simply $-2 = \lambda$. But if you do that, you lose any solutions with $y = 0$, and would need to go back and gather them separately! ■

You Try It

(4) Find the absolute extremes of $f(x, y, z) = x^2 - y + 2z$ subject to the constraint $x^2 + y^2 + z^2 = 8$. (Remember to keep your values exact. Come on now, you're going to have a computer solve your equations, the least you can do is keep your values as accurate as possible!)

We can also combine Lagrange Multipliers with routine location of critical points to find extremes over a full region and its boundary. In Sec. 14.2, we were fairly limited in the shapes of regions we could consider, but now we can broaden that a bit.

EX 5 Find the points in the region $2x^2 + 3y^2 \leq 6$ which locate the absolute extremes of $z = \ln(10 + xy)$ in that region.

In this scenario, the objective function is $z = \ln(10+xy)$ and the constraint is $2x^2+3y^2 \leq 6$. We can name the constraint function as $g(x, y) = 2x^2+3y^2$. First, let's just scan for regular critical points of the function $z = \ln(7+xy)$ to see if any lie inside the given region (i.e. if any satisfy the constraint). We have

$$\frac{\partial z}{\partial x} = \frac{y}{7 + xy} \quad \text{and} \quad \frac{\partial z}{\partial y} = \frac{x}{7 + xy}$$

and so both derivatives are zero at the origin, $(x, y) = (0, 0)$, and that critical point is inside the given region.

Now we use Lagrange Multipliers to seek possible locations of extremes of z over the boundary of the given region, i.e., we seek extremes of $z(x, y)$ subject to the constraint $g(x, y) = 6$. By setting $\nabla z = \lambda \nabla g$, we find

$$\langle \frac{y}{7 + xy}, \frac{x}{7 + xy} \rangle = \lambda \langle 4x, 6y \rangle$$

Matching the components on each side of this vector equation and tossing in the constraint itself gives the system of equations

$$\frac{y}{7+xy} = \lambda(4x)$$
$$\frac{x}{7+xy} = \lambda(6y)$$
$$2x^2 + 3y^2 = 6$$

Wolfram Alpha reports four solutions to this system:

$$(x,y) = \left(-\sqrt{\frac{3}{2}}, -1\right), \left(\sqrt{\frac{3}{2}}, -1\right), \left(-\sqrt{\frac{3}{2}}, 1\right), \left(\sqrt{\frac{3}{2}}, 1\right)$$

(Notice that these are four corners of a rectangle inscribed into the ellipse $g(x,y) = 6$.) Altogether, we now have five candidate points for extremes of $z(x,y) = \ln(7+xy)$ over the region $2x^2 + 3y^2 \leq 6$:

$$z(0,0) = \ln(7) \approx 1.946$$
$$z\left(-\sqrt{\frac{3}{2}}, -1\right) = \ln\left(7 + \sqrt{\frac{3}{2}}\right) \approx 2.107$$
$$z\left(\sqrt{\frac{3}{2}}, -1\right) = \ln\left(7 - \sqrt{\frac{3}{2}}\right) \approx 1.754$$
$$z\left(-\sqrt{\frac{3}{2}}, 1\right) = \ln\left(7 - \sqrt{\frac{3}{2}}\right) \approx 1.754$$
$$z\left(\sqrt{\frac{3}{2}}, 1\right) = \ln\left(7 + \sqrt{\frac{3}{2}}\right) \approx 2.107$$

Altogether, we find the absolute extremes on the boundary on the region. The absolute maximum of $z(x,y)$ over the region is found at two locations:

$$z\left(\sqrt{\frac{3}{2}}, -1\right) = z\left(-\sqrt{\frac{3}{2}}, 1\right) = \ln\left(7 - \sqrt{\frac{3}{2}}\right)$$

and the absolute minimum is found at two locations:

$$z\left(-\sqrt{\frac{3}{2}}, -1\right) = z\left(\sqrt{\frac{3}{2}}, 1\right) = \ln\left(7 + \sqrt{\frac{3}{2}}\right) \quad \blacksquare$$

You Try It

(5) Find the largest and smallest distances from points in or on the circle $x^2 + y^2 = 9$ in the xy-plane to the paraboloid $z = 1 + (x-1)^2 + (y-2)^2$. and minimized.

In EX 5, we came across a rectangle inscribed in an ellipse. Lagrange Multipliers can help determine fun information about such inscribed shapes. Give it a try!

You Try It

(6) Find the corners of the rectangle of greatest area that can be inscribed in the ellipse $9x^2 + 16y^2 = 144$, and give that maximum area.

Lagrange Multipliers — Problem List

Lagrange Multipliers — You Try It

These appeared above; solutions begin on the next page.

(1) Identify the objective function and the constraint in the following problem statements:

- The height of a rocket t seconds after launch is given by $f(t, u) = (u - 32)t^2/2$ feet, where u is a measure of thrust in ft^2/s. The fuel usage is limited by $u^2t = 10,000$. Find the value of u that maximizes the height that the rocket reaches.
- Find the points on the sort-of-sphere $x^4 + y^4 + z^4 = 1$ that are closest to the origin.

(2) Find the maximum value of $f(x, y) = 4x^2y$ subject to $x^2 + y^2 = 3$.
(3) The temperature at a point (x, y) on a metal plate is $T(x, y) = 4x^2 - 4xy + y^2$. The Calculus Ant walks around the plate on a circle of radius 5 centered at the origin. What are the highest and lowest temperatures encountered by the ant?
(4) Find the absolute extremes of $f(x, y, z) = x^2 - y + 2z$ subject to the constraint $x^2 + y^2 + z^2 = 8$. (Remember to keep your values exact. Come on now, you're going to have a computer solve your equations, the least you can do is keep your values as accurate as possible!)
(5) Find the largest and smallest distances from points in or on the circle $x^2 + y^2 = 9$ in the xy-plane to the paraboloid $z = 1 + (x - 1)^2 + (y - 2)^2$.
(6) Find the corners of the rectangle of greatest area that can be inscribed in the ellipse $9x^2 + 16y^2 = 144$, and give that maximum area.

Lagrange Multipliers — Practice Problems

Try these as you get the hang of the You Try It problems. Solutions to these problems are available in Sec. B.5.4.

(1) Find the maximum value of $f(x,y) = x^2 + y^3/3$ that can be achieved by points on the ellipse $2x^2 + y^2 = 6$, and the point(s) at which it occurs.
(2) Find the maximum value of $f(x,y,z) = x + 2y^2 + 2z$ that can be achieved by points on the sphere $x^2 + y^2 + z^2 = 4$, and the point(s) at which it occurs.
(3) To generate Fig. 13.21, I had to find the minimum and maximum values of $z = 1 + xy$ subject to the constraint $(x-3)^2 + (y-2)^2 = 0.25$. What were these values, and at what points did they occur?
(4) Find the point on the ellipse $x^2 + 2y^2 = 4$ that is closest to the point $(1,1)$.
(5) What is the largest volume of possible of a cylinder which is sized according to $\pi r^2 + h = 20$, and what are the dimensions which give that volume?
(6) Find the maximum value of $f(x,y,z,w) = 2x - 3y + 4z + w$ determined by points on the "hypersphere" $x^2 + y^2 + z^2 + w^2 = 3$.[3]
(7) Find the absolute extremes of $f(x,y) = xy(1-x-y) = xy - x^2y - xy^2$ anywhere on and inside the unit circle. (Compare to Practice Problem 2 in Sec. 14.2.)

Lagrange Multipliers — Challenge Problems

Try these problems to test your skills with the ideas in this section. Solutions to these problems are available in Sec. C.5.4.

(1) The galactic company that manufactures phasers (p), pulse rifles (r), and communicator badges (b) for the United Federation of Planets earns a profit of 10 quatloos for every phaser sold, 20 quatloos for every pulse rifle, and 5 quatloos for every communicator badge. The logistics of their operations require that the combined number of these items made each day is held strictly to $2p^2 + r^2 + 4b^2 = 10000$. How many of each item should they make each day to maximize profit?
(2) Find the maximum value of $f(x,y,z) = xy + 2yz - 3x + 3z$ that can be achieved anywhere on the unit sphere.

[3]Imagine a circle in 2D, then a sphere in 3D. Now imagine this surface in 4D. Easy!

(3) Find the point on the plane $2x + 3y + 4z = 12$ that is closest to the point $(5, 5, 5)$.

Lagrange Multipliers — You Try It — Solved

(1) Identify the objective function and the constraint in the following problem statements:

- The height of a rocket t seconds after launch is given by $f(t, u) = (u - 32)t^2/2$ feet, where u is a measure of thrust in ft^2/s. The fuel usage is limited by $u^2t = 10,000$. Find the value of u that maximizes the height that the rocket reaches.
- Find the points on the sort-of-sphere $x^4 + y^4 + z^4 = 1$ that are closest to the origin.

☐ In the first problem statement, we are asked to find the maximum height of the rocket, so the objective function is

$$f(t, u) = \frac{1}{2}(u - 32)t^2$$

Since the height is limited by fuel usage, the constraint is $u^2t = 10,000$. We could say the constraint function is $g(t, u) = u^2t$ and the constraint is the level curve of this function corresponding to $g(t, u) = 10,000$.

In the second problem statement, the magic word "closest" tells us we are hoping to locate a minimum distance. So, the objective function represents that distance. We could write this objective function as $f(x, y, z) = \sqrt{x^2 + y^2 + z^2}$, which represents the distance D from any point (x, y, z) to the origin. But as we have seen before, discovering a minimum of D^2 also reveals a minimum of D, and so we might simplify problem solving by seeking to minimize the objective function $f(x, y, z) = x^2 + y^2 + z^2$ instead. The constraint in this problem would be the equation of the sort-of-sphere $x^4 + y^4 + z^4 = 1$, and we'd set the constraint function as $g(x, y, z) = x^4 + y^4 + z^4$. ■

(2) Find the maximum value of $f(x, y) = 4x^2y$ subject to $x^2 + y^2 = 3$.

☐ The constraint is identified as $f(x, y) = 4x^2y$, and we set the constraint function as $g(x, y) = x^2 + y^2$. The relation $\nabla f(x, y) = \lambda g(x, y)$ gives a vector equation

$$\langle 8xy, 4x^2 \rangle = \lambda \langle 2x, 2y \rangle$$

which, along with the constraint $g(x,y) = 3$, leads to the system of equations

$$\begin{aligned} 8xy &= \lambda(2x) \\ 4x^2 &= \lambda(2y) \\ x^2 + y^2 &= 3 \end{aligned}$$

Wolfram Alpha gives six solution sets to these equations:

$$(-\sqrt{2},-1),\ (-\sqrt{2},1),\ (\sqrt{2},-1),\ (\sqrt{2},1),\ (0,-\sqrt{3}),\ (0,\sqrt{3})$$

which is our list of candidate points for the actual maximum. At these points, we can calculate

$$\begin{aligned} f(-\sqrt{2},-1) &= -8 \qquad & f(-\sqrt{2},1) &= 8 \\ f(\sqrt{2},-1) &= -8 & f(\sqrt{2},1) &= 8 \\ f(0,-\sqrt{3}) &= 0 & f(0,\sqrt{3}) &= 0 \end{aligned}$$

And so the maximum value of $f(x,y) = 4x^2y$ subject to $x^2 + y^2 = 3$ is attained at two points, where $f(-\sqrt{2},1) = f(\sqrt{2},1) = 8$. ■

---

(3) The temperature at a point (x,y) on a metal plate is $T(x,y) = 4x^2 - 4xy + y^2$. The Calculus Ant walks around the plate on a circle of radius 5 centered at the origin. What are the highest and lowest temperatures encountered by the ant?

□ Our objective function is $T(x,y) = 4x^2 - 4xy + y^2$ and our constraint is $g(x,y) = x^2 + y^2$ (because the constraint itself is $x^2 + y^2 = 25$, the circle of radius 5 centered at the origin). The relation $\nabla T(x,y) = \lambda \nabla g(x,y)$ gives the vector equation

$$\langle 8x - 4y, -4x + 2y \rangle = \lambda \langle 2x, 2y \rangle$$

which, with the constraint, leads to the system of equations

$$\begin{aligned} 8x - 4y &= \lambda(2x) \\ -4x + 2y &= \lambda(2y) \\ x^2 + y^2 &= 25 \end{aligned}$$

Wolfram Alpha gives four solution sets to these equations:

$$(-\sqrt{5},-2\sqrt{5}),\ (\sqrt{5},2\sqrt{5}),\ (-2\sqrt{5},\sqrt{5}),\ (2\sqrt{2},-\sqrt{5})$$

which is our list of candidate points for the extremes. The value of the objective function at each point is:

$$\begin{aligned} T(-\sqrt{5}, -2\sqrt{5}) &= 0 \\ T(\sqrt{5}, 2\sqrt{5}) &= 0 \\ T(-2\sqrt{5}, \sqrt{5}) &= 125 \\ T(2\sqrt{5}, -\sqrt{5}) &= 125 \end{aligned}$$

And so, our intrepid Calculus Ant experiences the minimum temperature of $T = 0$ at two locations and a maximum temperature of $T = 125$ at two locations. Let's hope the units are not Celsius. ■

---

(4) Find the absolute extremes of $f(x, y, z) = x^2 - y + 2z$ subject to the constraint $x^2 + y^2 + z^2 = 8$. (Remember to keep your values exact. Come on now, you're going to have a computer solve your equations, the least you can do is keep your values as accurate as possible!)

□ Our objective function is $f(x, y, z) = x^2 - y + 2z$ and our constraint is $g(x, y, z) = x^2 + y^2 + z^2$. The relation $\nabla f(x, y, z) = \lambda \nabla g(x, y, z)$ gives the vector equation

$$\langle 2x, -1, 2 \rangle = \lambda \langle 2x, 2y, 2z \rangle$$

which, with the constraint, leads to the system of equations

$$\begin{aligned} 2x &= \lambda(2x) \\ -1 &= \lambda(2y) \\ 2 &= \lambda(2z) \\ x^2 + y^2 + z^2 &= 8 \end{aligned}$$

Wolfram Alpha provides four solution sets for this system (approximated),

$$(x, y, z) = \left(-\frac{3\sqrt{3}}{2}, -\frac{1}{2}, 1\right), \left(\frac{3\sqrt{3}}{2}, -\frac{1}{2}, 1\right), \\ \left(0, 2\sqrt{\frac{2}{5}}, -4\sqrt{\frac{2}{5}}\right), \left(0, -2\sqrt{\frac{2}{5}}, 4\sqrt{\frac{2}{5}}\right)$$

which is our list of candidate points for the extremes. The value of the objective function at each point is:

$$f\left(-\frac{3\sqrt{3}}{2}, -\frac{1}{2}, 1\right) = \frac{37}{4}$$

$$f\left(\frac{3\sqrt{3}}{2}, -\frac{1}{2}, 1\right) = \frac{37}{4}$$

$$f\left(0, 2\sqrt{\frac{2}{5}}, -4\sqrt{\frac{2}{5}}\right) = -2\sqrt{10}$$

$$f\left(0, -2\sqrt{\frac{2}{5}}, 4\sqrt{\frac{2}{5}}\right) = 2\sqrt{10}$$

Since $37/4 > 2\sqrt{10}$, the absolute maximum is provided by the first two locations,

$$f\left(-\frac{3\sqrt{3}}{2}, -\frac{1}{2}, 1\right) = f\left(\frac{3\sqrt{3}}{2}, -\frac{1}{2}, 1\right) = \frac{37}{4}$$

and the absolute minimum is provided by

$$f\left(0, 2\sqrt{\frac{2}{5}}, -4\sqrt{\frac{2}{5}}\right) = -2\sqrt{10} \quad \blacksquare$$

---

(5) Find the largest and smallest distances from points in or on the circle $x^2 + y^2 = 9$ in the xy-plane to the vertex of the paraboloid $z = 1 + (x-1)^2 + (y-2)^2$.

□ To make everything match in three dimensions, note that we can write the equation of the constraint (inside or on the circle) as $g(x, y, z) = x^2 + y^2 + 0^2 \leq 9$. The objective function is a function that measures distance D (or D^2 as a surrogate) to the vertex of the paraboloid.

The vertex of the paraboloid is at $(1, 2, 1)$, because this is where we find $\partial z/\partial z = \partial z/\partial y = 0$. The point $(1, 2, 0)$ satisfies the constraint $x^2 + y^2 + 0^2 \leq 9$, and the distance from that point to the vertex of the paraboloid is $D = 1$. We need Lagrange Multipliers to trace the boundary of the circle and look for other candidate locations of extremes of distance to the vertex.

For this more general examination, we must build the constraint function; we'll seek to minimize and maximize D^2, where D is the distance from points $(x, y, 0)$ to the vertex $(1, 2, 1)$. This objective function is $f(x, y) = (x-1)^2 + (y-2)^2 + (0-1)^2$, or $D^2 = (x-1)^2 + (y-2)^2 + 1$ and our constraint is $x^2 + y^2 = 9$. Using $\nabla f(x, y) = \lambda \nabla g(x, y)$ and the constraint, we get the system,

$$\begin{aligned} 2(x-1) &= \lambda(2x) \\ 2(y-2) &= \lambda(2y) \\ x^2 + y^2 &= 9 \end{aligned}$$

which has solutions, per Wolfram Alpha, of $(x, y) = (3/\sqrt{5}, 6/\sqrt{5})$, $(-3/\sqrt{5}, -6/\sqrt{5})$. At our three points of interest, then, we compute the actual distance (squared) to $(1, 2, 1)$ and find:

$$\begin{aligned} D^2(1,2,0) &= (1-1)^2 + (2-2)^2 + 1 = 1 \\ D^2\left(\frac{3}{\sqrt{5}}, \frac{6}{\sqrt{5}}, 0\right) &= \left(\frac{3}{\sqrt{5}} - 1\right)^2 + \left(\frac{6}{\sqrt{5}} - 2\right)^2 + 1 = 15 - 6\sqrt{5} \\ &\approx 1.583 \\ D^2\left(-\frac{3}{\sqrt{5}}, -\frac{6}{\sqrt{5}}, 0\right) &= \left(1 - \frac{3}{\sqrt{5}}\right)^2 + \left(\frac{6}{\sqrt{5}} - 2\right)^2 + 1 = 15 + 6\sqrt{5} \\ &\approx 28.4 \end{aligned}$$

As it turns out, the point inside the circle was irrelevant and our extremes come from the boundary. Remember the values we just found are the *squares* of the actual distances, and so as far as distances from inside or on the circle $x^2 + y^2 = 9$ to the vertex of the paraboloid $z = 1 + (x-1)^2 + (y-2)^2$ go,

- the minimum distance $D_{min} \approx 1.26$ is found at the point $(3/\sqrt{5}, 6/\sqrt{5})$
- the minimum distance $D_{max} \approx 5.33$ is found at the point $(-3/\sqrt{5}, -6/\sqrt{5})$

Figure 17.6 shows the circle, the paraboloid, and dashed lines connecting these two points to the vertex. ■

---

(6) Find the corners of the rectangle of greatest area that can be inscribed in the ellipse $9x^2 + 16y^2 = 144$, and give that maximum area.

□ Whatever rectangle this is, it will be centered at the origin, and have opposite corners at some $(-x, -y)$ and (x, y); its width will be $2x$, and its height will be $2y$, so its area will be $4xy$. So we are trying to

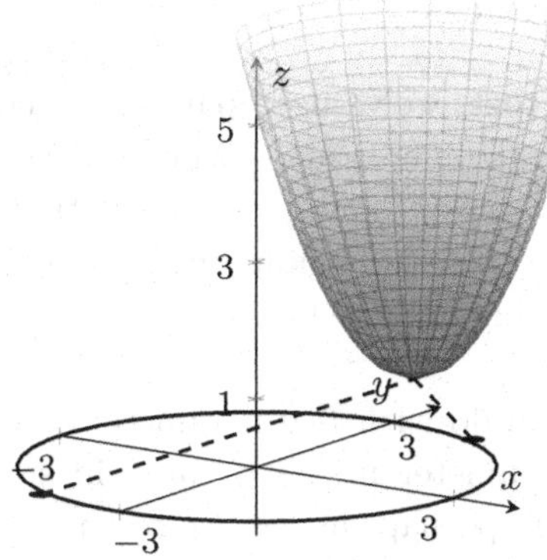

Fig. 17.6 The circle $x^2 + y^2 = 9$ and a nearby paraboloid.

maximize the function $f(x, y) = 4xy$ subject to the constraint $9x^2 + 16y^2 = 144$. With $g(x, y) = 9x^2 + 16y^2$, the relation $\nabla f(x, y)\lambda g(x, y)$ and the constraint itself give the system of equations

$$\begin{aligned} 4y &= \lambda(18x) \\ 4x &= \lambda(32y) \\ 9x^2 + 16y^2 &= 144 \end{aligned}$$

and sure enough, Wolfram Alpha reports four solutions which correspond to four corners of a rectangle:

$$\left(-2\sqrt{2}, \frac{3}{\sqrt{2}}\right), \left(2\sqrt{2}, -\frac{3}{\sqrt{2}}\right), \left(-2\sqrt{2}, -\frac{3}{\sqrt{2}}\right), \left(2\sqrt{2}, \frac{3}{\sqrt{2}}\right)$$

The width of this rectangle is $w = 2(2\sqrt{2}) = 4\sqrt{2}$ and the height is $h = 2(3/\sqrt{2}) = 6/\sqrt{2}$. The greatest area is

$$A = wh = 4\sqrt{2} \cdot \frac{6}{\sqrt{2}} = 24 \quad \blacksquare$$

17.5 Parametric Surfaces

At many places in this text, we've used parametric equations to generate curves in both 2D and 3D. The usual suspects have been circles — you know, the whole $x(t) = r\cos(t)$ and $y(t) = r\sin(t)$ for $0 \leq t \leq 2\pi$ thing. We don't need to go over this stuff again, but let's note how many different curves we could make using only one parameter, usually called t.

I wonder what we can do with *two* parameters? You might be able to guess. If use of one parameter makes a curve in space, which is a one-dimensional construction, perhaps use of two parameters will help generate a two-dimensional construction, which would be ... a surface!

At first, using a set of parametric equations to represent a surface seems complicated, but you have already won half the battle if you just remember your old friend $\sin^2\theta + \cos^2\theta = 1$. This identity is what links the pair of equations $x = 4\cos t$ and $y = 4\sin t$ to the circle $x^2 + y^2 = 16$. If we rewrite the two equations as $x/4 = \cos t$ and $y/4 = \sin t$, then we can combine them via

$$\left(\frac{x}{4}\right)^2 + \left(\frac{y}{4}\right)^2 = (\cos t)^2 + (\sin t)^2 = 1$$

so that

$$\frac{x^2}{16} + \frac{y^2}{16} = 1$$

or $x^2 + y^2 = 16$. Further, by playing games with the domain of t, we can select part or all of a curve. In the case of this circle of radius 4, a domain of $0 \leq t < 2\pi$ gives the full circle, whereas a smaller domain of $0 \leq t \leq \pi$ gives only the upper half.

With that in mind, consider the following set of equations with *two* parameters, t and s.

$$\begin{cases} x = 3\cos t \\ y = 3\sin t \\ z = \quad s \end{cases} \quad \text{with } 0 < s < 6 \quad ; \quad 0 \leq t < 2\pi$$

Would you believe me if I told you this represents a cylinder of radius 3, centered on the z-axis, from $z = 0$ to $z = 6$? You should, because it's true! The first two equations together give $x^2 + y^2 = 9$; as a set, we see that z can be any value between 0 and 6, and for all values of z, we see a circle of radius 3 in the horizontal plane.

Or, how about the parametric equations

$$\begin{cases} x = \sqrt{s}\cos t \\ y = \sqrt{s}\sin t \\ z = \quad s \end{cases} \quad \text{with } 0 < s < \infty \quad ; \quad 0 \leq t < 2\pi$$

Can you see why this represents the surface $z = x^2 + y^2$, and so is a paraboloid? (Hint: Compare $x^2 + y^2$ to z.)

With parametric equations for curves in 2D and 3D, we saw that we could list the equations separately, such as $x = f(t), y = g(t)$, or we could combine them into a single vector equation as $\mathbf{r}(t) = \langle f(t), g(t) \rangle$. The same can be done for parametric surfaces. For example, the equations for the paraboloid shown just above this paragraph could be regrouped into a vector equation as

$$\mathbf{r}(s,t) = \langle \sqrt{s}\cos t, \sqrt{s}\sin t, s \rangle \quad \text{for} \quad 0 < s < \infty, 0 \leq t < 2\pi$$

Of course, all these parametric equations and the rest in in Appendix A.2 which are based on trigonometric functions are only needed when we're trying to clip out a portion of a surface with circular or elliptical cross sections. If we want to clip out a portion of a surface that (a) passes the vertical line test, and (b) is set over a rectangular region, then the parametric equations don't need to be different than the rectangular form of the surface. Just like we can parameterize $y = f(x)$ as $\mathbf{r}(t) = \langle t, f(t) \rangle$ for $a \leq t \leq b$, we can also parameterize $z = f(x, y)$ as

$$\mathbf{r}(s,t) = \langle s, t, f(s,t) \rangle \quad \text{for} \quad a \leq s \leq b \,;\, c \leq t \leq d$$

For example, the hyperboloid $z = 3x^2 - 4y^2$ over the unit square can be set up as

$$\mathbf{r}(s,t) = \langle s, t, 3s^2 - 4t^2 \rangle \quad \text{for} \quad -1 \leq s \leq 1 \,;\, -1 \leq t \leq 1$$

Because this book is half supplemental textbook, half solution guide, half collection of handy references, and half the thing you put on a picnic table to hold down the tablecloth on a windy day, I have created Appendix A to contain information related to conic sections and quadric surfaces; the portion related to quadric surfaces contains descriptions of the parametric representation of those surfaces. Rather than repeat all that information in two places, I will just say that if you're interested in pursuing these concepts, go dive in to Appendix A. There are some problems given below here for you to practice with.

There are a couple of good reasons to pursue these ideas. One reason is just sheer curiosity. Some of the calculations to be done in the next Chapter (especially surface integrals) have other representations for surfaces posed in parametric form; while we won't see them in this text, perhaps you will encounter them in a different resource. Another reason is to prepare for "What's Next, Part 3!" As we get towards the end of our journey through multivariable calculus, you might be wondering, "OK, so if I like this stuff, where do I go next?" If you've been enjoying more recent topics such as gradients, multivariable optimization, arc length, curvature, and the Fresnel–Serret frame, then you should "level up" to a study of *differential geometry.* This is your gateway to concepts like tensors and manifolds. And, it is not just for pure math nerds; differential geometry is needed to gain a foothold in some advanced applied areas, such as the study of general relativity. This is highly recommended for those of you blending an interest in mathematics with an interest in physics or astrophysics.

Parametric Surfaces — Problem List

Parametric Surfaces — You Try It

These appeared above; solutions begin on the next page.

The manuscript for this text was prepared using the typsetting program LaTeX and a supplemental graphical package TikZ. To create a three-dimensional graph of a surface, TikZ requires the parametric equations for the surface. These You Try It problems relate to images found in this text.

(1) Figure 14.18 shows the cylinder $x^2+y^2=9$ for $0 \le z \le 8$ and the plane $y+z=5$ for $-3 \le x, y \le 3$. Provide the parametric equations for these two surfaces and give the appropriate bounds on the parameters.

(2) A figure you have seen in this text displayed the surface with parametric equations (in vector form),

$$\mathbf{r}(s,t) = \langle s, t, s^2 - t^2 \rangle \quad \text{for} \quad -1 \le s, t \le 1$$

Identify this surfaces by giving its "regular" expression in rectangular coordinates.

(3) Figure 15.26 shows the upper hemisphere of radius 4 centered at the origin. Provide the parametric equations for this surface and give the appropriate bounds on the parameters.

(4) A quadric surface has parametric equations $x(s,t) = \sqrt{s}\cos t$, $y(t) = \sqrt{s}\sin t$, and $z(t) = 5 - s$ for $0 \le s \le 5$ and $0 \le t < 2\pi$. Identify this surface by giving its "regular" expression in rectangular coordinates.

(5) Design Fig. 15.30, which shows the region between the cone $z^2 = x^2+y^2$ and the sphere $\rho = 1$. Give the parametric equations along with the bounds on the parameters required to clip the surfaces to the portions shown. Bonus: give the parametric equations of the curve shown at the intersection of the two surfaces.

Parametric Surfaces — Practice Problems

Try these as you get the hang of the You Try It problems. Solutions to these problems are available in Sec. B.5.5.

(1) Figure 15.29 shows the paraboloid $x = 5 - y^2 - z^2$ for $0 \le x \le 5$. Provide the parametric equations for this surface and give the appropriate bounds on the parameters.
(2) Figure C.9 shows the paraboloid $z = 2x^2 + y^2$ for $0 \le z \le 10$. Provide the parametric equations for this surface and give the appropriate bounds on the parameters.
(3) A quadric surface has the following parametric (vector) equations. Identify this surface by giving its "regular" expression in rectangular coordinates.

$$\mathbf{r}(s,t) = \langle \sqrt{9-s}\cos t, \sqrt{9-s}\sin t, s\rangle \quad \text{for} \quad 0 \le s \le 9, 0 \le t < 2\pi$$

(4) A quadric surface has the parametric equations $x = 1 + s + t, y = -s + t, z = 2s$, where s and t can be any real numbers. Identify this surface by giving its "regular" expression in rectangular coordinates.
(5) A quadric surface has the parametric equations $x = s\cos t, y = s^2, z = s\sin t$, for $0 \le s \le \pi$ and $0 \le t \le \pi$. Identify this surface by giving its "regular" expression in rectangular coordinates and appropriate bounds on those coordinates.
(6) Figure C.20 shows the hyperboloid $x^2 + 3y^2 - z^2 = 1$ between $z = -1$ and $z = 4$. Provide the parametric equations for this surface and give the appropriate bounds on the parameters.

Parametric Surfaces — Challenge Problems

Try these problems to test your skills with the ideas in this section. Solutions to these problems are available in Sec. C.5.5.

(1) Identify the parametric surface $\mathbf{r}(s,t) = \langle 3\sin t\cos s, 3\sin t\sin s, 3\cos t\rangle$ for $0 \le s \le \pi/2$ and $0 \le t \le \pi/2$.

(2) Give parametric equations for the portion of the plane $z = 2x + y$ that lies inside the cylinder $x^2 + y^2 = 1$.

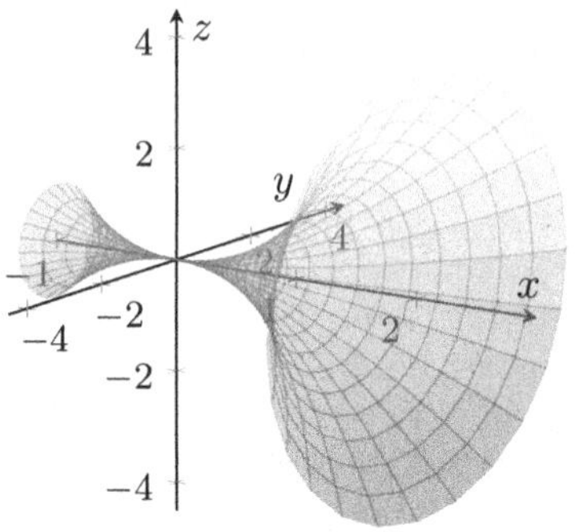

Fig. 17.7 $f(x) = x^2$ revolved around the x-axis on $[-1, 2]$.

(3) Do you remember surfaces of revolution? Those can be drawn parametrically, too! A figure in Chapter 9 (Volume 1), duplicated here as Fig. 17.7, shows the surface generated when the curve $f(x) = x^2$ is revolved around the x-axis on the interval $[-1, 2]$. Give parametric equations for this surface of revolution.

Parametric Surfaces — You Try It — Solved

(1) Figure 14.18 shows the cylinder $x^2+y^2 = 9$ for $0 \leq z \leq 8$ and the plane $y+z = 5$ for $-3 \leq x, y \leq 3$. Provide the parametric equations for these two surfaces and give the appropriate bounds on the parameters.

□ The equations for the cylinder need to create a circle of radius 3 parallel to the xy-plane at any z; they can be written as

$$\begin{cases} x = 3\cos t \\ y = 3\sin t \\ z = \quad s \end{cases} \quad \text{with} \quad 0 \leq s \leq 8 \quad ; \quad 0 \leq t < 2\pi$$

In vector form, we can write this (with the same bounds) as

$$\mathbf{r}(s,t) = \langle 3\cos t, 3\sin t, s \rangle$$

For the plane $y + z = 5$, we can just rename x and y then build z accordingly:

$$\mathbf{p}(s,t) = \langle s, t, 5 - t \rangle \quad \text{for} \quad -3 \leq s, t \leq 3 \quad \blacksquare$$

(2) A figure you have seen in this text displayed the surface with parametric equations (in vector form),

$$\mathbf{r}(s,t) = \langle s, t, s^2 - t^2 \rangle \quad \text{for} \quad -1 \le s, t \le 1$$

Identify this surfaces by giving its "regular" expression in rectangular coordinates.

□ This is a set of parametric equations built from direct replacement. We see that $x = s$ and $y = t$, so that $z = s^2 - y^2$ becomes the surface $z = y^2 - x^2$. The domain of x and y are $-1 \le x, y \le 1$. This happens to be Fig. 14.1. ■

(3) Figure 15.26 shows the upper hemisphere of radius 4 centered at the origin. Provide the parametric equations for this surface and give the appropriate bounds on the parameters.

□ By comparing the equations of a sphere found in Appendix A.2 to spherical coordinates, we can see that the parameter s plays the role of the azimuthal angle ϕ and the parameter t plays the role of the rotational angle θ. This means that to make a hemisphere, we should restrict s to $[0, \pi/2]$, while t can use the full $[0, 2\pi]$. To make our hemisphere have a radius of 4, then, we have:

$$\begin{cases} x = 4\sin s\cos t \\ y = 4\sin s\sin t \\ z = 4\cos s \end{cases} \quad \text{with} \quad 0 \le s \le \frac{\pi}{2} \quad ; \quad 0 \le t < 2\pi$$ ■

(4) A quadric surface has a (parametric) vector equation $\mathbf{r}(s,t) = \langle \sqrt{s}\cos t, \sqrt{s}\sin t, 5 - s \rangle$ for $0 \le s \le 5$ and $0 \le t < 2\pi$. Identify this surface by giving its "regular" expression in rectangular coordinates.

□ We can break this out as

$$\begin{cases} x = \sqrt{s}\cos t \\ y = \sqrt{s}\sin t \\ z = 5 - s \end{cases} \quad \text{with} \quad 0 \le s \le 5 \quad ; \quad 0 \le t < 2\pi$$

Even without reference to Appendix A.2, it's pretty clear that we can put the first two equations together as

$$x^2 + y^2 = s\cos^2 t + s\sin^2 t = s(\cos^2 t + \sin^2 t) = s$$

and since $z = 5 - s$, we've discovered that $z = 5 - (x^2 + y^2)$. These equations represent the (inverted) paraboloid $z = 5 - x^2 - y^2$ for $0 \le z \le 5$. ■

(5) Design Fig. 15.30, which shows the region between the cone $z^2 = x^2+y^2$ and the sphere $\rho = 1$. Give the parametric equations along with the bounds on the parameters required to clip the surfaces to the portions shown. Bonus: give the parametric equations of the curve shown at the intersection of the two surfaces.

□ The cone is just the standard "perfect diagonal" cone $\phi = \pi/4$ which, by Appendix A.2, has parametric equations

$$\begin{cases} x = s\cos t \\ y = s\sin t \\ z = \quad s \end{cases} \quad \text{with} \quad 0 \leq s \leq H \quad ; \quad 0 \leq t < 2\pi$$

where H is the height of the cone, clipped where it intersects the sphere. The sphere is the unit sphere,

$$\begin{cases} x = \sin u \cos v \\ y = \sin u \sin v \\ z = \quad \cos u \end{cases} \quad \text{with} \quad 0 \leq s \leq \frac{\pi}{4} \quad ; \quad 0 \leq v < 2\pi$$

To find both the height H of the cone and also the radius of the circle of intersection of the cone with the sphere, we have to find where these surfaces intersect. By combining $x^2+y^2+z^2 = 1$ with $z^2 = x^2+y^2$, we find $x^2+y^2 = 1/2$, which means that the z-coordinate of the intersection and the radius of the circle of intersection are both $1/\sqrt{2}$. The circle of intersection is

$$\mathbf{r}(t) = \langle \frac{1}{\sqrt{2}} \cos t, \frac{1}{\sqrt{2}} \sin t \rangle \quad \text{for} \quad 0 \leq t < 2\pi$$

We can also go back to the equations for the cone and set $H = 1/\sqrt{2}$. ■

Chapter 18

The Big Bang of Scalar and Vector Quantities

18.1 Line Integrals

Introduction

The last few topics were examples of how scalar and vector quantities play nice with each other. In the next few topics, we'll step up the interaction between them. For this first new topic, I have bad news. You may have thought that, having done single, double and triple integrals, you'd have exhausted all the varieties of integration and were done with it. But no, there are more! The three types of integrals you know so far, with their definitions and notation (the single integral notation is adjusted a bit to match the rest), are:

$$\int_I f(x)\,dx = \lim_{n\to\infty}\sum_{i=1}^{n} f(x_i^\star)\Delta x$$

$$\iint_R f(x,y)\,dA = \lim_{n,m\to\infty}\sum_{i=1}^{n}\sum_{j=1}^{m} f(x_i^\star, y_j^\star)\Delta A$$

$$\iiint_E f(x,y,z)\,dV = \lim_{n,m,p\to\infty}\sum_{i=1}^{n}\sum_{j=1}^{m}\sum_{k=1}^{p} f(x_i^\star, y_j^\star, z_k^\star)\Delta V$$

It may seem like there are no other kinds possible, but there are two more types of integration that are situated in between these — there's one type that's sort of a single integral but also sort of a double integral, and another that's sort of a double integral, but also sort of a triple integral. We'll take a look at the former in this section.

Line Integrals

Think back to the good old days in Calc I, when you learned about single integrals, $\int_a^b f(x)dx$. Here, you integrated along the x-axis, i.e. along a straight line. There was a function $f(x)$ somewhere above (or below) this line. You chopped the line into small pieces of size Δx, selected a representative point $x_i^\star$ from within each piece, and then found the function's value $f(x_i^\star)$ at each representative point. From this information, you created small rectangles, got their areas $f(x_i^\star)\Delta x$, and added the areas up. Then you asked the magic question, what happens as the number of rectangles goes to infinity?

You should also remember that we integrated with respect to y, too. So there was nothing special about the x-axis, since we did integrals in which the interval of integration could be either on a horizontal line (the x-axis) or a vertical line (the y-axis). Fundamentally, in a single integral, we specify a one-dimensional domain (interval) of integration, and the variable we assign to it is irrelevant.

Imagine taking this process of doing a single integral and embedding it in a three-dimensional coordinate system. Suppose we have a function or surface hovering over the xy-plane. Can't we select an interval on the x-axis, say $[a, b]$ (let's call it I) and do here exactly the same thing we do for a single integral? (Fig. 18.1 shows a one-dimensional integral process embedded along the x-axis — the trapezoid rule is being used.) We'd write the process like this:

$$\int_I f(x,0)\,dx = \lim_{n\to\infty}\sum_{i=1}^{n} f(x_i^\star,0)\Delta x$$

Or, what if we integrate over an interval from the y-axis:

$$\int_I f(0,y)\,dy = \lim_{m\to\infty}\sum_{j=1}^{n} f(0,y_j^\star)\Delta y$$

(Fig. 18.2 shows a one-dimensional integral process embedded along the y-axis.) Either of these are possible; and since we've ported the single integral into 3-space, why should we restrict our intervals of integration to the x- or y-axis? In 2D that was necessary, but in 3D, that's silly! Why can't our

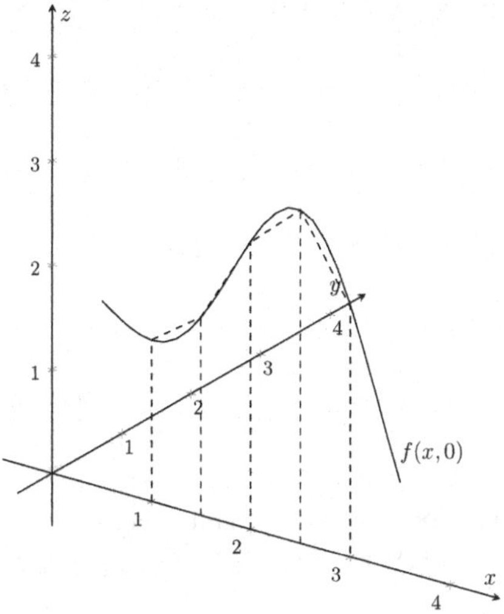

Fig. 18.1 A single integral along the x-axis in $\mathbb{R}^3$.

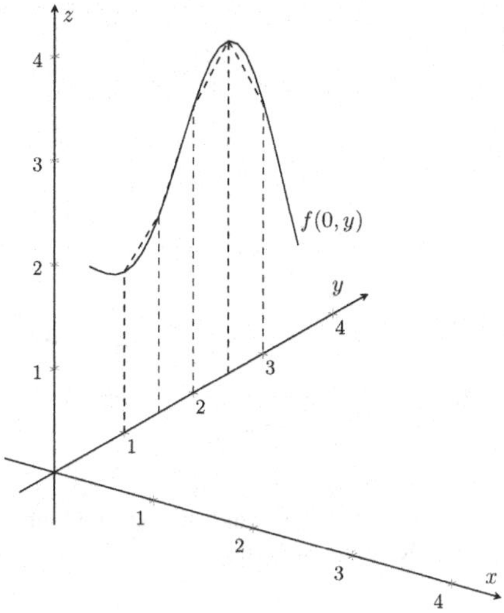

Fig. 18.2 A single integral along the y-axis in $\mathbb{R}^3$.

interval of integration I be taken from some line that's neither the x-axis nor the y-axis? Heck, the interval doesn't even have to be parallel to either axis: Figure 18.3 shows a one dimensional integral process embedded along the line $y - x$ in $\mathbb{R}^3$. Heck, the "interval" doesn't even have to be a straight line! The end result is that we can take any line or curve (*contour*) from the xy-plane and use it as a baseline for a one-dimensional integral process. Figure 18.4 shows this scenario. These are called *line integrals.*

Let's reach back to Sec. 14.3 and grab our list of delineated steps for a partitioning process, and apply it here for line integrals. We start with some contour C (continuous, but not necessarily straight) in the xy-plane, and a function $f(x, y)$ in space. We will have to adapt some of the terminology in the steps, but the overall ideas are still the same:

(1) **Select a region of integration from the domain of the function.** This will be our contour C.
(2) **Partition the region of integration into tiny pieces. Each tiny piece, or partition, will have its own size.** Since the contour C must be generated by parametric equations, using parameter t, the partitioning will divide $t = a$ to $t = b$ as $a \leq t_1 \leq$

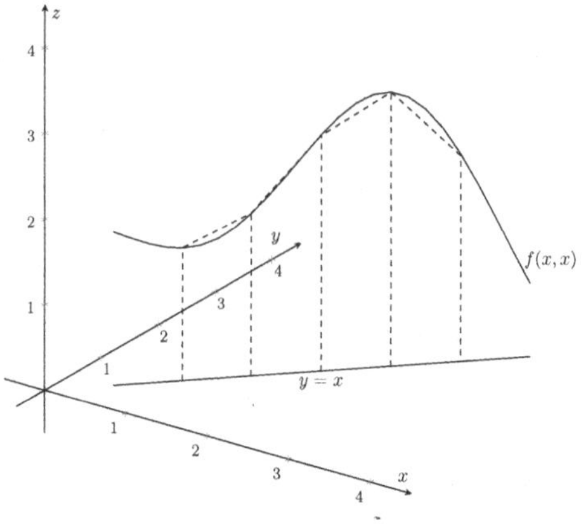

Fig. 18.3 A single integral along the line $y = x$ in $\mathbb{R}^3$.

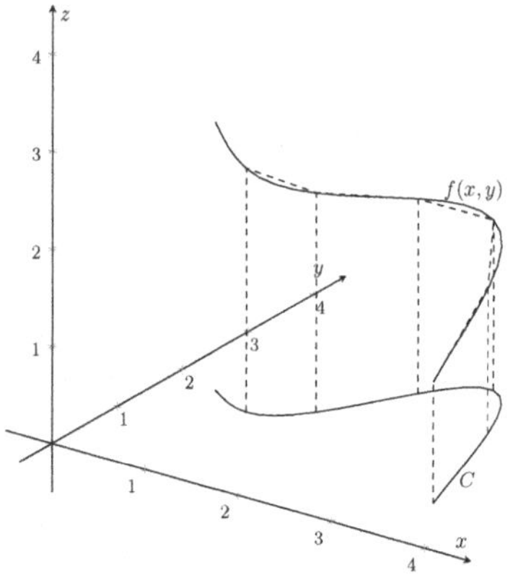

Fig. 18.4 A line integral along a contour C in $\mathbb{R}^3$.

$t_2 \leq \ldots \leq t_n = b$. The length of each little partitioned piece is described by its arc length s, or rather, we will upgrade what used to be Δx to Δs. (It is important to understand why we're not using Δt ... do you understand why?)

(3) **Within each individual partition, select a representative point.** Within each little Δs, a representative point must be referenced by both its x and y coordinates, so we will call this point $(x_i^\star, y_i^\star)$. Note that we only use one counter, i, not two. This is because there is only one representative point for each partition of the contour. Figure 18.5 shows a sample partitioned Δs in the xy-plane and its representative point; not shown in the figure is the function $f(x, y)$ hovering in space above the xy-plane.

(4) **Plug the coordinate(s) of that representative point into the function.** Our representative point gives the value $f(x_i^\star, y_i^\star)$.

(5) **Multiply the function's value at the representative point by the size of the partition.** Our "length times width" looks like $A_i = f(x_i^\star, y_i^\star)\,\Delta s$.

(6) **Sum these resulting contributions from all partitions.**

$$\sum_{i=1}^{n} f(x_i^\star, y_i^\star)\Delta s$$

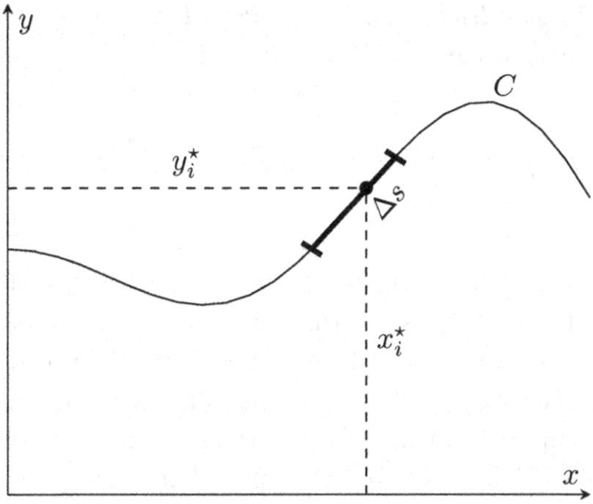

Fig. 18.5 Partitioning for a line integral.

(7) **Find the limit of this sum as the number of partitions goes to ∞.**

$$\lim_{n\to\infty}\sum_{i=1}^{n} f(x_i^\star, y_i^\star)\Delta s$$

(8) **Give this limit a name and symbol...**

Here's the part that's fun to think about! Are we building a single integral or a double integral? Do we want one integral sign or two? To answer this, we have to take a deliberate look at the integration process to notice something that may have been subtle before, but is now crucially important: *the number of integral signs we use does NOT come from the number of variables in the function.* We don't use two integral signs in a double integral because there are two variables in $f(x, y)$; similarly, we don't use three integral signs in a triple integral because there are three variables in $f(x, y, z)$. Rather, the number of integral signs reflects the dimensionality of the domain of integration. In a double integral, we integrate over dA — which evolved from the measures of ΔA in a two-dimensional domain. In a triple integral, we integrate over dV — which evolved from the measure of ΔV in a three-dimensional domain. Here, we're integrating over ds — and although this reflects a partition of a curve in the xy-plane, it is still inherently a one-dimensional construct; ds evolves from Δs. And so, we want to use one integral sign.

Definition 18.1. *The **line integral** (also called a contour integral) of the function $f(x, y)$ over a contour C is defined by:*

$$\int_C f(x,y)\,ds = \lim_{n\to\infty} \sum_{i=1}^{n} f(x_i^\star, y_i^\star)\Delta s$$

So what do you think a line integral computes? Look at Fig. 18.4 and be sure you see that each $f(x_i^\star, y_i^\star)\Delta s$ gives an estimate of the *area* of a standing rectangle; the total sum, then, gives an estimate of the total area of the "curtain" that hangs down from $f(x, y)$ over C. And the limit pushes that estimate into an exact value. (Picture that if we take scissors and cut along C, then go up and cut along the corresponding path on $f(x, y)$, we'll be cutting out a sheet that can be laid flat; it has area.)

Line integrals can compute something else, too. Recall that

- $\displaystyle\int_I (1)\,dx$ computes the length of the interval of integration I
- $\displaystyle\iint_R (1)\,dA$ computes the area of the region of integration R
- $\displaystyle\iiint_E (1)\,dV$ computes the volume of the region of integration E.

So what do you think $\displaystyle\int_C (1)\,ds$ computes?

Useful Fact 18.1. *Given a contour C in the xy-plane, the line integral $\displaystyle\int_C (1)\,ds$ computes the arc length of C.*

Although all this introductory stuff put the contour C in the xy-plane, that's not necessary. We could integrate $f(x, y)$ over a contour in the yz-plane or the xz-plane. Actually, if we stepped up a dimension, we could integrate some $f(x, y, z)$ over a contour C that roams freely through $\mathbb{R}^3$. Once we see how to lay out and solve line integrals, it will be very obvious how to change their placement or extend them to higher dimensions.

Evaluating line integrals really isn't that bad, because the first thing we do is convert them them into "regular" single integrals that reference the parameter t used to build the contour C. Line integrals can come in a few different formats, so let's look at them one at a time.

Line Integrals, Format 1: Arc Length Formulation

To solve a line integral of $f(x,y)$ formulated with the arc length element ds, you need to have the parametric equations that describe the contour C:

Useful Fact 18.2. *Given the parametric equations for the contour of integration, $C = (x(t), y(t))$: for $a \le t \le b$, we evaluate a line integral by opening it up into a single integral as follows:*

$$\int_C f(x,y)\,ds = \int_a^b f(x(t),y(t))\sqrt{(x'(t))^2 + (y'(t))^2}\,dt$$

This formula should make sense; the arc length element ds measures the arc length of a small partition of the curve. If you go back and look at the formula for the arc length of a curve in parametric form, it should be no surprise that $ds = \sqrt{(x'(t))^2 + (y'(t))^2}\,dt$. In all, this formulation means you need to:

(1) Find the parametric form of the curve C
(2) Compute the arc length differential $ds = \sqrt{(x'(t))^2 + (y'(t))^2}\,dt$
(3) Plug the parametric equations for x and y into the function $f(x,y)$
(4) Set up the total integral shown, using the first and last values of t from the parametric equations of C for the limits of integration

EX 1 Evaluate $\displaystyle\int_C 2 + x^2y\,ds$ where C is the upper half of the unit circle.

Described parametrically, the upper half of the unit circle is $x(t) = \cos t$, $y(t) = \sin t$ for $0 \le t \le \pi$. Using these parametric equations, we can get ready for the arc length formulation of the line integral:

$$\begin{aligned} x'(t) &= -\sin t \\ y'(t) &= \cos t \\ \sqrt{(x'(t))^2 + (y'(t))^2} &= \sqrt{\sin^2 t + \cos^2 t} = 1 \\ f(x(t),y(t)) &= 2 + (\cos t)^2(\sin t) \end{aligned}$$

Putting these together,

$$\begin{aligned} \int_C 2 + x^2y\,ds &= \int_a^b f(x(t),y(t))\sqrt{(x'(t))^2 + (y'(t))^2}\,dt \\ &= \int_0^\pi (2 + \cos^2 t \sin t)(1)\,dt = \left(2t - \frac{1}{3}\cos^3 t\right)\Big|_0^\pi \\ &= \left(2\pi + \frac{1}{3}\right) - \left(0 - \frac{1}{3}\right) = 2\pi + \frac{2}{3} \quad \blacksquare \end{aligned}$$

You Try It

(1) Evaluate $\int_C xy^4\, ds$, where C is the right half of the circle $x^2 + y^2 = 16$.

The contour of integration C can have many kinks and bends. You don't need to find a single set of parametric equations to take care of the entire contour at once, you can describe each segment with its own parametric equations. That is, if C consists of several individual contours $C_1, C_2, \ldots$ joined together, then

$$\int_C f(x,y)\, ds = \int_{C_1} f(x,y)\, ds + \int_{C_2} f(x,y)\, ds + \cdots$$

This will be important when we get to Sec. 18.4.

Line Integrals, Format 2: Cartesian Formulation

Line integrals aren't required to use the arc length element ds. Line integrals can also reference rectangular coordinates directly. This variety of line integrals appears as, for example,

$$\int_C P(x,y)\, dx + Q(x,y)\, dy \quad \text{or} \quad \int_C P(x,y)\, dx \quad \text{or} \quad \int_C Q(x,y)\, dy$$

Useful Fact 18.3. *To evaluate a line integral that references dx and dy directly, we use the parametric equations for the contour of integration, $C = (x(t), y(t))$: for $a \le t \le b$, to prepare $dx = x'(t)\, dt$ and / or $dy = y'(t)\, dt$ directly and construct the corresponding single integral with respect to t. For example,*

$$\int_C P(x,y)\, dx = \int_a^b P(x(t), y(t))\, x'(t)\, dt$$

Note that this format of a line integral is nice because it avoids the square root term in the arc length formulation. We'll see how to interpret line integrals of this form below, but for now, start looking closely at the general structure $\int_C P(x,y)\, dx + Q(x,y)\, dy$. Does that remind you of anything?

EX 2 Evaluate $\int_C \left(xy + \frac{1}{x}\right) dy$ where C is the arc of $y = x^2$ from (1,1) to (3,9).

A good parametric formulation of C is $x(t) = t, y(t) = t^2$ for $1 \leq t \leq 3$. From this, we can immediately get that $dy = 2t\,dt$. And so following the suggestion of Useful Fact 18.3,

$$\begin{aligned}\int_C \left(xy + \frac{1}{x}\right) dy &= \int_1^3 \left((t)(t^2) + \frac{1}{t}\right)(2t\,dt) = 2\int_1^3 (t^4 + 1)\,dt \\ &= 2\left(\frac{1}{5}t^5 + t\right)\Bigg|_1^3 = 2\left(\frac{1}{5}3^5 + 3\right) - 2\left(\frac{1}{5} + 1\right) \\ &= 2\left(\frac{242}{5} + 1\right) \quad \blacksquare\end{aligned}$$

You Try It

(2) Evaluate $\displaystyle\int_C xy\,dx + (x - y)\,dy$; where C is the line from (0,0) to (2,0) joined to the line from (2,0) to (3,2).

Okay, have you been thinking about the format of $\displaystyle\int_C P(x,y)\,dx + Q(x,y)\,dy$? If so, you should be getting suspicious. This is the first section in a Chapter titled for the big bang of scalar and vector quantities, but we've seen no vectors ... yet. Did you notice that the form $P(x,y)\,dx + Q(x,y)\,dy$ looks an awful lot like a dot product? Well, it is! We can write $P(x,y)\,dx + Q(x,y)\,dy$ as $\langle P(x,y), Q(x,y)\rangle \cdot \langle dx, dy\rangle$, and so:

$$\int_C P(x,y)\,dx + Q(x,y)\,dy = \int_C \langle P(x,y), Q(x,y)\rangle \cdot \langle dx, dy\rangle$$

And so, this brings us to the third format of line integrals, which involve a vector field $\mathbf{F}(x,y) = \langle P(x,y), Q(x,y)\rangle$.

Line Integrals, Format 3: Vector Formulation

Given a vector field $\mathbf{F}(x,y) = \langle P(x,y), Q(x,y)\rangle$ and a contour C given by the vector function $\mathbf{r}(t) = \langle (x(t), y(t)\rangle$ for $a \leq t \leq b$, we denote the integral of $\mathbf{F}$ along $\mathbf{r}(t)$ as:

$$\int_C \mathbf{F}(x,y) \cdot d\mathbf{r}$$

So, the bad news is that vectors have shown up in a big way. The good news is that to evaluate this, we're just going to convert it back to the previous form of a line integral.

Useful Fact 18.4. *Given a vector field* $\mathbf{F}(x,y) = \langle P(x,y), Q(x,y)\rangle$ *and a contour* C *given by the vector function* $\mathbf{r}(t) = \langle (x(t), y(t)\rangle$ *for* $a \le t \le b$, *we compute the integral of* $\mathbf{F}$ *along* $\mathbf{r}(t)$ *as follows:*

$$\int_C \mathbf{F}(x,y)\cdot d\mathbf{r} = \int_C \langle P(x,y), Q(x,y)\rangle \cdot \langle dx, dy\rangle = \int_C P(x,y)dx + Q(x,y)dy$$

EX 3 Evaluate $\int_C \mathbf{F}\cdot d\mathbf{r}$ for $\mathbf{F}(x,y) = \langle x^2, -xy\rangle$ and C the contour given by the vector function $\mathbf{r}(t) = \langle \cos t, \sin t\rangle, 0 \le t \le \frac{\pi}{2}$.

Since $\mathbf{r}(t) = \langle x(t), y(t)\rangle = \langle \cos t, \sin t\rangle$, we get $d\mathbf{r}(t) = \langle -\sin t, \cos t\rangle$. Plugging in the parametric form of C into the vector field $\mathbf{F}$, we get

$$\mathbf{F}(x,y) = \langle x^2, -xy\rangle = \langle \cos^2 t, -\cos t \sin t\rangle$$

So carrying out the dot product we get

$$\begin{aligned}\int_C \mathbf{F}\cdot d\mathbf{r} &= \int_C \langle \cos^2 t, -\cos t\sin t\rangle \cdot \langle -\sin t, \cos t\rangle\, dt \\ &= \int_0^{\pi/2} -\cos^2 t \sin t - \cos^2 t \sin t\, dt = -2\int_0^{\pi/2} \cos^2 t \sin t\, dt \\ &= \frac{2}{3}\cos^3 t\Big|_0^{\pi/2} = -\frac{2}{3} \quad \blacksquare\end{aligned}$$

You Try It

(3) Evaluate $\int_C \mathbf{F}\cdot d\mathbf{r}$ where $\mathbf{F}(x,y) = \langle x^2y^3, -y\sqrt{x}\rangle$ and $\mathbf{r}(t) = \langle t^2, -t^3\rangle$ for $0 \le t \le 1$.

So what do these vectorized line integrals mean, anyway? Well, I'm glad you asked. A long time ago, you probably learned that the work done by a constant force F to move an object a distance x is $W = Fx$. This equation assumes the direction of force is the same as the direction of motion. But a constant force in the same direction as motion is boring. How do we make it more interesting?

What if the direction of the force is NOT the same as the direction of motion? Then the work done by a constant force $\mathbf{F}$ in the direction of the vector $\mathbf{r}$ is $W = \mathbf{F}\cdot\mathbf{r}$. But this is still boring. Constant forces are dull, whether they're in the direction of motion or not.

So, now suppose we have a vector field that represents a *variable* force, and an object moving through the force field along a path C. If we consider C to be a vector function $\mathbf{r}(t)$, then along each tiny segment $d\mathbf{r}$ of the path C, the work done to move the object that small distance is $\mathbf{F}(x,y)\cdot d\mathbf{r}$. The total amount of work done moving the object along the whole path is ... you guessed it ...

$$\int_C \mathbf{F}(x,y)\cdot d\mathbf{r}$$

EX 4 Find the work done by the vector (force) field $\vec{F}(x,y,z) = \langle 3 + 2xy, x^2 - 3y^2, z^5\rangle$ when it moves an object around a complete circle in the plane $z = 3$ that has radius 5 and center $(0,0,3)$. Once you construct the integral, you are very likely to want to use tech to evaluate it.

First, we need the parametric equations for C. It's a circle of radius 5 centered at the origin, but up at $z = 3$. So we have

$$C = \mathbf{r}(t) = \langle 5\cos t, 5\sin t, 3\rangle \quad \text{for} \quad 0 \le t \le 2\pi$$

and with that, we also have $d\mathbf{r}(t) = \langle -5\sin t, 5\cos t, 0\rangle$. Next, we need to get $\mathbf{F}$ expressed in terms of t by using the component functions from C:

$$\begin{aligned}\mathbf{F} &= \langle 3 + 2xy, x^2 - 3y^2, z^5\rangle\\ &= \langle 3 + 2(5\cos t)(5\sin t), (5\cos t)^2 - 3(5\sin t)^2, (3)^5\rangle\\ &= \langle 3 + 50\cos t\sin t, 25\cos^2 t - 75\sin^2 t, 3^5\rangle\end{aligned}$$

Let's get $\mathbf{F}(x,y)\cdot d\mathbf{r}$ ready for the integral:

$$\begin{aligned}\mathbf{F}(x,y)\cdot d\mathbf{r} &= \langle 3 + 50\cos t\sin t, 25\cos^2 t - 75\sin^2 t, 3^5\rangle \cdot \langle -5\sin t, 5\cos t, 0\rangle\\ &= (3 + 50\cos t\sin t)(-5\sin t) + (25\cos^2 t - 75\sin^2 t)(5\cos t) + 3^5(0)\\ &= -15\sin t - 250\cos t\sin^2 t + 125\cos^3 t - 375\sin^2 t\cos t\\ &= -15\sin t - 625\sin^2 t\cos t + 125\cos^3 t\end{aligned}$$

And so the work integral is,

$$\int_C \mathbf{F}(x,y)\cdot d\mathbf{r} = \int_0^{2\pi} -15\sin t - 625\sin^2 t\cos t + 125\cos^3 t\, dt$$

Now, whether you use tech or just remembering how integrals of trig functions work, the integral of every term in that expression from $t = 0$ to $t = 2\pi$ is zero! And so,

$$W = \int_C \mathbf{F}(x,y)\cdot d\mathbf{r} = 0$$

That was a lot of work for nothing! ■

You Try It

(4) Repeat EX 4 except with $\mathbf{F}(x,y) = \langle x+1, y+2, z+3\rangle$.

Line Integrals — Problem List

Line Integrals — You Try It

These appeared above; solutions begin on the next page.

(1) Evaluate $\int_C xy^4\,ds$, where C is the right half of the circle $x^2+y^2=16$.
(2) Evaluate $\int_C xy\,dx+(x-y)\,dy$; where C is the line from (0,0) to (2,0) joined to the line from (2,0) to (3,2).
(3) Evaluate $\int_C \mathbf{F}\cdot d\mathbf{r}$ where $\mathbf{F}(x,y)=\langle x^2y^3,-y\sqrt{x}\rangle$ and $\mathbf{r}(t)=\langle t^2,-t^3\rangle$ for $0\le t\le 1$.
(4) Repeat EX 4 except with $\mathbf{F}(x,y)=\langle x+1,y+2,z+3\rangle$.

Line Integrals — Practice Problems

Try these as you get the hang of the You Try It problems. Solutions to these problems are available in Sec. B.6.1.

(1) Evaluate $\int_C ye^x\,ds$, where C is the line from (1,2) to (4,7).
(2) Evaluate $\int_C \sin x\,dx+\cos y\,dy$ where C that is the top half of $x^2+y^2=1$ from $(1,0)$ to $(-1,0)$ joined to the line from from $(-1,0)$ to $(-2,3)$.
(3) Evaluate $\int_C \mathbf{F}\cdot d\mathbf{r}$ where $\mathbf{F}(x,y,z)=\langle yz,xz,xy\rangle$ and $\mathbf{r}(t)=\langle t,t^2,t^3\rangle$ for $0\le t\le 2$.
(4) Evaluate $\int_C x^2z\,ds$, where C is the line segment from $(0,6,-1)$ to $(4,1,5)$.
(5) Evaluate $\int_C z\,dx+x\,dy+y\,dz$ along the contour C that is given by

$$x=t^2 \quad y=t^3 \quad z=t^2 \quad \text{for} \quad 0\le t\le 1$$

(6) Find the work done by $\mathbf{F}(x,y,z)=\langle z,y,-x\rangle$ around the contour $\mathbf{r}(t)=\langle t,\sin t,\cos t\rangle$ for $0\le t\le\pi$.
(7) *(Bonus! Following up Sec. 15.3 ...)* We expect $\int_C f(x,y)\,ds=0$ for which of the following combinations of function $f(x,y)$ and path of integration C?

I1) $f(x,y)=x^2y$ and C is the upper half of $2x^2+3y^2=6$?
I2) $f(x,y)=x^2y$ and C is the right half of $2x^2+3y^2=6$?
I3) $f(x,y)=xy^2$ and C is on $y=x^2-1$ from $(-1,0)$ to $(1,0)$?

Line Integrals — Challenge Problems

Try these problems to test your skills with the ideas in this section. Solutions to these problems are available in Sec. C.6.1.

(1) Evaluate $\int_C (x-4)(z-5)\,ds$, where C is the line segment from $(4,1,5)$ to $(0,6,-1)$.

(2) Evaluate $\int_C y\,dx + z\,dy + x\,dz$ along the contour C that is given by

$$x = t^3 \quad y = t^2 \quad z = t \quad \text{for} \quad 0 \le t \le 1$$

(3) Find the work done by $\mathbf{F}(x,y,z) = \langle -y, z^2, x\rangle$ around the contour $\mathbf{r}(t) = \langle \sin t, t, \cos t\rangle$ for $0 \le t \le 2\pi$.

(4) *(Bonus! Following up Sec. 15.3 ...)* We expect $\int_C \mathbf{F}(x,y) \cdot d\vec{r} = 0$ for which of the following combinations of vector field $\mathbf{F}(x,y)$ and oriented path of integration C?

I4) $\mathbf{F}(x,y) = \langle y^3, x^3\rangle$ and C is the circle of radius 2, oriented counterclockwise

I5) $\mathbf{F}(x,y) = \langle x/y, x+y\rangle$ and C is the cardioid $r = 2 + \cos\theta$, oriented counterclockwise

I6) $\mathbf{F}(x,y) = \langle (x+y)^2, x^2y^2\rangle$ and C follows $y = 1/(x^2+1)$ from $x = -2$ to $x = 2$.

Line Integrals — You Try It — Solved

---

(1) Evaluate $\int_C xy^4\,ds$, where C is the right half of the circle $x^2+y^2=16$.

☐ Note that this involves the first type of line integral, i.e. one posed in terms of arc length. We can describe the contour C as

$$x=4\cos t \qquad y=4\sin t \qquad \text{for } -\frac{\pi}{2}\le t\le\frac{\pi}{2}$$

so that

$$\sqrt{[x'(t)]^2+[y'(t)]^2}=\sqrt{(-4\sin t)^2+(4\cos t)^2}=\sqrt{16}=4$$

and

$$\begin{aligned}\int_C xy^4\,ds &= \int_a^b f(x(t),y(t))\sqrt{[x'(t)]^2+[y'(t)]^2}\,dt\\ &= \int_{-\pi/2}^{\pi/2}(4\cos t)(4\sin t)^4(4)\,dt=\int_{-\pi/2}^{\pi/2}4^6\sin^4 t\cos t\,dt\\ &= \frac{8192}{5}=1638.4 \quad ■\end{aligned}$$

---

(2) Evaluate $\int_C xy\,dx+(x-y)\,dy$; where C is the line from (0,0) to (2,0) joined to the line from (2,0) to (3,2).

☐ Note that this line integral does not involve the arc length parameter, so we'll just convert everything directly to parametric form. The contour C has two parts: the line from (0,0) to (2,0) (call this C_1), and then from (2,0) to (3,2) (call this C_2). The contours are

$$C_1:\quad x=2t\quad y=0\quad \text{for}\quad 0\le t\le 1\to dx=2\,dt\quad dy=0$$

$$C_2:\quad x=2+t\quad y=2t\quad \text{for}\quad 0\le t\le 1\to dx=dt\quad dy=2\,dt$$

The integrals are:

$$\int_{C_1} xy\,dx+(x-y)\,dy=\int_0^1(2t)(0)(2\,dt)+(2t-0)(0)=0$$

$$\begin{aligned}\int_{C_2} xy\,dx+(x-y)\,dy &= \int_0^1(2+t)(2t)(dt)+(2+t-2t)(2\,dt)\\ &= \int_0^1[2t(2+t)+2(2-t)]\,dt=\frac{17}{3}\end{aligned}$$

Putting them together,

$$\int_C xy\,dx+(x-y)\,dy=\int_{C_1} xy\,dx+(x-y)\,dy+\int_{C_2} xy\,dx+(x-y)\,dy=\frac{17}{3} \quad ■$$

(3) Evaluate $\displaystyle\int_C \mathbf{F} \cdot d\mathbf{r}$ where $\mathbf{F}(x, y) = \langle x^2y^3, -y\sqrt{x}\rangle$ and $\mathbf{r}(t) = \langle t^2, -t^3\rangle$ for $0 \le t \le 1$.

□ From $\mathbf{r}(t)$ we have

$$x = t^2 \to dx = 2t\,dt$$
$$y = -t^3 \to dy = -3t^2\,dt$$

so that

$$\begin{aligned}\int_C \mathbf{F} \cdot d\mathbf{r} &= \int_C x^2y^3\,dx - y\sqrt{x}\,dy \\ &= \int_0^1 (t^2)^2(-t^3)^3(2t\,dt) - (-t^3)\sqrt{t^2}(-3t^2\,dt) \\ &= \int_0^1 (-2t^14 - 3t^6)\,dt = -\frac{59}{105} \quad \blacksquare\end{aligned}$$

(4) Repeat EX 4 except with $\mathbf{F}(x, y) = \langle x+1, y+2, z+3\rangle$.

□ In EX 4, we used the contour

$$C = \mathbf{r}(t) = \langle 5\cos t, 5\sin t, 3\rangle \quad \text{for} \quad 0 \le t \le 2\pi$$

for which $d\mathbf{r}(t) = \langle dx, dy, dz\rangle = \langle -5\sin t, 5\cos t, 0\rangle$. If we expand $\mathbf{F}$ using the component functions from C, we get

$$\mathbf{F}(x, y) = \langle x+1, y+2, z+3\rangle = \langle 5\cos t + 1, 5\sin t + 2, 6\rangle$$

So,

$$\begin{aligned}\int_C \mathbf{F}(x, y) \cdot d\mathbf{r} &= \int_0^{2\pi} \langle 5\cos t + 1, 5\sin t + 2, 6\rangle \cdot \langle -5\sin t, 5\cos t, 0\rangle\,dt \\ &= \int_0^{2\pi} (5\cos t + 1)(5\sin t) + (5\sin t + 2)(5\cos t) + 6(0)\,dt \\ &= \int_0^{2\pi} 25\sin t\cos t + 5\sin t + 25\sin t\cos t + 10\cos t\,dt \\ &= \int_0^{2\pi} 50\sin t\cos t + 5\sin t + 10\cos t\,dt \\ &= 0\end{aligned}$$

Once again, around a closed contour C, $W = \displaystyle\int_C \mathbf{F}(x, y) \cdot d\mathbf{r} = 0$. Is this a pattern? I guess we have something to investigate in the next section. ■

18.2 Conservative Vector Fields

Introduction

This topic will tie up some threads that have been introduced in previous sections: line integrals, vector fields, gradients, and curls. Near the end of Sec. 16.4, there was an open question: We can compute the gradient of a scalar function $f(x,y)$, $\nabla f = \langle f_x, f_y \rangle$, and that gradient is a vector field. But what about the reverse? If we grab just any old vector field $\mathbf{F} = \langle P(x,y), Q(x,y) \rangle$, is it guaranteed to be the gradient of some scalar function? Well, the answer is no, it's not always guaranteed, but there are indicators for when it is guaranteed.

When a vector field is known to be the gradient of a scalar function, it is called a *conservative vector field*, and the scalar function that produces the vector fiend as its gradient is called the *potential function* for that vector field. As familiar examples, you may have encountered gravitational potentials and electrostatic potentials in a physics course. They are named such because their gradients are the vector fields we know as the gravitational field and electrostatic field.

Identifying Conservative Vector Fields

To formalize some terms given just above,

Definition 18.2. *A vector field* $\mathbf{F}$ *is* ***conservative*** *if there is a scalar function* f *such that* $\mathbf{F} = \nabla f$. *When* $\nabla f = \mathbf{F}$, *we call* f *the* ***potential function*** *for* $\mathbf{F}$.

We can test whether or not a given vector field $\mathbf{F}$ is conservative:

Useful Fact 18.5. *A 2D vector field* $\mathbf{F}(x,y) = \langle P(x,y), Q(x,y) \rangle$ *is conservative if* $\partial P/\partial y = \partial Q/\partial x$.

A 3D vector field $\mathbf{F}(x,y,z) = \langle P(x,y,z), Q(x,y,z), R(x,y,z) \rangle$ *is conservative if* $\nabla \times \mathbf{F} = \mathbf{0}$.

Both of these are easy tests, and in fact they are the same test. You might think, "Hey, you can't do a curl of a two-dimensional vector field!" But you can indeed do that, you just have to cheat by adding a third component of zero. If you upgrade a 2D vector field $\mathbf{F}(x,y) = \langle P(x,y), Q(x,y) \rangle$ to a 3D vector field $\mathbf{F}(x,y,z) = \langle P(x,y), Q(x,y), 0 \rangle$ and find its curl, you will find that the curl is $\langle 0, 0, Q_x - P_y \rangle$. So checking if $\nabla \times \mathbf{F} = \mathbf{0}$ is the same as checking if $P_y = Q_x$. Go ahead, try it!

So, why does this test work? Well, you'll sort of have to take it on faith. Challenge Problem 3 of Sec. 16.4 asked you to show that $\nabla \times (\nabla f)$ is always $\mathbf{0}$. In other words, you proved the statement "If $\mathbf{F}$ is the gradient of a scalar function f, then $\nabla \times \mathbf{F} = \mathbf{0}$." Now consider that statement along with two others:

A: If $\mathbf{F}$ is conservative, then $\nabla \times \mathbf{F} = \mathbf{0}$.
B: If $\nabla \times \mathbf{F} \neq \mathbf{0}$, then $\mathbf{F}$ is not conservative.
C: If $\nabla \times \mathbf{F} = \mathbf{0}$, then $\mathbf{F}$ is conservative.

The truth of Statement A implies the truth of Statement B. But Statement C is still just a hunch. Formally moving it from a hunch to a proven statement is not something we're ready to do. (Maybe we'll see it later? Maybe not? I'll keep you in suspense.) But, we're still going to use Statement C as if it's been proven.

EX 1 Is the vector field $\mathbf{F}(x, y) = \langle 3 + 2xy, x^2 - 3y^2 \rangle$ conservative?

Matching $\mathbf{F}$ to the form $\langle P, Q \rangle$, we have $P(x, y) = 3 + 2xy$ and $Q(x, y) = x^2 - 3y^2$, and so also $P_y = 2x$ and $Q_x = 2x$. Useful Fact 18.5 then confirms that since $P_y = Q_x$, the vector field is conservative. ■

EX 2 Is the vector field $\mathbf{F}(x, y, z) = \langle 3 + 2xy, x^2 - 3y^2, z^5 \rangle$ conservative?

Matching $\mathbf{F}$ to the form $\langle P, Q, R \rangle$, we have $P(x, y, z) = 3+2xy$, $Q(x, y, z) = x^2 - 3y^2$, and $R(x, y, z) = z^5$. Since this is a 3D vector field, Useful Fact 18.5 directs is to look at its curl:

$$\text{curl}(\mathbf{F}) = \begin{vmatrix} \mathbf{i} & \mathbf{j} & \mathbf{k} \\ \frac{\partial}{\partial x} & \frac{\partial}{\partial y} & \frac{\partial}{\partial z} \\ 3 + 2xy & x^2 - 3y^2 & z^5 \end{vmatrix} = 0\mathbf{i} - 0\mathbf{j} + (2x - 2x)\mathbf{k} = \mathbf{0}$$

Since $\nabla \times \mathbf{F} = \mathbf{0}$, this is a conservative vector field. ■

Finding the Potential Function for a Conservative Vector Field

Once we have determined that a vector field is conservative, we can systematically search for its potential function. In a sense, searching for a potential function is the multivariable analog of finding an antiderivative. The vector field we have is a gradient (made of derivatives) of some mystery function, and we're trying to find its ... "antigradient"? As with antiderivatives, potential functions are not unique and can differ by an arbitrary constant. After all, if we know that $\nabla f = \mathbf{F}$, then certainly $\nabla(f + C) = \mathbf{F}$ as well.

Here's how we can find a potential function, at least in the 2D case. We take advantage of the fact that if $\mathbf{F}(x, y) = \langle P(x, y), Q(x, y)\rangle$ is the gradient of some $f(x, y)$, then $f_x = P$ and $f_y = Q$. The overall strategy is to take one of the two relationships ($f_x = P$ or $f_y = Q$) and learn as much as we can via a regular old antiderivative, and then tailor our result to match the other relationship. So,

- STEP 1: Pick either component of $\mathbf{F}$, i.e. $P(x, y)$ or $Q(x, y)$. Find its antiderivative with respect to the appropriate variable; that is, find the antiderivative of P with respect to x, or the antiderivative of Q with respect to y. But there's a catch to this: in these antiderivatives, we have to attach an arbitrary *function* of the other variable, not just an arbitrary constant. Here's why: Suppose we have $P(x, y) = f_x(x, y) = 3x^2$. Since f is a function of x AND y, then the collection of valid antiderivatives of this function with respect to x include things like $x^3 + y$, $x^3 + \sin(y)$, $x^3 + e^{y^2}$, and so on — the derivatives of every one of these functions with respect to x is just $3x^2$. So we must write that the antiderivative of $f_x(x, y) = x^2$ with respect to x is $x^3 + g(y)$, where $g(y)$ is any arbitrary function of y alone.

- STEP 2: Knowing that the other component (P or Q) must be the derivative of the function f with respect to the other variable, you should determine what the arbitrary function must look like; it should be pretty obvious on a case-by-case basis, but this step may take a bit of trial and error.

EX 3 In EX 1, we found that $\mathbf{F}(x, y) = \langle 3+2xy, x^2-3y^2\rangle$ is conservative. Find a potential function for this vector field.

The potential function for this vector field will be some scalar function $f(x, y)$ such that $\mathbf{F} = \langle P, Q\rangle = \langle f_x, f_y\rangle$. Let's follow the steps above to find it.

STEP 1: Picking the first component of **F**, i.e. $P(x, y) = 3 + 2xy$, we know that whatever our potential function $f(x, y)$ is, it generates the derivative $f_x = 3 + 2xy$. Therefore, at worst, $f(x, y)$ itself will look like the antiderivative of this expression with respect to x. That is, $f(x, y) = 3x + x^2y + g(y)$ (where we've installed $g(y)$, an arbitrary function of y).

STEP 2: We also know that the other component of **F**, i.e. $Q(x, y) = x^2 - 3y^2$ must be the same as f_y. But in the last step, we found *most* of $f(x, y)$ already. So we have two hooks into

- From **F** itself, we know that $f_y = x^2 - 3y^2$.
- From our half-built potential function we found in STEP 1, $f(x, y) = 3x + x^2y + g(y)$, we would have $f_y = x^2 + g'(y)$.

Comparing these two representations of f_y gives $x^2 - 3y^2 = x^2 + g'(y)$, and now we know $g'(y) = -3y^2$. This means that $g(y)$ itself is $g(y) = -y^3 + C$, and so:

$$f(x, y) = 3x + x^2y + g(y) = 3x + x^2y - y^3 + C$$

You can verify that the gradient of this function is indeed the vector field **F** we started with. ■

Note that you may not need to follow this formal procedure. If you know that **F** is conservative, then you may be able to get at its potential function almost through trial and error. Knowing that the first component of **F** must be f_x, and the second component must be f_y, and (if 3D) the third component must be f_z, you might be able to puzzle out $f(x, y, z)$ just based on that information. I particularly recommend this strategy for 3D conservative vector fields, although you can extend the formal procedure described above if you need to.

You Try It

(1) Determine if the vector field $\mathbf{F}(x, y) = \langle 6x + 5y, 5x + 4y \rangle$ is conservative. If it is, find a potential function for it.
(2) Determine if the vector field $\mathbf{F}(x, y) = \langle xe^y, ye^x \rangle$ is conservative. If it is, find a potential function for it.
(3) Determine if the vector field $\mathbf{F}(x, y, z) = \langle yz, xz, xy \rangle$ is conservative. If it is, find a potential function for it.

Line Integrals With Conservative Vector Fields

Here is another analogy between potential functions and antiderivatives. Everyone remembers The Fundamental Theorem of Calculus, which says that as long as f obeys certain properties, then we can use its antiderivative F to compute definite integrals:

$$\int_a^b f(x)dx = F(b) - F(a)$$

Here is an upgrade:

Theorem 18.1. *If* $\mathbf{F}(x, y)$ *is a conservative vector field with potential function* $f(x, y)$*, and* C *is a piecewise smooth contour from* (x_1, y_1) *to* (x_2, y_2)*, then*

$$\int_C \mathbf{F}(x, y) \cdot d\mathbf{r} = f(x_2, y_2) - f(x_1, y_1)$$

This is called the ***Fundamental Theorem for Line Integrals****.*

There is one important fact inherent in that formula; if $\mathbf{F}$ is conservative, then the line integral $\int_C \mathbf{F} \cdot d\mathbf{r}$ is **path independent**. That is, the value of the line integral only depends on the first and last points on the contour C; how the contour gets from the first to last point does not matter.

A slightly more "vectorized" version of Theorem 18.1 relies on the fact that we often have the contour C given via its vector equation form: $C = \mathbf{r}(t)$ for $a \leq t \leq b$; in this case, the Theorem might read this way:

If $\mathbf{F}(x, y)$ *is a conservative vector field with potential function* $f(x, y)$*, and* C *is a piecewise smooth contour* $\mathbf{r}(t)$ *for* $a \leq t \leq b$*, then*

$$\int_C \mathbf{F}(x, y) \cdot d\mathbf{r} = f(\mathbf{r}(b)) - f(\mathbf{r}(a))$$

Be ready to describe the endpoints of the contour in either format.

FFT: You can probably imagine what is means for a contour C to be *closed*. It means the contour starts and ends at the same point (think of a circle, square, triangle, etc). The Fundamental Theorem for Line Integrals gives us a definitive conclusion about the value of $\int_C \mathbf{F} \cdot d\mathbf{r}$ when C is a piecewise smooth and closed contour. Do you know what that value is?

EX 4 Evaluate $\int_C \mathbf{F} \cdot d\mathbf{r}$ if $\mathbf{F}(x, y) = \langle ye^x + \sin y, e^x + x\cos y \rangle$ and C is a contour that starts at (0,0), goes counterclockwise in a quarter circular arc, slants diagonally at 45°, and then follows a parabola to the point (π, π).

Wow, what a horrible contour! Luckily, it will turn out that the contour itself won't matter, only its endpoints. Looking at $\mathbf{F}$, we see that $P = ye^x + \sin y$ and $Q = e^x + x\cos y$, so that $P_y = e^x + \cos y$ and $Q_x = e^x + \cos y$. Since $P_y = Q_x$, then $\mathbf{F}$ is conservative and the Fundamental Theorem for Line Integrals will apply. But we need a potential function for $\mathbf{F}$ to use that Theorem. Again looking at $\mathbf{F}$ and knowing that it has a potential function $f(x, y)$, we need

$$f_x = ye^x + \sin y \qquad \text{and} \qquad f_y = e^x + x\cos y$$

I think we can determine without much ado that $f(x, y) = ye^x + x\sin y + C$. Therefore, with the first and last points on the messy contour being $(0, 0)$ and (π, π), we have by the Fundamental Theorem for Line Integrals,

$$\begin{aligned}\int_C \mathbf{F} \cdot d\mathbf{r} &= f(x_2, y_2) - f(x_1, y_1) = f(\pi, \pi) - f(0, 0) \\ &= (\pi e^\pi + \pi \sin \pi) - (0e^0 + 0\sin 0) = \pi e^\pi \ \blacksquare\end{aligned}$$

You Try It

(4) Evaluate $\int_C \mathbf{F} \cdot d\mathbf{r}$ where $\mathbf{F}(x, y) = \langle x^3y^4, x^4y^3 \rangle$ and C is the curve given by $\mathbf{r}(t) = \langle \sqrt{t}, 1 + t^3 \rangle$ for $0 \le t \le 1$.

We've been using Useful Fact 18.5 without a solid proof behind it, and I know you're not going to let me get away with that twice, so I'll show you why Theorem 18.1 works. Surprisingly, it isn't that bad! Here are two ingredients to start with:

- First, the conditions of the theorem require that the vector field $\mathbf{F}$ is conservative, meaning it's the gradient of a scalar function:

$$\mathbf{F} = \nabla f = \left\langle \frac{\partial f}{\partial x}, \frac{\partial f}{\partial y} \right\rangle$$

- Second, the contour C can be given parametrically via $\mathbf{r}(t) = \langle x(t), y(t) \rangle$ for $a \le t \le b$, so

$$d\mathbf{r} = \left\langle \frac{dx}{dt}, \frac{dy}{dt} \right\rangle$$

So when we build the integral and expand the dot product $\mathbf{F} \cdot \mathbf{r}(t)$, we get

$$\int_C \mathbf{F} \cdot d\mathbf{r} = \int_a^b \left(\frac{\partial f}{\partial x} \frac{dx}{dt} + \frac{\partial f}{\partial y} \frac{dy}{dt} \right) dt$$

Now take out the integrand and look at it on its own:

$$\frac{\partial f}{\partial x} \frac{dx}{dt} + \frac{\partial f}{\partial y} \frac{dy}{dt}$$

That should look *really* familiar! It's a chain rule expression, built for the derivative with respect to t of a function $f(x(t), y(t))$, i.e. the contour integral itself can be reset as:

$$\int_C \mathbf{F} \cdot d\mathbf{r} = \int_a^b \frac{d}{dt} f(x(t), y(t))\, dt$$

which by the single variable fundamental theorem of calculus is just the difference between the values of $f(x, y)$ evaluated at $t = a$ and $t = b$:

$$\int_C \mathbf{F} \cdot d\mathbf{r} = f(x(b), y(b)) - f(x(a), y(a))$$

which is the formula given in Theorem 18.1 except with slightly different referencing to the endpoints.

Summary

Here is a summary of everything we now know about conservative vector fields and their line integrals. The following statements are *equivalent*, meaning that if one of them is known to be true, then they ALL are true:

(1) $\mathbf{F}$ is conservative.
(2) $\mathbf{F}$ is the gradient of some scalar potential function f, i.e. $\mathbf{F} = \nabla f$.
(3) $\int_C \mathbf{F} \cdot \mathbf{dr}$ is path independent; if p_1 and p_2 represent the starting and ending points on the contour C, then $\int_C \mathbf{F} \cdot d\mathbf{r} = f(p_2) - f(p_1)$.
(4) $\int_C \mathbf{F} \cdot \mathbf{dr} = 0$ along ALL piecewise-smooth closed contours C.
(5) For a 2D vector field $\mathbf{F}(x, y) = \langle P(x, y), Q(x, y) \rangle$, $P_y(x, y) = Q_x(x, y)$. For a 3D vector field $\mathbf{F}(x, y, z)$, $\nabla \times \mathbf{F} = \mathbf{0}$

Conservative Vector Fields — Problem List

Conservative Vector Fields — You Try It

These appeared above; solutions begin on the next page.

(1) Determine if the vector field $\mathbf{F}(x, y) = \langle 6x+5y, 5x+4y\rangle$ is conservative. If it is, find a potential function for it.
(2) Determine if the vector field $\mathbf{F}(x, y) = \langle xe^y, ye^x\rangle$ is conservative. If it is, find a potential function for it.
(3) Determine if the vector field $\mathbf{F}(x, y, z) = \langle yz, xz, xy\rangle$ is conservative. If it is, find a potential function for it.
(4) Evaluate $\int_C \mathbf{F} \cdot d\mathbf{r}$ where $\mathbf{F}(x, y) = \langle x^3y^4, x^4y^3\rangle$ and C is the curve given by $\mathbf{r}(t) = \langle \sqrt{t}, 1 + t^3\rangle$ for $0 \le t \le 1$.

Conservative Vector Fields — Practice Problems

Try these as you get the hang of the You Try It problems. Solutions to these problems are available in Sec. B.6.2.

(1) Determine if the vector field $\mathbf{F}(x, y) = \langle x^3 + 4xy, 4xy - y^3\rangle$ is conservative. If it is, find a potential function for it.
(2) Determine if the vector field $\mathbf{F}(x, y) = \langle e^y, xe^y\rangle$ is conservative. If it is, find a potential function for it.
(3) Determine if the vector field $\mathbf{F}(x, y, z) = \langle 3z^2, \cos y, 2xz\rangle$ is conservative. If it is, find a potential function for it.
(4) Evaluate $\int_C \mathbf{F} \cdot d\mathbf{r}$ where $\mathbf{F}(x, y) = \left\langle \frac{y^2}{1 + x^2}, 2y \tan^{-1}(x) \right\rangle$ and C is the curve given by $\mathbf{r}(t) = \langle t^2, 2t\rangle$ for $0 \le t \le 1$.
(5) Determine if the vector field $\mathbf{F}(x, y) = \langle 1+2xy+\ln x, x^2\rangle$ is conservative. If it is, find a potential function for it.
(6) Determine if the vector field $\mathbf{F}(x, y, z) = \langle e^z, 1, xe^z\rangle$ is conservative. If it is, find a potential function for it.
(7) Use the Fundamental Theorem for Line Integrals to evaluate $\int_C \mathbf{F} \cdot d\mathbf{r}$ where $\mathbf{F}(x, y, z) = \langle 2xz + y^2, 2xy, x^2 + 3z^2\rangle$, and C is the curve given by $\mathbf{r}(t) = \langle t^2, t + 1, 2t - 1\rangle$ for $0 \le t \le 1$.

Conservative Vector Fields — Challenge Problems

Try these problems to test your skills with the ideas in this section. Solutions to these problems are available in Sec. C.6.2.

(1) Determine if the vector field $\mathbf{F}(x,y) = \langle y^2 + e^x + xe^x, 2xy\rangle$ is conservative. If it is, find a potential function for it.
(2) Determine if the vector field $\mathbf{F}(x,y,z) = \langle 3x^2, -\cos(y), 2xz\rangle$ is conservative. If it is, find a potential function for it.
(3) Use the Fundamental Theorem for Line Integrals to evaluate $\int_C \mathbf{F}\cdot d\mathbf{r}$ where $\mathbf{F}(x,y,z) = \langle 10x+3y+yz, 3x+20y+xz, xy\rangle$ and C is the curve given by

$$\mathbf{r}(t) = \left\langle t^{5/2} - 1, \sqrt{t+3}, \sin\left(\frac{\pi}{2}t\right)\right\rangle \quad \text{for} \quad 0 \le t \le 1$$

Conservative Vector Fields — You Try It — Solved

(1) Determine if the vector field $\mathbf{F}(x,y) = \langle 6x+5y, 5x+4y\rangle$ is conservative. If it is, find a potential function for it.

□ Matching this to the form $\mathbf{F}(x,y) = \langle P(x,y), Q(x,y)\rangle$, we have

$$P = 6x + 5y \to \frac{\partial P}{\partial y} = 5$$
$$Q = 5x + 4y \to \frac{\partial Q}{\partial x} = 5$$

Since $\partial P/\partial y = \partial Q/\partial x$ then the vector field is conservative. So there is a function $f(x,y)$ such that $\nabla f = \mathbf{F}$.

We know from $\mathbf{F}$ that for this function, $f_x = 6x+5y$ and $f_y = 5x+4y$. Based on f_x, we know that at worst, $f(x,y) = 3x^2 + 5xy + g(y)$ where $g(y)$ is some unknown function of y.

With $f(x,y)$ in this form, $f_y = 5x + g'(y)$. But since we know $f_y = 5x + 4y$ we have that $g'(y) = 4y$ and $g(y) = 2y^2 + C$. So, $f(x,y) = 3x^2 + 5xy + 2y^2 + C$. ■

(2) Determine if the vector field $\mathbf{F}(x,y) = \langle xe^y, ye^x\rangle$ is conservative. If it is, find a potential function for it.

☐ Matching this to the form $\mathbf{F}(x,y) = \langle P(x,y), Q(x,y)\rangle$, we have

$$P = xe^y \to \frac{\partial P}{\partial y} = xe^y$$

$$Q = ye^x \to \frac{\partial Q}{\partial x} = ye^x$$

Since $\partial P/\partial y \neq \partial Q/\partial x$ then the vector field is not conservative, and there is no function $f(x,y)$ such that $\nabla f = \mathbf{F}$. ■

---

(3) Determine if the vector field $\mathbf{F}(x,y,z) = \langle yz, xz, xy\rangle$ is conservative. If it is, find a potential function for it.

☐ Since $\nabla \times \mathbf{F} = \mathbf{0}$ then $\mathbf{F}$ is conservative, and so there is a scalar function f such that $\mathbf{F} = \nabla f$. Since that's true, we know $f_x = yz$, $f_y = xz$, and $f_z = xy$. Without any further ado, it's pretty easy to figure out that $f(x,y,z) = xyz + C$. ■

---

(4) Evaluate $\displaystyle\int_C \mathbf{F} \cdot d\mathbf{r}$ where $\mathbf{F}(x,y) = \langle x^3y^4, x^4y^3\rangle$ and C is the curve given by $\mathbf{r}(t) = \langle \sqrt{t}, 1+t^3\rangle$ for $0 \le t \le 1$.

☐ A quick check will show that $\mathbf{F}$ is conservative. Therefore, we can evaluate the integral using the Fundamental Theorem for Line Integrals. The potential function for this vector field is a function $f(x,y)$ such that $f_x = x^3y^4$ and $f_y = x^4y^3$. Based on f_x, we know that at worst, $f(x,y) = x^4y^4/4 + g(y)$ where $g(y)$ is some unknown function of y. With $f(x,y)$ in this form, we get $f_y = x^4y^3 + g'(y)$. But since we know $f_y = x^4y^3$ we have that $g'(y) = 0$ and $g(y) = C$. Let's choose $C = 0$ (do you know why we can??) so that

$$f(x,y) = \frac{1}{4}x^4y^4$$

Next we need $\int_C \mathbf{F} \cdot d\mathbf{r}$ where $\mathbf{r}(t) = \langle \sqrt{t}, 1+t^3\rangle$ for $0 \le t \le 1$. The values of $\mathbf{r}(t)$ and $f(x,y)$ at the endpoints of this curve, i.e. at $t=0$ and $t=1$, are

$$\mathbf{r}(1) = \langle 1, 2\rangle \to f(1,2) = \frac{1}{4}(1)(2)^4 = 4$$

$$\mathbf{r}(0) = \langle 0, 1\rangle \to f(0,1) = \frac{1}{4}(0)(1)^4 = 0$$

So by the Fundamental Theorem for Line Integrals, we have

$$\int_C \mathbf{F} \cdot d\mathbf{r} = f(1,2) - f(0,1) = 4 \quad ■$$

18.3 Surface Integrals

Introduction

Let's insert our newest discovery, line integrals, into the hierarchy of integral types; line integrals live in a zone between single and double integrals:

$$\int_a^b f(x)\,dx = \lim_{n\to\infty}\sum_{i=1}^{n} f(x_i^\star)\Delta x$$

$$\to \int_C f(x,y)\,ds = \lim_{n\to\infty}\sum_{i=1}^{n} f(x_i^\star, y_i^\star)\Delta s$$

$$\iint Rf(x,y)\,dA = \lim_{n,m\to\infty}\sum_{i=1}^{n}\sum_{j=1}^{m} f(x_i^\star, y_j^\star)\Delta A$$

$$\iiint Ef(x,y,z)\,dV = \lim_{n,m,p\to\infty}\sum_{i=1}^{n}\sum_{j=1}^{m}\sum_{k=1}^{p} f(x_i^\star, y_j^\star, z_k^\star)\Delta V$$

Are you getting nervous about that spot between double and triple integrals? So far, we've used straight lines and curves as domains of integration (single and line integrals); we've used a flat 2D region from the xy-plane as a domain of integration (double integrals); we've used a 3D solid volume from $\mathbb{R}^3$ as a domain of integration (triple integrals). Are there any options left? You bet! We haven't used a *surface* in $\mathbb{R}^3$ as a domain of integration. Just as line integrals fit in between single and double integrals, *surface integrals* will fit in between double integrals and triple integrals.

Surface Integrals: Definition

Consider a function $f(x,y,z)$, which — as you remember — can't be plotted on our usual sets of axes, and a region of integration E in 3-space. We already know how to do the triple integral of f over E: it's $\iiint_E f(x,y,z)\,dV$. But let S be the part or all of the *surface* of the solid region E. We can also integrate f over S — this is a surface integral. To develop it, we'll just follow our standard partitioning procedure.

(1) **Select a region of integration from the domain of the function.** This is the surface S.

(2) **Partition the region of integration into tiny pieces.** First, we have to visualize how the partitioning works. Imagine a grid in the

xy-plane, just as for a double integral, into i steps along the x-axis and j steps along the y-axis, so that there are a total of ij individual grid calls. Each grid cell is a rectangle with size $\Delta A = \Delta x \Delta y$. Now imagine that grid projected upwards (or downwards) onto the surface S. Each grid cell ΔA projects onto a small patch on the surface; this patch will not necessarily be a rectangle, but it will have area ΔS. Each grid cell down in the xy-plane may very well have the same size ΔA, but there's no guarantee that every grid cell on the surface will have the same size, so we need to specify the area of each as ΔS_{ij} (where the subscript ij lands us in one specific patch on the surface). The projection of ΔA onto ΔS_{ij} is shown in Fig. 18.6.

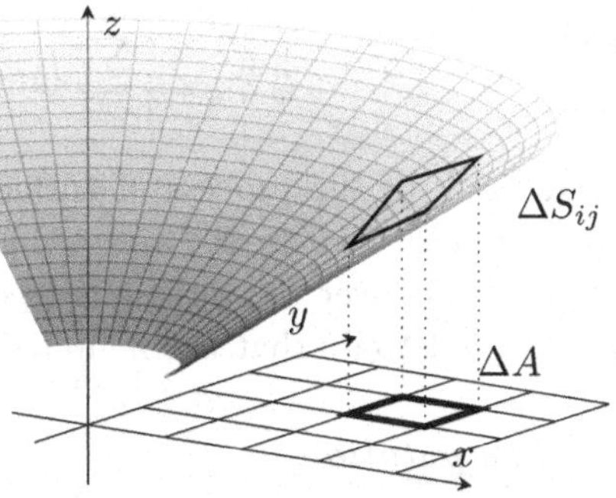

Fig. 18.6 Partitioning for a surface integral; dA vs dS_{ij}.

(3) **Within each partition, select a representative point.** Each grid cell in the xy-plane (and thus also on the surface) is characterized by a point $(x_i^\star, y_j^\star)$. The corresponding patch of S above each grid cell can be assigned one z-coordinate from that patch. Therefore we'll call our representative point $(x_i^\star, y_j^\star, z_{ij}^\star)$.
(4) **Plug the coordinate(s) of that representative point into the function.** We get $f(x_i^\star, y_j^\star, z_{ij}^\star)$.
(5) **Multiply the function's value at the representative point by the size of the partition.** This product will be $f(x_i^\star, y_j^\star, z_{ij}^\star)\Delta S_{ij}$.
(6) Sum these resulting contributions from all partitions.

$$\sum_{i=1}^{n}\sum_{j=1}^{m} f(x_i^\star, y_j^\star, z_{ij}^\star)\Delta S_{ij}$$

(7) **Find the limit of this sum as the number of partitions goes to ∞.**

$$\lim_{n,m\to\infty} \sum_{i=1}^{n}\sum_{j=1}^{m} f(x_i^\star, y_j^\star, z_{ij}^\star)\Delta S_{ij}$$

(8) **Give this limit a name and symbol.** Remember from our discussion about line integrals that the number of integral signs is not due to the number of variables in the function being integrated (which here is $f(x, y, z)$), but rather a reflection of how many dimensions we have in the domain or region of integration. In a surface integral, the term ΔS_{ij} is the area of a patch of the surface, and so is two-dimensional. Therefore, we will need two integral signs on a surface integral:

$$\iint_S f(x, y, z)\, dS = \lim_{n,m\to\infty} \sum_{i=1}^{n}\sum_{j=1}^{m} f(x_i^\star, y_j^\star, z_{ij}^\star)\Delta S_{ij}$$

All in all, you can think of the process of surface integration as like having an ant walk over a surface in $\mathbb{R}^3$ and accumulating values of a function $f(x, y, z)$ as he walks over the designated portion of the surface. Now, why an ant would want to do that is left up to your imagination.

Surface Integrals: Computation

When we defined line integrals, we didn't really learn a new way to compute an integral, we just learned how to convert the new category of integral into one we already knew how to do. There were two basic versions of line integrals (the rectangular version and the "vectorized" version), but either way, we converted the line integral as a regular single integral in terms of a parameter t.

We'll do the same kind of thing for surface integrals. Some of the expressions are going to look pretty nasty, but fundamentally, we don't need a new method of determining an integral, we'll just learn how to convert a surface integral $\iint_S f(x, y, z)\, dS$ into a regular double integral, $\iint_S (\cdot) dA$. Once we've done that, we can use all the tools we have available for evaluating double integrals — including changing to polar coordinates!

You will need to be on your toes when dealing with notation in surface integrals, because we need to keep track of *two* different functions:

(1) The function that's being integrated is usually designated $f(x, y, z)$
(2) The surface over which the integration is taking place is usually designated $z = g(x, y)$. This means, for example, that if the surface of integration is a portion of the plane $2x + 3y + z = 1$, we will refer to it as $z = g(x, y) = 1 - 2x - 3y$.

The conversion of the dS in the surface integral to the regular dA of a double integral hinges on an idea whose development involves lots of neat things like tangent planes and cross products. The end result is that if we have a grid in the xy-plane where the cells have size $\Delta x \Delta y$, and a surface $z = g(x, y)$ is hovering over this grid, then the corresponding partitions of S have surface area

$$\Delta S_{ij} \approx \sqrt{[g_x(x_i, y_j)]^2 + [g_y(x_i, y_j)]^2 + 1}\,\Delta A$$

Therefore, we can adapt our definition of the surface integral to:

$$\begin{aligned}\iint_S f(x, y, z)\, dS &= \lim_{n,m\to\infty} \sum_{i=1}^{n} \sum_{j=1}^{m} f(x_i^\star, y_j^\star, z_{ij}^\star)\Delta S_{ij}\\ &= \lim_{n,m\to\infty} \sum_{i=1}^{n} \sum_{j=1}^{m} f(x_i^\star, y_j^\star, z_{ij}^\star)\sqrt{[g_x(x_i, y_j)]^2 + [g_y(x_i, y_j)]^2 + 1}\,\Delta A\\ &= \iint_D f(x, y, z)\sqrt{(g_x)^2 + (g_y)^2 + 1}\, dA\end{aligned}$$

But we can't do a double integral of a function with z's still floating around in it, and so we must restrict the function $f(x, y, z)$ only to points from the surface $z = g(x, y)$. Ultimately, we design surface integrals for evaluation as "regular" double integrals as follows:

Useful Fact 18.6. *Given a well-behaved function $f(x, y, z)$ and smooth surface $z = g(x, y)$,*

$$\iint_S f(x, y, z)\, dS = \iint_D f(x, y, g(x, y))\sqrt{(g_x)^2 + (g_y)^2 + 1}\, dA$$

The procedure is broken down like this:

- If it's not already given this way, write the equation of the surface of integration in the form $z = g(x, y)$.
- From the equation of the surface of integration, compute g_x, g_y, and create the term $\sqrt{(g_x)^2 + (g_y)^2 + 1}$.

- If $f(x,y,z)$ has z's in it, replace z with the equation of the surface $z = g(x,y)$.
- Assemble the integral shown in Useful Fact 18.6. The domain D of the resulting double integral is just the "shadow" in the xy-plane of the surface S.
- Decide whether rectangular or polar coordinates are more appropriate to solve the integral.
- Solve it.

Isn't that easy???

EX 1 Evaluate $\iint_S x^2z^2\,dS$ where S is the part of the cone $z^2 = x^2 + y^2$ that lies between the planes $z = 1$ and $z = 3$.

The surface of integration is shown in Fig. 18.7. The dot on the surface is our ant walking around on the cone, collecting up values of the function $f(x,y,z) = x^2z^2$.

To get started with the calculation, we must rewrite the surface of integration as $z = g(x,y) = \sqrt{x^2+y^2}$. From this, we get:

$$g_x = \frac{x}{\sqrt{x^2+y^2}} \quad ; \quad g_y = \frac{y}{\sqrt{x^2+y^2}}$$

and so

$$\sqrt{(g_x)^2 + (g_y)^2 + 1} = \sqrt{\frac{x^2}{x^2+y^2} + \frac{y^2}{x^2+y^2} + 1} = \sqrt{2}$$

The function which is known as $f(x,y,z) = x^2z^2$ everywhere in $\mathbb{R}^3$ can be rewritten more specifically for this surface as

$$f(x,y,g(x,y)) = x^2(z)^2 = x^2(\sqrt{x^2+y^2})^2 = x^2(x^2+y^2)$$

For our double integral, we need a suitable region of integration D; this is just the portion of the xy-plane which is directly under the surface of integration. This region is shown in Fig. 18.8. D is the region in the xy-plane between the circles $x^2 + y^2 = 1$ and $x^2 + y^2 = 9$. And now we have all the pieces we need to assemble the surface integral:

$$\begin{aligned}\iint_S x^2z^2\,dS &= \iint_D f(x,y,g(x,y))\sqrt{(g_x)^2+(g_y)^2+1}\,dA \\ &= \iint_D x^2(x^2+y^2)\sqrt{2}\,dA\end{aligned}$$

This is an integral for which polar coordinates is quite appropriate, due to the circular nature of D and the presence of $x^2 + y^2$ in the integrand. Since D is between two concentric circles of radius 1 and 3,

$$\begin{aligned}\iint_S x^2 z^2\, dS &= \iint_D x^2(x^2 + y^2)\sqrt{2}\, dA \\ &= \sqrt{2}\int_0^{2\pi}\int_1^3 (r\cos\theta)^2 (r^2)\, r\, drd\theta \\ &= \sqrt{2}\int_0^{2\pi}\int_1^3 r^5 \cos^2\theta\, drd\theta\end{aligned}$$

Because setting up the correct integral is 90% of the battle in one of these problems, it's totally fair to head for tech to get the integral evaluated. In this case, we can find that

$$\iint_S x^2 z^2\, dS = \sqrt{2}\int_0^{2\pi}\int_1^3 r^5 \cos^2\theta\, drd\theta = \frac{364\sqrt{2}\pi}{3} \quad \blacksquare$$

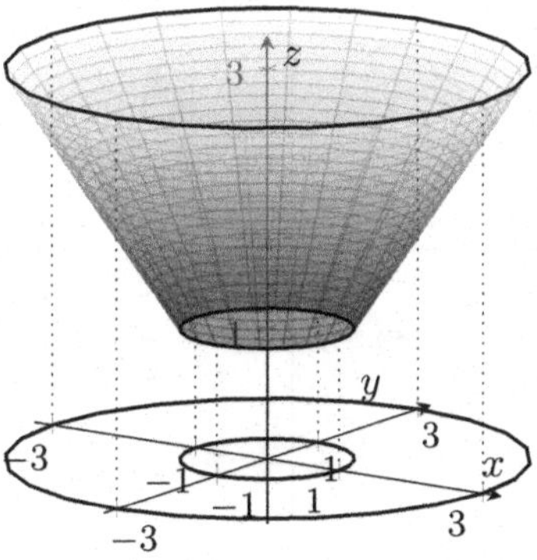

Fig. 18.7 The cone $z = r$, $1 \leq z \leq 3$, with EX 1.

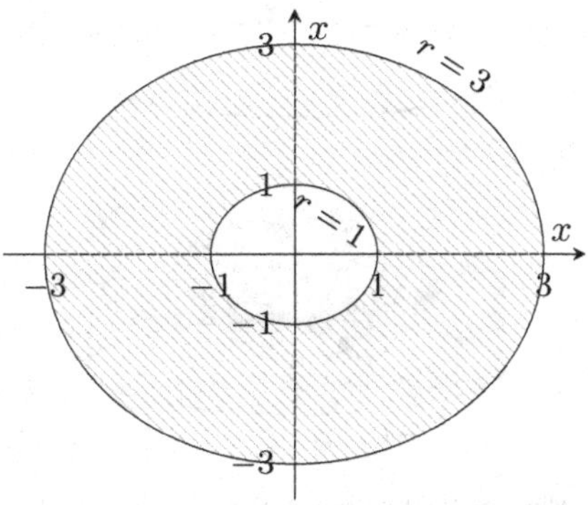

Fig. 18.8 The support region D for the surface in Fig. 18.7.

EX 2 Evaluate $\iint_S z\, dS$ where S is the part of the paraboloid $z = x^2 + y^2$ inside the cylinder $x^2 + y^2 = 1$.

From the surface of integration $z = g(x, y) = x^2 + y^2$, we get:

$$g_x = 2x \qquad ; \qquad g_y = 2y$$

and so

$$\sqrt{(g_x)^2 + (g_y)^2 + 1} = \sqrt{4x^2 + 4y^2 + 1}$$

The function $f(x,y,z)=z$ can be rewritten for this surface as

$$f(x,y,g(x,y))=x^2+y^2$$

The region D in the xy-plane below our surface comes from the intersection of the paraboloid and the cylinder; this occurs when $x^2+y^2=1$. Therefore D is the unit circle. The surface and the associated region in the xy-plane are shown in Figs. 18.9 and 18.10. And now we have all the pieces we need to assemble the surface integral (polar coordinates seem useful here):

$$\begin{aligned}\iint_S z\,dS &= \iint_D f(x,y,g(x,y))\sqrt{(g_x)^2+(g_y)^2+1}\,dA\\ &= \iint_D (x^2+y^2)\sqrt{4x^2+4y^2+1}\,dA = \iint_D r^2\sqrt{4r^2+1}\,dA\\ &= \int_0^{2\pi}\int_0^1 r^2\sqrt{4r^2+1}\,r\,drd\theta = \int_0^{2\pi}\int_0^1 r^3\sqrt{4r^2+1}\,drd\theta\\ &= \frac{\pi}{60}(25\sqrt{5}+1) \quad \blacksquare\end{aligned}$$

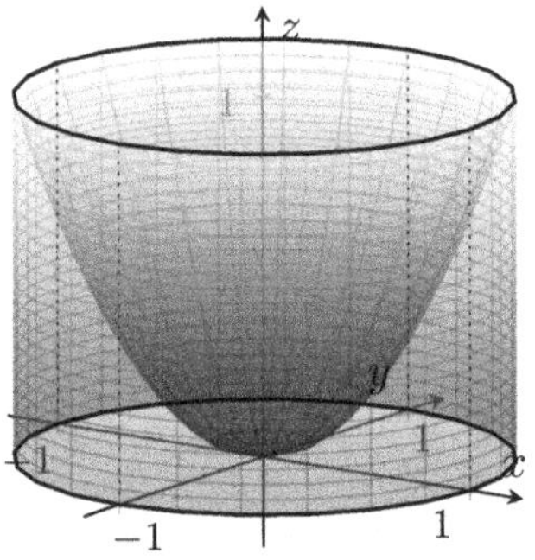

Fig. 18.9 The paraboloid $z=r^2$ inside the cylinder $r=1$, with EX 2.

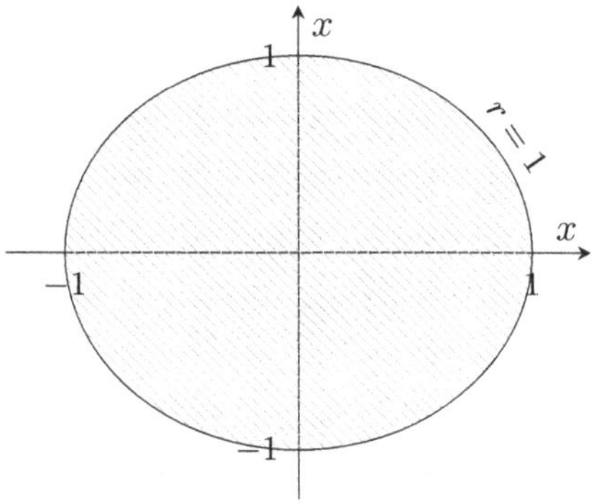

Fig. 18.10 The support region D for the surface in Fig. 18.9.

You Try It

(1) Find $\iint_S x^2yz\,dS$ where the surface S is the portion of the plane $z=1+2x+3y$ above $0\le x\le 3$ and $0\le y\le 2$

Computing Surface Areas

We can use surface integrals to compute surface area. The integrals above are *not* computations of surface area; rather, they are the result of the process where we travel over a surface S and sample / integrate values of some function $f(x, y, z)$ from locations on that surface. This isn't the same as computing the actual area of the surface S.

However, we've laid plenty of clues to how we can compute the area of S. Recall this growing inventory of calculations:

- $\int_I (1)\, dx$ computes the length of the interval of integration I
- $\int_C (1)\, ds$ computes the length of the contour of integration C
- $\iint_R (1)\, dA$ computes the area of the region of integration R
- $\iiint_E (1)\, dV$ computes the volume of the region of integration E.

Take a crazy guess as to how we can compute the area of a surface S.

Useful Fact 18.7. *The surface area of a smooth surface S can be calculated by:*

$$A_S = \iint_S (1)\, dS$$

EX 3 Find the surface area of the part of the paraboloid $z = x^2 + y^2$ under the plane $z = 9$.

From the surface of integration $z = g(x, y) = x^2 + y^2$, we get:

$$g_x = 2x \quad ; \quad g_y = 2y$$

and so

$$\sqrt{(g_x)^2 + (g_y)^2 + 1} = \sqrt{4x^2 + 4y^2 + 1}$$

The function of integration is just $f(x, y, z) = 1$, since we want to compute a surface area. The region of integration comes from the intersection of the paraboloid with the plane $z = 9$, which happens when $x^2 + y^2 = 9$; so, D is a circle of radius 3. (If you want an image for this, go look at Figs. 18.9 and 18.10 and imagine the paraboloid there extended up to $z = 9$.) Using polar coordinates again,

$$\begin{aligned} A_S &= \iint_S (1)\, dS = \iint_S (1)\sqrt{(g_x)^2 + (g_y)^2 + 1}\, dA \\ &= \iint_S (1)\sqrt{4x^2 + 4y^2 + 1}\, dA = \iint_S \sqrt{4x^2 + 4y^2 + 1}\, dA \\ &= \int_0^{2\pi} \int_0^3 \sqrt{4r^2 + 1}\, r\, dr d\theta = \frac{\pi}{6}(37\sqrt{37} - 1) \quad \blacksquare \end{aligned}$$

You Try It

(2) Find the surface area of the cylinder $y^2+z^2=9$ above the rectangle with vertices (0,0), (4,0), (0,2) and (4,2).

Surface Integrals: Vector Version

Remember how line integrals started as something harmless like $\int_C f(x,y)ds$ but, when we found we could involve vectors, they changed into something more scary: $\int_C \mathbf{F}\cdot d\mathbf{r}$? The same thing is about to happen with surface integrals. Given a 3D vector field $\mathbf{F}(x,y,z)$, we can perform a surface integral that looks like this (which is very similar to the vector version of a line integral):

$$\iint_S \mathbf{F}\cdot d\mathbf{S}$$

This expression should confuse you. After all, we know the scalar quantity dS evolved from ΔS, a small patch of our surface, with area. But what does $d\mathbf{S}$ mean when it's a vector? Can you just take a surface element and turn it into a vector?? Well, yes, it's just notation. Think of each patch of our surface S with area ΔS as an itty bitty piece of a plane; thus, each patch has a unit normal vector $\mathbf{n}$ associated with it. (These vectors are the same as the normal vectors that describe the tangent plane at any point on the surface). The "vector" form of dS is really just $\mathbf{n}\,dS$. Therefore, the vector surface integral might also look like this:

$$\iint_S \mathbf{F}\cdot\mathbf{n}\,dS$$

Before worrying about how to compute it, think for a minute about what this means. If $\mathbf{F}$ is a vector that represents some sort of flow — like fluid flow or electrostatic current. If $\mathbf{n}$ happens to be a *unit* vector perpendicular to the surface, then $\mathbf{F}\cdot\mathbf{n}$ is the component of $\mathbf{F}$ perpendicular to the surface. And when we integrate that over an entire surface, then we are measuring the total outward flux due to $\mathbf{F}$ across the surface. This is a very powerful thing to be able to compute.

Now, to compute this "flux integral", we must convert it into a regular double integral. Here's how.

Back in Sec. 16.5, we encountered Useful Fact 16.14, which said that a vector normal to a surface $g(x, y)$ at a point $(x_0, y_0, z_0$ could be given by $\langle -g_x(x, y), -g_y(x, y), 1\rangle$.[1] And so a *unit* normal to a surface $z = g(x, y)$ at any point is

$$\mathbf{n} = \frac{\langle -g_x(x, y), -g_y(x, y), 1\rangle}{\sqrt{[g_x(x, y)]^2 + [g_y(x, y)]^2 + 1}}$$

With this, we can expand the flux integral. If we name the components of $\mathbf{F}$ to be P, Q, R and borrow the extended version of dS from above, $dS = \sqrt{[g_x(x, y)]^2 + [g_y(x, y)]^2 + 1}\, dA$, then

$$\begin{aligned}
\iint_S \mathbf{F} \cdot d\mathbf{S} &= \iint_S \mathbf{F} \cdot \mathbf{n}\, dS \\
&= \iint_D \left(\langle P, Q, R\rangle \cdot \frac{\langle -g_x, -g_y, 1\rangle}{\sqrt{[g_x]^2 + [g_y]^2 + 1}} \right) \sqrt{[g_x]^2 + [g_y]^2 + 1}\, dA \\
&= \iint_D (\langle P, Q, R\rangle \cdot \langle -g_x, -g_y, 1\rangle)\, dA \\
&= \iint_D \left(-P\frac{\partial g}{\partial x} - Q\frac{\partial g}{\partial y} + R \right) dA
\end{aligned}$$

It certainly was fortunate that the disgusting term $\sqrt{[g_x]^2 + [g_y]^2 + 1}$ has canceled out!

There is one last detail to pin down: back in Useful Fact 16.14, we noted *two* gradient-type vectors which would be perpendicular to a surface — one pointing inwards, and one pointing outwards. The one we've selected is the one designed to point *outwards*; when we take a surface and select to use *outward* normal vectors, we have assigned the surface to be **positively orientated**.

Useful Fact 18.8. *Given a vector field* $\mathbf{F} = \langle P(x, y, z), Q(x, y, z), R(x, y, z)\rangle$ *and a surface* $z = g(x, y)$ *that is positively oriented and smooth, we have*

$$\iint_S \mathbf{F} \cdot d\mathbf{S} = \iint_D \left(-P\frac{\partial g}{\partial x} - Q\frac{\partial g}{\partial y} + R \right) dA$$

[1] The name of the surface there was f instead of g, but that's no big deal, right?

What if the surface is not oriented positively? Do remember back in Calculus I when you learned single integrals, and we had the formula

$$\int_b^a f(x)\,dx = -\int_a^b f(x)\,dx$$

If our interval of integration was "backwards", we would just reverse it at the cost of introducing a negative sign. Wouldn't it be great if that was all we had to do here to account for a surface given with negative orientation instead of positive?

To summarize, here's the procedure. Given a vector field $\mathbf{F} = \langle P, Q, R\rangle$ and positively oriented surface $z = g(x, y)$,

(1) Determine g_x and g_y from the equation of the surface.
(2) Get the components P, Q, and R from the vector field $\mathbf{F}$.
(3) If there are any instances of z in P, Q, and R, plan to replace them with the surface equation $z = g(x, y)$.
(4) Determine the region of integration D in the xy-plane that corresponds to the surface of integration.
(5) Assemble and solve the integral as shown in Useful Fact 18.8.
(6) If the surface happened to be negatively oriented, change the sign of your result.

EX 4 Evaluate $\iint_S \mathbf{F}\cdot d\mathbf{S}$ where $\mathbf{F} = \langle xze^y, -xze^y, z\rangle$ and S is the part of the plane $x+y+z = 1$ in the first octant, with positive orientation.

Let's write the surface as $z = g(x, y) = 1-x-y$, from which we get $g_x = -1$ and $g_y = -1$. Then using the components of $\mathbf{F}$, $P = xze^y$, $Q = -xze^y$, and $R = z$, we can begin to assemble the flux integral.

$$\begin{aligned}\iint_S \mathbf{F}\cdot d\mathbf{S} &= \iint_D \left(-P\frac{\partial g}{\partial x} - Q\frac{\partial g}{\partial y} + R\right) dA \\ &= \iint_D (-xze^y(-1) - (-xze^y)(-1) + z)\; dA \\ &= \iint_D (xze^y - xze^y + z)\; dA = \iint_D (z)\, dA\end{aligned}$$

Note that an instance of z has survived, so we are going to need to replace that with the equation of the surface $z = 1-x-y$. Also, we need to identify the region of integration D. But this is easy; we need the "shadow" of the plane $x + y + z = 1$ in the first quadrant of the xy-plane; this is found (by setting $z = 0$) the be line $x + y = 1$. The plane $z = g(x, y)$ and the subsequent domain of integration D are shown in Figs. 18.11 and 18.12. (The normal vector $\mathbf{n}$ in Fig. 18.11 looks ridiculously large, but on the scale of the figure, that really is a length of 1!) Our region of integration D is the region in the first quadrant below the line $x + y = 1$, and so we can continue with the integral:

$$\iint_S \mathbf{F} \cdot \mathbf{n}\, dS = \iint_D (z)\, dA = \int_0^1 \int_0^{1-x} (1 - x - y)\, dy dx = \frac{1}{6} \quad \blacksquare$$

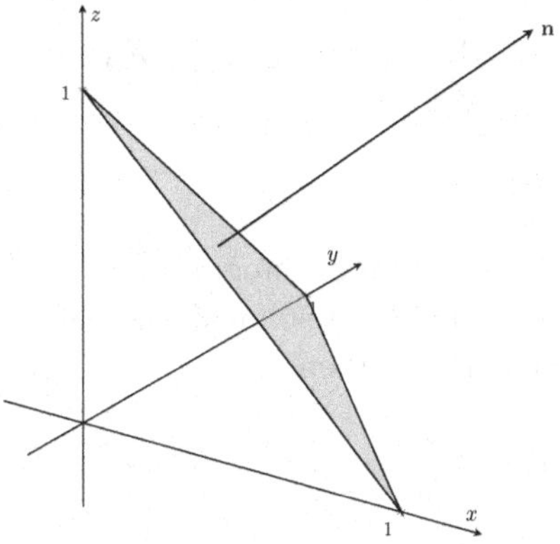

Fig. 18.11 The plane $x + y + z = 1$ in the first octant, w/ EX 4.

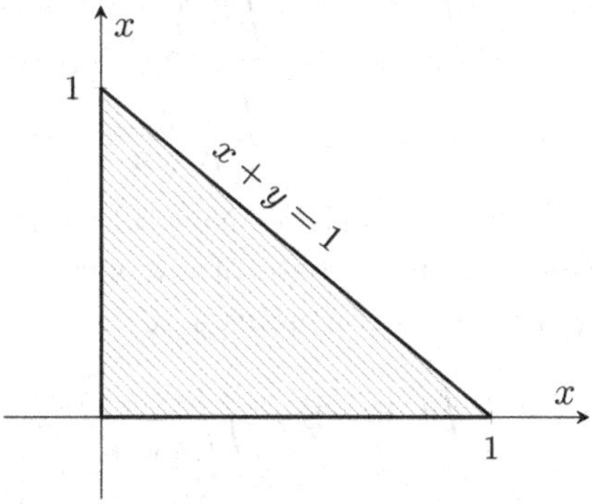

Fig. 18.12 The support region D for the surface in Fig. 18.11.

EX 5 Evaluate $\iint_S \mathbf{F} \cdot d\mathbf{S}$ where $\mathbf{F} = \langle x, -z, y \rangle$ and S is the part of the sphere $x^2 + y^2 + z^2 = 4$ in the first octant, with normal vectors pointing toward the origin.

First, since the normal vectors on S point to the origin, they are not pointing outwards from the sphere, so this is a negatively oriented surface. From the equation of the surface (sphere), we can write $z = g(x, y) = \sqrt{4 - x^2 - y^2}$, and so

$$g_x = \frac{-x}{\sqrt{4 - x^2 - y^2}} \quad ; \quad g_y = \frac{-y}{\sqrt{4 - x^2 - y^2}}$$

Then with $P = x$, $Q = -z$ and $R = y$, we have

$$\begin{aligned}\iint_S \mathbf{F} \cdot d\mathbf{S} &= \iint_D \left(-P\frac{\partial g}{\partial x} - Q\frac{\partial g}{\partial y} + R\right) dA \\ &= \iint_D \left(-x \cdot \frac{-x}{\sqrt{4-x^2-y^2}} - (-z)\frac{-y}{\sqrt{4-x^2-y^2}} + y\right) dA \\ &= \iint_D \left(\frac{x^2}{\sqrt{4-x^2-y^2}} - \frac{yz}{\sqrt{4-x^2-y^2}} + y\right) dA\end{aligned}$$

Wow, this looks terrible! But look what happens when we replace z with the equation of the surface, $z = \sqrt{4-x^2-y^2}$:

$$\begin{aligned}\iint_S \mathbf{F} \cdot d\mathbf{S} &= \iint_D \left(\frac{x^2}{\sqrt{4-x^2-y^2}} - \frac{y\sqrt{4-x^2-y^2}}{\sqrt{4-x^2-y^2}} + y\right) dA \\ &= \iint_D \frac{x^2}{\sqrt{4-x^2-y^2}}\, dA\end{aligned}$$

That's much better. Now all we need is the region of integration D, but this is just the base of our surface (sphere) in the first quadrant, i.e. a quarter circle of radius 4. Using polar coordinates, then,

$$\begin{aligned}\iint_S \mathbf{F} \cdot d\mathbf{S} &= \iint_D \frac{x^2}{\sqrt{4-x^2-y^2}}\, dA = \int_0^{\pi/2} \frac{r^2\cos^2\theta}{\sqrt{4-r^2}} r\, dr d\theta \\ &= \int_0^{\pi/2} \frac{r^3\cos^2\theta}{\sqrt{4-r^2}}\, dr d\theta = \frac{4\pi}{3} \quad \blacksquare\end{aligned}$$

You Try It

(3) Find $\iint_S \mathbf{F} \cdot d\mathbf{S}$ where $\mathbf{F} = \langle xy, yz, zx\rangle$ and the surface S is the paraboloid $z = 4 - x^2 - y^2$ over $0 \le x \le 1$ and $0 \le y \le 1$, oriented positively.

Surface Integrals — Problem List

Surface Integrals — You Try It

These appeared above; solutions begin on the next page.

(1) Find $\iint_S x^2yz\, dS$ where the surface S is the portion of the plane $z = 1 + 2x + 3y$ above $0 \le x \le 3$ and $0 \le y \le 2$.
(2) Find the surface area of the cylinder $y^2 + z^2 = 9$ above the rectangle with vertices (0,0), (4,0), (0,2) and (4,2).
(3) Find $\iint_S \mathbf{F} \cdot d\mathbf{S}$ where $\mathbf{F} = \langle xy, yz, zx \rangle$ and the surface S is the paraboloid $z = 4 - x^2 - y^2$ over $0 \le x \le 1$ and $0 \le y \le 1$, oriented positively.

Surface Integrals — Practice Problems

Try these as you get the hang of the You Try It problems. Solutions to these problems are available in Sec. B.6.3.

(1) Find $\iint_S xy\, dS$ where the surface S is the triangular region with vertices P(1,0,0), Q(0,2,0) and R(0,0,2).
(2) Find the surface area of the portion of the paraboloid $z = 4 - x^2 - y^2$ above the xy-plane.
(3) Find $\iint_S \mathbf{F} \cdot d\mathbf{S}$ where $\mathbf{F} = \langle xy, 4x^2, yz \rangle$ and the surface S is $z = xe^y$ over $0 \le x \le 1$ and $0 \le y \le 1$, oriented positively.
(4) Find $\iint_S yz\, dS$ where the surface S is the part of the plane $x+y+z = 1$ in the first octant.
(5) Find the surface area of the part of the hyperbolic paraboloid $z = y^2 - x^2$ between the cylinders $x^2 + y^2 = 1$ and $x^2 + y^2 = 4$.
(6) Find $\iint_S \mathbf{F} \cdot d\mathbf{S}$ where $\mathbf{F} = \langle x, y, z^4 \rangle$ and the surface S is $z = \sqrt{x^2 + y^2}$ under $z = 1$ oriented negatively.
(7) *(Bonus! Following up Sec. 15.3 ...)* We expect $\iint Sf(x, y, z)\, dS = 0$ for which of the following combinations of function $f(x, y, z)$ and surfaces of integration S?

I1) $f(x, y) = e^{x^2+y^2+z^2}$ and S is the upper half of the unit sphere
I2) $f(x, y) = ze^{x^2+y^2}$ and S is the right half of the unit sphere?

I3) $f(x, y) = \sin(x)\cos(yz)$ and S is the paraboloid $z = x^2 + y^2$ from $z = 0$ to $z = 1$?

Surface Integrals — Challenge Problems

Try these problems to test your skills with the ideas in this section. Solutions to these problems are available in Sec. C.6.3.

(1) Find $\iint_S y^2(3 - z)\, dS$ where the surface S is the part of the plane $x + y + z = 3$ in the first octant.
(2) Find the surface area of the portion of the surface $z = 4 - x^2 - y^2$ over the region between $x^2 + y^2 = 2$, $x^2 + y^2 = 9$, and $y = 0$.
(3) Find $\iint_S \mathbf{F} \cdot d\mathbf{S}$ where $\mathbf{F} = \langle x, y, z^2 \rangle$ and the surface S is the inverted cone $z = 4 - \sqrt{x^2 + y^2}$ above the xy-plane (oriented positively).
(4) *(Bonus! Following up Sec. 15.3 ...)* We expect $\iint S\mathbf{F}(x, y, z) \cdot d\mathbf{S} = 0$ for which of the following combinations of vector field $\mathbf{F}(x, y, z)$ and oriented surface S?

I4) $\mathbf{F}(x, y, z) = \langle xy, xyz, 0 \rangle$ and S is the inverted paraboloid $z = 4 - x^2 - y^2$ oriented outwards
I5) $\mathbf{F}(x, y, z) = \langle x + y + z, e^{-xyz}, \cos(x)\sin(y) \rangle$ and S is the upwards oriented plane $z = 6$ over $\{(x, y) : -\pi \le x \le \pi, -\pi \le y \le \pi\}$
I6) $\mathbf{F}(x, y, z) = \langle x, y, z \rangle$ and S is the unit sphere oriented outwards

Surface Integrals — You Try It — Solved

---

(1) Find $\displaystyle\iint_S x^2yz\,dS$ where the surface S is the portion of the plane $z = 1+2x+3y$ above $0 \le x \le 3$ and $0 \le y \le 2$.

□ From the equation of the surface $z(x,y)$ we have $g_x = 2$ and $g_y = 3$. The region D is the rectangle $0 \le x \le 3$, $1 \le y \le 2$. So using the scalar function version of the surface integral,

$$\begin{aligned}\iint_S x^2yz\,dS &= \iint_D x^2yz\sqrt{g_x^2+g_y^2+1}\,dA \\ &= \int_0^3\int_0^2 x^2y(1+2x+3y)\sqrt{2^2+3^2+1}\,dydx \\ &= \sqrt{14}\int_0^3\int_0^2 x^2y(1+2x+3y)\,dydx = 171\sqrt{14} \quad \blacksquare\end{aligned}$$

---

(2) Find the surface area of the cylinder $y^2+z^2 = 9$ above the rectangle with vertices (0,0), (4,0), (0,2) and (4,2).

□ We can write the function as $z = \sqrt{9-y^2}$. So, with $g_x = 0$ and $g_y = -y/\sqrt{9-y^2}$,

$$\sqrt{g_x^2+g_y^2+1} = \sqrt{\frac{y^2}{9-y^2}+1} = \sqrt{\frac{9}{9-y^2}} = \frac{3}{\sqrt{9-y^2}}$$

and then,

$$\begin{aligned}A_S &= \iint_S (1)\,dS = \iint_S (1)\sqrt{g_x^2+g_y^2+1}\,dA = \int_0^4\int_0^2 \frac{3}{\sqrt{9-y^2}}\,dydx \\ &= \int_0^4 3\sin^{-1}\left(\frac{y}{3}\right)\Big|_0^2 dx = \int_0^4 3\sin^{-1}\left(\frac{2}{3}\right)dx \\ &= 3x\sin^{-1}\left(\frac{2}{3}\right)\Big|_0^4 = 12\sin^{-1}\left(\frac{2}{3}\right) \quad \blacksquare\end{aligned}$$

---

(3) Find $\displaystyle\iint_S \mathbf{F}\cdot d\mathbf{S}$ where $\mathbf{F} = \langle xy, yz, zx\rangle$ and the surface S is the paraboloid $z = 4-x^2-y^2$ over $0 \le x \le 1$ and $0 \le y \le 1$, oriented positively.

□ Matching the surface to the form $z = g(x,y)$ we have $g = 4-x^2-y^2$ and

$$\frac{\partial g}{\partial x} = -2x \quad ; \quad \frac{\partial g}{\partial y} = -2y$$

Matching to the form $\mathbf{F} = \langle P, Q, R \rangle$, we have

$$P = xy \quad ; \quad Q = yz = y(4 - x^2 - y^2) \quad ; \quad R = xz = x(4 - x^2 - y^2)$$

So using the vector version of the surface integral,

$$\begin{aligned}
\iint_S \mathbf{F} \cdot d\mathbf{S} &= \iint_D \left(-P\frac{\partial g}{\partial x} - Q\frac{\partial g}{\partial y} + R \right) dA \\
&= \iint_D (-xy(-2x) - (yz)(-2y) + xz) \; dA \\
&= \int_0^1 \int_0^1 \left(2x^2y + 2y^2(4 - x^2 - y^2) + x(4 - x^2 - y^2)\right) \; dydx \\
&= \frac{713}{180} \quad \blacksquare
\end{aligned}$$

18.4 Green's Theorem

Introduction

Given the choice between a double integral and a line integral, which would you rather compute? To me, double integrals are often a bit more straightforward. For example, a line integral around a rectangular region of integration is annoying, because you have to do it in 4 parts (one per side). But a double integral over the interior of that same rectangle is actually pretty easy (since the limits in rectangular coordinates would be constant).

In math, we are sometimes offered the chance to convert something we don't want to evaluate into something that might be a bit easier. Integration by parts is an example of this: we can convert an integral we can't do into one we can. In this section and the next two, we explore conversion equations for higher order integrals. Often, the importance of the conversion equation is not in the use of it, but in terms of what it means physically. The first conversion equation we'll see is called Green's Theorem, and it offers a chance to convert some line integrals (in both scalar and vector form) into double integrals that may be easier. Green's Theorem is actually the direct analog of integration by parts for double integrals, but that won't be obvious at all at our level of exploration.

Some Preliminary Terms

Here are some terms that are either new or reminders. A contour C is

- **positively oriented** if it is traversed in the counterclockwise direction
- **piecewise smooth** if each segment does not have breaks, jumps, etc.
- **simple** if it doesn't intersect itself
- **closed** if it starts and ends at the same point

Green's Theorem, which we'll see below, is going to require a contour that is all of the above properties: positively-oriented, piecewise smooth, simple, and closed. This essentially means the boundary must connect all the way around, close up, and not contain more than one "blob". The direction we traverse the contour should be counterclockwise. A circle is piecewise smooth, simple, and closed. A figure-eight is not; do you know which property it fails to have?

Green's Theorem: Scalar Version

Theorem 18.2. *Let C be a positively oriented, piecewise smooth, simple closed curve in $\mathbb{R}^2$, and let D be the region surrounded by C. If P and Q have continuous partial derivatives in and around D, then*

$$\oint_C Pdx + Qdy = \iint_D \left(\frac{\partial Q}{\partial x} - \frac{\partial P}{\partial y}\right) dA$$

This is called ***Green's Theorem.***

In other words, if you don't feel like doing a line integral around the contour C, you can convert it to a regular double integral over the region enclosed by C. Notice that there's a little hoop in the integral sign: $\oint$. This is added to the integral sign to indicate that C is closed.

Here are a couple of follow-ups to the theorem:

(1) The theorem requires that the contour C be positively oriented. A negatively oriented contour can still be used, but we have to change the sign of the result. This is similar changing the sign on a regular single integral if we reverse the order from $\int_a^b f(x)\,dx$ to $\int_b^a f(x)\,dx$.

(2) Although a washer / annulus / donut shape is technically not simple, we can still use Green's Theorem there.[2]

OK, let's try it!

EX 1 Find $\oint_C xy\,dx + x^2y^3\,dy$ for C being the triangle with vertices $(0,0)$, $(1,0)$, $(1,2)$, oriented positively.

If we wanted to do this as a line integral, we'd have to do it in three pieces, one for each side of the triangle. Ick. Fortunately, this contour is piecewise-smooth, simple, and closed, so we can use Green's Theorem. Matching the integral to the form $\oint_C Pdx + Qdy$, we have

$$P = xy \to \frac{\partial P}{\partial y} = x$$

$$Q = x^2y^3 \to \frac{\partial Q}{\partial x} = 2xy^3$$

[2]Because ... reasons. Take Advanced Calculus or Complex Analysis!

The triangle goes from (0,0) to (1,0), then up to (1,2) — so it's the region under the line $y = 2x$ from $x = 0$ to $x = 1$ — see Fig. 18.13. So by Green's Theorem (Theorem 18.2),

$$\begin{aligned}\oint_C xy\,dx + x^2y^3\,dy &= \iint_D \left(\frac{\partial Q}{\partial x} - \frac{\partial P}{\partial y}\right) dA \\ &= \int_0^1 \int_0^{2x} (2xy^3 - x)\,dydx = \int_0^1 \left(\frac{xy^4}{2} - xy\right)\Big|_0^{2x} dx \\ &= \int_0^1 \left(\frac{x(16x^4)}{2} - x(2x)\right) dx = \int_0^1 (8x^5 - 2x^2)\,dx \\ &= \left(\frac{8x^6}{6} - \frac{2x^3}{3}\right)\Big|_0^1 = \frac{2}{3} \quad \blacksquare\end{aligned}$$

You Try It

(1) Find $\oint_C e^y\,dx + 2xe^y\,dy$ where C is the boundary of the square with corners (0,0) and (1,1).

(2) Find $\oint_C \frac{1}{x^2+y^2}\,dx + \frac{1}{x^2+y^2}\,dy$ where C is the boundary of the region inside the circles $x^2 + y^2 = 1$ and $x^2 + y^2 = 4$ (take C to be positively oriented).

Green's Theorem: Vector Version

You should have been a bit worried about the form of the line integral involved in Theorem 18.2. After all, it's why we introduced the vector form of line integrals. That kind of fun can be had here, too. Given a vector field $\mathbf{F} = \langle P(x,y), Q(x,y)\rangle$ and a contour C given by the vector function $\mathbf{r}(t) = \langle x(t), y(t)\rangle$, remember that these two things are the same:

$$\int_C P(x,y)\,dx + Q(x,y)\,dy = \int_C \mathbf{F}\cdot d\mathbf{r}$$

And so ...

Useful Fact 18.9. *Green's Theorem can also be posed as follows: Let the contour $C = \mathbf{r}(t)$, for $a \le t \le b$, be a positively oriented, piecewise smooth, simple closed curve in $\mathbb{R}^2$, and let D be the region surrounded by C. If $\mathbf{F}(x,y)$ is a vector field whose components $P(x,y)$ and $Q(x,y)$ have continuous partial derivatives, then*

$$\oint_C \mathbf{F}\cdot d\mathbf{r} = \iint_D \left(\frac{\partial Q}{\partial x} - \frac{\partial P}{\partial y}\right) dA$$

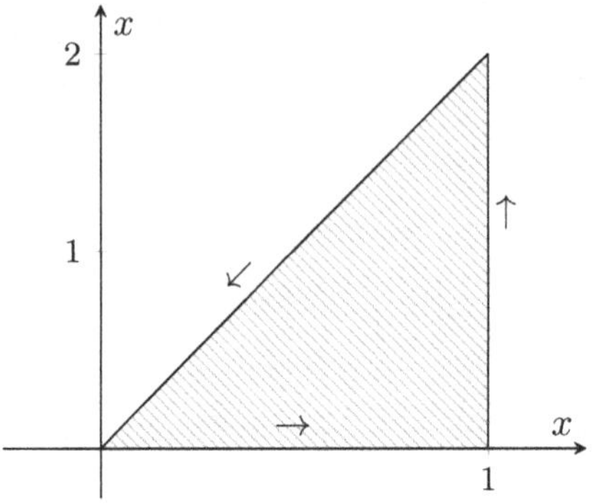

Fig. 18.13 The triangle between (0, 0), (1, 0), and (1, 2).

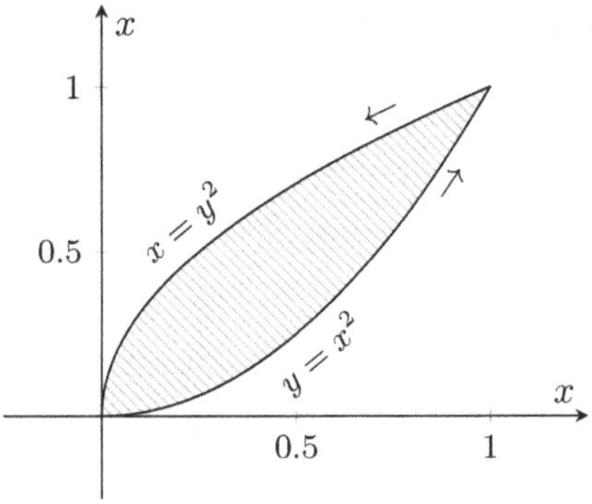

Fig. 18.14 The region between $y = x^2$ and $x = y^2$.

Let's try this one now.

EX 2 Find $\oint_C \mathbf{F} \cdot d\mathbf{r}$ where $\mathbf{F}(x, y) = \langle y + e^{\sqrt{x}}, 2x + \cos y^2 \rangle$ and C is the boundary of the region enclosed by $y = x^2$ and $x = y^2$, oriented positively.

The components of $\mathbf{F}$ are nasty enough we might not be able to do this in standard line integral form. But C is piecewise-smooth, simple, and closed, and so we can convert the line integral into a double integral that might be better. From the components of the vector field $\langle P, Q \rangle$ we get

$$P = y + e^{\sqrt{x}} \to \frac{\partial P}{\partial y} = 1$$
$$Q = 2x + \cos(y^2) \to \frac{\partial Q}{\partial x} = 2$$

To describe the region D bounded by C, we need to rewrite the second curve as $y = \sqrt{x}$, and recognize that for $0 \leq x \leq 1$, the curve $y = x^2$ is actually below $y = \sqrt{x}$ — see Fig. 18.14. So by Useful Fact 18.9,

$$\begin{aligned}\oint_C \mathbf{F} \cdot d\mathbf{r} &= \iint_D \left(\frac{\partial Q}{\partial x} - \frac{\partial P}{\partial y} \right) dA = \int_0^1 \int_{x^2}^{\sqrt{x}} (2-1)\, dy dx \\ &= \int_0^1 (\sqrt{x} - x^2)\, dx \\ &= \left(\frac{2}{3} x^{3/2} - \frac{1}{3} x^3 \right) \Big|_0^1 = \frac{1}{3}\end{aligned}$$

That was much easier than the line integral version! ■

You Try It

(3) Find $\oint_C \mathbf{F} \cdot d\mathbf{r}$ where $\mathbf{F}(x, y) = \langle \sqrt{x} + y^3, x^2 + \sqrt{y} \rangle$ and C is the arc of $y = \sin x$ from (0,0) to $(\pi, 0)$ and then straight back to (0,0).

FFT: If we know that our vector field $\mathbf{F}$ is *conservative*, what does Green's Theorem (as posed in Useful Fact 18.9) tell us about the value of $\oint_C \mathbf{F} \cdot d\mathbf{r}$ over *any* simple closed contour C?

Green's Theorem: More Vector Versions and Interpretation

This section does not culminate in any more problems, it's just a discussion of another two forms of Green's Theorem that help us understand its physical importance. The vector version just above is fine for computation, but remains a bit unsatisfactory. We converted the left side of Green's Theorem to vector form, but not the right side; we sort of have apples and oranges. Is there something we can do to the right side to make it look more "vectory"? Yes!

Remember that we can fool the curl into acting on a 2D vector field by writing it as $\langle P(x, y), Q(x, y), 0 \rangle$. When we do this, we get

$$\nabla \times \mathbf{F} = \nabla \times \langle P(x, y), Q(x, y), 0 \rangle = \ldots = \left(\frac{\partial Q}{\partial x} - \frac{\partial P}{\partial y} \right) \mathbf{k}$$

This looks a lot like what we have on the right side of Green's Theorem, except this expression has a $\mathbf{k}$ stuck to it. But we can get rid of it quickly. Since we know that $\mathbf{k}$ is a unit vector, then we also know that $\mathbf{k} \cdot \mathbf{k} = 1$. Therefore,

$$\nabla \times \mathbf{F} \cdot \mathbf{k} = \left(\frac{\partial Q}{\partial x} - \frac{\partial P}{\partial y} \right) \mathbf{k} \cdot \mathbf{k} = \left(\frac{\partial Q}{\partial x} - \frac{\partial P}{\partial y} \right)$$

and this is now precisely what we have on the right side of Green's Theorem. So, we can make Green's Theorem more "vectory" by writing it like this:

- Let $C = vbfr(t)$, be a positively oriented, piecewise smooth, simple closed curve in the plane, and let D be the region surrounded by C.

If $\mathbf{F}(x, y)$ is a vector field whose components $P(x, y)$ and $Q(x, y)$ have continuous partial derivatives, then

$$\oint_C \mathbf{F} \cdot d\mathbf{r} = \iint_D \nabla \times \mathbf{F} \cdot \mathbf{k}\, dA \tag{18.1}$$

The point of this rewrite of the vector version of Green's Theorem is to make it more tidy, more consistent, and to make sure you're getting good at interpreting such a soup of scalar and vector concepts.

We can create one more version of the Green's Theorem formula by bringing in a few more ideas. This cascade of changes is rather fun, so while you can certainly take these items on faith, I encourage you to verify them for yourself.

(1) An identity involving dot and cross products is $\mathbf{a} \cdot (\mathbf{b} \times \mathbf{c}) = (\mathbf{a} \times \mathbf{b}) \cdot \mathbf{c}$.
(2) Recall that we can find unit tangent vectors $\mathbf{T}$ at any point along a contour $\mathbf{r}(t)$. These unit tangent vectors are related to unit normal vectors $\mathbf{n}$ by $\mathbf{T} = \mathbf{k} \times \mathbf{n}$, where $\mathbf{k}$ is the usual unit vector $\mathbf{k} = \langle 0, 0, 1 \rangle$.
(3) Along a contour $C = \mathbf{r}(t)$, we can write the differential element $d\mathbf{r}$ as $\mathbf{T} ds$, where s is the arc length parameter along C.
(4) Given a vector field $\mathbf{F} = \langle P, Q \rangle$ embedded in $\mathbb{R}^3$ as $\langle P, Q, 0 \rangle$, we have

$$(\nabla \times \mathbf{F}) \cdot \mathbf{k} = \nabla \cdot (\mathbf{F} \times \mathbf{k})$$

All of these together allow us to make two modifications:

$$\oint_C \mathbf{F} \cdot d\mathbf{r} = \oint_C \mathbf{F} \cdot \mathbf{T}\, ds = \oint_C \mathbf{F} \cdot (\mathbf{k} \times \mathbf{n})\, ds = \oint_C (\mathbf{F} \times \mathbf{k}) \cdot \mathbf{n}\, ds$$

and

$$\iint_D \nabla \times \mathbf{F} \cdot \mathbf{k}\, dA = \iint_D \nabla \cdot (\mathbf{F} \times \mathbf{k})\, dA$$

Therefore the previous version of Green's Theorem in (18.1),

$$\oint_C \mathbf{F} \cdot d\mathbf{r} = \iint_D \nabla \times \mathbf{F} \cdot \mathbf{k}\, dA$$

can be converted into this:

$$\oint_C (\mathbf{F} \times \mathbf{k}) \cdot \mathbf{n}\, ds = \iint_D \nabla \cdot (\mathbf{F} \times \mathbf{k})\, dA$$

Now since $\mathbf{F} \times \mathbf{k}$ is itself a vector field, we can just rename it, say as $\mathbf{G} = \mathbf{F} \times \mathbf{k}$, and so we have this new version of Green's Theorem:

$$\oint_C \mathbf{G} \cdot \mathbf{n}\, ds = \iint_D \nabla \cdot \mathbf{G}\, dA \tag{18.2}$$

This version nicely illuminates a physical interpretation of Green's Theorem. Let's think in terms of fluid flow, and say $\mathbf{G}$ is a vector field containing flux vectors. At any location on the boundary of C, $\mathbf{G} \cdot \mathbf{n}$ measures the amount of fluid flowing across the boundary at that point — we have called this the **flux** of $\mathbf{G}$ across C at the point. Therefore, the left side of the new Green's Theorem formula, $\oint_C \mathbf{G} \cdot \mathbf{n}\, ds$, is the *total* flow outward across the whole (closed) boundary C of the domain D. If everything in the flow system is balanced out, then flow entering D must equal flow leaving D, and the total flux of $\mathbf{G}$ across C will be zero — what goes in equals what comes out. But what if we find that the total flux is not zero? Then something is out of balance, and flow is being added to or removed from the flow field. Where is it coming from or going to?

Remember that the divergence of $\mathbf{G}$ at any point is the amount of fluid gained or lost by the flow field at that point. This is not a measure of how much flow is moving past the point, but rather how much flow is being added to (or taken from) the flow field itself at the point. So the right side of Green's Theorem is a measure of the *total* divergence of the flow field over the entire domain D.

Together, this version of Green's Theorem says that the total amount of fluid gained or lost over the inside of the domain (divergence) must equal the imbalance in the total amount of flow in or out across the boundary. Physically, that makes perfect sense. If you are watching a 2D flow field across, say, a circle, and you note a larger total of flow leaving the circle than there was entering the circle, where is that extra flow coming from? It is a consequence of the divergence of $\mathbf{G}$ inside the circle.

If a vector field has no divergence, i.e. if $\nabla \cdot \mathbf{G} = 0$, then flow is always balanced in that flow field; whatever flow enters any domain D will be exactly the same amount as flow leaving D.

It's always very nice when you look at an ugly mathematical expression and realize it's nothing more than a very efficient statement of the obvious!

From Eq. (18.2) onward, we could have recognized that since **G** was just a random name assigned to a vector field, we could have reset it to **F**, but that may not have been advisable since **G** evolved from $\nabla \times \mathbf{F}$. However, at this point, let's go ahead and reset it to **F** as we pose this list of all the different versions of Green's Theorem we have seen:

$$\oint_C P dx + Q dy = \iint_D \left(\frac{\partial Q}{\partial x} - \frac{\partial P}{\partial y} \right) dA$$

$$\oint_C \mathbf{F} \cdot d\mathbf{r} = \iint_D \left(\frac{\partial Q}{\partial x} - \frac{\partial P}{\partial y} \right) dA$$

$$\oint_C \mathbf{F} \cdot d\mathbf{r} = \iint_D \nabla \times \mathbf{F} \cdot \mathbf{k}\, dA$$

$$\oint_C \mathbf{F} \cdot \mathbf{n}\, ds = \iint_D \nabla \cdot \mathbf{F}\, dA$$

Green's Theorem — Problem List

Green's Theorem — You Try It

These appeared above; solutions begin on the next page.

(1) Find $\oint_C e^y\, dx + 2xe^y\, dy$ where C is the boundary of the square with corners (0,0) and (1,1).

(2) Find $\oint_C \frac{1}{x^2+y^2}\, dx + \frac{1}{x^2+y^2}\, dy$ where C is the boundary of the region inside the circles $x^2 + y^2 = 1$ and $x^2 + y^2 = 4$ (take C to be positively oriented).

(3) Find $\oint_C \mathbf{F} \cdot d\mathbf{r}$ where $\mathbf{F}(x, y) = \langle \sqrt{x} + y^3, x^2 + \sqrt{y} \rangle$ and C is the arc of $y = \sin x$ from (0,0) to $(\pi, 0)$ and then straight back to (0,0).

Green's Theorem — Practice Problems

Try these as you get the hang of the You Try It problems. Solutions to these problems are available in Sec. B.6.4.

(1) Find $\oint_C x^2y^2\, dx + 4xy^3\, dy$ where C is the boundary of the triangle with corners traversed in order (0,0) to (1,3) to (0,3) and back to (0,0).

(2) Find $\oint_C \mathbf{F} \cdot d\mathbf{r}$ where $\mathbf{F}(x,y) = \langle y^2 \cos x, x^2 + 2y \sin x \rangle$ and C is the triangle from (0,0) to (2,6) to (2,0) and back to (0,0).

(3) Find $\oint_C xe^{-2x}\, dx + (x^4 + 2x^2y^2)\, dy$ where C is the boundary of the region inside the circles $x^2 + y^2 = 4$ and $x^2 + y^2 = 9$.

(4) Find $\oint_C \mathbf{F} \cdot d\mathbf{r}$ where $\mathbf{F}(x,y) = \langle e^x + x^2 y, e^y - xy^2 \rangle$ and C is the clockwise perimeter of $x^2 + y^2 = 25$.

Green's Theorem — Challenge Problems

Try these problems to test your skills with the ideas in this section. Solutions to these problems are available in Sec. C.6.4.

(1) Find $\oint_C x^2 e^{-3x}\, dx + \frac{2}{3}(x^2 + y^2)^3\, dy$ where C is the boundary of the region inside the circles $x^2 + y^2 = 2$ and $x^2 + y^2 = 9$ (take C to be positively oriented).

(2) Find $\oint_C \mathbf{F} \cdot d\mathbf{r}$ where

$$\mathbf{F}(x,y) = \langle e^{-x} + \frac{1}{2}x^3y^2, \sin(y) - \frac{1}{2}x^2y^3 \rangle$$

and C is the perimeter of $x^2 + y^2 = 16$, oriented clockwise.

(3) The integral $\oint_C \mathbf{F} \cdot d\mathbf{r}$ gives the *circulation* of the vector field $\mathbf{F}$ around the contour C. Use Green's Theorem to find the circulation of $\mathbf{F}(x,y) = \langle y, -x \rangle$ around a circle $x^2 + y^2 = a^2$ (where a is the constant radius of the circle). Take the boundary as being oriented positively.

Green's Theorem — You Try It — Solved

---

(1) Find $\oint_C e^y\,dx + 2xe^y\,dy$ where C is the boundary of the square with corners (0,0) and (1,1).

☐ Matching this integral to the form $\oint_C P\,dx + Q\,dy$, we have

$$P = e^y \to \frac{\partial P}{\partial y} = e^y$$
$$Q = 2xe^y \to \frac{\partial Q}{\partial x} = 2e^y$$

Since C is closed and positively oriented, Green's Theorem applies and

$$\begin{aligned}\oint_C e^y\,dx - 2xe^y\,dy &= \iint_D \left(\frac{\partial Q}{\partial x} - \frac{\partial P}{\partial y}\right) dA \\ &= \int_0^1\int_0^1 (2e^y - e^y)\,dydx = \int_0^1\int_0^1 e^y\,dydx \\ &= \int_0^1 (e-1)\,dx = e-1 \quad \blacksquare\end{aligned}$$

---

(2) Find $\oint_C \frac{1}{x^2+y^2}\,dx + \frac{1}{x^2+y^2}\,dy$ where C is the boundary of the region inside the circles $x^2+y^2=1$ and $x^2+y^2=4$ (take C to be positively oriented).

☐ Matching this integral to the form $\oint_C P\,dx + Q\,dy$, we have

$$P = \frac{1}{x^2+y^2} \to \frac{\partial P}{\partial y} = -\frac{2y}{(x^2+y^2)^2}$$
$$Q = \frac{1}{x^2+y^2} \to \frac{\partial Q}{\partial x} = -\frac{2x}{(x^2+y^2)^2}$$

Let's assemble $Q_x - P_y$ before building the integral:

$$\frac{\partial Q}{\partial x} - \frac{\partial P}{\partial y} = \frac{-2x+2y}{(x^2+y^2)^2}$$

Since our region of integration is between circles $r=1$ and $r=2$, let's convert to polar coordinates:

$$\frac{\partial Q}{\partial x} - \frac{\partial P}{\partial y} = \frac{-2r\cos\theta + 2r\sin\theta}{(r^2)^2} = \frac{2r(\sin\theta - \cos\theta)}{r^4} = 2\,\frac{\sin\theta - \cos\theta}{r^3}$$

and Green's Theorem gives

$$\oint_C \frac{1}{x^2+y^2}\,dx + \frac{1}{x^2+y^2}\,dy = \iint_D \left(\frac{\partial Q}{\partial x} - \frac{\partial P}{\partial y}\right) dA$$
$$= \int_0^{2\pi}\int_1^2 \left(2\,\frac{\sin\theta - \cos\theta}{r^3}\right) r\,drd\theta$$
$$= 2\int_0^{2\pi}\int_1^2 \frac{\sin\theta - \cos\theta}{r^2}\,drd\theta = 0$$

FFT: Can you think of a physical meaning for this result? ■

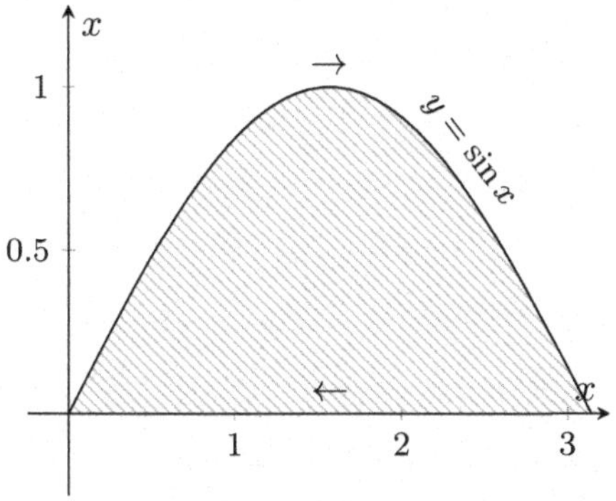

Fig. 18.15 Following $y = \sin(x)$ from $x = 0$ to $x = \pi$ and back to $(0, 0)$.

(3) Find $\oint_C \mathbf{F}\cdot d\mathbf{r}$ where $\mathbf{F}(x,y) = \langle \sqrt{x} + y^3, x^2 + \sqrt{y}\rangle$ and C is the arc of $y = \sin x$ from (0,0) to $(\pi, 0)$ and then straight back to (0,0).

□ Matching the vector field to the form $\mathbf{F}(x,y) = \langle P(x,y), Q(x,y)\rangle$, we have

$$P = \sqrt{x} + y^3 \to \frac{\partial P}{\partial y} = 3y^2$$

$$Q = x^2 + \sqrt{y} \to \frac{\partial Q}{\partial x} = 2x$$

Note that C is the area under $y = \sin x$ from $x = 0$ to $x = \pi$, but as described it is negatively oriented (clockwise) — see Fig. 18.15. Green's Theorem applies, but we have to change the sign to account for the orientation:

$$\begin{aligned}\oint_C \mathbf{F}\cdot d\mathbf{r} = \oint_C P\,dx + Q\,dy &= -\iint_D \left(\frac{\partial Q}{\partial x} - \frac{\partial P}{\partial y}\right) dA \\ &= -\int_0^{\pi}\int_0^{\sin x} (2x - 3y^2)\,dydx = -\frac{2}{3}(3\pi - 2) \\ &= \frac{4}{3} - 2\pi \quad \blacksquare\end{aligned}$$

18.5 The Divergence Theorem

Introduction

Green's Theorem allowed us to convert some line integrals into easier double integrals. Even though we bumped up the number of dimensions in the integral (from 1 to 2), we actually created problems that are (hopefully) easier to solve. You might think that if there's a way to convert a 1D line integral to a 2D double integral, there might be a way to convert a 2D surface integral into a 3D triple integral. And there is — the Divergence Theorem (which is actually an upgrade of Green's Theorem). Its name tells you what to expect as one of the ingredients.

This particular subject is like the first flat relaxing portion on a breath-taking roller coaster: after you've gone down the first hill and through the first set of corkscrew turns and inverted loops, you get to slow down and take a breath.

The Divergence Theorem

Theorem 18.3. *Let E be a simple solid region and let S be the boundary of E, given with outward (positive) orientation. Let $\mathbf{F}$ be a vector field whose components have continuous partial derivatives in and around E. Then*

$$\iint_S \mathbf{F} \cdot d\mathbf{S} = \iiint_E \nabla \cdot \mathbf{F}\, dV \tag{18.3}$$

This is called the Divergence Theorem.

In other words, given a surface (flux) integral that's too gross to compute directly, you can do a triple integral of the divergence of the flux vector instead (thus the name "Divergence Theorem"). The volume of integration in the triple integral is simply the volume surrounded by the surface S. The triple integral can be done in rectangular, spherical, or cylindrical coordinates.

If Eq. (18.3) looks familiar, then you must have read all the way to the end of Sec. 18.4 ... good job! Equation (18.3) is an upgrade of Eq. 18.2, and the physical interpretation of the Divergence Theorem is almost identical to that Green's Theorem as posed in Eq. (18.2). The left side of (18.3) measures the total amount of flux crossing the entire surface S. In a balanced system, in which flow into the volume E equals the flow out of the volume, the net flux across the surface should be 0. If it's not zero, then the added or lost flow must be accounted for by the divergence of the flow field within the volume. This divergence is measured on the right side.

You might be disappointed, but that's about all there is to say for now about the Divergence Theorem!

EX 1 Find the total flux of the vector field $F(x, y, z) = \langle z, y, x \rangle$ over the unit sphere.

This is asking for the value of the surface integral $\iint_S \mathbf{F} \cdot d\mathbf{S}$. Well, no thanks. I'll take a triple integral, instead. Since the unit sphere is simple and closed, we can use the Divergence Theorem. The divergence of $\mathbf{F}$ is

$$\nabla \cdot \mathbf{F} = \frac{\partial}{\partial x}(z) + \frac{\partial}{\partial y}(y) + \frac{\partial}{\partial z}(x) = 1$$

When we use the divergence theorem, we're going to want a triple integral over the unit sphere: so spherical coordinates will be useful.

$$\iint_S \mathbf{F} \cdot d\mathbf{S} = \iiint_E \nabla \cdot \mathbf{F}\, dV = \int_0^{2\pi} \int_0^{\pi} \int_0^1 (1)\rho^2 \sin\phi \, d\rho d\phi d\theta = \frac{4\pi}{3} \quad \blacksquare$$

EX 2 Evaluate $\iint_S \mathbf{F} \cdot d\mathbf{S}$ for $\mathbf{F}(x, y, z) = \langle x^3, y^3, z^3 \rangle$ and S is the surface of the solid bounded by $x^2 + y^2 = 1$, $z = 0$, $z = 2$.

Since the solid of integration (a cylinder) is simple and closed, we can use the Divergence Theorem. The divergence of $\mathbf{F}$ is

$$\nabla \cdot \mathbf{F} = \frac{\partial}{\partial x}(x^3) + \frac{\partial}{\partial y}(y^3) + \frac{\partial}{\partial z}(z^3) = 3(x^2 + y^2 + z^2)$$

When we use the divergence theorem, we're going to want a triple integral over the cylinder, so cylindrical coordinates will be useful.

$$\iint_S \mathbf{F} \cdot d\mathbf{S} = \iiint_E \nabla \cdot \mathbf{F}\, dV = \iiint_E 3(x^2 + y^2 + z^2)\, dV$$
$$= \int_0^{2\pi} \int_0^1 \int_0^2 3(r^2 + z^2)\, r \, dz dr d\theta = 11\pi \quad \blacksquare$$

You Try It

(1) Find $\iint_S \mathbf{F}\cdot d\mathbf{S}$ for the vector field $\mathbf{F} = \langle e^x \sin y, e^x \cos y, yz^2\rangle$ where S is the surface of the region E that is the rectangular box $0 \le x \le 1$, $0 \le y \le 1$, $0 \le z \le 2$.

(2) Compute $\iint_S \mathbf{F}\cdot d\mathbf{S}$ for the vector field $\mathbf{F} = \langle 3xy^2, xe^z, z^3\rangle$ where S is the surface of the cylinder $y^2 + z^2 = 1$ between $x = -1$ and $x = 2$.

Divergence Theorem — Problem List

Divergence Theorem — You Try It

These appeared above; solutions begin on the next page.

(1) Find $\iint_S \mathbf{F}\cdot d\mathbf{S}$ for the vector field $\mathbf{F} = \langle e^x \sin y, e^x \cos y, yz^2\rangle$ where S is the surface of the region E that is the rectangular box $0 \le x \le 1$, $0 \le y \le 1$, $0 \le z \le 2$.

(2) Compute $\iint_S \mathbf{F}\cdot d\mathbf{S}$ for the vector field $\mathbf{F} = \langle 3xy^2, xe^z, z^3\rangle$ where S is the surface of the cylinder $y^2 + z^2 = 1$ between $x = -1$ and $x = 2$.

Divergence Theorem — Practice Problems

Try these as you get the hang of the You Try It problems. Solutions to these problems are available in Sec. B.6.5.

(1) Find $\iint_S \mathbf{F}\cdot d\mathbf{S}$ for the vector field $\mathbf{F} = \langle x^2z^3, 2xyz^3, xz^4\rangle$ where S is the surface of the region E that is the rectangular box $-1 \le x \le 1$, $-2 \le y \le 2$, $-3 \le z \le 3$.

(2) Evaluate $\iint_S \mathbf{F}\cdot d\mathbf{S}$ for $\mathbf{F}(x, y, z) = \langle xy, (y^2 + e^{xz^2}), \sin(xy)\rangle$ and S is the surface of the region bounded by $z = 1 - x^2$, $z = 0$, $y = 0$, $y = 2$.

(3) Find $\iint_S \mathbf{F}\cdot d\mathbf{S}$ for the vector field $\mathbf{F} = \langle x^3y, -x^2y^2, -x^2yz\rangle$ where S is the surface of the hyperboloid $x^2 + y^2 - z^2 = 1$ between $z = -2$ and $z = 2$.

(4) Find $\iint_S \mathbf{F}\cdot d\mathbf{S}$ for the vector field $\mathbf{F} = \langle x^2y, xy^2, 2xyz\rangle$ where S is the surface of the tetrahedron formed by the plane $x + 2y + z = 2$ in the first octant.

Divergence Theorem — Challenge Problems

Try these problems to test your skills with the ideas in this section. Solutions to these problems are available in Sec. C.6.5.

(1) Find $\iint_S \mathbf{F} \cdot d\mathbf{S}$ for the vector field $\mathbf{F} = \langle -x^3y + yz, x^2y^2 + e^x, x^2yz \rangle$ where S is the surface of the hyperboloid $x^2 + 3y^2 - z^2 = 1$ between $z = -1$ and $z = 4$.

(2) Find $\iint_S \mathbf{F} \cdot d\mathbf{S}$ for the vector field $\mathbf{F} = \langle x^2yz + e^y, xy^2z + \sin(xz), \pi - xyz^2 \rangle$ where S is the surface of the tetrahedron formed by the plane $2x + 2y + z = 4$ in the first octant.

(3) As noted, the integral $\iint_S \mathbf{F} \cdot d\mathbf{S}$ gives the *flux* of the vector field $\mathbf{F}$ across the surface S. Use the Divergence Theorem to find the flux of a vector field $\mathbf{F}(x, y, z) = \langle bz - cy, cx - az, ay - bx \rangle$ across ANY smooth closed surface S in $\mathbf{R}^3$ (where a, b, c are constants).

The Divergence Theorem — You Try It — Solved

---

(1) Find $\iint_S \mathbf{F} \cdot d\mathbf{S}$ for the vector field $\mathbf{F} = \langle e^x \sin y, e^x \cos y, yz^2 \rangle$ where S is the surface of the region E that is the rectangular box $0 \leq x \leq 1$, $0 \leq y \leq 1$, $0 \leq z \leq 2$.

□ The divergence of $\mathbf{F}$ is

$$\begin{aligned}\nabla \cdot \mathbf{F} &= \frac{\partial}{\partial x}(e^x \sin y) + \frac{\partial}{\partial y}(e^x \cos y) + \frac{\partial}{\partial z}(yz^2) \\ &= e^x \sin y - e^x \sin y + 2yz = 2yz\end{aligned}$$

Then by the divergence theorem,

$$\iint_S \mathbf{F} \cdot d\mathbf{S} = \iiint_E \nabla \cdot \mathbf{F}\, dV = \int_0^1 \int_0^1 \int_0^2 (2yz)\, dz dy dx = 2 \quad \blacksquare$$

---

(2) Compute $\iint_S \mathbf{F} \cdot d\mathbf{S}$ for the vector field $\mathbf{F} = \langle 3xy^2, xe^z, z^3 \rangle$ where S is the surface of the cylinder $y^2 + z^2 = 1$ between $x = -1$ and $x = 2$.

□ The divergence of $\mathbf{F}$ is

$$\nabla \cdot \mathbf{F} = \frac{\partial}{\partial x}(3xy^2) + \frac{\partial}{\partial y}(xe^z) + \frac{\partial}{\partial z}(z^3) = 3y^2 + 3z^2 = 3(y^2 + z^2)$$

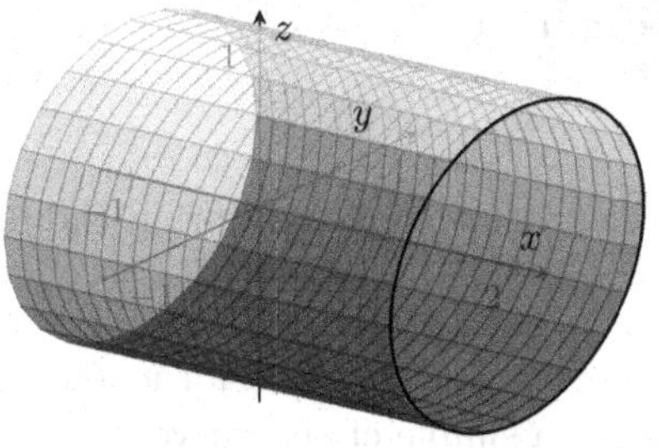

Fig. 18.16 The cylinder $y^2 + z^2 = 1$ for $-1 \leq x \leq 2$.

Note that the region E is a cylinder opening along the x-axis — see Fig. 18.16. It will be convenient to use "sideways" cylindrical coordinates — where r and θ are set in the yz-plane, and x is the remaining rectangular coordinate (like z is normally).

$$\iint_S \mathbf{F} \cdot d\mathbf{S} = \iiint_E \nabla \cdot \mathbf{F}\, dV = \iiint_E 3(y^2 + z^2)\, dV$$
$$= \int_0^{2\pi} \int_0^1 \int_{-1}^2 3(r^2)\, r\, dx dr d\theta = \frac{9\pi}{2} \quad \blacksquare$$

18.6 Stokes' Theorem

Introduction

Here are a few new ideas that came up during the discussion of Green's Theorem:

- The idea of "positive" orientation for a contour
- The definition of a simple closed curve
- The idea of converting a work or flux integral into a different form to make it easier to evaluate

Green's Theorem is fine, I suppose, if you like being restricted to the xy-plane. But intrepid mathematical souls such as yourself like their integrals in 3D settings, so let's move there. Here's an example of where Stokes' Theorem will be in action. Suppose we passed a slanted plane through a paraboloid, as shown in Fig. 18.17. Their intersection forms a tilted and ellipse-ish contour in $\mathbb{R}^3$. Suppose we wanted to do a work integral $\int_C \mathbf{F} \cdot d\mathbf{r}$ over that contour (do you remember those from Sec. 18.1?). Green's Theorem isn't sufficient any more because we're not in the xy-plane. This is where Stokes' Theorem takes over. Stokes' Theorem is the worst of the three conversion formulas (Green's, Divergence, Stokes') to visualize, so of course we've saved it for last.[3]

Oh, and just to be sure you've noticed: this theorem is named after George Stokes (1819–1903), and so it is properly referred to as Stokes' Theorem. It is not Stoke's Theorem. Don't do that.

Preliminaries

First, let's work on some terms. The main ingredients of Stokes' Theorem are a vector field $\mathbf{F}$, an oriented surface S, and a simple closed positively oriented boundary curve C (alias $\mathbf{r}(t)$) of the surface. Two of these terms are new: the idea of a boundary curve may be somewhat self explanatory, but what the heck is an oriented surface?

Imagine taking a regular surface S with a closed end, like a paraboloid, cone, or ellipsoid, and crossing through it with a plane in such a way that

[3]Here's a secret: Green's Theorem is actually a special case of Stokes' Theorem.

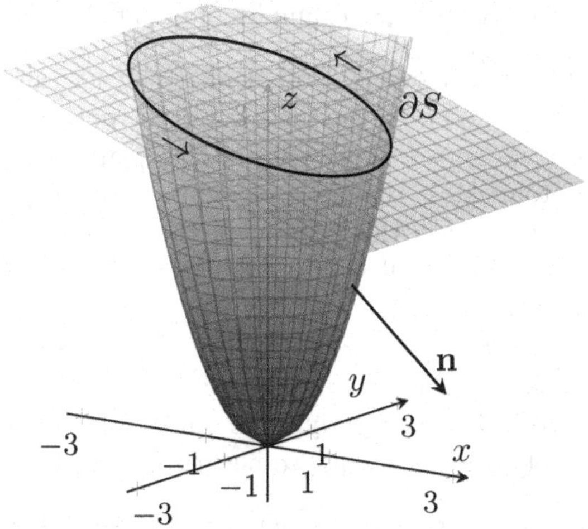

Fig. 18.17 Ingredients of a positively oriented surface.

the intersection forms a closed contour. As we assigned positive and negative orientations to contours, we'll do the same with the surface — but sometimes this can be a bit tricky to visualize. The contour at the intersection of the plane and surface S is called the boundary contour, ∂S; in this context, the symbol ∂ means "boundary of" instead of being a partial derivative symbol. Here are two pieces of information we need:

- Remember from Sec. 16.5 that if the equation of the surface is $z = g(x, y)$, then we can find normal vectors on the surface as $\langle g_x, g_y, -1\rangle$.
- We assign positive orientation to ∂S by designing its vector equation $\mathbf{r}(t)$ such that we traverse it counterclockwise.

So now, let's pretend S and ∂S are really big, so that we can walk around ∂S counterclockwise, with our right hand pointing in the direction of the normal vectors of S; if the interior of the surface is to our left, then the surface itself is called **positively oriented**.

This scenario can be seen in Fig. 18.17, where as we traverse ∂S counterclockwise, with the normal vectors of S pointing outwards, the surface itself is to the left. So, this surface is positively oriented. Together, the surface and contour provide a "right" direction and a "wrong" direction, just like there is a right and wrong direction when you try to screw a cap onto the top of a bottle.

Here are some other simple examples of oriented surfaces.

- If we take the sideways paraboloid $x = y^2 + z^2$ and slice through it with the plane $x = 4$, the surface S is the portion of the paraboloid between its vertex at $x = 0$ and the plane $x = 4$. The bounding contour ∂S is the circle of radius 2 now capping the surface, where we traverse this contour in the counterclockwise (positive) direction. FFT: Can you design parametric equations for this contour?
- If we take the sphere $x^2 + y^2 + z^2 = 9$ and slice through it with the plane $z = 1 + x$, the surface S is the spherical cap above the plane; this plane is slightly tilted, and so the bounding contour ∂S (positively oriented) will also be tilted, and is not a perfect circle.

There are other subtle ways of defining an oriented surface. Consider the cylinder $x^2 + y^2 = 4$ and its intersection with the plane $z = 3$. Neither end of the cylinder is closed, and in fact the cylinder extends infinitely in both directions from the plane. So neither the part of the cylinder above the plane or below the plane generates a surface with one closed end, and another end bounded by a contour. But, remember that a plane is a surface, too. The cylinder is a bit of misdirection; it does not have to become part of the surface. Rather, it just tells us where to truncate the surface that's the plane itself. In other words, our oriented surface can be the circular portion of the plane $z = 3$ inside the cylinder, with a bounding contour generated from the intersection of the plane with the cylinder.

As with Green's Theorem and the Divergence Theorem, Stokes' Theorem can be thought of as way to take one integral which may be difficult or impossible to set up or solve, and convert it into another one that works better. When we used Green's Theorem and the Divergence Theorem, we only used them in one direction; that is, for example, we only used Green's Theorem to convert a line integral into a double integral ... while it's certainly possible to use the Green's Theorem formula to convert a double integral to a line integral, it is rarely used that way. Also, we only used the Divergence Theorem to convert a surface integral to a triple integral, we never used it to convert a triple integral into a surface integral. Stokes' Theorem, though, gets used both ways: to convert a surface integral into a contour integral *and* vice versa.

Stokes' Theorem

Theorem 18.4. *Let S be an oriented piecewise-smooth surface that is bounded by a simple, closed, piecewise smooth boundary curve C with positive orientation. Let $\mathbf{F}$ be a vector field whose components have continuous partial derivatives in and around S. Then*

$$\int_{\partial S} \mathbf{F} \cdot d\mathbf{r} = \iint_S \nabla \times \mathbf{F} \cdot d\mathbf{S}$$

As noted above, we can use this in both direction. Given a contour integral on the left, we can convert it to a surface intergal, and vice versa. To do the former, we can construct a more streamlined formula for use. Stokes' Theorem gets us this far:

$$\int_C \mathbf{F} \cdot d\mathbf{r} = \iint_S \nabla \times \mathbf{F} \cdot d\mathbf{S}$$

but then Useful Fact 18.8 in Sec. 18.3 can be adapted for the surface integral of $\nabla \times \mathbf{F} \cdot d\mathbf{r}$ rather than $\mathbf{F} \cdot d\mathbf{r}$ as follows:

$$\iint_S \nabla \times \mathbf{F} \cdot d\mathbf{S} = \iint_D \left(-P\frac{\partial g}{\partial x} - Q\frac{\partial g}{\partial y} + R \right) dA$$

where we have to be really careful to remember that in this case, P, Q, and R are the components of the *curl* of $\mathbf{F}$, not $\mathbf{F}$ itself. Altogether:

Useful Fact 18.10. *A more utilitarian pathway through Stokes' Theorem is that, for all the conditions posed in Theorem 18.4, we have:*

$$\int_C \mathbf{F} \cdot d\mathbf{r} = \iint_D \left(-P\frac{\partial g}{\partial x} - Q\frac{\partial g}{\partial y} + R \right) dA \tag{18.4}$$

where $\langle P, Q, R\rangle = \nabla \times \mathbf{F}$, C is the boundary of the the surface $z = g(x, y)$, and the domain D is the planar support of the surface.

Converting Line Integrals to Surface Integrals

Here are examples of conversion of a line integral to a surface integral, i.e. using Stokes' Theorem as shown in Theorem 18.4 "from left to right".

EX 1 Use Stokes' Theorem to evaluate $\int_C \mathbf{F} \cdot d\mathbf{r}$ where $\mathbf{F}(x, y, z) = \langle x + y, y + z, x - z\rangle$ and C is the intersection of the paraboloid $z = 2x^2 + 2y^2$ with the plane $z = 8$, oriented positively. Then, evaluate the surface integral.

The overall flow of the problem starts with, "Hey, let's use Stokes' Theorem to convert our line integral into a surface integral." The price of being able to convert a line integral into a surface integral is that **F** is replaced by $\nabla \times \mathbf{F}$. Once we have a surface integral of the curl of **F**, we refer to Sec. 18.3, where we said, "Hey, if you have a surface integral to do, you're going to evaluate it by swapping it with a regular double integral based on the components of the curl and the derivatives of the surface equation." That double jump, line integral to surface integral, and surface integral to double integral, is encoded into (18.4). When it's all set up, we never actually have to see the surface integral itself.

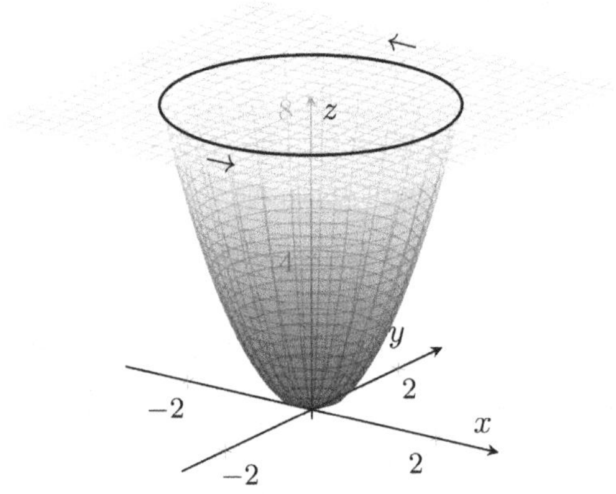

Fig. 18.18 $z = 2x^2 + 2y^2$ and $z = 8$.

The paraboloid, plane, and resulting contour ∂S are shown in Fig. 18.18. Here are a few ingredients we're going to need to implement Eq. (18.4):

- With $\mathbf{F} = \langle x + y, y + z, x - z\rangle$, we can get $\nabla \times \mathbf{F} = \langle P, Q, R\rangle = \langle -1, -1, -1\rangle$ (details omitted, finding a curl should be trivial at this point).
- Our surface is $z = g(x, y) = x^2 + y^2$ and so $g_x = 2x$, $g_y = 2y$.

We can now build the integrand in advance; this helps us decide what coordinate system to use for D:

$$-P\frac{\partial g}{\partial x} - Q\frac{\partial g}{\partial y} + R = -(-1)2x - (-1)(2y) + (-1) = 2x + 2y - 1$$

The equation $2x + 2y - 1$ doesn't immediately make us think of polar coordinates, but given the information about S and ∂S in the problem statement, the domain of integration in the xy-plane is going to be a circle of radius 2 (the intersection of $z = 2x^2 + 2y^2$ with $z = 8$ is the circle $x^2 + y^2 = 4$). So polar coordinates are going to be best, even though the integrand will look icky:

$$\begin{aligned}\int_C \mathbf{F} \cdot d\mathbf{r} &= \iint_D \left(-P\frac{\partial g}{\partial x} - Q\frac{\partial g}{\partial y} + R\right) dA = \iint_D (2x + 2y - 1)\ dA \\ &= \int_0^{2\pi} \int_0^2 (2r\cos\theta + 2r\sin\theta - 1)\ r\, dr d\theta = 4\pi \quad \blacksquare\end{aligned}$$

EX 2 Evaluate $\int_C \mathbf{F} \cdot d\mathbf{r}$ for $\mathbf{F}(x, y, z) = \langle xy, yz, xz\rangle$ and C is the triangle with vertices $(1, 0, 0)$, $(0, 1, 0)$, and $(0, 0, 1)$, oriented counterclockwise as viewed from above.

Since we're given a contour C and need to come up with a corresponding surface S, we must decide on a surface which has C as a boundary curve. With no other information at hand, we can pick *any* surface as long as it has C as its boundary curve. The simplest thing to do is recognize that the three given points determine the plane $x + y + z = 1$, which is itself a surface. So we can say that C is the boundary of the portion of the plane $z = g(x, y) = 1 - x - y$ in the first octant. So to implement Stokes' Theorem via Useful Fact 18.10, we need the following items:

- Since we determined our surface to be $z = g(x, y) = 1 - x - y$, then $g_x = -1$ and $g_y = -1$.
- Since $\mathbf{F}(x, y, z) = \langle xy, yz, xz\rangle$, then $\nabla \times \mathbf{F} = \langle -y, -z, -x\rangle$ (details omitted). Therefore, matching to the form $\nabla \times \mathbf{F} = \langle P, Q, R\rangle$, we have

  $$P = -y \quad ; \quad Q = -z \quad ; \quad R = -x$$

  Now, we can't have z's floating around in our integral, so we have to use the equation of our surface to write $Q = -z = -(1-x-y) = x + y - 1$.
- Since S is determined to be the plane $z = 1 - x - y$ in the first octant, the corresponding region of integration D in the xy-plane is the triangle between the lines $x + y = 1$, $x = 0$, and $y = 0$.

Now we have all the information we need to load up (18.4):

$$\begin{aligned}\int_C \mathbf{F}\cdot d\mathbf{r} &= \iint_S \nabla\times\mathbf{F}\cdot d\mathbf{S} = \iint_D \left(-P\frac{\partial g}{\partial x} - Q\frac{\partial g}{\partial y} + R\right) dA\\ &= \iint_D (-(-y)(-1) - (x+y-1)(-1) + (-x))\, dA\\ &= \iint_D (-y + x + y - 1 - x)\, dA\\ &= \int_0^1 \int_0^{-x+1} (-1)\, dydx = -\frac{1}{2}\end{aligned}$$

An interpretation of this is that the total work done by the force field **F** around the contour C is $-1/2$. ■

EX 3 Find the work done by $\mathbf{F}(x,y,z) = \langle -y^2, x, z^2\rangle$ around the intersection of the plane $y+z=2$ and the cylinder $x^2+y^2=1$, oriented counterclockwise as viewed from above.

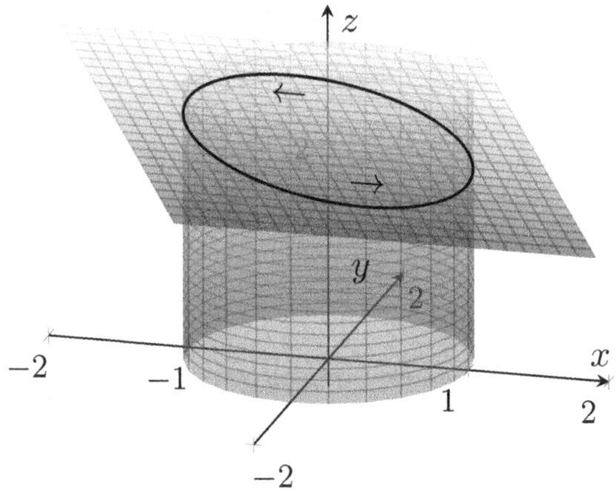

Fig. 18.19 $x^2+y^2=1$ and $y+z=2$.

The problem statement means that we need to evaluate $\int_C \mathbf{F}\cdot d\mathbf{r}$. There is a "gotcha" in this problem, which is that we can't use the cylinder as our surface S, because we don't know where the bottom of the cylinder would be. So like in EX 2, we can rely on "hey, a plane is a perfectly good surface, too!" Our surface can be the portion of the plane $y+z=2$ which gets cut out by the given cylinder; thus, S will be the portion of $y + z = 2$ above the unit circle — see Fig. 18.19. The ingredients we need for (18.4) are:

- Our surface (plane) can be written $z = g(x, y) = 2 - y$ and so $g_x = 0$ and $g_y = -1$.
- $\nabla \times \mathbf{F} = \langle 0, 0, 2y + 1\rangle$ (details omitted). Matching to the form $\nabla \times \mathbf{F} = \langle P, Q, R\rangle$, we have

$$P = 0 \quad ; \quad Q = 0 \quad ; \quad R = 2y + 1$$

(We don't need to modify these any further since z hasn't shown up in P, Q, or R.)
- Since our region of integration D is the unit circle, then we should plan on using polar coordinates for our upcoming double integral.

Altogether, we have for (18.4) — in rectangular coordinates —

$$\begin{aligned}\int_C \mathbf{F} \cdot d\mathbf{r} = \iint_S \nabla \times \mathbf{F} \cdot d\mathbf{S} &= \iint_D \left(-P\frac{\partial g}{\partial x} - Q\frac{\partial g}{\partial y} + R\right) dA \\ &= \iint_D (-(0)(0) - (0)(-1) + (2y + 1))\, dA \\ &= \iint_D (2y + 1)\, dA \cdots\end{aligned}$$

Then converting this integral to polar coordinates, we continue:

$$\begin{aligned}\cdots &= \int_0^{2\pi} \int_0^1 (2(r \sin\theta) + 1)\, r\, dr d\theta \\ &= \int_0^{2\pi} \int_0^1 (2r^2 \sin\theta + r)\, dr d\theta = \pi \quad \blacksquare\end{aligned}$$

You Try It

(1) Compute $\int_C \mathbf{F} \cdot d\mathbf{r}$ for the vector field $\mathbf{F} = \langle x^2, y^2, x^2 + y^2\rangle$ where C is the lower half of the hemisphere $x^2 + y^2 + z^2 = 1$ capped by the xy-plane.

(2) Compute $\int_C \mathbf{F} \cdot d\mathbf{r}$ for the vector field $\mathbf{F} = \langle x + y^2, y + z^2, z + x^2\rangle$ where C is the triangle with vertices (1,0,0), (0,1,0), and (0,0,1).

Converting Surface Integrals to Line Integrals

If we flip Stokes' Theorem around, it looks like this:

$$\iint_S \nabla \times \mathbf{F} \cdot d\mathbf{S} = \int_{\partial S} \mathbf{F} \cdot d\mathbf{r}$$

So given a nasty surface integral, perhaps we can convert it to a nicer line integral. Remember that once the line integral is set up, we'll have to parameterize the contour C to evaluate it!

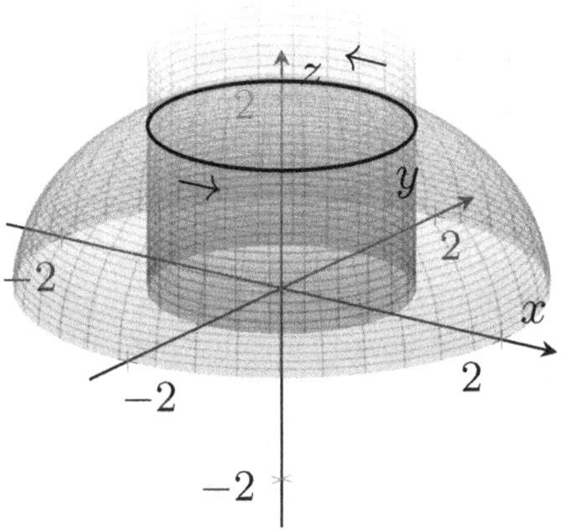

Fig. 18.20 Sphere $\rho = 2$ and cylinder $r = 1$.

EX 4 Evaluate $\iint_S \nabla \times \mathbf{F} \cdot d\mathbf{S}$ for $\mathbf{F}(x, y, z) = \langle xz, yz, xy \rangle$ and S is the part of the sphere $x^2 + y^2 + z^2 = 4$ that is inside the cylinder $x^2 + y^2 = 1$ and above the xy-plane.

To use Stokes' Theorem we have to come up with a boundary curve $C = \partial S$ for this surface. But the intersection of the sphere with the cylinder will be a circle, and that circle is found (by merging the two equations) at $1 + z^2 = 4$, or $z = \sqrt{3}$. Further, the circle of intersection is cut out of the sphere by the cylinder, and so the circle of intersection is a unit circle centered on the z-axis and parallel to the xy-plane — see Fig. 18.20. Because we are going to evaluate a line integral around $C = \mathbf{r}(t)$, we need to ensure C is positively oriented, regardless of what's happening with the sphere and cylinder. We would be evaluating the same line integral for any intersection of two surfaces which produces C as the bounding contour of the intersection. All together, the boundary curve ∂S can be expressed parametrically as

$$\begin{aligned} x &= \cos t && \to dx = -\sin t dt \\ y &= \sin t && \to dy = \cos t dt \\ z &= \sqrt{3} && \to dz = 0 \end{aligned}$$

for $0 \leq t \leq 2\pi$. With these equations, we produce $d\mathbf{r} = \langle -\sin t, \cos t, 0\rangle\, dt$. By Stokes Theorem we convert the surface integral into a line integral. In the process of forming the line integral, we plug the parametric forms of x, y and z into the vector field $\mathbf{F}$:

$$\begin{aligned}\iint_S (\nabla \times \mathbf{F}) \cdot d\mathbf{S} &= \int_{\partial S} \mathbf{F} \cdot d\mathbf{r} = \int_0^{2\pi} \langle xz, yz, xy\rangle \cdot \langle -\sin t, \cos t, 0\rangle\, dt \\ &= \int_0^{2\pi} \langle \sqrt{3}\cos t, \sqrt{3}\sin t, \cos t \sin t\rangle \cdot \langle -\sin t, \cos t, 0\rangle\, dt \\ &= \int_0^{2\pi} \left(-\sqrt{3}\sin t \cos t + \sqrt{3}\sin t \cos t + 0\right)\, dt = 0 \quad \blacksquare\end{aligned}$$

EX 5 Evaluate $\iint_S \nabla \times \mathbf{F} \cdot d\mathbf{S}$ for $\mathbf{F}(x, y, z) = \langle x^2yz, yz^2, z^2e^{xy}\rangle$ and S is the part of the sphere $x^2 + y^2 + z^2 = 5$ that lies above $z = 1$ (and S is positively oriented).

I'm glad we have Stokes' Theorem available, because I really don't want to compute the curl of that vector field. To use Stokes' Theorem we have to come up with a boundary curve $C = \partial S$ for this surface. But the intersection of the sphere with the $z = 1$ is the circle $x^2 + y^2 = 4$. Therefore, the boundary curve ∂S can be expressed parametrically as

$$\begin{aligned} x &= 2\cos t &&\rightarrow dx = -2\sin t\, dt \\ y &= 2\sin t &&\rightarrow dy = 2\cos t\, dt \\ z &= 1 &&\rightarrow dz = 0 \end{aligned}$$

for $0 \leq t \leq 2\pi$. With these equations, we produce $d\mathbf{r} = \langle -2\sin t, \cos t, 0\rangle\, dt$. By Stokes Theorem we convert the surface integral into a line integral. In the process of forming the line intergal, we plug the parametric forms of x, y and z into the vector field $\mathbf{F}$:

$$\begin{aligned}\iint_S (\nabla \times \mathbf{F}) \cdot d\mathbf{S} &= \int_{\partial S} \mathbf{F} \cdot d\mathbf{r} = \int_0^{2\pi} \langle x^2yz, yz^2, z^2e^{xy}\rangle \cdot \langle -2\sin t, 2\cos t, 0\rangle\, dt \\ &= \int_0^{2\pi} \langle 8\cos^2 t \sin t, 2\sin t, e^{4\cos t \sin t}\rangle \cdot \langle -\sin t, \cos t, 0\rangle\, dt \\ &= \int_0^{2\pi} -8\cos^2 t \sin^2 t + 2\sin t \cos t + 0\, dt = -2\pi \quad \blacksquare\end{aligned}$$

You Try It

(2) Compute $\iint_S (\nabla \times \mathbf{F}) \cdot d\mathbf{S}$ for the vector field $\mathbf{F} = \langle x^2e^{yz}, y^2e^{xz}, z^2e^{xy}\rangle$ where S is the hemisphere $x^2 + y^2 + z^2 = 4$ for $z \geq 0$.

(3) Compute $\iint_S \nabla \times \mathbf{F} \cdot d\mathbf{S}$ for the vector field $\mathbf{F} = \langle x^2, y^2, z^2\rangle$ where S is the unit cube with opposite corners at $(0, 0, 0)$ and $(1, 1, 1)$.

Stokes' Theorem — Problem List

Stokes' Theorem — You Try It

These appeared above; solutions begin on the next page.

(1) Compute $\int_C \mathbf{F} \cdot d\mathbf{r}$ for the vector field $\mathbf{F} = \langle x^2, y^2, x^2 + y^2\rangle$ where C is the lower half of the hemisphere $x^2 + y^2 + z^2 = 1$ capped by the xy-plane.

(2) Compute $\int_C \mathbf{F} \cdot d\mathbf{r}$ for the vector field $\mathbf{F} = \langle x + y^2, y + z^2, z + x^2\rangle$ where C is the triangle with vertices (2,0,0), (0,1,0), and (0,0,2).

(3) Compute $\iint_S (\nabla \times \mathbf{F}) \cdot d\mathbf{S}$ for the vector field $\mathbf{F} = \langle x^2e^{yz}, y^2e^{xz}, z^2e^{xy}\rangle$ where S is the hemisphere $x^2 + y^2 + z^2 = 4$ for $z \geq 0$.

(4) Compute $\iint_S \nabla \times \mathbf{F} \cdot d\mathbf{S}$ for the vector field $\mathbf{F} = \langle x^2, y^2, z^2\rangle$ where S is the unit cube with opposite corners at $(0, 0, 0)$ and $(1, 1, 1)$.

Stokes' Theorem — Practice Problems

Try these as you get the hang of the You Try It problems. Solutions to these problems are available in Sec. B.6.6.

(1) Compute $\int_C \mathbf{F} \cdot d\mathbf{r}$ for the vector field $\mathbf{F} = \langle yz, 2xz, e^{xy}\rangle$ where C is the boundary of the cylinder $x^2 + y^2 = 16$ at $z = 5$.

(2) Compute $\iint_S (\nabla \times \mathbf{F}) \cdot d\mathbf{S}$ for the vector field $\mathbf{F} = \langle yz, xz, xy\rangle$ where S is the paraboloid $z = 9 - x^2 - y^2$ above $z = 5$.

(3) Compute $\int_C \mathbf{F} \cdot d\mathbf{r}$ for the vector field $\mathbf{F} = \langle e^{-x}, e^x, e^z\rangle$ where C is the boundary of the plane $2x + y + 2z = 2$ in the first octant, traversed counterclockwise.

(4) Find the work done by the vector field $\mathbf{F} = \langle x^2y^3z, \sin(xyz), xyz\rangle$ around the bounding contour of the cone $y^2 = x^2 + z^2$ between $y = 0$ and $y = 3$ with normal vectors oriented outwards.

Stokes' Theorem — Challenge Problems

Try these problems to test your skills with the ideas in this section. Solutions to these problems are available in Sec. C.6.6.

(1) Compute $\int_C \mathbf{F} \cdot d\mathbf{r}$ for the vector field $\mathbf{F} = \langle (1+y)z, (1+z)x, (1+x)y\rangle$ where C is the boundary of the plane $2x+2y+z = 4$ in the first octant, traversed counterclockwise.

(2) Compute $\iint_S (\nabla \times \mathbf{F}) \cdot d\mathbf{S}$ for the vector field $\mathbf{F} = \langle z^2, -3xy, x^3y^3\rangle$ where S is the top part of the inverted paraboloid $z = 5 - x^2 - y^2$ above the plane $z = 1$, oriented positively.

(3) The integral $\oint_C \mathbf{F} \cdot d\mathbf{r}$ gives the *circulation* of the vector field $\mathbf{F}$ around the contour C. Let g be the function $g(x, y, z) = xe^z \sin(y)$ and let $\mathbf{F}(x, y, z) = \nabla g$. Let the contour C be the intersection of the plane $x + y + z = 5$ and the cylinder $x^2 + y^2 = 11$, oriented positively. What is the total circulation due to $\mathbf{F}$ around C? (Hint: There's an easy way, and there's a hard way...)

Stokes' Theorem — You Try It — Solved

---

(1) Compute $\int_C \mathbf{F} \cdot d\mathbf{r}$ for the vector field $\mathbf{F} = \langle x^2, y^2, x^2 + y^2 \rangle$ where C is the lower half of the hemisphere $x^2 + y^2 + z^2 = 1$ capped by the xy-plane.

□ Our surface is $z = -\sqrt{1 - x^2 - y^2}$, with bounding contour ∂S being is the unit circle in the xy-plane, oriented clockwise. From the equation of the surface, we have

$$\frac{\partial g}{\partial x} = \frac{x}{\sqrt{1 - x^2 - y^2}} \quad \text{and} \quad \frac{\partial g}{\partial y} = \frac{y}{\sqrt{1 - x^2 - y^2}}$$

Let's build the integrand of (18.4) before creating the whole integral; this may help determine the best coordinate system.

$$\begin{aligned} -P\frac{\partial g}{\partial x} - Q\frac{\partial g}{\partial y} + R &= -(2y) \cdot \frac{x}{\sqrt{1 - x^2 - y^2}} - (-2x) \cdot \frac{y}{\sqrt{1 - x^2 - y^2}} + 0 \\ &= \frac{-2xy}{\sqrt{1 - x^2 - y^2}} + \frac{2xy}{\sqrt{1 - x^2 - y^2}} = \frac{-2xy + 2xy}{\sqrt{1 - x^2 - y^2}} = 0 \end{aligned}$$

So it turns out we don't need to choose a coordinate system! For the given conditions,

$$\oint_C \mathbf{F} \cdot d\mathbf{r} = \iint D \left(-P\frac{\partial g}{\partial x} - Q\frac{\partial g}{\partial y} + R \right) dA = \iint D\,(0)\, dA = 0 \quad \blacksquare$$

---

(2) Compute $\oint_C \mathbf{F} \cdot d\mathbf{r}$ for the vector field $\mathbf{F} = \langle x + y^2, y + z^2, z + x^2 \rangle$ where C is the triangle with vertices (2,0,0), (0,1,0), and (0,0,2).

□ The contour is also known as the boundary of the portion of the plane $x + 2y + z = 2$ in the first octant. We'll use Stokes' Theorem to convert this to a surface integral involving $\nabla \times \mathbf{F}$, which is $\nabla \times \mathbf{F} = \langle -2z, -2x, -2y \rangle$ (details omitted).

Matching the surface to the form $z = g(x, y)$ we have $g = 2 - x - 2y$ and

$$\frac{\partial g}{\partial x} = -1 \quad ; \quad \frac{\partial g}{\partial y} = -2$$

Matching to the form $\nabla \times \mathbf{F} = \langle P, Q, R \rangle$, we have

$$P = -2z = -2(2 - x - 2y) = -4 + 2x + 4y \quad ; \quad Q = -2x \quad ; \quad R = -2y$$

Using Eq. (18.4) to leapfrog from the line integral to a double integral, and then apply the information we just collected. Let's build the integrand first:

$$-P\frac{\partial g}{\partial x} - Q\frac{\partial g}{\partial y} + R = (4-2x-4y)(-1) + 2x(-2) - 2y$$
$$= -4 + 2x + 4y - 4x - 2y = -4 - 2x + 2y$$

The region of integration D will be the area under the line $x + 2y = 2$ in the first quadrant.

$$\oint_C \mathbf{F}\cdot d\mathbf{r} = \iint S\nabla\times\mathbf{F}\cdot d\mathbf{S} = \iint D\left(-P\frac{\partial g}{\partial x} - Q\frac{\partial g}{\partial y} + R\right)\,dA$$
$$= \iint D\,(-4-2x+2y)\,dA$$
$$= \int_0^1\int_0^{2-2y}(2y-2x-4)\,dxdy = -\frac{14}{3} \quad \blacksquare$$

---

(3) Compute $\iint_S \nabla\times\mathbf{F}\cdot d\mathbf{S}$ for the vector field $\mathbf{F} = \langle x^2e^{yz}, y^2e^{xz}, z^2e^{xy}\rangle$ where S is the hemisphere $x^2+y^2+z^2 = 4$ for $z \geq 0$.

☐ The boundary curve ∂S of this surface is the circle $x^2+y^2 = 4$ (with $z = 0$) and can be expressed as

$$x = 2\cos t \quad \to dx = -2\sin t\,dt$$
$$y = 2\sin t \quad \to dy = 2\cos t\,dt$$
$$z = 0 \quad \to dz = 0$$

for $0 \leq t \leq 2\pi$. By Stokes' Theorem we convert the surface integral into a line integral:

$$\iint_S(\nabla\times\mathbf{F})\cdot d\mathbf{S} = \oint_{\partial S}\mathbf{F}\cdot d\mathbf{r}$$
$$= \oint_{\partial S}(x^2e^{yz})\,dx + (y^2e^{xz})\,dy + (z^2e^{xy})\,dz$$
$$= \int_0^{2\pi}(2\cos t)^2e^0(-2\sin t\,dt) + (2\sin t)^2e^0(2\cos t\,dt) + 0$$
$$= \int_0^{2\pi}(8\sin^2 t\cos t - 8\cos^2 t\sin t)\,dt$$
$$= \left(\frac{8}{3}\sin^3 t + \frac{8}{3}\cos^3 t\right)\Bigg|_0^{2\pi} = 0 \quad \blacksquare$$

(4) Compute $\iint_S \nabla \times \mathbf{F} \cdot d\mathbf{S}$ for the vector field $\mathbf{F} = \langle x^2, y^2, z^2 \rangle$ where S is the unit cube with opposite corners at $(0,0,0)$ and $(1,1,1)$.

☐ The boundary curve ∂S of this surface is the square at $z = 1$ with corners at $(0,0,1)$, $(1,0,1)$, $(1,1,1)$, and $(0,1,1)$. To set this up with parametric equations, we need to do one edge at a time; while this requires four sets of equations, it's still better than a surface integral over a surface with six sides! Fortunately, the edges of ∂S are simple:

$$\begin{aligned}
L1\,(0,0,1) \to (1,0,1):\quad & x = t, y = 0, z = 1\,(0 \le t \le 1)\\
& dx = dt, dy = 0, dz = 0\\
L2\,(1,0,1) \to (1,1,1):\quad & x = 1, y = t, z = 1\,(0 \le t \le 1)\\
& dx = 0, dy = dt, dz = 0\\
L3\,(1,1,1) \to (0,1,1):\quad & x = 1-t, y = 1, z = 1\,(0 \le t \le 1)\\
& dx = -dt, dy = 0, dz = 0\\
L4\,(0,1,1) \to (0,0,0):\quad & x = 0, y = 1-t, z = 1\,(0 \le t \le 1)\\
& dx = 0, dy = -dt, dz = 0
\end{aligned}$$

By Stokes' Theorem we convert the surface integral into a line integral:

$$\iint_S (\nabla \times \mathbf{F}) \cdot d\mathbf{S} = \oint_{\partial S} \mathbf{F} \cdot d\mathbf{r}$$

which in turn must be split into four parts; in each part, we pull together information from above into the appropriate $d\mathbf{r} = \langle dx, dy, dx \rangle$:

$$\oint_{\partial S} \mathbf{F} \cdot d\mathbf{r} = \int_{L1} \mathbf{F} \cdot d\mathbf{r}_1 + \ldots + \int_{L4} \mathbf{F} \cdot d\mathbf{r}_4$$

where

$$\begin{aligned}
\int_{L1} \mathbf{F} \cdot d\mathbf{r}_1 &= \int_{L1} \langle t^2, 0^2, 1^2 \rangle \cdot \langle dt, 0, 0 \rangle = \int_0^1 t^2\, dt = \frac{1}{3}\\
\int_{L2} \mathbf{F} \cdot d\mathbf{r}_2 &= \int_{L2} \langle 0^2, t^2, 1^2 \rangle \cdot \langle 0, dt, 0 \rangle = \int_0^1 t^2\, dt = \frac{1}{3}\\
\int_{L3} \mathbf{F} \cdot d\mathbf{r}_3 &= \int_{L3} \langle (1-t)^2, 1^2, 1^2 \rangle \cdot \langle -dt, 0, 0 \rangle = -\int_0^1 (1-t)^2\, dt = -\frac{1}{3}\\
\int_{L4} \mathbf{F} \cdot d\mathbf{r}_4 &= \int_{L4} \langle 0, (1-t)^2, 1^2 \rangle \cdot \langle 0, -dt, 0 \rangle = -\int_0^1 (1-t)^2\, dt = -\frac{1}{3}
\end{aligned}$$

And so

$$\oint_{\partial S} \mathbf{F} \cdot d\mathbf{r} = \frac{1}{3} + \frac{1}{3} - \frac{1}{3} - \frac{1}{3} = 0$$

Wow, that was a lot of work for nothing! ■

Appendix A

Conics, Quadrics, and Parametrics

A.1 Conic Sections

Conic Sections

Let's break down the phrase "conic section" by focusing on the first part, *conic*. Imagine two sugar cones (you know, the ice cream cones with the circular top and pointy bottoms that are much yummier than those mealy cake cones) that are touching pointy end to pointy end, with the circular tops pointing in opposite directions. This is a full cone that has both halves. Now imagine this full cone is a bit saggy, so that the cross sections are actually ellipses rather than perfect circles. When we make flat slices through this cone, we generate conic sections. A slice through the cone parallel to one of the open ends makes an ellipse. A diagonal slice that only intersects one half of the cone makes a parabola. A slice that goes up and down through both halves makes a hyperbola (one half of the hyperbola comes from each half of the cone).

When we draw these cross sections in the 2D coordinate plane, we get the familiar conic sections; all conic sections are somehow buried in the expression

$$ax^2 + by^2 + cx + dy + e = 0 \tag{A.1}$$

Let's go through each conic section, see the general form of its equation, and the parametric equations that can also generate it. The very scattered details here are not intended to be comprehensive, but just enough to remind you of what's going on — and, frankly, to be a resource for me as I generate lots of the figures in this text!

Lines

Technically, lines are conic sections. We get lines from (A.1) when $a = 0$ and $b = 0$, leaving us with the familiar $cx + dy + e = 0$.

Line segments can be generated parametrically by

$$\begin{cases} x = c_1 t + c_2 \\ y = c_3 t + c_4 \end{cases} \quad t_1 < t < t_2$$

If we set $t_1 = -\infty$ and $t_2 = \infty$, then we get the entire line. Figure A.1 shows the line segment generated by

$$\begin{cases} x = 4t - 2 \\ y = \ \ t + 3 \end{cases} \quad -2 < t < 3$$

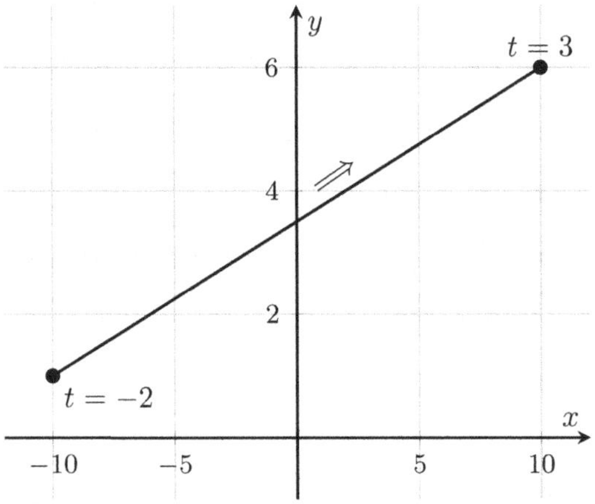

Fig. A.1 The (directed) parametric curve $x = 4t - 2, y = t + 3$ for $-2 \le t \le 3$.

Ellipses

Of course, this category also includes circles. We get ellipses from (A.1) when a and b have the same sign. Assuming $a > 0$ and $b > 0$, we have $ax^2 + by^2 + cx + dy + e = 0$. (If $a < 0$ and $b < 0$ then we can just multiply both sides of (A.1) by -1.) After completing the square and renaming constants, we get this into the form $A(x - x_0)^2 + B(y - y_0)^2 = C$, or,

$$\frac{(x - x_0)^2}{D^2} + \frac{(y - y_0)^2}{E^2} = 1 \tag{A.2}$$

In the case that $a = b$ in the original format, we'd get to

$$(x - x_0)^2 + (y - y_0)^2 = R^2$$

which is the familiar equation of a circle with center (x_0, y_0) and radius R.

Let's keep it simple and restrict ourselves to the case where $x_0 = 0$ and $y_0 = 0$. In that case, an ellipse (or circle) in the form (A.2) reduces to

$$\frac{x^2}{D^2} + \frac{y^2}{E^2} = 1 \tag{A.3}$$

and we can represent this ellipse (or circle, when $D = E$) parametrically by:

$$\begin{cases} x = D\cos\theta \\ y = E\sin\theta \end{cases} \quad 0 < \theta < 2\pi$$

(When we use these to eliminate the parameter in (A.3), the expression reduces to $\cos^2\theta + \sin^2\theta = 1$.) There are other possible parametric forms as well. Figure A.2 shows the ellipse generated by

$$\begin{cases} x = 4\cos\theta \\ y = \frac{3}{2}\sin\theta \end{cases} \quad 0 < \theta < 2\pi$$

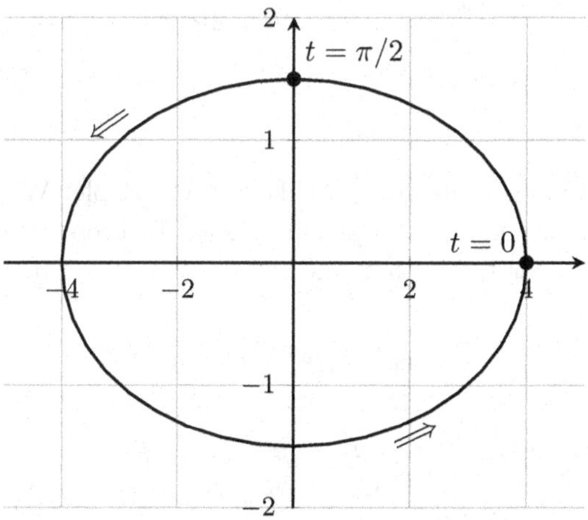

Fig. A.2 The (directed) parametric curve $x = 4\cos\theta, y = \frac{3}{2}\sin\theta$.

Parabolas

We get parabolas from (A.1) when either $a = 0$ or $b = 0$. Again, to keep things simple, we'll restrict ourselves to parabolas that are either horizontal or vertical in the xy-plane. The common forms of these are either

$$(x-h)^2 = 4p(y-k) \qquad \text{or} \qquad (y-k)^2 = 4p(x-h)$$

where the vertex is located at (h, k) in both. The left hand case represents the "usual" case of a parabola opening up or down; the latter case represents a parabola opening either left or right. The parameter p has to do with how wide the parabola is, and if you want to follow that trail, go read about the "directrix" of a parabola.

The best way to represent a parabola parametrically is to solve it for the variable that's not squared, and apply parametrics directly. For example, if we rearrange the equation of a parabola opening upwards or downwards as $y = f(x)$, then we represent the parabola as

$$\begin{cases} x = t \\ y = f(t) \end{cases} \qquad -\infty < t < \infty$$

(or restrict t for only a portion of the parabola). The "sideways" parabola $x = g(y)$ would look similar, with $y = t$ and $x = g(t)$. Figure A.3 shows the parabola generated by

$$\begin{cases} x = t \\ y = -2t^2 + \frac{3}{2} \end{cases} \qquad -2 < t < 2$$

Hyperbolas

These are the conic sections we *really* like to keep simple. We get hyperbolas from (A.1) when a and b have opposite signs. To keep things simple, let's ignore the first order terms in x and y ($c = d = 0$); that will lead to the form

$$Ax^2 - By^2 + C = 0$$

or

$$\frac{x^2}{D^2} - \frac{y^2}{E^2} = 1 \tag{A.4}$$

We can represent this hyperbola parametrically by:

$$\begin{cases} x = D\sec\theta \\ y = E\tan\theta \end{cases} \qquad \theta_1 < \theta < \theta_2$$

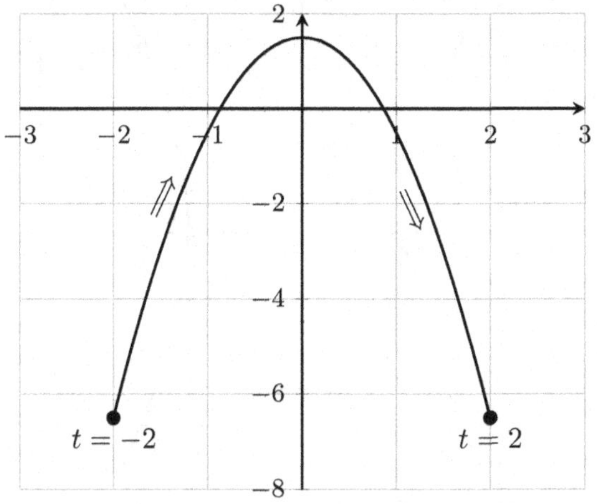

Fig. A.3 The (directed) parametric curve $x = t, y = -2t^3 + \frac{3}{2}$.

(When we use these to eliminate the parameter in (A.4), the expression reduces to $\sec^2\theta - \tan^2\theta = 1$.) There are other possible parametric forms as well. Figure A.4 shows the hyperbola generated by

$$\begin{cases} x = 2\sec\theta \\ y = 4\tan\theta \end{cases} \quad -1 < t < 1$$

These parametric equations only generate half of the hyperbola (the solid curve). We have to use symmetry if we want to see the other half (dashed):

$$\begin{cases} x = -2\sec\theta \\ y = \ \ 4\tan\theta \end{cases} \quad -1 < t < 1$$

A.2 Quadric Surfaces

Quadric Surfaces

If you are here because of Sec. 13.3, please just look at the discussions of rectangular coordinates associated with quadric surfaces. The parts related to parametric representation won't make sense until you've visited Chapter 15.

In Appendix A.1, we saw how conic sections are all somehow buried in the expression

$$Ax^2 + By^2 + Cx + Dy + E = 0$$

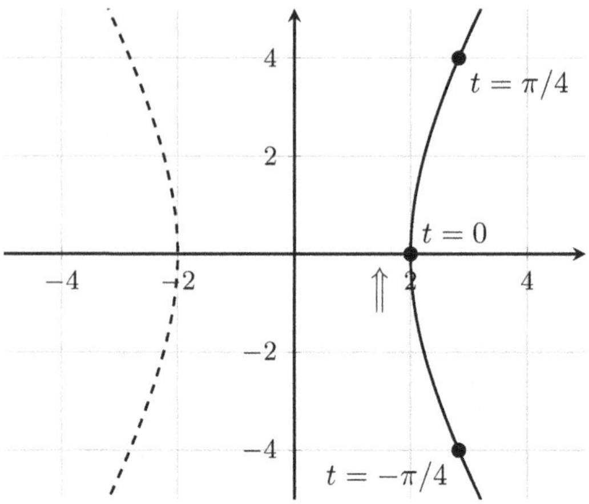

Fig. A.4 The (directed) parametric curve $x = 2\sec\theta, y = 4\tan\theta$.

When we extend that one more dimension by adding the coordinate z, we get

$$Ax^2 + By^2 + Cz^2 + Dx + Ey + Fz + G = 0 \tag{A.5}$$

Many three-dimensional surfaces can be generated by variations of (A.5); here, we'll see the "greatest hits". There will be some representative images here, but you are encouraged to go play with the resources you have and make plots of your own: change the magnitude of a constant, or switch the sign of a term to see what the effects are. The web site Wolfram Alpha is a good resource and is fairly intuitive to use. For example, typing in "`plot x^2 - y^2 + z^2 = 1`" will get you started on hyperboloids (see below.)

Try not to giggle as you see that the names of many 3D surfaces end in "-oid", making them sound like medical afflictions.

Planes

We get planes from (A.5) when $A = 0$, $B = 0$, and $C = 0$, leaving us with $Dx + Ey + Fz + G = 0$.

To build a parametric representation of a plane, we need two vectors in the plane. These could come from three points on the plane, $P = (p_x, p_y, p_x)$, $Q = (q_x, q_y, q_z)$, and $R = (r_x, r_y, r_z)$, with the two vectors being **PQ** and **RQ**. Then the full plane is traced out by the vector $\mathbf{v} = s \cdot \mathbf{PQ} + t \cdot \mathbf{RQ}$, where the parameters s and t are any real numbers. In this case, we have:

$$\mathbf{v} = [s(p_x - q_x) + t(r_x - q_x), s(p_y - q_y) + t(r_y - q_y), s(p_z - q_z) + t(r_z - q_z)]$$

or

$$\begin{cases} x = (p_x - q_x)s + (r_x - q_x)t \\ y = (p_y - q_y)s + (r_y - q_y)t \\ z = (p_z - q_z)s + (r_z - q_z)t \end{cases} \quad -\infty < s < \infty, \quad -\infty < t < \infty$$

Figure A.5 shows the plane containing the points $P(1, 2, 3)$, $Q(0, -3, 2)$, and $R(-2, 4, 3)$.

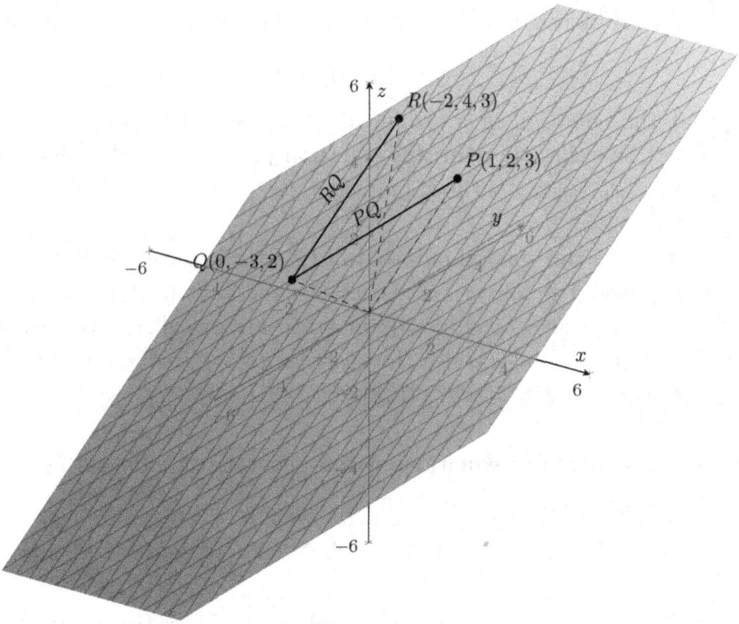

Fig. A.5 Plane with points $P(1, 2, 3)$, $Q(0, -3, 2)$, $R(-2, 4, 3)$.

Ellipsoids / Spheres

From this point forwards, we are going to simplify things by treating only surfaces that have not been translated — so that the origin remains the anchor point. Some simple translations might be noted on a case-by-case basis.

The category of "ellipsoids" also includes spheres. Equation (A.5) gives an ellipsoid when it can be rearranged to

$$\frac{x^2}{a^2} + \frac{y^2}{b^2} + \frac{z^2}{c^2} = 1 \tag{A.6}$$

In the case that $a = b = c$, that common equal denominator can be renamed r^2, and we get

$$x^2 + y^2 + z^2 = r^2$$

which is the familiar equation of a sphere with radius r centered at the origin. This shape has an easy translation, and we can place the center at some (x_0, y_0, z_0) like so:

$$(x - x_0)^2 + (y - y_0)^2 + (z - z_0)^2 = r^2$$

Recall that the equation of a graphical object states what all the points on the object have in common; this equation just collects all points whose distance from (x_0, y_0, z_0) is the same value, r.

When we have the general ellipsoid in (A.6), the direction in which the axis of the ellipsoid is longer can be configured by varying a, b, c. FFT: Given that traces of ellipsoids in both horizontal and vertical cross sections are ellipses, can you use those traces to locate the longest axis of an ellipsoid in which $c > b > a$?

One set of parametric equations which generate an ellipsoid are

$$\begin{cases} x = a\cos t \sin s \\ y = b \sin t \sin s \\ z = \quad c\cos s \end{cases} \quad \text{with} \quad 0 \le s < \pi \quad ; \quad 0 \le t < 2\pi$$

These should remind you a lot of spherical coordinates; the parameter t is related to the rotational angle that might otherwise be named θ, and s is related to the azimuthal angle also known as ϕ.

Figure A.6 shows the ellipsoid generated by parametric equations

$$\begin{cases} x = \quad \cos s \sin t \\ y = 1.25 \sin s \sin t \\ z = \quad 1.5 \cos s \end{cases} \quad \text{with} \quad 0 \le s < \pi \quad ; \quad 0 \le t < 2\pi$$

which, in rectangular form, is

$$x^2 + \frac{y^2}{1.25^2} + \frac{z^2}{1.5^2} = 1$$

Is this figure consistent with your digestion of the "Food For Thought" given above?

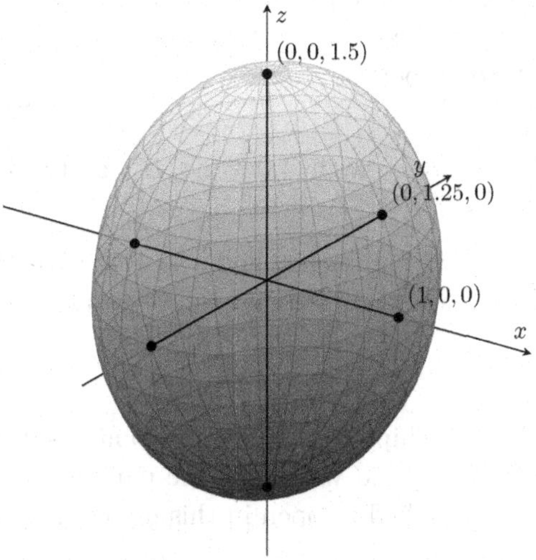

Fig. A.6 Sample ellipsoid with $a = 1$, $b = 1.25$, $c = 1.5$.

(Elliptical) Paraboloids

We get paraboloids from (A.5) when one of a, b, or c is zero. In that case, (A.5) can be reorganized — through completing the square and shifting constants — into forms like

$$z = ax^2 + by^2 \quad \text{or} \quad y = ax^2 + cz^2 \quad \text{or} \quad x = by^2 + cz^2 \tag{A.7}$$

Each version dictates a different direction in which the paraboloid opens. The former, where we have $z = f(x, y)$ is a paraboloid opening upwards, and traces of such a paraboloid would be ellipses in a horizontal cross sections, or parabolas in vertical cross sections. A paraboloid in which traces are ellptical are called an *elliptic paraboloid* (I know, crazy, right?).

This shape is easy to translate so that the vertex is at any (x_0, y_0, z_0):

$$z - z_0 = a(x - x_0)^2 + b(y - y_0)^2$$
$$y - y_0 = a(x - x_0)^2 + c(z - z_0)^2$$
$$x - x_0 = b(y - y_0)^2 + c(z - z_0)^2$$

In order to form parametric equations of a/an (elliptic) paraboloid, we need to decide how much of the paraboloid to show; this is fixed by stating its "height" H. Then, an example set of parametric equations for an upright elliptic paraboloid would be:

$$\begin{cases} x = a\sqrt{s}\cos t \\ y = b\sqrt{s}\sin t \\ z = \quad s \end{cases} \quad \text{with} \quad 0 \le s \le H \quad ; \quad 0 \le t < 2\pi$$

Note that t is an angular parameter, and s is a linear distance. Reconstituted, these equations would correspond to the parabola

$$z = \frac{x^2}{a^2} + \frac{y^2}{b^2}$$

Figure A.7 shows an ellipsoid paraboloid opening sideways around the y-axis. FFT: How could we shuffle the parametric equations shown above to allow the paraboloid to open in this direction?

`https://mathworld.wolfram.com/EllipticParaboloid.html`

Hyperboloid of One Sheet

The name implies that there is a hyperboloid of two sheets, and there is (that's next). We get hyperboloids of one sheet from (A.5) when one of a, b, or c has a different sign than the others, such that we can reduces it to the form

$$\frac{x^2}{a^2} + \frac{y^2}{b^2} - \frac{z^2}{c^2} = 1 \tag{A.8}$$

This is one of three possible arrangements; the axis of the variable whose term is negative is the centerline around which the hyperboloid opens. The hyperboloid represented by (A.8) will be centered on the z-axis. When $a \neq b$, we can call this an elliptic hyperboloid of one sheet, since traces perpendicular to the centerline will be ellipses; if $a = b$, those traces are circles. Traces in the other two coordinate directions will be hyperbolas.

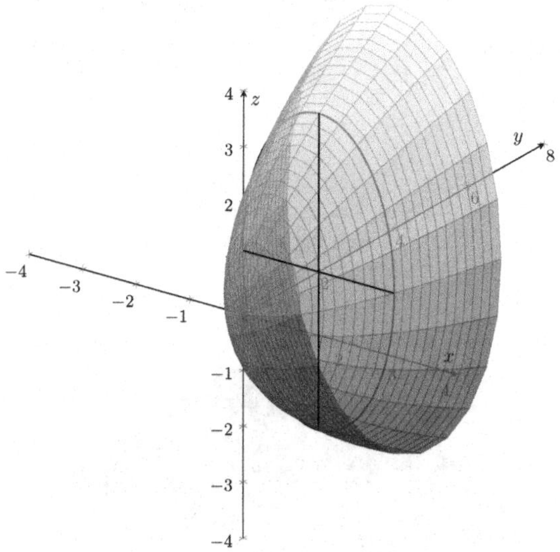

Fig. A.7 An elliptic paraboloid opening around the positive y-axis.

A hyperboloid of one sheet can be represented parametrically by:

$$\begin{cases} x = a\sqrt{1+s^2}\cos t \\ y = b\sqrt{1+s^2}\sin t \\ z = \quad cs \end{cases} \quad \text{with} \quad 0 \le s \le H \quad ; \quad 0 \le t < 2\pi$$

These equations would have to be rearranged if the hyperboloid was centered on the x- or y-axis. Figure A.8 shows a unit ($a = b = c = 1$) hyperboloid of one sheet opening sideways around the z-axis. In the event $a \neq b$, we can call this an elliptic hyperboloid, since the traces perpendicular to the centerline are ellipses.

https://mathworld.wolfram.com/EllipticHyperboloid.html

Hyperboloid of Two Sheets

This shape is what you get if you create a surface of revolution with a regular 2D hyperbola, by rotating both halves of the hyperbola around its centerline. Hyperboloids of two sheets come from (A.5) when we allow a, b, or c to have different signs, such that we can get

$$\frac{x^2}{a^2} + \frac{y^2}{b^2} - \frac{z^2}{c^2} = -1 \qquad (A.9)$$

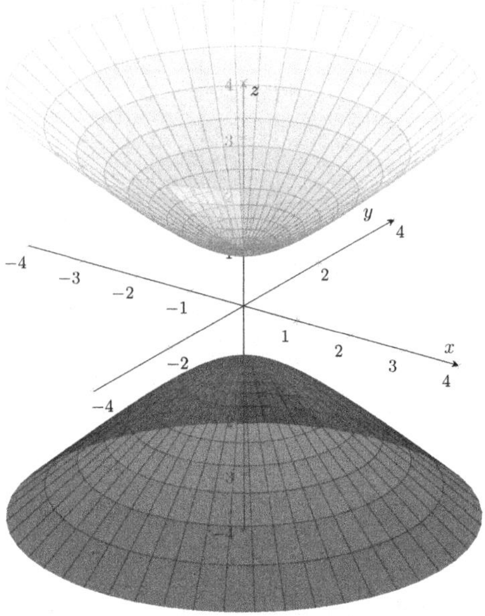

Fig. A.8 A elliptic hyperboloid of two sheets.

Note the right hand side of −1; the difference between (A.9) and (A.8) is that when the right side is −1, there will be z values for which the equation is undefined — thus breaking the "nuclear cooling tower" shape of the hyperboloid of one sheet into two parts.

Having set this up to compare (A.9) to (A.7), it's worth noting that we can also arrange this to have a right side of 1:

$$\frac{z^2}{c^2} - \frac{x^2}{a^2} - \frac{y^2}{b^2} = 1 \tag{A.10}$$

When given a quadric equation in general form (A.5), we can simplify until the right side is a constant of 1 — then, if two quadratic terms on the left are negative, we have a hyperboloid of two sheets; if only one is negative, we have a hyperboloid of one sheet.

A hyperboloid of two sheets can be represented parametrically by:

$$\begin{cases} x = a\sinh s\cos t \\ y = b\sinh s\sin t \\ z = \ \pm c\cosh s \end{cases} \quad \text{with } -\infty < s < \infty \quad ; \quad 0 \leq t < \pi$$

(As usual, these equations would have to be rearranged if the hyperboloid was centered on the x- or y-axis.) Because $\cosh(u)$ is always positive, we must repeat the generation of the parametric surface twice — once with $+c$ and once with $-c$ — to get both halves of the hyperboloid. Figure A.9 shows a unit ($a = b = c = 1$) hyperboloid of two sheets opening around the z-axis.

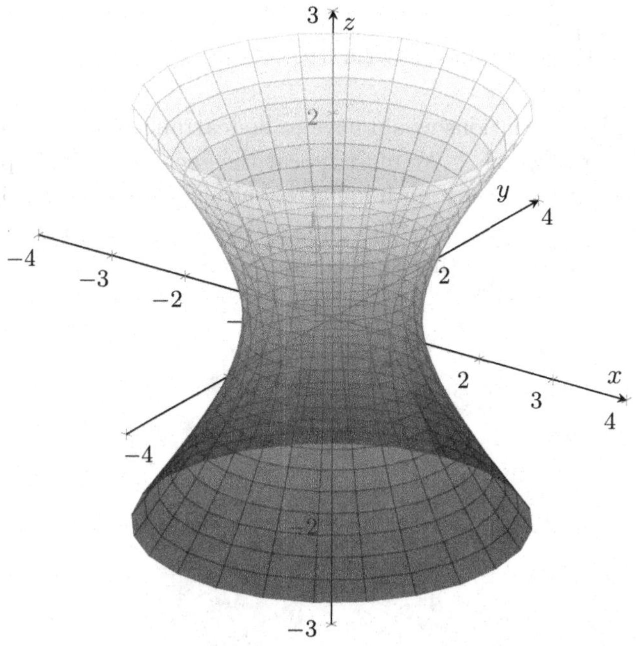

Fig. A.9 A elliptic hyperboloid of one sheet.

https://mathworld.wolfram.com/Two-SheetedHyperboloid.html

(Elliptical) Cone

Cones are generated by (A.5) when we can arrange it to eliminate the free constant term G, and get

$$z^2 = \frac{x^2}{a^2} + \frac{y^2}{b^2} \tag{A.11}$$

As written, this equation generates two symmetric halves of the cone joined at the vertex because we can solve for both $z = +\sqrt{f(x,y)}$ and $z = -\sqrt{f(x,y)}$. As written, (A.11) gives a cone centered on the z-axis, in which horizontal cross sections would reveal either elliptical ($a \neq b$) or circular ($a = b$) cross sections. FFT: How do we realigning the cone along the other axes?

One half of an (elliptical) cone can be represented parametrically by:

$$\begin{cases} x = as\cos t \\ y = bs\sin t \qquad \text{with } 0 < s < \infty \quad ; \quad 0 \leq t < \pi \\ z = \quad s \end{cases}$$

(As usual, these equations would have to be rearranged if the cone was centered on the x or y-axis.) We must repeat the generation of the parametric surface again for $z = -u$ if both halves of the cone are needed. Figure A.10 shows the cone $z^2 = x^2/4 + y^2/9$. Can you design the parametric equations I had to use to generate that surface graphically?

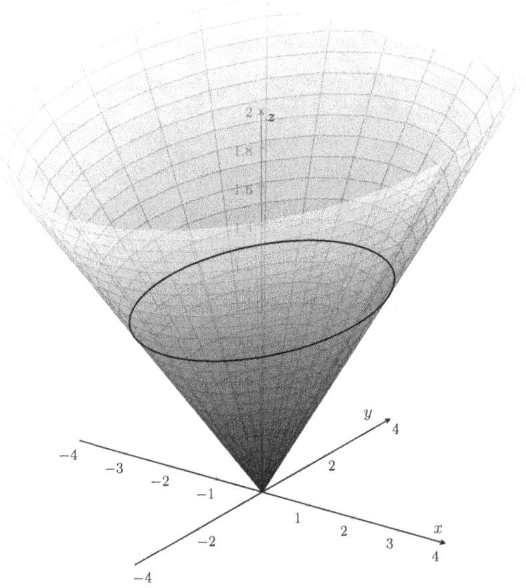

Fig. A.10 An elliptic cone opening around the positive z-axis.

https://mathworld.wolfram.com/Cone.html

Appendix B

Solutions to All Practice Problems

B.1 Chapter 13: Practice Problem Solutions

B.1.1 *Life in Three-Dimensions — Practice — Solved*

(1) What region is described by the expression $y \geq 0$ in 2D and 3D?

□ In 2D, $y \geq 0$ describes the half-plane of Quadrants 1 and 2, everything on and above the x-axis. In 3D, this region is a half space, everything on and "above" the xz-plane. ■

(2) What region is described by the expression $1 < x^2 + y^2 + z^2 \leq 25$

□ This region is between two concentric spheres: outside the unit sphere, but on and inside a sphere of radius 5 centered at the origin. ■

(3) Is the triangle defined by the points $P(1,1,0)$, $Q(2,4,1)$, and $R(-1,-1,3)$ a right triangle?

□ Let's find the squares of the lengths of the sides:

$$|PQ|^2 = (2-1)^2 + (4-1)^2 + (1-0)^2 = 11$$
$$|QR|^2 = (-1-2)^2 + (-1-4)^2 + (3-1)^2 = 38$$
$$|PR|^2 = (-1-1)^2 + (-1-1)^2 + (3-0)^2 = 17$$

Since no arrangement of the lengths of these sides follows $c^2 = a^2 + b^2$, the triangle is not a right triangle. ■

B.1.2 *Multivariable Functions — Practice — Solved*

---

(1) For $f(x, y) = \ln(x + y - 1)$, find $f(e, 1)$ and $f(4x, x + 1)$, and describe the domain and range of $f(x, y)$.

□ We have $f(e, 1) = \ln(e + 1 - 1) = \ln(e) = 1$. Also, $f(4x, x + 1) = \ln(4x + (x + 1) - 1) = \ln(5x)$. For the domain of $f(x, y)$, all (x, y) values are allowed except those that make $x + y - 1 \leq 0$. That is, we must have $x + y > 1$, i.e. $y > 1 - x$. This is everything above the line $y = 1 - x$ in the xy-plane. Since all real numbers are possible as output from the natural log function, the range is all reals, $\mathbb{R}$. ■

---

(2) Describe the domain and range of $f(x, y) = \sqrt[3]{x} + \sqrt[4]{y}$.

□ In this function x can be anything, but y must be non-negative. So the domain is, in fancy set notation, $\{(x, y) : x \in \mathbb{R}, y \geq 0\}$. This is all of Quadrants 1 and 2 in the xy-plane, including the x-axis. The range of this function is all reals (the range of the cube root is all reals). ■

---

(3) Describe the domain and range of $f(x, y, z) = \ln(16 - 4x^2 - 4y^2 - z^2)$.

□ We must have $16 - 4x^2 - 4y^2 - z^2 > 0$, i.e. $4x^2 + 4y^2 + z^2 < 16$. This almost looks like a sphere, but it isn't (the coefficients of x, y and z are not all the same). It turns out that this is the interior of an ellipsoid, which we'll learn about in the next topic. The range of this function is $(-\infty, \ln 16]$. ■

---

(4) Find the equations of the line segment starting at the point $(2, 5, -1)$ and ending at $(3, 3, 0)$.

□ How about $x = 2 + t, y = 5 - 2t, z = -1 + t$ for $0 \leq t \leq 1$? ■

---

(5) Describe the curve given by the parametric equations $x = 3 - t, y = 1 + 2t, z = -t$ for $-1 \leq t \leq 1$.

□ At $t = -1$, we have the point $(4, -1, 1)$ and at $t = 1$ we have the point $(2, 3, -1)$. All functions are linear in t. Therefore, this is the line segment from the point $(4, -1, 1)$ to the point $(2, 3, -1)$. ■

---

(6) Do the lines $x = 1+2t, y = 2-t, z = -t$ and $x = -s, y = 1-2s, z = 1+s$ share a point?

□ We can make the x and y coordinates match by forcing $1 + 2t = -s$ and $2 - t = 1 - 2s$. From the first equation, we have $s = -1 - 2t$. Plugging into the second, we get $2 - t = 1 - 2(-1 - 2t)$, or $t = -1/5$ ... and so also $s = -3/5$. Passing these values to the z coordinates we get (1) $z = -1/5$ and (2) $z = 1 - 3/5 = 2/5$. The z coordinate can not be made the same while the x and y coordinates are also the same. The lines do not share a point. ■

B.1.3 *3D Surfaces — Practice — Solved*

(1) Find the equation of the plane containing the points $(1, 0, 1)$, $(0, 1, 1)$ and $(1, 1, 0)$.

□ Plugging each point into the general form for a plane and simplifying, we get:

$$\begin{aligned} a(1) + b(0) + c(1) + d = 0 \quad &\rightarrow \quad a + c + d = 0 \\ a(0) + b(1) + c(1) + d = 0 \quad &\rightarrow \quad b + c + d = 0 \\ a(1) + b(1) + c(0) + d = 0 \quad &\rightarrow \quad a + b + d = 0 \end{aligned}$$

We have 3 equations, but 4 unknowns. Let's let d be the "extra" unknown, and choose it to be a convenient value — such as -1. With $d = -1$, we then get the three equations

$$\begin{aligned} a + c &= 1 \\ b + c &= 1 \\ a + b &= 1 \end{aligned}$$

The solution to this system is $a = 1/2$, $b = 1/2$, and $c = 1/2$, so that the equation of the plane becomes

$$\frac{1}{2}x + \frac{1}{2}y + \frac{1}{2}z - 1 = 0$$

or multiplying by 2,

$$x + y + z - 2 = 0$$

This can also be rewritten $z = -x - y + 2$. ■

(2) Identify the surface $x^2 + y^2 + z^2 = 4x - 2y$. Give at least two pieces of identifying information that distinguishes this surface from others of the same type.

□ Completing the square on the given equation,

$$\begin{aligned} x^2 + y^2 + z^2 &= 4x - 2y \\ (x^2 - 4x + (-2)^2) + (y^2 + 2y + (1)^2) + z^2 &= 0 + (-2)^2 + (1)^2 \\ (x^2 - 4x + 4) + (y^2 + 2y + 1) + z^2 &= 4 + 1 \\ (x - 2)^2 + (y + 1)^2 + (z - 0)^2 &= 5 \end{aligned}$$

So this is a sphere with center $(2, -1, 0)$ and radius $\sqrt{5}$. ■

(3) Find the equation of a sphere that has center $(3, 8, 1)$ and passes through the point $(4, 3, -1)$.

□ A sphere with center $(3, 8, 1)$ that passes through $(4, 3, -1)$ has a radius equal to the distance between those points. In fact, we'll only need r^2, so:

$$\begin{aligned} r^2 &= (x_2 - x_1)^2 + (y_2 - y_1)^2 + (z_2 - z_1)^2 \\ &= (4-3)^2 + (3-8)^2 + (-1-1)^2 \\ &= 30 \end{aligned}$$

So the sphere's equation is

$$(x-3)^2 + (y-8)^2 + (z-1)^2 = 30 \quad \blacksquare$$

(4) Describe the 3D surface $x^2 - y^2 = 1$. Give identifying information that supports your description.

□ The surface $x^2 - y^2 = 1$ is a hyperbola in the xy-plane. Then z is unrestricted, so we get an hyperbolic cylinder extending / opening in the z-direction. ■

(5) Identify the surface $z = x^2 - y^2$ using traces. Provide at least one trace parallel to each each of the xy-, yz-, and xz-planes.

□ In the planes $x = 0$, $y = 0$, and $z = 0$, we get traces $z = -y^2$, $z = x^2$ and $x^2 - y^2 = 0$, which are two parabolas and the pair of lines $y = \pm x$. In any plane $x = c$ we get $z = -y^2 + c^2$ which is a parabola. In any plane $y = c$ we get $z = x^2 - c^2$ which is another parabola. In any plane $z = c$ we get $x^2 - y^2 = c$ which is a hyperbola. This combination of parabolic and hyperbolic traces (cross sections) indicates a hyperbolic paraboloid. ■

(6) Complete the squares on $4x^2 + y^2 + 4z^2 - 4y - 24z + 36 = 0$ and identify it using traces.

□ If we group each variable and complete the squares, we get

$$\begin{aligned} 4x^2 + y^2 + 4z^2 - 4y - 24z + 36 &= 0 \\ 4x^2 + (y^2 - 4y + 4) + 4(z^2 - 6z + 9) &= -36 + 4 + 36 \\ 4x^2 + (y-2)^2 + 4(z-3)^2 &= 4 \\ x^2 + \frac{(y-2)^2}{4} + (z-3)^2 &= 1 \end{aligned}$$

Traces in all directions (set $x = c$, $y = c$, $z = c$) are all ellipses, so this is an ellipsoid. However, the center is $(0, 2, 3)$ rather than $(0, 0, 0)$. ■

---

(7) Identify the surface $z^2 = 4x^2 + 9y^2 + 36$ using traces. Provide at least one trace parallel to each each of the xy-, yz-, and xz-planes.

□

- In any plane such that $z^2 < 36$, this surface does not exist.
- In any plane such that $z^2 > 36$, traces are ellipses.
- In the two planes $z = 6$ and $z = -6$, this surface is only a point, $(0, 0, \pm 6)$.
- In any plane $x = c$ or $y = c$, traces are hyperbolas

With elliptical and hyperbolic cross sections, and with the surface not existing at all for $-6 < z < 6$, we must have a hyperboloid of two sheets; it opens along the z-axis. ■

---

(8) Does the contour plot in Fig. B.1 represent $z = \sin(x)\cos(y)$ or $z = \sin(y)\cos(x)$? Why?

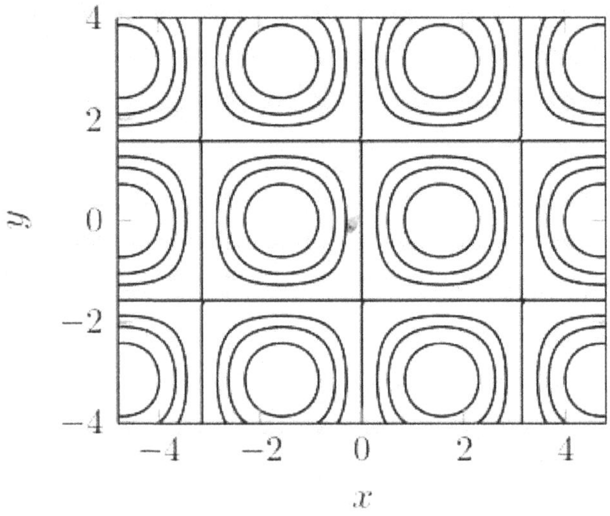

Fig. B.1 Contours for $z = \sin x \cos y$ or $z = \sin y \cos x$?

□ Contours of the function $z = \sin(x)\cos(y)$ are of the form $\sin(x)\cos(y) = c$. We would generate the contour corresponding to $c = 0$ at every x-coordinate that is a multiple of π and every y-coordinate that is an odd multiple of $\frac{\pi}{2}$. Contours of the function $z = \sin(y)\cos(x)$ are of the form $\sin(y)\cos(x) = c$. We would generate the contour corresponding to $c = 0$ at every x-coordinate that is an odd multiple of $\pi/2$ and every y-coordinate that is a multiple of π. In the figure, we see a contour forming at $x = -\pi, 0, \pi$ and $y = -\pi/2, \pi/2$. This would be the contour $c = 0$ and we have the first case, the plot represents $z = \sin(x)\cos(y)$. ■

(9) Present the level curves of $z = 3x^2 - 2y^2$ for $z = -2, -1, 0, 1, 2$ as a 2D contour plot.

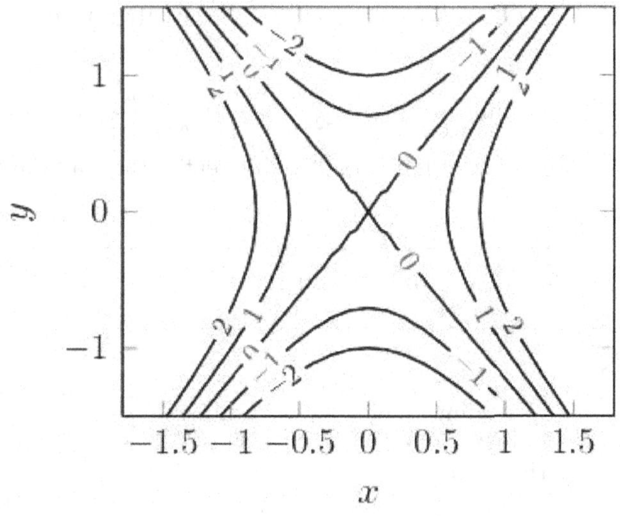

Fig. B.2 Some level curves of $z = 3x^2 - 2y^2$ (w/ PP 9).

□ Each level curve for $z = \pm 1, \pm 2$ is of the form $3x^2 - 2y^2 = c$ fpr $c \neq 0$, and so is a hyperbola. For $z = 0$, the level curve is a pair of lines, $3x^2 = 2y^2$ or $y = \pm\sqrt{3/2}x$. These level curves are shown in Fig. B.2. ■

B.1.4 *Limits of MV Functions — Practice — Solved*

---

(1) Investigate the limit $\displaystyle\lim_{(x,y)\to(0,0)} \frac{xy}{\cos(xy)}$.

☐ This is an "easy one". As $(x, y) \to (0, 0)$, we have $xy \to 0$ and so the overall limit becomes

$$\lim_{(x,y)\to(0,0)} \frac{xy}{\cos(xy)} = \lim_{(x,y)\to(0,0)} \frac{0}{\cos(0)} = \lim_{(x,y)\to(0,0)} \frac{0}{1} = 0 \quad \blacksquare$$

---

(2) Investigate the limit $\displaystyle\lim_{(x,y)\to(0,0)} \frac{x^2 + \sin^2 y}{2x^2 + y^2}$.

☐ Along the path $x = 0$,

$$\lim_{(x,y)\to(0,0)} \frac{x^2 + \sin^2 y}{2x^2 + y^2} = \lim_{(x,y)\to(0,0)} \frac{\sin^2 y}{y^2} = \lim_{(x,y)\to(0,0)} \left(\frac{\sin y}{y}\right)^2 = 1$$

(Hopefully you remember that as y approaches 0, the function $\sin y/y$ approaches 1. Or, you can invoke L-Hopital's Rule.) Along the path $y = 0$,

$$\lim_{(x,y)\to(0,0)} \frac{x^2 + \sin^2 y}{2x^2 + y^2} = \lim_{(x,y)\to(0,0)} \frac{x^2}{2x^2} = \frac{1}{2}$$

So we get two different limits on two different paths, and the limit does not exist. ■

---

(3) Investigate the limit $\displaystyle\lim_{(x,y)\to(0,0)} \frac{xy^4}{x^2 + y^8}$.

☐ Along the path $x = 0$,

$$\lim_{(x,y)\to(0,0)} \frac{xy^4}{x^2 + y^8} = \lim_{(x,y)\to(0,0)} \frac{0}{y^8} = 0$$

Along the path $x = y^4$,

$$\lim_{(x,y)\to(0,0)} \frac{xy^4}{x^2 + y^8} = \lim_{(x,y)\to(0,0)} \frac{y^4 y^4}{(y^4)^2 + y^8} = \lim_{(x,y)\to(0,0)} \frac{y^8}{2y^8} = \frac{1}{2}$$

So we get two different limits on two different paths, and the limit does not exist. ■

---

(4) Where is the function $f(x, y) = \dfrac{x - y}{1 + x^2 + y^2}$ continuous?

☐ There are no locations where this function "goes bad". It is continuous everywhere in the xy-plane. ■

B.1.5 *Partial Derivatives — Practice — Solved*

(1) Find both first order derivatives of $z = y \ln x$.

$$\square \quad \frac{\partial z}{\partial x} = \frac{y}{x} \qquad \text{and} \qquad \frac{\partial z}{\partial y} = \ln x \quad \blacksquare$$

(2) Find both first order derivatives of $u = te^{w/t}$.

$$\square \quad u_w = t\left(\frac{1}{t}\right)e^{w/t} = e^{w/t}$$

$$u_t = e^{w/t} + t\left(\frac{-w}{t^2}\right)e^{w/t} = e^{w/t}\left(1 - \frac{w}{t}\right) \quad \blacksquare$$

(3) Find all first order derivatives of $f(x,y,z) = x^2e^{yz}$.

$$\square \quad f_x = 2xe^{yz} \quad ; \quad f_y = x^2ze^{yz} \quad ; \quad f_z = x^2ye^{yz} \quad \blacksquare$$

(4) Find all first order derivatives of $w = \sqrt{r^2 + s^2 + t^2}$.

□ There are three first order derivatives:

$$\frac{\partial w}{\partial r} = \frac{r}{\sqrt{r^2 + s^2 + t^2}}$$

$$\frac{\partial w}{\partial s} = \frac{s}{\sqrt{r^2 + s^2 + t^2}}$$

$$\frac{\partial w}{\partial t} = \frac{t}{\sqrt{r^2 + s^2 + t^2}} \quad \blacksquare$$

(5) If $f(x,y) = \sin(2x + 3y)$, what is $f_y(-6,4)$?

□ Finding the derivative f_y and plugging in $(-6,4)$:

$$f_y = 3\cos(2x + 3y) \qquad \rightarrow \qquad f_y(-6,4) = 3\cos(-12 + 12) = 3 \quad \blacksquare$$

(6) Find all second order derivatives of $f(x,y) = \ln(3x + 5y)$.

□ Starting with first derivatives,

$$f_x = \frac{3}{3x + 5y} \qquad \text{and} \qquad f_y = \frac{5}{3x + 5y}$$

so that

$$f_{xx} = \frac{-9}{(3x + 5y)^2} \,;\; f_{xy} = f_{yx} = \frac{-15}{(3x + 5y)^2} \,;\; f_{yy} = \frac{-25}{(3x + 5y)^2} \quad \blacksquare$$

(7) If $f(r,s,t) = r\ln(rs^2t^3)$, what are f_{rss} and f_{rst}?

☐ It's easier if we expand the function using properties of the natural log:

$$\begin{aligned} f(r,s,t) &= r\ln r + 2r\ln s + 3r\ln t \\ f_r &= \ln r + 1 + 2\ln s + 3\ln t \\ f_{rs} &= \frac{2}{s} \end{aligned}$$

$$f_{rss} = -\frac{2}{s^2} \quad ; \quad f_{rst} = 0 \quad \blacksquare$$

(8) Does $u = x^2 - y^2$ satisfy Laplace's Equation?

☐ We have $u_{xx} = 2$ and $u_{yy} = -2$, so $u_{xx} + uyy = 0$, and the function DOES satisfy Laplace's equation. ■

(9) For $z = 3x^2 - 2y^2$, compute $z_x(1,0)$, $z_y(1,0)$, $z_x(1,1)$, $z_y(1,1)$, $z_x(0,1)$ and $z_y(0,1)$. In PP 9 of Sec. 13.3, you presented some level curves of this function. On which level curves (specified by associated value of z) will we find the points $(1,0)$, $(1,1)$, and $(0,1)$?

☐ Since $z_x = 6x$, then $z_x(1,0) = 6$, $z_x(1,1) = 6$, and $z_x(0,1) = 0$.

Since $z_y = -4y$, then $z_y(1,0) = 0$, $z_y(1,1) = -4$, and $z_y(0,1) = -4$.

The point $(1,0)$ falls on the level curve for $z = 3$; the point $(1,1)$ falls on the level curve for $z = 1$; the point $(0,1)$ falls on the level curve for $z = -2$. The level curve for $z = 3$ was not originally displayed in PP 9 of Sec. 13.3, but it has been added to Fig. B.3 along with these the points given here. ■

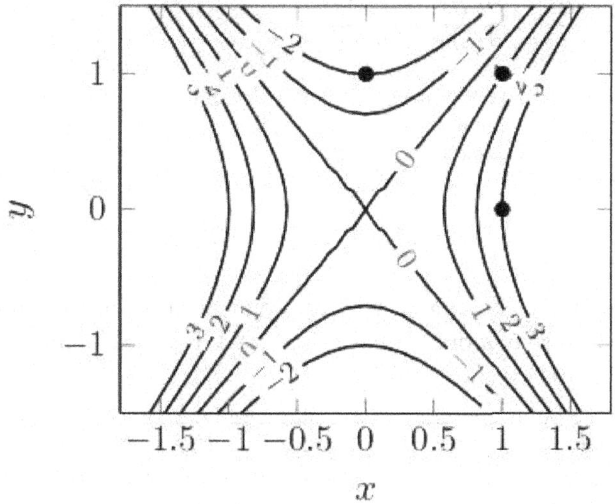

Fig. B.3 Some level curves of $z = 3x^2 - 2y^2$ (w/ PP 9).

B.2 Chapter 14: Practice Problem Solutions

B.2.1 *The Chain Rule — Practice — Solved*

---

(1) If $z = f(x(t), y(t)) = \sqrt{x^2 + y^2}$ with $x(t) = e^{2t}$ and $y(t) = e^{-2t}$, what is $\dfrac{\partial z}{\partial t}$?

☐ z depends on x and y, which in turn depend on t. So the proper chain rule and its completion are:

$$\begin{aligned}\frac{\partial z}{\partial t} &= \frac{\partial z}{\partial x}\frac{\partial x}{\partial t} + \frac{\partial z}{\partial y}\frac{\partial y}{\partial t} \\ &= \frac{x}{\sqrt{x^2+y^2}}(2e^{2t}) + \frac{y}{\sqrt{x^2+y^2}}(-2e^{-2t}) \\ &= \frac{2}{\sqrt{x^2+y^2}}(xe^{2t} - ye^{-2t}) \quad \blacksquare\end{aligned}$$

---

(2) If $z = f(x(s,t), y(s,t)) = e^{xy}\tan y$ with $x(s,t) = s + 2t$ and $y(s,t) = \dfrac{s}{t}$, find both $\partial z/\partial s$ and $\partial z/\partial t$.

□
$$\frac{\partial z}{\partial s} = \frac{\partial z}{\partial x}\frac{\partial x}{\partial s} + \frac{\partial z}{\partial y}\frac{\partial y}{\partial s}$$
$$= (ye^{xy}\tan y)(1) + (xe^{xy}\tan y + e^{xy}\sec^2 y)\left(\frac{1}{t}\right)$$
$$\frac{\partial z}{\partial t} = \frac{\partial z}{\partial x}\frac{\partial x}{\partial t} + \frac{\partial z}{\partial y}\frac{\partial y}{\partial t}$$
$$= (ye^{xy}\tan y)(2) + (xe^{xy}\tan y + e^{xy}\sec^2 y)\left(\frac{-s}{t^2}\right)$$

This is a great example of why we don't always want the final answer written in terms of only s and t, and it's OK to leave in the intermediate variables. ■

---

(3) If $w = w(x, y, z)$ with $x = x(t, u)$, $y = y(t, u)$, and $z = z(t, u)$, write chain rule expressions for all possible first derivatives of w with respect to the independent variables.

□ There will be two first derivatives of w; these derivatives are with respect to t, u going through the intermediate variables of x, y, z. The chain rule formulations are:

$$\frac{\partial w}{\partial t} = \frac{\partial w}{\partial x}\frac{\partial x}{\partial t} + \frac{\partial w}{\partial y}\frac{\partial y}{\partial t} + \frac{\partial w}{\partial z}\frac{\partial z}{\partial t}$$
$$\frac{\partial w}{\partial u} = \frac{\partial w}{\partial x}\frac{\partial x}{\partial u} + \frac{\partial w}{\partial y}\frac{\partial y}{\partial u} + \frac{\partial w}{\partial z}\frac{\partial z}{\partial u} \quad ■$$

---

(4) If $u = \sqrt{r^2 + s^2}$ where $r = y + x\cos t$ and $s = x + y\sin t$, find the values of all possible first derivatives of u with respect to the independent variables when $x = 1, y = 2, t = 0$.

□ There are three first partials with respect to x, y, t, going through the intermediate variables of r and s. We want them for $x = 1, y = 2, t = 0$. First, note that for these values we have

$$r = 2 + 1\cos(0) = 3 \qquad \text{and} \qquad s = 1 + 2\sin(0) = 1$$

so that $\sqrt{r^2 + s^2} = \sqrt{10}$, and we get the following derivative values:

$$\frac{\partial u}{\partial r} = \frac{r}{\sqrt{r^2 + s^2}} = \frac{3}{\sqrt{10}} \qquad \text{and} \qquad \frac{\partial u}{\partial s} = \frac{s}{\sqrt{r^2 + s^2}} = \frac{1}{\sqrt{10}}$$

Further, derivatives of r with respect to x, y, and t are:

$$\frac{\partial r}{\partial x} = \cos(t) = 1 \quad ; \quad \frac{\partial r}{\partial y} = 1 \quad ; \quad \frac{\partial r}{\partial t} = -x\sin(t) = 0$$

and for s,

$$\frac{\partial s}{\partial x} = 1 \quad ; \quad \frac{\partial s}{\partial y} = \sin(t) = 0 \quad ; \quad \frac{\partial s}{\partial t} = y\cos(t) = 2$$

By the Chain Rule, then,

$$\begin{aligned}
\frac{\partial u}{\partial x} &= \frac{\partial u}{\partial r}\frac{\partial r}{\partial x} + \frac{\partial u}{\partial s}\frac{\partial s}{\partial x} \\
&= \frac{3}{\sqrt{10}}(1) + \frac{1}{\sqrt{10}}(1) = \frac{4}{\sqrt{10}} \\
\frac{\partial u}{\partial y} &= \frac{\partial u}{\partial r}\frac{\partial r}{\partial y} + \frac{\partial u}{\partial s}\frac{\partial s}{\partial y} \\
&= \frac{3}{\sqrt{10}}(1) + \frac{1}{\sqrt{10}}(0) = \frac{3}{\sqrt{10}} \\
\frac{\partial u}{\partial t} &= \frac{\partial u}{\partial r}\frac{\partial r}{\partial t} + \frac{\partial u}{\partial s}\frac{\partial s}{\partial t} \\
&= \frac{3}{\sqrt{10}}(0) + \frac{1}{\sqrt{10}}(2) = \frac{2}{\sqrt{10}} \quad \blacksquare
\end{aligned}$$

---

(5) If $z = f(\alpha(s,t), \beta(s,t)) = \sin\alpha\tan\beta$ with $\alpha(s,t) = 3s + t$ and $\beta(s,t) = s - t$, find the first derivatives of z with respect to the independent variables.

□ We have that z is ultimately a function of s and t, going through the intermediate variables α and β. So,

$$\begin{aligned}
\frac{\partial z}{\partial s} &= \frac{\partial z}{\partial \alpha}\frac{\partial \alpha}{\partial s} + \frac{\partial z}{\partial \beta}\frac{\partial \beta}{\partial s} \\
&= (\cos\alpha\tan\beta)(3) + (\sin\alpha\sec^2\beta)(1) = 3\cos\alpha\tan\beta + \sin\alpha\sec^2\beta \\
\frac{\partial z}{\partial t} &= \frac{\partial z}{\partial \alpha}\frac{\partial \alpha}{\partial t} + \frac{\partial z}{\partial \beta}\frac{\partial \beta}{\partial t} \\
&= (\cos\alpha\tan\beta)(1) + (\sin\alpha\sec^2\beta)(-1) = \cos\alpha\tan\beta - \sin\alpha\sec^2\beta \quad \blacksquare
\end{aligned}$$

---

(6) If $u = u(s,t)$ with $s = s(w,x,y,z)$ and $t = t(w,x,y,z)$, write chain rule expressions for all possible first derivatives of u with respect to the independent variables.

□ There will be four first derivatives of u; these derivatives are with respect to w, x, y, z going through the intermediate variables of s, t.

The chain rule formulations are:

$$\frac{\partial u}{\partial w} = \frac{\partial u}{\partial s}\frac{\partial s}{\partial w} + \frac{\partial u}{\partial t}\frac{\partial t}{\partial w}$$
$$\frac{\partial u}{\partial x} = \frac{\partial u}{\partial s}\frac{\partial s}{\partial x} + \frac{\partial u}{\partial t}\frac{\partial t}{\partial x}$$
$$\frac{\partial u}{\partial y} = \frac{\partial u}{\partial s}\frac{\partial s}{\partial y} + \frac{\partial u}{\partial t}\frac{\partial t}{\partial y}$$
$$\frac{\partial u}{\partial z} = \frac{\partial u}{\partial s}\frac{\partial s}{\partial z} + \frac{\partial u}{\partial t}\frac{\partial t}{\partial z} \quad \blacksquare$$

---

(7) If $R = \ln(u^2 + v^2 + w^2)$ where $u = x + 2y$, $v = 2x - y$, and $w = 2xy$, find the values of all possible first derivatives of R with respect to x and y when $x = 1, y = 1$.

□ There are two first partials with respect to x, y, going through the intermediate variables u, v, w. We want them for $x = 1, y = 1$. First, note that for these values we have

$$u = 1 + 2(1) = 3$$
$$v = 2(1) - 1 = 1$$
$$w = 2(1)(1) = 2$$

So with these values,

$$\begin{aligned}
\frac{\partial R}{\partial x} &= \frac{\partial R}{\partial u}\frac{\partial u}{\partial x} + \frac{\partial R}{\partial v}\frac{\partial v}{\partial x} + \frac{\partial R}{\partial w}\frac{\partial w}{\partial x} \\
&= \frac{2u}{u^2+v^2+w^2}(1) + \frac{2v}{u^2+v^2+w^2}(2) + \frac{2w}{u^2+v^2+w^2}(2y) \\
&= \frac{6}{14}(1) + \frac{4}{14}(2) + \frac{4}{14}(2) \quad = \quad \frac{18}{14} \quad = \quad \frac{9}{7}
\end{aligned}$$

$$\begin{aligned}
\frac{\partial R}{\partial y} &= \frac{\partial R}{\partial u}\frac{\partial u}{\partial y} + \frac{\partial R}{\partial v}\frac{\partial v}{\partial y} + \frac{\partial R}{\partial w}\frac{\partial w}{\partial y} \\
&= \frac{2u}{u^2+v^2+w^2}(2) + \frac{2v}{u^2+v^2+w^2}(-1) + \frac{2w}{u^2+v^2+w^2}(2x) \\
&= \frac{6}{14}(2) + \frac{4}{14}(-1) + \frac{4}{14}(2) \quad = \quad \frac{18}{14} \quad = \quad \frac{9}{7} \quad \blacksquare
\end{aligned}$$

B.2.2 *Optimization — Practice — Solved*

(1) Find and characterize the critical points of $f(x, y) = x^3y + 12x^2 - 8y$.

□ We have $f_x = 3x^2y + 24x$ and $f_y = x^3 - 8$. The equation $f_y = 0$ is easier to solve, and we have $f_y = 0$ only at $x = 2$. Handing that back to $f_x = 0$ gives us $3(2)^2y + 24(2) = 0$, or $12y = -48$, or $y = -4$. So, the point $(x, y) = (2, -4)$ is the only critical point. Getting ready for the second derivative test via Useful Fact 14.1, we have:

$$f_{xx} = 6xy + 24 \to f_{xx}(2,-4) = -24$$
$$f_{yy} = 0 \to f_{yy}(2,-4) = 0$$
$$f_{xy} = 3x^2 \to f_{xy}(2,-4) = 12$$
$$D(2,-4) = f_{xx}(2,-4)f_{yy}(2,-4) - [f_{xy}(2,-4)]^2$$
$$(-24)(0) - (12)^2 = -144$$

Since $D(2,-4) < 0$, the critical point is a saddle point. ■

(2) Find and characterize the critical points of $f(x, y) = xy(1 - x - y) = xy - x^2y - xy^2$.

□ We have $f_x = y - 2xy - y^2 = y(1 - 2x - y)$ and $f_y = x - x^2 - 2xy = x(1 - x - 2y)$. Let's start unpacking $f_x = 0$ and $f_y = 0$.

We have $f_x = 0$ when $x = 0$ and when $y = 1 - 2x$. We have $f_y = 0$ when $y = 0$ and $x = 1 - 2y$. There are three combinations here:

- One combination is when $x = 0$ and $y = 0$.
- A second combination is when $x = 0$ and so $y = 1 - 2(0) = 1$.
- A third combination is when $y = 0$ and so $x = 1 - 2(0) = 1$.

Altogether, $f_x = 0$ and $f_y = 0$ together at the points $(0, 0), (0, 1), (1, 0)$. And, we also must look for simultaneous solutions to $f_x = 0$ and $f_y = 0$ via

$$1 - 2x - y = 0$$
$$1 - x - 2y = 0$$

Twice the first minus the second gives $1 - 3x = 0$, or $x = 1/3$. That value leads to $y = 1/3$ as well. So $(x, y) = (1/3, 1/3)$ is another critical point. There are four critical points altogether. Getting ready for the second derivative test via Useful Fact 14.1, we have:

$$f_{xx} = -2y \qquad f_{yy} = -2x \qquad f_{xy} = 1 - 2x - 2y$$

and in general,

$$D(x,y) = f_{xx}(x,y)f_{yy}(x,y) - [f_{xy}(x,y)]^2 = 4xy - (1-2x-2y)^2$$

- $D(0,0) = -1 < 0$ so $(0,0)$ locates a saddle point.
- $D(0,1) = -1 < 0$ so $(0,1)$ locates a saddle point.
- $D(1,0) = -1 < 0$ so $(1,0)$ locates a saddle point.
- $D(1/3,1/3) = 1/3 > 0$; since also $f_{xx} < 0$ here, $(1/3,1/3)$ locates a local max. ■

---

(3) Find the absolute extremes of $f(x,y) = 3+xy-x-2y$ over the domain D that's a triangle with vertices P(1,0), Q(5,0) and R(1,4).

☐ Are there extremes over the interior of D? With $f_x = y-1$ and $f_y = x-2$, we have $f_x = f_y = 0$ at (2,1) which is indeed in the interior of D.

Let edge L1 be the line segment PR. On L1, $x = 1$ and the function reduces to $f(1,y) = 2-y$. Since this is linear, there are no extremes along L1 itself.

Let edge L2 be the line segment PQ. On L2, $y = 0$ and the function reduces to $f(x,0) = 3-x$. Since this is linear, there are no extremes along L2 itself.

Let edge L3 be the line segment QR. This is the line $y = -x+5$, and the function reduces to

$$f_{L3}(x) = 3 + x(-x+5) - x - 2(-x+5) = -x^2 + 6x - 7$$

which has a critical point at $x = 3$ (and so $y = 2$).

At the two critical points we've found so far, we have

$$f(2,1) = 1 \quad ; \quad f(3,2) = 2$$

Then also at the vertices:

$$f(1,0) = 2 \quad ; \quad f(1,4) = -2 \quad ; \quad f(5,0) = -2$$

Comparing the five candidate locations for absolute extremes (two critical points and three vertices), we see absolute maximums of $f(3,2) = f(1,0) = 2$ and absolute minimums of $f(1,4) = f(5,0) = -2$. ■

(4) Find the minimum distance d from the point (1,2,3) to the plane $x - y + z = 4$ and the point on the plane where this distance occurs.

☐ To make the calculations easier, we'll minimize d^2 (whatever minimizes d^2 also minimizes d). The distance between (1,2,3) and any point in the universe is our objective function, which is to be minimized:

$$d^2 = (x-1)^2 + (y-2)^2 + (z-3)^2$$

Our constraint is that we are only interested in points on the plane $x-y+z=4$, i.e. for which $z = 4-x+y$. This constrains our objective function down to two independent variables:

$$d^2 = (x-1)^2 + (y-2)^2 + (1-x+y)^2$$

Treating the right hand side as the function $f(x,y)$ we want to minimize, let's get some derivatives:

$$\begin{aligned} f_x &= 2(x-1) - 2(1-x+y) = 4x - 2y - 4 \\ f_y &= 2(y-2) + 2(1-x+y) = -2x + 4y - 2 \end{aligned}$$

Now $f_x = 0$ gives $4x - 2y - 4 = 0$ or $y = 2x - 2$. Also, $f_y = 0$ gives $-2x + 4y - 2 = 0$, or $-x + 2y - 1 = 0$. Merging that with $y = 2x - 2$ gives

$$\begin{aligned} -x + 2(2x-2) - 1 &= 0 \\ 3x - 5 &= 0 \\ x &= \frac{5}{3} \end{aligned}$$

Passing that back to $y = 2x - 2$ gives $y = 4/3$. So our critical point is located by $(x,y) = (5/3, 4/3)$. To confirm this is a minimum (although we don't really need to, because what would be the maximum?), we head for the second derivative test:

$$f_{xx} = 4 \quad ; \quad f_{yy} = 4 \quad ; \quad f_{xy} = -2$$

and use them to build:

$$D(x,y) = f_{xx}f_{yy} - [f(x,y)]^2 = 12$$

Since $D(x,y) > 0$ and $f_{xx} > 0$ everywhere, any critical point is a local minimum. (See, I told you the critical point is a minimum.) Finally, we get that the point on $z = 4 - x + y$ with $(x,y) = (5/3, 4/3)$ gives $z = 11/3$ and so the point on the plane $x-y+z=4$ closest to the point

(1,2,3) is $(x, y, z) = (5/3, 4/3, 11/3)$. The actual minimum distance is

$$d_{min} = \sqrt{\left(\frac{5}{3} - 1\right)^2 + \left(\frac{4}{3} - 2\right)^2 + \left(\frac{11}{3} - 3\right)^2} = \frac{2}{\sqrt{3}} \quad \blacksquare$$

---

(5) What is the largest volume that can be contained in a rectangular box (with a lid!) that has a total surface area of $64\,cm^2$.

□ Let the dimensions of the box be x, y, z (length, width, height). Our objecive function, which we want to maximize, is for volume, V: $V = xyz$. Our constraint comes from the surface area which must be a fixed value. Since we have a lid, there are two faces of each possible size xy, yz, and xz, and the total surface area fixed to 64 is given by $2xy + 2xz + 2yz = 64$, or $xy + xz + yz = 32$. We will use information about the surface area to constrain V to only two independent variables. Let's solve $xy + xz + yz = 32$ for z,

$$z = \frac{32 - xy}{x + y}$$

hand that to the volume $V = xyz$, and find derivatives:

$$\begin{aligned} V = xyz &= xy\left(\frac{32 - xy}{x + y}\right) \\ V_x &= \frac{-y^2(x^2 + 2xy - 32)}{(x + y)^2} \\ V_y &= \frac{-x^2(y^2 + 2xy - 32)}{(x + y)^2} \end{aligned}$$

Now, immediately we can see we'll have $V_x = 0$ and $V_y = 0$ when $x = 0$ and $y = 0$, but that would make a pretty silly box. We need to hope we get better information from the other parts:

$$\begin{aligned} V_x &= 0 \rightarrow x^2 + 2xy - 32 = 0 \\ V_y &= 0 \rightarrow y^2 + 2xy - 32 = 0 \end{aligned}$$

Solving systems of non-linear equations is no fun. You can do it by hand by solving $V_x = 0$ for y, then plugging that in to $V_y = 0$, and then performing some excruciating algebra. Being more judicious in how we use our time, let's consult Wolfram Alpha, which reports that the non-zero and real solution to this system is $x = y = 8/\sqrt{6}$. These values lead back to $z = 8/\sqrt{6}$ as well. So, the maximum volume occurs when the box is a cube, with dimensions

$$x = y = z = \frac{8}{\sqrt{6}}$$

The actual maximum volume is,

$$V = \left(\frac{8}{\sqrt{6}}\right)^3 \approx 135\,cm^3 \quad \blacksquare$$

---

(6) Find and characterize the critical points of $f(x, y) = x^2 + y^2 + \dfrac{1}{x^2y^2}$.

□ We have

$$f_x = 2x - \frac{2}{x^3y^2} \quad \text{and} \quad f_y = 2y - \frac{2}{x^2y^3}$$

Solving for where $f_x = 0$ and $f_y = 0$,

$$\begin{aligned}
2x - \frac{2}{x^3y^2} = 0 \ \text{ and } \ 2y - \frac{2}{x^2y^3} &= 0 \\
2x^4y^2 - 2 = 0 \ \text{ and } \ 2x^2y^4 - 2 &= 0 \\
x^4y^2 = 1 \ \text{ and } \ x^2y^4 &= 1 \\
x^4y^2 - x^2y^4 &= 0 \\
x^2y^2(x^2 - y^2) &= 0 \\
(x, y) &= (\pm 1, \pm 1)
\end{aligned}$$

So, there are 4 possible critical points. (Remember, (0,0) is not a critical point since it's not in the domain of the function.) Let's build the second derivative test function $D(x, y)$ in general:

$$f_{xx} = \frac{6}{x^4y^2} + 2 \quad ; \quad f_{yy} = \frac{6}{x^2y^4} + 2 \quad ; \quad f_{xy} = \frac{4}{x^3y^3}$$

$$\begin{aligned}
D(x, y) &= f_{xx}(x, y)f_{yy}(x, y) - [f_{xy}(x, y)]^2 \\
&= \left(\frac{6}{x^4y^2} + 2\right)\left(\frac{6}{x^2y^4} + 2\right) - \left(\frac{4}{x^3y^3}\right)^2 \\
&= \frac{36}{x^6y^6} + \frac{12}{x^4y^2}\frac{12}{x^2y^4} + 4 - \frac{16}{x^6y^6} \\
&= \frac{20}{x^6y^6} + \frac{12}{x^4y^2}\frac{12}{x^2y^4} + 4
\end{aligned}$$

So, $D(x,y) > 0$ everywhere. Also $f_{xx}(x,y) > 0$ everywhere. So all critical points are local minimums. ■

---

(7) Find the absolute extremes of $f(x,y) = 4x + 6y - x^2 - y^2$ over the domain D with $0 \leq x \leq 4, 0 \leq y \leq 5$.

□ Are there extremes over the interior of D? With $f_x = 4 - 2x$ and $f_y = 6 - 2y$, we have $f_x = f_y = 0$ at (2,3) which is indeed in the interior of D.

Let edge L1 be the left boundary of D, where $x = 0$. Here, the function reduces to $f(0,y) = 6y - y^2$. There is a critical point here at $y = 3$, so $(0,3)$ is a candidate extreme of $f(x,y)$.

Let edge L2 be the right boundary of D, where $x = 4$. Here, the function again reduces to $f(4,y) = 6y - y^2$. There is a critical point here at $y = 3$, so $(4,3)$ is a candidate extreme of $f(x,y)$.

Let edge L3 be the lower boundary of D, where $y = 0$. Here, the function reduces to $f(x,0) = 4x - x^2$. There is a critical point here at $x = 2$, so $(2,0)$ is a candidate extreme of $f(x,y)$.

Let edge L4 be the upper boundary of D, where $y = 5$. Here, the function reduces to $f(x,5) = 4x - x^2 + 5$. There is a critical point here at $x = 2$, so $(2,5)$ is a candidate extreme of $f(x,y)$.

At the critical point we found on the interior, we have $f(2,3) = 13$.

At the candidate extremes we found on the boundaries, we have these values of the function:

$$f(0,3) = 9 \quad , \quad f(4,3) = 9 \quad , \quad f(2,0) = 4 \quad , \quad f(2,5) = 5$$

At the vertices,

$$f(0,0) = 0 \quad , \quad f(4,0) = 0 \quad , \quad f(0,5) = 5 \quad \text{and} \quad f(4,5) = 5$$

Comparing the critical points and vertices, we see an absolute maximum of 13 at (2,3), and absolute minimums of 0 at two points: (0,0), and (4,0). ■

(8) What dimensions should we use for a rectangular aquarium of volume 12,000 cm^3 if we want to minimize its cost, given that the base costs 5 times more per unit area than the sides. (There is no lid on the aquarium.)

□ Our objective function (which is to be minimized) will be the cost function. Let the dimensions of the tank be x, y, z (length, width, height). Assume there is no top; the base has area xy. Let p be the cost per unit area of the glass sides, then the cost of the base is $5p(xy)$; there are two sides of dimensions yz and xz, so the total cost to be minimized is

$$C = 5pxy + 2pyz + 2pxz$$

Our constraint is that we need to have, specifically, $V = 12,000$. Since $V = xyz$ and $V = 12,000$, our constraint is:

$$12,000 = xyz \qquad \text{or} \qquad z = \frac{12,000}{xy}$$

We can use this to constrain our objective function down to two independent variables:

$$\begin{aligned} C &= 5pxy + 2pyz + 2pxz \\ &= 5pxy + 2py\left(\frac{12,000}{xy}\right) + 2px\left(\frac{12,000}{xy}\right) \\ &= 5pxy + \frac{24,000p}{x} + \frac{24,000p}{y} \end{aligned}$$

Now let's start some derivative action:

$$C_x = 0 \quad \rightarrow \quad 5py - \frac{24,000p}{x^2} = 0$$

$$C_y = 0 \quad \rightarrow \quad 5px - \frac{24,000p}{y^2} = 0$$

Let's start solving these two equations by dividing $5p$:

$$C_x = 0 \rightarrow y - \frac{4800}{x^2} = 0$$

$$C_y = 0 \rightarrow x - \frac{4800}{y^2} = 0$$

and then merge them by using the common term 4800. From $C_x = 0$, we get $4800 = yx^2$. Passing that to $C_y = 0$, we have

$$x - \frac{x^2}{y} = 0$$

$$x\left(1 - \frac{x}{y}\right) = 0$$

which holds for $x = 0$ or $x = y$. The $x = 0$ case is not useful. The other case tells us that BOTH $C_x = 0$ and $C_y = 0$ when $x = y$. For $x = y$, the equation $C_x = 0$ becomes

$$4800 = y(y)^2$$

$$y = \sqrt[3]{4800}$$

So, the minimum cost occurs when

$$x = y = \sqrt[3]{4800} = 4\sqrt[3]{75} \quad ; \quad z = \frac{12000}{xy} = 10\sqrt[3]{75}$$

Note that the actual value of p never became an issue. ■

B.2.3 *Double Integrals — Practice — Solved*

(1) Evaluate $\int_1^4 \int_0^2 x + \sqrt{y}\, dxdy$.

☐ This integral is ready to be evaluated with no prep work:

$$\int_1^4 \int_0^2 x + \sqrt{y}\, dxdy = \int_1^4 \left[\left(\frac{1}{2}x^2 + x\sqrt{y} \right) \Big|_0^2 \right] dy$$
$$= \int_1^4 (2 + 2\sqrt{y})\, dy = \left(2y + \frac{4}{3}y^{3/2} \right) \Big|_1^4$$
$$= 8 + \frac{4}{3}(8) - 2 - \frac{4}{3} = \frac{46}{3} \quad \blacksquare$$

(2) Evaluate $\iint_R \frac{xy^2}{x^2+1}\, dA$ where R is the region $\{(x, y) : 0 \le x \le 1, -3 \le y \le 3\}$.

☐ Let's choose the order of integration as $dxdy$ (rather than $dydx$) to get at x first and remove the fraction.

$$\iint_R \frac{xy^2}{x^2+1}\, dA = \int_{-3}^3 \int_0^1 \frac{xy^2}{x^2+1}\, dxdy$$
$$= \int_{-3}^3 \left(\frac{1}{2}y^2 \ln(x^2+1) \right) \Big|_0^1 dy = \int_{-3}^3 \left(\frac{1}{2}y^2 (\ln 2) \right) dy$$
$$= \frac{\ln 2}{6} \left(y^3 \right) \Big|_{-3}^3 = \frac{54 \ln 2}{6} = 9 \ln 2 \quad \blacksquare$$

(3) Find the volume under $z = 4 + x^2 - y^2$ over the region $R = \{(x, y) : -1 \le x \le 1, 0 \le y \le 2\}$.

☐ We use the bounds on x and y to set the limits of integration:

$$\int_{-1}^1 \int_0^2 (4 + x^2 - y^2)\, dydx = \int_{-1}^1 \left(4y + x^2 y - \frac{1}{3}y^3 \right) \Big|_0^2 dx$$
$$= \int_{-1}^1 \left(\frac{16}{3} + 2x^2 \right) dx$$
$$= \left(\frac{16}{3}x + \frac{2}{3}x^3 \right) \Big|_{-1}^1 = 12 \quad \blacksquare$$

---

(4) Evaluate $\iint_D e^{y^2}\,dA$, where D is the region $\{(x,y) : 0 \le x \le y, 0 \le y \le 1\}$.

□ We have to integrate with respect to x first, since the boundaries of x are not constant:

$$\iint_D e^{y^2}\,dA = \int_0^1 \int_0^y e^{y^2}\,dxdy = \int_0^1 \left(xe^{y^2}\right)\Big|_0^y dy = \int_0^1 ye^{y^2}\,dy$$

$$= \frac{1}{2}e^{y^2}\Big|_0^1 = \frac{1}{2}(e-1) \quad ■$$

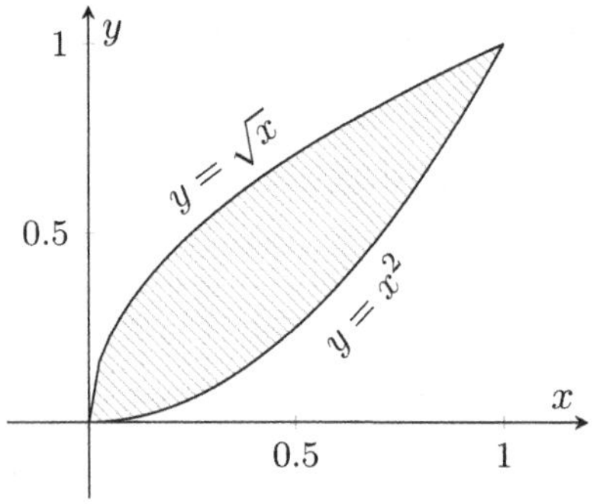

Fig. B.4 Region of integration for PP 5.

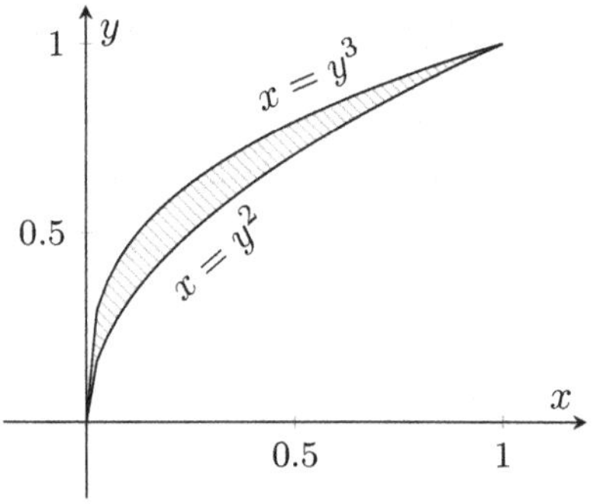

Fig. B.5 Region of integration for PP 6.

---

(5) Evaluate $\iint_D (x+y)\,dA$ where D is the region between $y = \sqrt{x}$ and $y = x^2$.

□ Since these curves intersect at (0,0) and (1,1) and $y = x^2$ is below the other on that interval, the region D bounded by them is the area under $y = \sqrt{x}$ and over $y = x^2$ from $x = 0$ to $x = 1$. This region is shown in Fig. B.4. We have to integrate with respect to y first, since the boundaries of y are not constant:

$$\iint_D (x+y)\,dA = \int_0^1 \int_{x^2}^{\sqrt{x}} (x+y)\,dydx = \int_0^1 \left(xy + \frac{1}{2}y^2\right)\Big|_{x^2}^{\sqrt{x}} dx$$

$$= \int_0^1 \left(x^32 + \frac{1}{2}x - x^3 - \frac{1}{2}x^4\right) dx$$

$$= \left(\frac{2}{5}x^52 + \frac{1}{4}x^2 - \frac{1}{4}x^4 - \frac{1}{10}x^5\right)\Big|_0^1$$

$$= \frac{2}{5} - \frac{1}{10} = \frac{3}{10} \quad ■$$

(6) Evaluate $\iint_D (2x + y^2)\, dA$, where D is the region between $x = y^2$ and $x = y^3$.

□ Since these curves intersect at (0,0) and (1,1) and $x = y^3$ is "below" (i.e. closer to the y-axis than) the other on that interval, the region D bounded by them is the area "under" $x = y^2$ and "over" $x = y^3$ from $y = 0$ to $y = 1$. This region is shown in Fig. B.5. To find the volume under the surface $z = 2x + y^2$, we integrate with respect to x first, since the boundaries of x are not constant:

$$\begin{aligned}\iint_D (2x + y^2)\, dA &= \int_0^1 \int_{y^3}^{y^2} (2x + y^2)\, dx dy \\ &= \int_0^1 (x^2 + xy^2)\Big|_{y^3}^{y^2} dy = \int_0^1 (2y^4 - (y^6 + y^5))\, dy \\ &= \left(\frac{2}{5}y^5 - \frac{1}{7}y^7 - \frac{1}{6}y^6\right)\Big|_0^1 = \frac{2}{5} - \frac{1}{7} - \frac{1}{6} = \frac{19}{210} \quad ■\end{aligned}$$

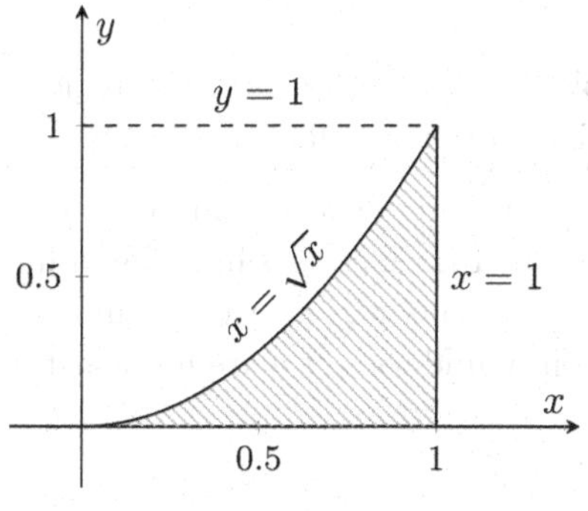

Fig. B.6 Region of integration for PP 7.

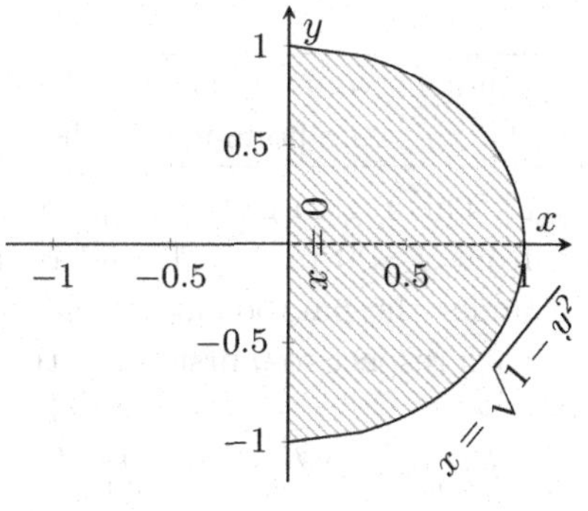

Fig. B.7 Region of integration for PP 8.

(7) Write the equivalent integral in which the order of integration is reversed: $\int_0^1 \int_{\sqrt{y}}^1 \sqrt{x^3 + 1}\, dx dy$.

□ We note that the limits of integration describe the region to the right of the curve $x = \sqrt{y}$ and left of $x = 1$, from $y = 0$ to $y = 1$. This is the same as the region between above the line $y = 0$ and below $y = x^2$

from $x = 0$ to $x = 1$. This region is shown in Fig. B.6. So, we have

$$\int_0^1 \int_{\sqrt{y}}^1 \sqrt{x^3+1}\, dxdy = \int_0^1 \int_0^{x^2} \sqrt{x^3+1}\, dydx$$

This can now be solved, whereas the original could not. ■

---

(8) Evaluate $\iint_D xy^2\, dA$, where D is the region bounded by $x = 0$ and $x = \sqrt{1-y^2}$.

□ The region D is the right half of the unit circle, i.e. $D = \{(x, y) : 0 \leq x \leq \sqrt{1-y^2}, -1 \leq y \leq 1\}$ and is shown in Fig. B.7. We have to integrate with respect to x first, since the boundaries of x are not constant:

$$\begin{aligned}\iint_D xy^2\, dA &= \int_{-1}^1 \int_0^{\sqrt{1-y^2}} (xy^2)\, dxdy \\ &= \int_{-1}^1 \left(\frac{1}{2}x^2y^2\right)\Bigg|_0^{\sqrt{1-y^2}} dy = \int_{-1}^1 \left(\frac{1}{2}(1-y^2)y^2\right) dy \\ &= \left(\frac{1}{6}y^3 - \frac{1}{10}y^5\right)\Bigg|_{-1}^1 = 2\left(\frac{1}{6} - \frac{1}{10}\right) = \frac{2}{15} \quad ■\end{aligned}$$

---

(9) Find the volume under the paraboloid $z = x^2 + 3y^2$ over the region in the xy-plane bounded by the the lines $y = 1, x = 0$, and $y = x$.

□ This region in the xy-plane is the area under $y = 1$ and over $y = x$ from $x = 0$ to $x = 1$, and is shown in Fig. B.8. To find the volume under the paraboloid $z = x^2 + 3y^2$ over this region, we can integrate with respect to y first, since the given boundaries of y are not constant:

$$\begin{aligned}\iint_D (x^2 + 3y^2)]\, dA &= \int_0^1 \int_x^1 (x^2 + 3y^2)\, dydx \\ &= \int_0^1 (x^2y + y^3)\Bigg|_x^1 dx = \int_0^1 \left((x^2+1) - 2x^3\right) dx \\ &= \left(\frac{1}{3}x^3 + x - \frac{1}{2}x^4\right)\Bigg|_0^1 = \frac{1}{3} + 1 - \frac{1}{2} = \frac{5}{6}\end{aligned}$$

The same answer can be obtained by reordering the limits of integration,

$$\iint_D (x^2 + 3y^2) dA = \int_0^1 \int_0^y (x^2 + 3y^2) dydx = \frac{5}{6} \quad ■$$

(10) Evaluate $\int_0^1 \int_{x^2}^1 x^3 \sin(y^3)\, dydx$.

□ The integral cannot be solved with the current order of integration, so we have to reverse it. We note that the limits of integration describe below the line $y = 1$ and above $y = x^2$ from $x = 0$ to $x = 1$. This is the same as the region to the right of the line $x = 0$ and right of $x = \sqrt{y}$ from $y = 0$ to $y = 1$. The region is shown in Fig. B.9. So, we have

$$\int_0^1 \int_{x^2}^1 x^3 \sin(y^3)\, dydx = \int_0^1 \int_0^{\sqrt{y}} x^3 \sin(y^3)\, dxdy$$

$$= \int_0^1 \left(\frac{1}{4}x^4 \sin(y^3)\right)\Big|_0^{\sqrt{y}} dy = \int_0^1 \left(\frac{1}{4}y^2 \sin(y^3)\right) dy$$

$$= \left(-\frac{1}{12}\cos(y^3)\right)\Big|_0^1 = \frac{1}{12}(1 - \cos 1) \quad \blacksquare$$

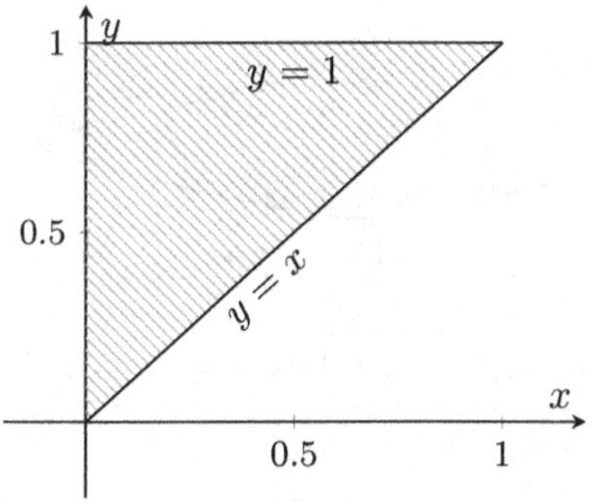

Fig. B.8 Region of integration for PP 9.

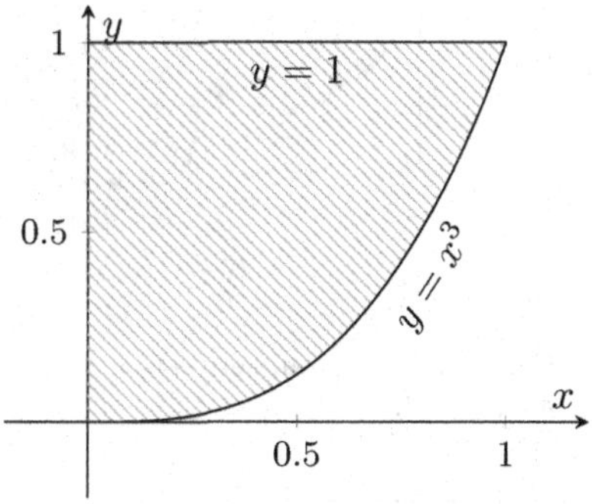

Fig. B.9 Region of integration for PP 10.

B.2.4 *Triple Integrals — Practice — Solved*

(1) Evaluate $\int_0^1 \int_x^{2x} \int_0^y (2xyz)\, dzdydx$ by hand.

□ $$\int_0^1 \int_x^{2x} \int_0^y (2xyz)\, dzdydx = \int_0^1 \int_x^{2x} xyz^2 \Big|_0^y dydx$$
$$= \int_0^1 \int_x^{2x} xy^3\, dydx = \int_0^1 \frac{1}{4}xy^4 \Big|_x^{2x} dx$$
$$= \int_0^1 \frac{1}{4}x(16x^4 - x^4)\, dx = \int_0^1 \frac{15}{4}x^5\, dx = \frac{15}{24} = \frac{5}{8} \quad ■$$

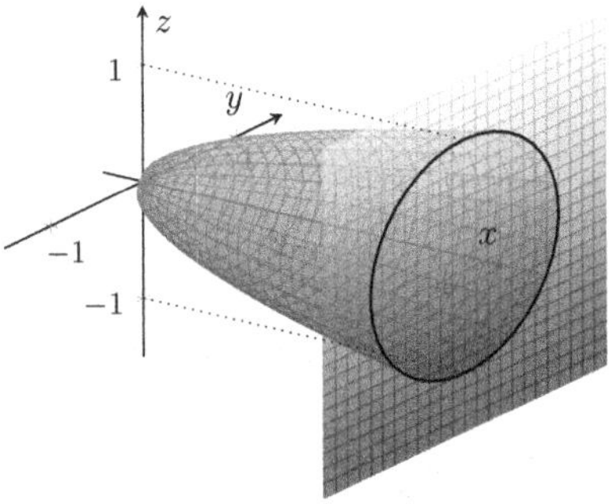

Fig. B.10 Region between $x = 4y^2 + 4z^2$, $x = 4$.

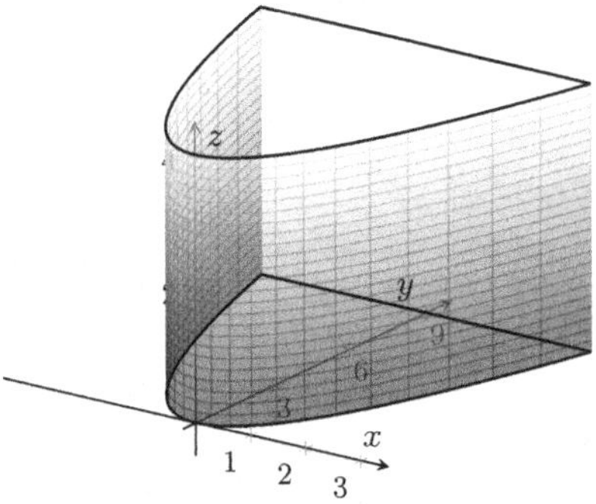

Fig. B.11 Region between $y = x^2$, $y = 9$, $z = 0$, $z = 4$.

(2) Evaluate $\iiint_E x\, dV$, where E is the region between the paraboloid $x = 4y^2 + 4z^2$ the plane $x = 4$.

□ The region in question is shown in Fig. B.10. The paraboloid $x = 4y^2 + 4z^2$ intersects the plane $x = 4$ in the circle $4y^2 + 4z^2 = 4$, i.e. $y^2+z^2 = 1$. The paraboloid opens sideways, along the x-axis; the region E between the paraboloid and the plane $x = 4$ is to the right of the paraboloid and left of $x = 4$. This defines limits on x: $4y^2+4z^2 \le x \le 4$. The cross section of the intersection of the paraboloid and plane in the yz-plane is the circle $y^2 + z^2 = 1$, which defines limits on y and z. Let's call those limits $-\sqrt{1-z^2} \le y \le \sqrt{1-z^2}$ and $-1 \le z \le 1$. We must set up limits of integration so that variables are progressively eliminated; the final limits must be constants. So, x must go first, then y and finally z: So,

$$\iiint_E x\,dV = \int_{-1}^{1}\int_{-\sqrt{1-z^2}}^{\sqrt{1-z^2}}\int_{4y^2+4z^2}^{4} x\,dxdydz = \frac{16\pi}{3} \quad \blacksquare$$

---

(3) Find the the volume of the solid bounded by the parabolic cylinder $y = x^2$ and the planes $z = 0$, $z = 4$, and $y = 9$.

□ The region in question is shown in Fig. B.11. The parabolic cylinder $y = x^2$ is the curve $y = x^2$ in the xy-plane, extended both ways along the z-axis.

We have perfectly good limits on y and z already: $0 \leq z \leq 4$ and $x^2 \leq y \leq 9$. To get limits on x, we can examine the intersection of the curves $y = x^2$ and $y = 9$, which are $x = \pm 3$.

There are multiple possible orders of integration, but we do need to ensure that we integrate with respect to y before x. Since we're computing the volume of a solid region, we integrate the function $F(x, y, z) = 1$. So how about this arrangement,

$$V = \iiint_E dV = \int_0^4\int_{-3}^{3}\int_{x^2}^{9} (1)\,dydzdz = 144 \quad \blacksquare$$

---

(4) Describe and sketch the 3D region whose volume is being computed in the following integral:

$$\int_{-3}^{3}\int_{-\sqrt{9-x^2}}^{\sqrt{9-x^2}}\int_{x^2+y^2-9}^{9-x^2-y^2} (1)\,dzdydx$$

□ The limits on z indicate a lower limit of $z = x^2 + y^2 - 9$, which is a paraboloid that has a vertex at $(0, 0, -9)$; the upper limit is $z = 9 - x^2 - y^2$, which is an inverted paraboloid with vertex $(0, 0, 9)$. These two paraboloids intersect in the circle $x^2 + y^2 = 3$ in the xy-plane; this entire circle is consistent with the bounds on the x and y integrals. So this integral computes the volume of the solid above the paraboloid $z = x^2 + y^2 - 9$ and below the paraboloid $z = 9 - x^2 - y^2$. The region in question is shown in Fig. B.12. ■

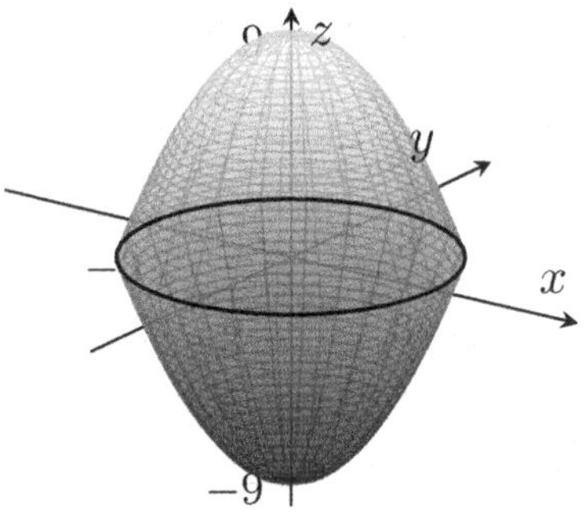

Fig. B.12 Region between $z = 9 - x^2 - z^2$, $z = x^2 + y^2 - 9$.

(5) Evaluate (by hand) $\iiint_E yz\cos(x^5)\,dV$, where E is defined as the region $E = \{(x, y, z) : 0 \leq x \leq 1; 0 \leq y \leq x; x \leq z \leq 2x\}$.

□ We must set up limits of integration so that variables are progressively eliminated; the final limits must be constants. Either y or z can go first, but x must go last. So,

$$\iiint_E yz\cos(x^5)dV = \int_0^1 \int_0^x \int_x^{2x} yz\cos(x^5)\,dzdydx$$

$$= \int_0^1 \int_0^x \frac{1}{2}yz^2\cos(x^5)\Big|_x^{2x} dydx = \int_0^1 \int_0^x \frac{3}{2}x^2y\cos(x^5)\,dydx$$

$$= \int_0^1 \frac{3}{4}x^2y^2\cos(x^5)\Big|_0^x dx = \int_0^1 \frac{3}{4}x^4\cos(x^5)\,dx$$

$$= \frac{3}{20}\sin(x^5)\Big|_0^1 = \frac{3}{20}\sin(1)$$

(6) Find the volume of region between the paraboloid $y = x^2 + z^2$ and the plane $y = 16$.

□ The region in question is shown in Fig. B.13. The paraboloid opens around the y-axis, so the "floor" of this solid is to the paraboloid, and the "roof" is the plane $y = 16$. This gives limits on y as: $x^2 + z^2 \leq y \leq 16$. With y pinned down, we get limits on x and z by exploring the region of the xz-plane used by this solid.

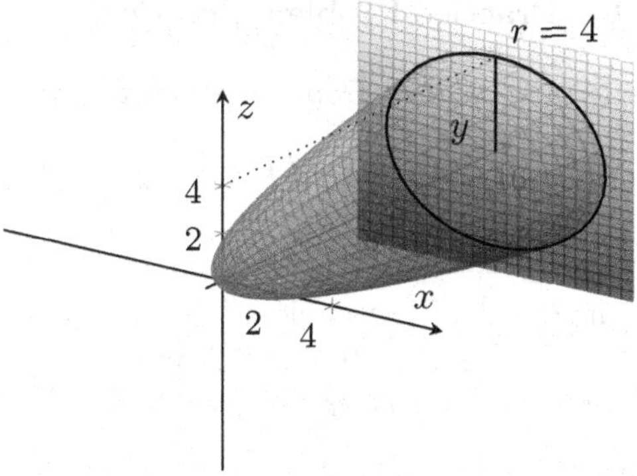

Fig. B.13 Region between $y = x^2 + z^2$, $y = 16$.

The intersection of the solid with the plane $y = 4$ is $x^2 + z^2 = 16$. This is a circle of radius 4 in the xz-plane and defines limits on x and z. We can set $-\sqrt{16 - z^2} \leq x \leq \sqrt{16 - z^2}$ and $-4 \leq x \leq 4$.

We must order limits of integration so that variables are progressively eliminated; the final limits must be constants. So, y must go first, then z and finally x: So,

$$V = \iiint E dV = \int_{-4}^{4} \int_{-\sqrt{16-z^2}}^{\sqrt{16-z^2}} \int_{x^2+z^2}^{16} dydxdz = 128\pi$$

If you don't like sideways paraboloids and regions of integration, then a different way to tackle this would be to say that the volume of this region would be the same as that between $z = x^2 + y^2$ and $z = 16$. That would be exactly the same shape and same size, just opening upwards. If you make such an adjustment in any problems, be sure to note that up front in your solution, so that your reader does not have to reverse-engineer your limits of integration, which might look incorrect at first!

■

B.3 Chapter 15: Practice Problem Solutions

B.3.1 *Double Ints in Polar Coords — Practice — Solved*

(1) Use polar coordinates to evaluate $\iint_R \sqrt{4-x^2-y^2}\,dA$, where R is the region $\{(x,y) : x^2+y^2 \leq 4, x \geq 0\}$.

☐ R is the inner half of the circle $x^2+y^2 = 4$ to the right of the y-axis, and so corresponds to $0 \leq r \leq 2$ and $-\pi/2 \leq \theta \leq \pi/2$. Since $x^2+y^2 = r^2$, and $dA = r\,drd\theta$ in polar coordinates,

$$\begin{aligned}\iint_R \sqrt{4-x^2-y^2}\,dA &= \int_{-\pi/2}^{\pi/2}\int_0^2 \sqrt{4-r^2}r\,drd\theta\\ &= \int_{-\pi/2}^{\pi/2}\left(-\frac{1}{3}(4-r^2)^{3/2}\right)\Big|_0^2\,d\theta = \int_{-\pi/2}^{\pi/2}\frac{8}{3}\,d\theta\\ &= \left(\frac{8}{3}\theta\right)\Big|_{-\pi/2}^{\pi/2} = \frac{8\pi}{3} \quad \blacksquare\end{aligned}$$

(2) Find the area inside the curve $r = 4+3\cos\theta$.

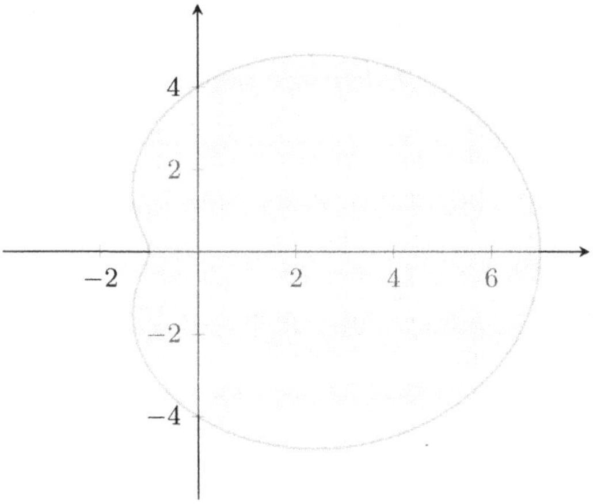

Fig. B.14 The region inside $r = 4+3\cos\theta$.

☐ Since this is an area problem, we need to describe the given region entirely with the limits of integration. A graph of this region is shown in Fig. B.14. The r variable is described explicitly by the function; to go all the way around the curve we use $0 \leq \theta \leq 2\pi$. So, the area enclosed is

$$\iint_D dA = \int_0^{2\pi} \int_0^{4+3\cos\theta} r\,drd\theta = \int_0^{2\pi} \left(\frac{1}{2}r^2\right)\Big|_0^{4+3\cos\theta} d\theta$$
$$= \int_0^{2\pi} \frac{1}{2}(16 + 24\cos\theta + 9\cos^2\theta)\,d\theta$$
$$= (8\theta + 12\sin\theta)\Big|_0^{2\pi} + \frac{9}{2}\int_0^{2\pi} \cos^2\theta\,d\theta$$
$$= 16\pi + 0 + \frac{9}{2}\left(\frac{1}{2}\theta + \frac{1}{4}\sin 2\theta\right)\Big|_0^{2\pi}$$
$$= 16\pi + \frac{9\pi}{2} = \frac{41\pi}{2} \quad \blacksquare$$

---

(3) Find the volume inside the sphere $x^2 + y^2 + z^2 = 16$ and outside the cylinder $x^2 + y^2 = 4$.

□ Note that this 3D region involves a solid both above and below the xy-plane! Figures B.15 and B.16 show the full 3D region as well as the corresponding polar region in the 2D plane. The bounds in the xy-plane of this solid are the intersections of the sphere and cylinder with the xy-plane, i.e. inside the circle $x^2 + y^2 = 16$ and outside the circle $x^2 + y^2 = 4$. The easiest thing to do here is to find the volume under the hemisphere $z = \sqrt{16 - x^2 - y^2} = \sqrt{16 - r^2}$ and outside the cylinder, then multiply by two. Thus, in polar coordinates we want the volume under $z = \sqrt{16 - r^2}$ over the region in the xy-plane with bounds of $2 \le r \le 4$ and $0 \le \theta \le 2\pi$:

$$\iint_D \sqrt{16 - r^2}\,dA = \int_0^{2\pi} \int_2^4 \sqrt{16 - r^2} \cdot r\,drd\theta$$
$$= \int_0^{2\pi} \left(-\frac{1}{3}(16 - r^2)^{3/2}\right)\Big|_2^4 d\theta = \int_0^{2\pi} \frac{1}{3} 12^{3/2}\,d\theta$$
$$= \frac{2\pi}{3}(2\sqrt{3})^3 = \frac{2\pi}{3}(24\sqrt{3}) = 16\pi\sqrt{3}$$

The entire volume is then $32\pi\sqrt{3}$. ■

---

(4) Convert this integral into polar coordinates: $\displaystyle\int_{-a}^{a} \int_0^{\sqrt{a^2-y^2}} (x^2 + y^2)^{3/2}\,dxdy$.

□ The limits of integration describe the right half of a circle of radius a (since y goes from $-a$ to a) — which is the region $0 \le r \le a$ and $-\pi/2 \le$

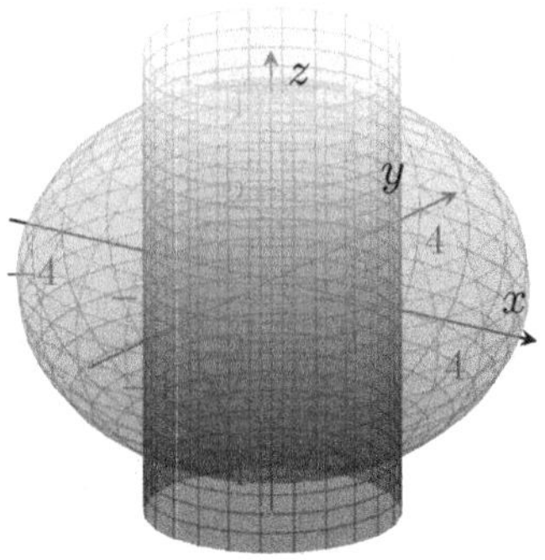

Fig. B.15 Inside $x^2+y^2+z^2=16$ and outside $x^2+y^2=4$.

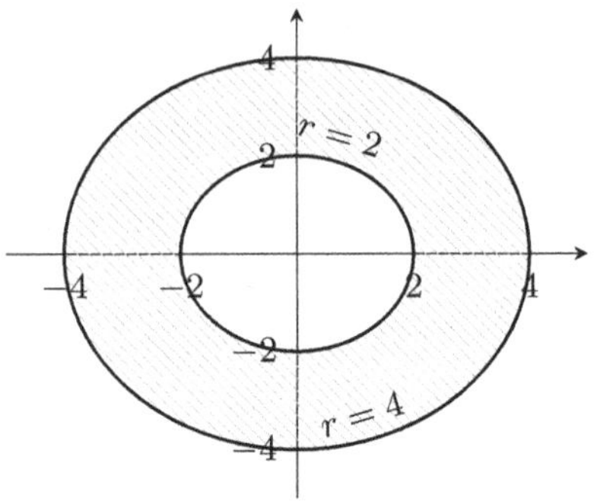

Fig. B.16 Between $x^2+y^2=4$ and $x^2+y^2=16$.

$\theta \le \pi/2$. Also, the function converts as $(x^2+y^2)^{3/2} = (r^2)^{3/2} = r^3$ and the area element $dydx$ becomes $r\,drd\theta$:

$$\int_{-a}^{a}\int_{0}^{\sqrt{a^2-y^2}} (x^2+y^2)^{3/2}\,dxdy = \int_{-\pi/2}^{\pi/2}\int_{0}^{a} r^3\cdot r\,drd\theta \quad \blacksquare$$

(5) Find the area outside $r=2$ and inside $r=4\sin\theta$.

□ This region is shown in Fig. B.17. The curves $r=4\sin\theta$ and $r=2$ intersect when $4\sin\theta=2$, i.e. when $\sin\theta=1/2$, which is at $\theta=\pi/6, 5\pi/6$. The area outside $r=2$ and inside $r=4\sin\theta$ is then

$$\begin{aligned}\iint_R dA &= \int_{\pi/6}^{5\pi/6}\int_{2}^{4\sin\theta} r\,drd\theta = \int_{\pi/6}^{5\pi/6}\left(\frac{1}{2}r^2\right)\Big|_2^{4\sin\theta}\,d\theta\\ &= \int_{\pi/6}^{5\pi/6}(8\sin^2\theta-2)\,d\theta = ((4\theta-2\sin 2\theta)-2\theta)\Big|_{\pi/6}^{5\pi/6}\\ &= (2\theta-2\sin 2\theta)\Big|_{\pi/6}^{5\pi/6} = \frac{5\pi}{3}-2\sin\frac{5\pi}{3}-\frac{\pi}{3}+2\sin\frac{\pi}{3}\\ &= \frac{5\pi}{3}+2\frac{\sqrt{3}}{2}-\frac{\pi}{3}+2\frac{\sqrt{3}}{2} = \frac{4\pi}{3}+2\sqrt{3} \quad \blacksquare\end{aligned}$$

(6) Find the volume between the paraboloid $z=10-3x^2-3y^2$ and the plane $z=4$.

□ Figures B.18 and B.19 show the full 3D region as well as the corresponding polar region in the 2D plane. The paraboloid intersects the plane $z=4$ where $10-3x^2-3y^2=4$, i.e. where $x^2+y^2=2$. So the

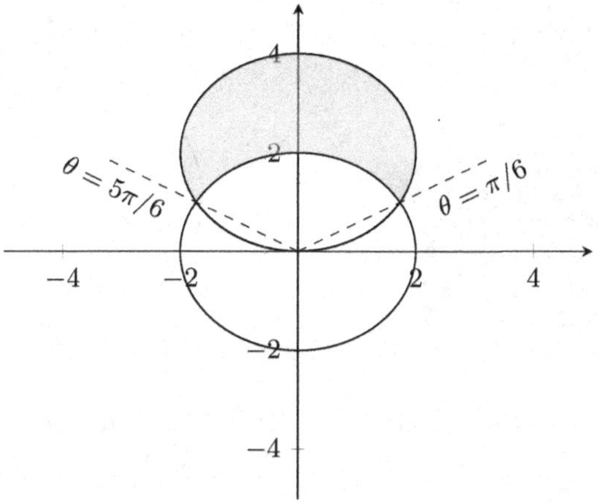

Fig. B.17 The region outside $r = 2$ and inside $r = 4\sin\theta$.

volume between the paraboloid and the plane is below the paraboloid (which in polar coordinates is $z = 10 - 3r^2$) and above the plane $z = 4$; the volume sits over the region $0 \le r \le \sqrt{2}$ and $0 \le \theta \le 2\pi$. So,

$$\begin{aligned}\iint_R [(10 - 3r^2) - 4]\, dA &= \int_0^{2\pi} \int_0^{\sqrt{2}} (6 - 3r^2) \cdot r\, dr d\theta \\ &= \int_0^{2\pi} \left(3r^2 - \frac{3}{4}r^4\right) \Big|_0^{\sqrt{2}} d\theta \\ &= \int_0^{2\pi} (3)\, d\theta = 6\pi \quad \blacksquare\end{aligned}$$

---

(7) Convert this integral into polar coordinates: $\displaystyle\int_0^2 \int_{-\sqrt{4-y^2}}^{\sqrt{4-y^2}} x^2 y^2\, dx dy.$

□ The limits of integration describe the upper half of a circle of radius 2 (since y goes from 0 to 2) — which is the region $0 \le r \le 2$ and $0 \le \theta \le \pi$. Also, recalling that $x = r\cos\theta$ and $y = r\sin\theta$, we have

$$\begin{aligned}\int_0^2 \int_{-\sqrt{4-y^2}}^{\sqrt{4-y^2}} x^2 y^2\, dx dy &= \int_0^{\pi} \int_0^2 (r\cos\theta)^2 (r\sin\theta)^2 \cdot r\, dr d\theta \\ &= \int_0^{\pi} \int_0^2 r^5 \cos^2\theta \sin^2\theta\, dr d\theta \quad \blacksquare\end{aligned}$$

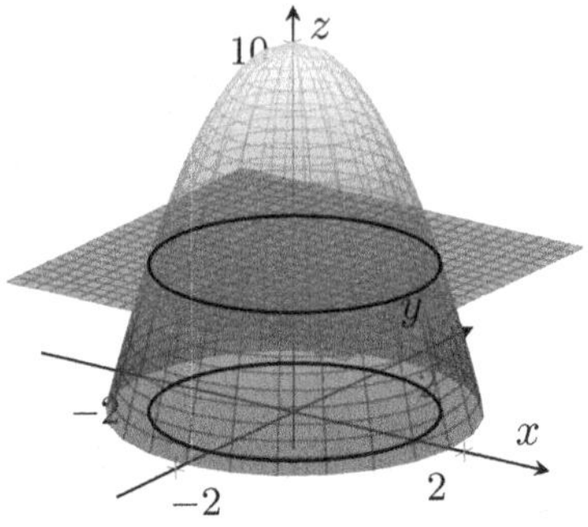

Fig. B.18 Under the paraboloid $z = 10 - 3x^2 - 3y^2$, above the plane $z = 4$.

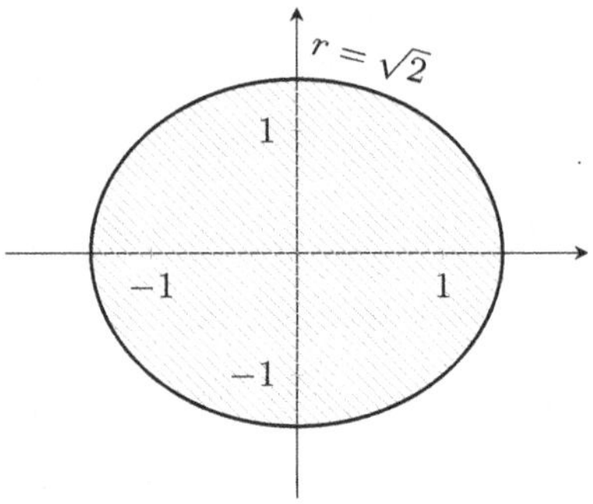

Fig. B.19 Intersection of $z = 10 - 3x^2 - 3y^2$ and $z = 4$.

B.3.2 *Cylindrical and Spherical Coords — Practice — Solved*

(1) Locate the point whose cylindrical coordinates are $(1, 3\pi/2, 2)$ and give its rectangular coordinates.

Because $\theta = 3\pi/2$, the radial angle takes us from the positive x-axis, three-fourths of the way around to the negative y-axis. The cylindrical coordinates $(r, \theta, z) = (1, 3\pi/2, 2)$ give rectangular coordinates

$$x = 1\cos\left(\frac{3\pi}{2}\right) = 0 \quad ; \quad y = 1\sin\left(\frac{3\pi}{2}\right)$$

and of course, $z = 2$. So this is the rectangular point $(x, y, z) = (0, -1, 2)$. ■

(2) What are the cylindrical coordinates of the point whose rectangular coordinates are $(3, 3, -2)$?

□ With rectangular coordinates $(3, 3, -2)$, we have $z = -2$ immediately, and then use Eq. (15.1) to give

$$r = \sqrt{x^2 + y^2} = \sqrt{3^2 + 3^2} = 3\sqrt{2}$$
$$\theta = \tan^{-1}\frac{y}{x} = \tan^{-1}\frac{3}{3} = \pi/4$$

so this is the cylindrical point $(r, \theta, z) = (3\sqrt{2}, \pi/4, -2)$. ■

(3) What surface is described in cylindrical coordinates by the equation $r^2 + z^2 = 25$?

□ Since $r^2 = x^2 + y^2$, the equation can be written in rectangular coordinates as $x^2 + y^2 + z^2 = 25$; the surface is is a sphere of radius 5 centered at the origin. ■

(4) Convert the cylindrical equation $r^2 - 2z^2 = 4$ into rectangular coordinates, and identify the kind of surface it is.

□ Let's convert the r^2 term using $r^2 = x^2 + y^2$:

$$r^2 - 2z^2 = 4$$
$$x^2 + y^2 - 2z^2 = 4$$
$$\frac{x^2}{4} + \frac{y^2}{4} - \frac{z^2}{2} = 1$$

This is a hyperboloid of one sheet (see Appendix A.2). ■

(5) Locate the point whose spherical coordinates are $(\rho, \theta, \phi) = (5, \pi, \pi/2)$ and give its rectangular coordinates.

□ We convert from spherical to rectangular using Eq. (15.4):

$$x = \rho \sin\phi \cos\theta = 5 \sin\left(\frac{\pi}{2}\right) \cos(\pi) = 5(1)(-1) = -5$$
$$y = \rho \sin\phi \sin\theta = 5 \sin\left(\frac{\pi}{2}\right) \sin(\pi) = 5(1)(0) = 0$$
$$z = \rho \cos\phi = 5 \cos\left(\frac{\pi}{2}\right) = 0$$

So this is the rectangular point $(x, y, z) = (-5, 0, 0)$. ■

(6) What are the spherical coordinates of the point whose rectangular coordinates are $(0, \sqrt{3}, 1)$?

□ Rectangular coords $(x, y, z) = (0, \sqrt{3}, 1)$ lead to spherical:

$$\rho = \sqrt{x^2 + y^2 + z^2} = \sqrt{0 + 3 + 1} = 2$$
$$\cos\phi = \frac{z}{\rho} = \frac{1}{2} \to \phi = \pi/3$$
$$\cos\theta = \frac{x}{\rho \sin\phi} = \frac{0}{2\sin(\pi/3)} = 0 \to \theta = \pi/2$$

So this is the spherical point $(2, \pi/2, \pi/3)$. ■

(7) What surface is described by the spherical equation $\rho = 3$?

☐ This is a spherical equation describing all points whose ρ coordinate (distance from the origin) is 3. This, it is a sphere of radius 3 centered on the origin. ■

(8) Convert the equation $x^2 + y^2 + z^2 = 2$ into spherical coordinates.

☐ With the direct application of a conversion equation from (15.3), we get $\rho^2 = 2$, or $\rho = \sqrt{2}$. This is very simple, but it's here to help you avoid the common error of saying this is $\rho = 2$. It's not. We are matching the following two equations,

$$x^2 + y^2 + z^2 = 2$$
$$x^2 + y^2 + z^2 = \rho^2$$

Pay close attention to your right hand sides! ■

(9) What are the spherical coordinates of the point whose cylindrical coordinates are $(r, \theta, z) = (\sqrt{6}, \pi/4, \sqrt{2})$?

☐ We don't have any direct conversion equations from cylindrical to spherical coordinates, but we can patch some things together. First, the θ coordinate is the same in both cylindrical and spherical coordinates, so we don't have to convert that one.

We can put together $r^2 = x^2 + y^2$ (rectangular to cylindrical) and $\rho^2 = x^2 + y^2 + z^2$ (rectangular to spherical) to get $\rho^2 = r^2 + z^2$, and so

$$\rho^2 = r^2 + z^2 = 6 + 2 = 8 \qquad \to \qquad \rho = 2\sqrt{2}$$

And also, from the spherical-to-rectangular equations, we know that $z = \rho \cos\phi$, or:0

$$\cos\phi = \frac{z}{\rho} = \frac{\sqrt{2}}{2\sqrt{2}} = \frac{1}{2} \qquad \to \qquad \phi = \frac{\pi}{3}$$

So in spherical coordinates, the point is $(\rho, \theta, \phi) = (2\sqrt{2}, \pi/4, \pi/3)$. ■

(10) What are the cylindrical coordinates of the point whose spherical coordinates are $(\rho, \theta, \phi) = (2\sqrt{2}, 3\pi/2, \pi/2)$?

☐ Remember that θ is the same in spherical and cylindrical coordinates, so we don't need to convert that coordinate. Also, both θ and ϕ

put us right on (rectangular) coordinate axes, so this shouldn't be too bad.

Let's get the rectangular z-coordinate, since that gets used in cylindrical, too. Directly from Eq. (15.4):

$$z = \rho\cos\phi = 2\sqrt{2}\cos\left(\frac{\pi}{2}\right) \qquad \to \qquad z = 0$$

It's also easy to piece together an equation for ρ:

$$\rho^2 = x^2 + y^2 + z^2 = r^2 + z^2 = r^2 + 0 \qquad \to \qquad \rho = r = 2\sqrt{2}$$

So in cylindrical coordinates, the point is $(r, \theta, z) = (2\sqrt{2}, 3\pi/2, 0)$. ■

---

(11) Convert the rectangular equation $y^2 + z^2 = 1$ into both cylindrical and spherical coordinates.

□ Since z is already a cylindrical coordinate, we just need to convert y. From the basic $y = r\sin\theta$, we get: $r^2\sin^2\theta + z^2 = 1$.

In spherical coordinates, we can really just directly substitute from (15.3):

$$y^2 + z^2 = (\rho\sin\theta\sin\phi)^2 + (\rho\cos\phi)^2$$

so the equation becomes

$$\rho^2\sin^2\theta\sin^2\phi + \rho^2\cos^2\phi = 1$$

It's not very exciting either way. ■

B.3.3 *Triple Ints in Cyl. & Spher. Coords — Practice — Solved*

---

(1) Evaluate $\iiint_E x\,dV$, where E is the region bounded by the planes $z = 0$ and $z = x + y + 5$, and the cylinders $x^2 + y^2 = 4$ (i.e. $r^2 = 4$) and $x^2 + y^2 = 9$.

□ Call me crazy, but the presence of two cylinders makes me think that it'll be best to use cylindrical coordinates. The planes $z = 0$ and $z = x+y+5$ form the bottom and top of the region. We designed the bounds on this region, which is shown in Fig. B.20, back in Practice Problem 9a of Sec. 15.2. The plane $z = x+y+5$ becomes $z = r\cos\theta + r\sin\theta + 5$, and doesn't get any better — but remember, we're trading some minor inconveniences like that for a lot more conveniences in other limits of integration. The cylinders $x^2 + y^2 = 4$ (i.e. $r^2 = 4$) and $x^2 + y^2 = 9$ (i.e. $r^2 = 9$) are nested cylinders, and form the sides of this region. This region used in the xy-plane between the cylinders is inside two concentric circles (an annulus / washer shape). Thus, we have

$$0 \le z \le r\cos\theta + r\sin\theta + 5 \quad ; \quad 2 \le r \le 3 \quad ; \quad 0 \le \theta \le 2\pi$$

and the integral is (with the volume element $dV = rdzdrd\theta$ in cylindrical coordinates):

$$\iiint_E xdV = \int_0^{2\pi}\int_2^3\int_0^{r\cos\theta + r\sin\theta + 5} (r\cos\theta)rdzdrd\theta = \frac{65\pi}{4} \quad \blacksquare$$

---

(2) Evaluate $\iiint_E e^{\sqrt{x^2+y^2+z^2}}\,dV$, where E is the region inside the sphere $x^2 + y^2 + z^2 = 9$ in the first octant.

□ The sphere $x^2+y^2+z^2 = 9$ is known as $\rho = 3$ in spherical coordinates. Since we're in the first octant, we restrict θ to $0 \le \theta \le \pi/2$ and ϕ to $0 \le \phi \le \pi/2$. The volume element in spherical coordinates is $dV = \rho^2\sin\phi d\rho d\phi d\theta$. So,

$$\begin{aligned}\iiint_E e^{\sqrt{x^2+y^2+z^2}}\,dV &= \int_0^{\pi/2}\int_0^{\pi/2}\int_0^3 e^\rho\rho^2\sin\phi\,d\rho d\phi d\theta \\ &= \frac{\pi}{2}(5e^3 - 2) \quad \blacksquare\end{aligned}$$

(3) Evaluate the following integral by first converting it to the most appropriate coordinate system:

$$\int_0^1 \int_0^{\sqrt{1-y^2}} \int_{x^2+y^2}^{\sqrt{x^2+y^2}} (xyz)\, dzdxdy$$

☐ From the given limits of integration on x and y, we can recognize that in the xy-plane, our region of integration is the portion of the unit circle in the first quadrant. From the limits on z, we go from the paraboloid $z = x^2 + y^2$ to the upper part of the cone $z = \sqrt{x^2+y^2}$ (a.k.a. $z^2 = x^2 + y^2$). This region is shown in Fig. B.21.

Although a cone often suggests spherical coordinates, the equation of a paraboloid does not convert nicely into spherical coordinates. So we'll try cylindrical. In cylindrical coordinates, this solid is bounded by $z = r^2$ (paraboloid), $z = r$ (cone), $0 \le r \le 1$, and $0 \le \theta \le \pi/2$ (to stay in the first quadrant). Remember that for cylindrical coordinates, the volume element is $dV = r\, dzdrd\theta$. Therefore, we have

$$\begin{aligned}
\int_0^1 \int_0^{\sqrt{1-y^2}} \int_{x^2+y^2}^{\sqrt{x^2+y^2}} (xyz)\, dzdxdy \\
&= \int_0^{\pi/2} \int_0^1 \int_{r^2}^{r} (r\cos\theta)(r\sin\theta)(z)\, r\, dzdrd\theta \\
&= \int_0^{\pi/2} \int_0^1 \int_{r^2}^{r} r^3 z \cos\theta \sin\theta\, dzdrd\theta \\
&= \frac{1}{96} \quad \blacksquare
\end{aligned}$$

(4) The region above the cone $z^2 = x^2 + y^2$ and below the sphere $x^2 + y^2 + z^2 = 5$ is shaped like an ice cream cone. What is the volume of this region?

☐ Recall that to find the volume of any 3D region E, we solve the triple integral $\iiint_E (1) dV$. With both a sphere and a cone involved, we should use spherical coordinates. We designed a region very much like this in EX 5c of Sec. 15.2, although with a sphere of radius 1. But updating for the larger sphere of radius $\sqrt{5}$ is simple. The cone $z^2 = x^2 + y^2$ is the perfect-diagonal cone with equation $\phi = \pi/4$. The sphere is also known as $\rho = \sqrt{5}$. The region is shown in Fig. B.22, and the variables

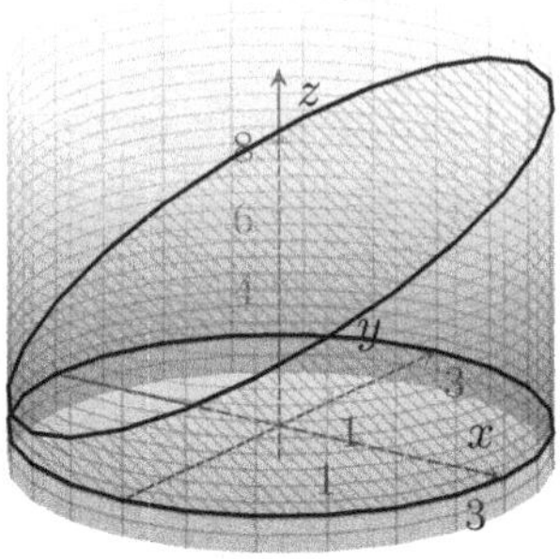

Fig. B.20 $x^2 + y^2 = 9$ between $z = 0$, $z = x + y + 5$.

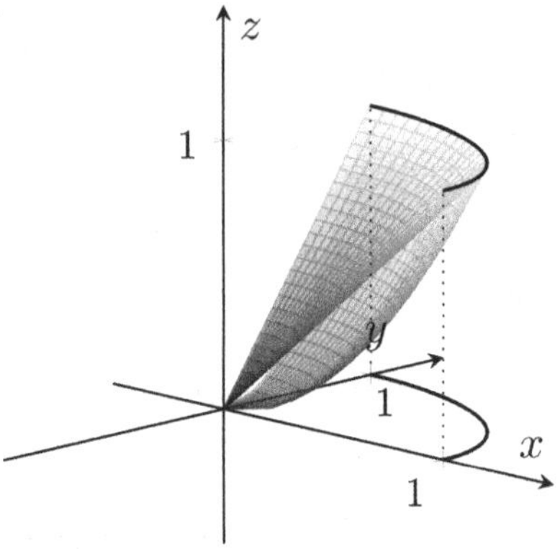

Fig. B.21 Above $z = r^2$, below $z = r$, in the first octant.

are bound as follows:

$$0 \le \rho \le \sqrt{5} \quad ; \quad 0 \le \phi \le \frac{\pi}{4} \quad ; \quad 0 \le \theta \le 2\pi$$

Since the volume element in spherical coordinates is $dV = \rho^2 \sin\phi d\rho d\phi d\theta$, we get:

$$\iiint_E (1)\, dV = \int_0^{2\pi} \int_0^{\pi/4} \int_0^{\sqrt{5}} (1)\rho^2 \sin\phi \, d\rho d\phi d\theta$$
$$= \frac{\pi}{3}\left(10\sqrt{5} - 5\sqrt{10}\right) \quad \blacksquare$$

---

(5) Find the volume of the region common to (inside) both the cylinder $x^2 + y^2 = 1$ and the sphere $x^2 + y^2 + z^2 = 4$.

□ Recall that to find the volume of any 3D region E, we solve the triple integral $\iiint_E (1)dV$. The given region E of integration is inside both the cylinder $x^2 + y^2 = 1$ (i.e. $r^2 = 1$) and the sphere $x^2 + y^2 + z^2 = 4$ (i.e. $r^2 + z^2 = 4$). The sphere encloses the cylinder, and so the bounds of this region are the upper and lower portions of the sphere within the cylinder. Figure B.23 shows this region which looks like a tennis ball can with rounded top and bottom; the spherical caps are from the sphere. Thus, we have

$$-\sqrt{4 - r^2} \le z \le \sqrt{4 - r^2} \quad ; \quad 0 \le r \le 1 \quad ; \quad 0 \le \theta \le 2\pi$$

and the volume is (with the volume element $dV = r\, dz dr d\theta$ in cylindrical coordinates):

$$V = \iiint_E (1)dV = \int_0^{2\pi} \int_0^1 \int_{-\sqrt{4-r^2}}^{\sqrt{4-r^2}} r dz dr d\theta = \frac{32\pi}{3} - 4\pi\sqrt{3} \quad \blacksquare$$

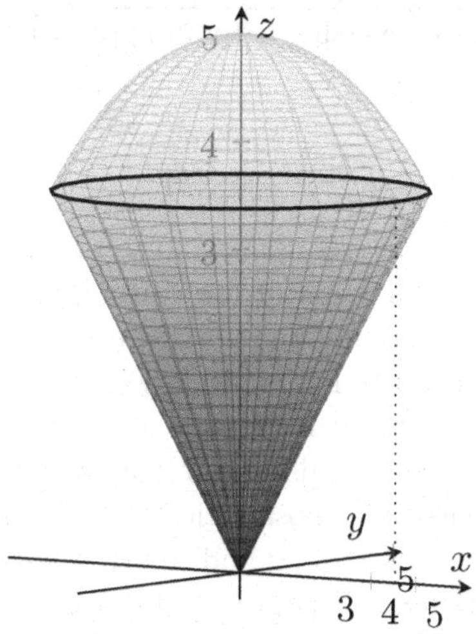

Fig. B.22 Above $z^2 = x^2 + y^2$, inside $\rho = 5$.

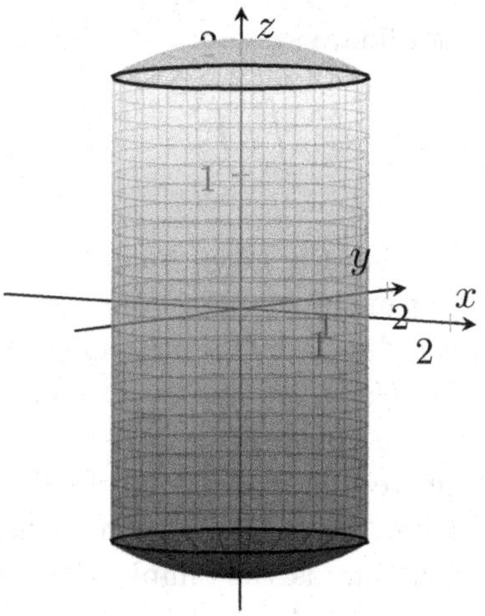

Fig. B.23 Volume common to both $x^2 + y^2 = 1$ and $\rho = 2$.

---

(6) Evaluate $\displaystyle\iiint_E xyz\, dV$, where E is the region between between the spheres $\rho = 2$ and $\rho = 4$ and above the cone $\phi = \pi/3$.

□ The given region E of integration is between the spheres $\rho = 2$ and $\rho = 4$ and above the cone $\phi = \pi/3$. Bounds on ρ are given by the spheres. The region above the cone is $0 \leq \phi \leq \pi/3$ (remember, ϕ is measured from the top down). And, since there are no other restrictions, $0 \leq \theta \leq 2\pi$. Also, converting the given function,

$$xyz = (\rho \sin \phi \cos \theta)(\rho \sin \phi \sin \theta)(\rho \cos \phi) = \rho^3 \sin^2 \phi \cos \phi \sin \theta \cos \theta$$

The volume element in spherical coordinates is $dV = \rho^2 \sin \phi d\rho d\phi d\theta$. So, the triple integral becomes,

$$\int_0^{2\pi} \int_0^{\pi/3} \int_2^4 (\rho^3 \sin^2 \phi \cos \phi \sin \theta \cos \theta)\, \rho^2 \sin \phi d\rho d\phi d\theta = 0$$

Note that 0 is a perfectly valid result, since this integral does NOT compute a volume! ■

(7) Evaluate the following integral by first converting it to an appropriate coordinate system:

$$\int_0^3 \int_0^{\sqrt{9-y^2}} \int_{\sqrt{x^2+y^2}}^{\sqrt{18-x^2-y^2}} (x^2+y^2+z^2)\, dzdxdy$$

□ From the given limits of integration on x and y, we can recognize that in the xy-plane, our region of integration is the portion of a circle of radius 3 in the first quadrant. From the limits on z, we go from $z = \sqrt{x^2+y^2}$ to $z = \sqrt{18-x^2-y^2}$, that is from the cone $z^2 = x^2+y^2$ to the sphere $x^2+y^2+z^2 = 18$. Since bounds of our region are a sphere and a cone, spherical coordinates are most logical. So, we're under the top half of the sphere $\rho = \sqrt{18} = 3\sqrt{2}$ and above a cone; this cone is the simple 45-degree wide cone, so it goes from $\phi = 0$ to $\phi = \pi/4$. And, we're over the first quadrant in the xy-plane, so $0 \leq \theta \leq \pi/2$. Remember that for spherical coordinates, the volume element is $dV = \rho^2 \sin\phi\, d\rho d\phi d\theta$.

$$\begin{aligned}
&\int_0^3 \int_0^{\sqrt{9-y^2}} \int_{\sqrt{x^2+y^2}}^{\sqrt{18-x^2-y^2}} (x^2+y^2+z^2)\, dzdxdy \\
&\qquad = \int_0^{\pi/2} \int_0^{\pi/4} \int_0^{3\sqrt{2}} (\rho^2)\rho^2 \sin\phi d\rho d\phi d\theta \\
&\qquad = \int_0^{\pi/2} \int_0^{\pi/4} \int_0^{3\sqrt{2}} \rho^4 \sin\phi d\rho d\phi d\theta \\
&\qquad = \frac{486\pi}{5}(\sqrt{2}-1)
\end{aligned}$$

B.4 Chapter 16: Practice Problem Solutions

B.4.1 *Vector Basics — Practice — Solved*

(1) If $\mathbf{a} = 2\mathbf{i} - 3\mathbf{j}$ and $\mathbf{b} = \mathbf{i} + 5\mathbf{j}$, what are $|\mathbf{a}|$, $\mathbf{a} + \mathbf{b}$, $\mathbf{a} - \mathbf{b}$, $2\mathbf{a}$, $3\mathbf{a} + 4\mathbf{b}$?

$$\begin{aligned} \square \quad |\mathbf{a}| &= \sqrt{4+9} = \sqrt{13} \\ \mathbf{a} + \mathbf{b} &= 3\mathbf{i} + 2\mathbf{j} \\ \mathbf{a} - \mathbf{b} &= \mathbf{i} - 8\mathbf{j} \\ 2\mathbf{a} &= 4\mathbf{i} - 6\mathbf{j} \\ 3\mathbf{a} + 4\mathbf{b} &= (6+4)\mathbf{i} + (-9+20)\mathbf{j} = 10\mathbf{i} + 11\mathbf{j} \quad \blacksquare \end{aligned}$$

(2) Find a unit vector in the same direction as $8\mathbf{i} - \mathbf{j} + 4\mathbf{k}$.

□ The length of $8\mathbf{i} - \mathbf{j} + 4\mathbf{k}$ is $\sqrt{64+1+16} = 9$, so a unit vector in the same direction is

$$\frac{8}{9}\mathbf{i} - \frac{1}{9}\mathbf{j} + \frac{4}{9}\mathbf{k} \quad \blacksquare$$

(3) Find a vector of length 6 in the same direction as $\mathbf{v} = \langle -2, 4, 2 \rangle$.

□ The vector $\mathbf{v} = \langle -2, 4, 2 \rangle$ has length $|\mathbf{v}| = \sqrt{4+16+4} = 2\sqrt{6}$, so a unit vector in the same direction is

$$\frac{\mathbf{v}}{|\mathbf{v}|} = \left\langle -\frac{2}{2\sqrt{6}}, \frac{4}{2\sqrt{6}}, \frac{2}{2\sqrt{6}} \right\rangle = \left\langle -\frac{1}{\sqrt{6}}, \frac{2}{\sqrt{6}}, \frac{1}{\sqrt{6}} \right\rangle$$

and therefore a vector in the same direction having length 6 is

$$6\frac{\mathbf{v}}{|\mathbf{v}|} = \left\langle -\frac{6}{\sqrt{6}}, \frac{6 \cdot 2}{\sqrt{6}}, \frac{6}{\sqrt{6}} \right\rangle = \langle -\sqrt{6}, 2\sqrt{6}, \sqrt{6} \rangle \quad \blacksquare$$

(4) If $\mathbf{v} = \langle v_1, v_2 \rangle$ and $c > 0$, prove that $|c\mathbf{v}| = c|\mathbf{v}|$, i.e. prove that the length of the vector $c\mathbf{v}$ is c times the length of $\mathbf{v}$. (You can use specific examples to test out the process, but in the end, you should prove the general rule for any vector $\mathbf{v} = \langle v_1, v_2 \rangle$.)

□ Since $c\mathbf{v} = \langle cv_1, cv_2 \rangle$, its length is

$$|c\mathbf{v}| = \sqrt{(cv_1)^2 + (cv_2)^2} = \sqrt{c^2(v_1^2 + v_2^2)} = c\sqrt{v_1^2 + v_2^2}$$

But the square root on the right is known as $\sqrt{v_1^2 + v_2^2} = |\mathbf{v}|$, and so $|c\mathbf{v}| = c|\mathbf{v}|$. ■

(5) If $\mathbf{v} = \langle v_1, v_2 \rangle$, prove that $\left| \frac{\mathbf{v}}{|\mathbf{v}|} \right| = 1$. (You can use specific examples to test out the process, but in the end, you should prove the general rule for any vector $\mathbf{v} = \langle v_1, v_2 \rangle$.)

☐ Since $|\mathbf{v}| = \sqrt{v_1^2 + v_2^2}$, then

$$\frac{\mathbf{v}}{|\mathbf{v}|} = \frac{1}{|\mathbf{v}|}\mathbf{v} = \frac{1}{\sqrt{v_1^2 + v_2^2}}\langle v_1, v_2 \rangle = \left\langle \frac{v_1}{\sqrt{v_1^2 + v_2^2}}, \frac{v_1}{\sqrt{v_1^2 + v_2^2}} \right\rangle$$

and so

$$\begin{aligned} \left| \frac{\mathbf{v}}{|\mathbf{v}|} \right| &= \sqrt{\left(\frac{v_1}{\sqrt{v_1^2 + v_2^2}} \right)^2 + \left(\frac{v_2}{\sqrt{v_1^2 + v_2^2}} \right)^2} \\ &= \sqrt{\frac{v_1^2}{v_1^2 + v_2^2} + \frac{v_2}{v_1^2 + v_2^2}} \\ &= \sqrt{\frac{v_1^2 + v_2^2}{v_1^2 + v_2^2}} = 1 \quad \blacksquare \end{aligned}$$

B.4.2 *Dot and Cross Products — Practice — Solved*

(1) What is the angle between $\mathbf{a} = \langle 6, -3, 2 \rangle$ and $\mathbf{b} = \langle 2, 1, -2 \rangle$?

☐ We have $\mathbf{a} \cdot \mathbf{b} = 5$, $|\mathbf{a}| = 7$ and $|\mathbf{b}| = 3$ so that

$$\begin{aligned} \mathbf{a} \cdot \mathbf{b} &= |\mathbf{a}||\mathbf{b}| \cos\theta \\ 5 &= (7)(3)\cos\theta \\ \cos\theta &= \frac{5}{21} \quad \Rightarrow \quad \theta \approx 76^\circ \quad \blacksquare \end{aligned}$$

(2) Are the vectors $\mathbf{a} = \langle -1, 2, 5 \rangle$ and $\mathbf{b} = \langle 3, 4, -1 \rangle$ parallel, perpendicular, or neither?

☐ We have $\mathbf{a} \cdot \mathbf{b} = 0$, so that the vectors are perpendicular. ■

(3) Are the vectors $\mathbf{a} = \langle 2, 6, -4 \rangle$ and $\mathbf{b} = \langle -3, -9, 6 \rangle$ parallel, perpendicular, or neither?

☐ We have $\mathbf{a} \cdot \mathbf{b} = -84$, $|\mathbf{a}| = \sqrt{56} = 2\sqrt{14}$, and $|\mathbf{b}| = \sqrt{126} = 3\sqrt{14}$.

So,

$$\begin{aligned}\mathbf{a}\cdot\mathbf{b} &= |\mathbf{a}||\mathbf{b}|\cos\theta\\ -84 &= (2\sqrt{14}(3\sqrt{14})\cos\theta\\ \cos\theta &= -\frac{84}{6(14)} = -1 \to \theta = \pi\end{aligned}$$

Since the angle between them is π, the vectors are parallel (although pointing in the opposite direction). ■

(4) If $\mathbf{v} = \langle -1, -2, 2\rangle$ and $\mathbf{w} = \langle 3, 3, 4\rangle$, what are $\text{comp}_{\mathbf{v}}\mathbf{w}$ and $\text{proj}_{\mathbf{v}}\mathbf{w}$?

□ We want the component and projection of $\mathbf{w}$ along $\mathbf{v}$. We need $\mathbf{v}\cdot\mathbf{w} = -1$ and $|\mathbf{v}| = 3$ so that

$$\text{comp}_{\mathbf{v}}\mathbf{w} = \frac{\mathbf{v}\cdot\mathbf{w}}{|\mathbf{v}|} = \frac{-1}{3}$$

which gets used in:

$$\begin{aligned}\text{proj}_{\mathbf{v}}\mathbf{w} = \text{comp}_{\mathbf{v}}\mathbf{w}\left(\frac{\mathbf{v}}{|\mathbf{v}|}\right) &= \left(\frac{-1}{3}\right)\left(\frac{1}{3}\right)\langle -1, -2, 2\rangle\\ &= \frac{-1}{9}\langle -1, -2, 2\rangle = \left\langle \frac{1}{9}, \frac{2}{9}, -\frac{2}{9}\right\rangle\end{aligned}$$ ■

(5) If $\mathbf{a} = \langle 1, -1, 1\rangle$ and $\mathbf{b} = \langle 1, 1, 1\rangle$, demonstrate that $\mathbf{a}\times\mathbf{b}$ is perpendicular to both $\mathbf{a}$ and $\mathbf{b}$.

□ We have

$$\begin{aligned}\mathbf{c} = \mathbf{a}\times\mathbf{b} = \begin{vmatrix}\mathbf{i} & \mathbf{j} & \mathbf{k}\\ 1 & -1 & 1\\ 1 & 1 & 1\end{vmatrix} &= \begin{vmatrix}-1 & 1\\ 1 & 1\end{vmatrix}\mathbf{i} - \begin{vmatrix}1 & 1\\ 1 & 1\end{vmatrix}\mathbf{j} + \begin{vmatrix}1 & -1\\ 1 & 1\end{vmatrix}\mathbf{k}\\ &= (-1-1)\mathbf{i} - (1-1)\mathbf{j} + (1-(-1))\mathbf{k} = -2\mathbf{i} + 2\mathbf{k}\end{aligned}$$

And then we have

$$\begin{aligned}\mathbf{c}\cdot\mathbf{a} &= (-2)(1) + (0)(-1) + (2)(1) = 0\\ \mathbf{c}\cdot\mathbf{b} &= (-2)(1) + (0)(1) + (2)(1) = 0\end{aligned}$$

so that $\mathbf{a}\times\mathbf{b}$ is orthogonal to both $\mathbf{a}$ and $\mathbf{b}$. ■

(6) Which of the following are legitimate expressions? (a) $\mathbf{a}\cdot(\mathbf{b}\cdot\mathbf{c})$, (b) $\mathbf{a}\times(\mathbf{b}\cdot\mathbf{c})$, and (c) $(\mathbf{a}\times\mathbf{b})\cdot(\mathbf{c}\times\mathbf{d})$.

□ (a) is not valid, the cross product of a vector $\mathbf{a}$ and a scalar $\mathbf{b}\cdot\mathbf{c}$ is not defined.

(b) is not valid, the cross product of a vector $\mathbf{a}$ and a scalar $\mathbf{b}\cdot\mathbf{c}$ is not defined.

(c) is fine, $\mathbf{a}\times\mathbf{b}$ and $\mathbf{c}\times\mathbf{d}$ are both vectors, so their dot product is defined. ■

---

(7) If $\mathbf{a} = \langle 2, -3, 1\rangle$ and $\mathbf{b} = \langle 1, 6, -2\rangle$, what are $\text{comp}_{\mathbf{a}}\mathbf{b}$ and $\text{proj}_{\mathbf{a}}\mathbf{b}$?

□ We have $\mathbf{a}\cdot\mathbf{b} = -18$ and $|\mathbf{a}| = \sqrt{14}$ so that

$$\text{comp}_{\mathbf{a}}\mathbf{b} = \frac{\mathbf{a}\cdot\mathbf{b}}{|\mathbf{a}|} = \frac{-18}{\sqrt{14}}$$

and with that,

$$\begin{aligned}\text{proj}_{\mathbf{a}}\mathbf{b} &= \left(\frac{\mathbf{a}\cdot\mathbf{b}}{|\mathbf{a}|}\right)\left(\frac{\mathbf{a}}{|\mathbf{a}|}\right) = \left(\frac{-18}{\sqrt{14}}\right)\left(\frac{1}{\sqrt{14}}\right)\langle 2, -3, 1\rangle \\ &= \frac{-9}{7}\langle 2, -3, 1\rangle = \left\langle -\frac{18}{7}, \frac{27}{7}, -\frac{9}{7}\right\rangle = -\frac{18}{7}\mathbf{i} + \frac{27}{7}\mathbf{j} - \frac{9}{7}\mathbf{k}\end{aligned}$$ ■

---

(8) Find a unit vector orthogonal to both $\mathbf{i}+\mathbf{j}$ and $\mathbf{i}+\mathbf{k}$.

□ Let's find just any vector orthogonal to both, then scale it to a unit vector. A vector orthogonal to both of these vectors will come from their cross product. We can find that $\langle 1, 1, 0\rangle \times \langle 1, 0, 1\rangle = \langle 1, -1, -1\rangle$; the length of this vector is $\sqrt{3}$. Therefore a unit vector orthogonal to the two given vectors is

$$\left\langle \frac{1}{\sqrt{3}}, -\frac{1}{\sqrt{3}}, -\frac{1}{\sqrt{3}}\right\rangle$$ ■

---

(9) We know that the dot product of a vector with itself is related to the length of the vector. What is special about the cross product of a vector with itself? Test this out for the general case $\mathbf{v} = \langle v_1, v_2, v_3\rangle$ and show all the details to support your conclusion.

□ We have

$$\begin{aligned}\mathbf{v}\times\mathbf{v} &= \begin{vmatrix} \mathbf{i} & \mathbf{j} & \mathbf{k} \\ v_1 & v_2 & v_3 \\ v_1 & v_2 & v_3 \end{vmatrix} = \begin{vmatrix} v_2 & v_3 \\ v_2 & v_3 \end{vmatrix}\mathbf{i} - \begin{vmatrix} v_1 & v_3 \\ v_1 & v_3 \end{vmatrix}\mathbf{j} + \begin{vmatrix} v_1 & v_1 \\ v_2 & v_2 \end{vmatrix}\mathbf{k} \\ &= (v_2v_3 - v_2v_3)\,\mathbf{i} - (v_1v_3 - v_1v_3)\,\mathbf{j} + (v_1v_2 - v_1v_2)\,\mathbf{k} \\ &= 0\mathbf{i} + 0\mathbf{j} + 0\mathbf{k}\end{aligned}$$

The cross product of a vector with itself always gives the zero vector $\mathbf{0}$! ■

B.4.3 *Vector Functions — Practice — Solved*

(1) Find the vector and parametric equations of the line segment joining the points $P(1,0,1)$ and $Q(2,3,1)$.

☐ This vector function has initial vector $\mathbf{r}_0 = \langle 1,0,1\rangle$ and parallel vector $\mathbf{r}_1 = \mathbf{PQ} = \langle 1,3,0\rangle$ so that its equation is

$$\mathbf{r}_0 + t\mathbf{r}_1 = \langle 1,0,1\rangle + t\,\langle 1,3,0\rangle = \langle 1+t, 3t, 1\rangle$$

To restrict this to the line segment that starts at P and ends at Q, we need $0 \le t \le 1$. We can then write the equation of this line segment as the vector function

$$\mathbf{r}(t) = \langle 1+t, 3t, 1\rangle \quad \text{for} \quad 0 \le t \le 1$$

or using the parametric equations

$$x = 1+t \;\; ; \;\; y = 3t \;\; ; \;\; z = 1 \quad \text{for} \quad 0 \le t \le 1 \;\; \blacksquare$$

(2) Evaluate $\displaystyle\lim_{t\to 1}\left(\sqrt{t+3}\,\mathbf{i} + \frac{t-1}{t^2-1}\mathbf{j} + \frac{\tan t}{t}\mathbf{k}\right)$.

☐ Notice that the middle component simplifies to $1/(t+1)$, so that

$$\begin{aligned}&\lim_{t\to 1}\left(\sqrt{t+3}\,\mathbf{i} + \frac{t-1}{t^2-1}\mathbf{j} + \frac{\tan t}{t}\mathbf{k}\right)\\ &= \lim_{t\to 1}\left(\sqrt{t+3}\,\mathbf{i} + \frac{1}{t+1}\mathbf{j} + \frac{\tan t}{t}\mathbf{k}\right)\\ &= 2\,\mathbf{i} + \frac{1}{2}\mathbf{j} + \tan(1)\,\mathbf{k} \;\; \blacksquare\end{aligned}$$

(3) If $\mathbf{r}(t) = (at\cos 3t)\,\mathbf{i} + (b\sin^3 t)\,\mathbf{j} + (c\cos^3 t)\,\mathbf{k}$, what is $\mathbf{r}'(t)$?

☐ By direct evaluation (with a product rule and a couple of chain rules),

$$\mathbf{r}'(t) = (a\cos 3t - 3at\sin 3t)\mathbf{i} + (3b\sin^2 t\cos t)\mathbf{j} - (3c\cos^2 t\sin t)\mathbf{k} \;\; \blacksquare$$

(4) Identify the curve given by $\mathbf{r}(t) = e^t\,\mathbf{i} + e^{-t}\,\mathbf{j}$ and find a vector tangent to the curve at $t = 0$.

☐ By eliminating t from $x(t) = e^t, y(t) = e^{-t}$, we find that this is the hyperbola $y = 1/x$. Since $\mathbf{r}'(t) = e^t\mathbf{i} - e^{-t}\mathbf{j}$, then a vector tangent to the curve at $t = 0$ is $\mathbf{r}'(0) = \mathbf{i} - \mathbf{j}$. ■

(5) Find the vector equation of the line tangent to $\mathbf{r}(t) = \langle t^2-1, t^2+1, t+1\rangle$ at $\langle -1, 1, 1\rangle$.

☐ Note that the vector $\langle -1, 1, 1\rangle$ is given by $t = 0$. We need:

$$\mathbf{r}'(t) = \langle 2t, 2t, 1\rangle$$
$$\mathbf{r}'(0) = \langle 0, 0, 1\rangle$$

So we have an initial vector $\mathbf{r}_0 = \langle -1, 1, 1\rangle$ and a parallel vector $\mathbf{r}'(0)$, so the vector equation of the tangent line is

$$\mathbf{r}_0 + t\mathbf{r}'(0) = \langle -1, 1, 1\rangle + t\,\langle 0, 0, 1\rangle = \langle -1, 1, 1+t\rangle \quad \blacksquare$$

(6) Find the vector equation of the line tangent to $\mathbf{r}(t) = \langle \ln t, 2\sqrt{t}, t^2\rangle$ at $\langle 0, 2, 1\rangle$.

☐ Note that $\langle 0, 2, 1\rangle$ is given by $t = 1$. We need:

$$\mathbf{r}'(t) = \left\langle \frac{1}{t}, \frac{1}{\sqrt{t}}, 2t \right\rangle$$
$$\mathbf{r}'(1) = \langle 1, 1, 2\rangle$$

So we have an initial vector $\mathbf{r}_0 = \langle 0, 2, 1\rangle$ and a parallel vector $\mathbf{r}'(1)$, and so the vector equation of the line is

$$\mathbf{r}(t) = \mathbf{r}_0 + t\,\mathbf{r}'(1) = \langle 0, 2, 1\rangle + t\,\langle 1, 1, 2\rangle = \langle t, 2+t, 1+2t\rangle \quad \blacksquare$$

(7) Based on Useful Fact 16.10, you may be suspicious that there could be an expression like this, under the right circumstances:

$$\frac{d}{dt}\,\mathbf{p}(t) \times \mathbf{r}(t) = \mathbf{p}'(t) \times \mathbf{r}(t) + \mathbf{p}(t) \times \mathbf{r}'(t)$$

Test this relation on the vector functions $\mathbf{p}(t) = \langle 1, t, t^2\rangle$ and $\mathbf{r}(t) = \langle t^2, t, 1\rangle$ by constructing both sides separately and comparing the results.

☐ Let's work on the left side first. Since

$$\mathbf{p}(t) \times \mathbf{r}(t) = \langle t - t^3, t^4 - 1, t - t^3\rangle$$

(details omitted, you can do it!), we have

$$\frac{d}{dt}\,\mathbf{p}(t) \times \mathbf{r}(t) = \langle 1 - 3t^2, 4t^3, 1 - 3t^2\rangle$$

To build the right side, we build $\mathbf{p}'(t) = \langle 0, 1, 2t\rangle$ and $\mathbf{r}'(t) = \langle 2t, 1, 0\rangle$, so that

$$\begin{aligned}\mathbf{p}'(t) \times \mathbf{r}(t) &= \langle 0, 1, 2t\rangle \times \langle t^2, t, 1\rangle = \langle 1 - 2t^2, 2t^3, -t^2\rangle \\ \mathbf{p}(t) \times \mathbf{r}'(t) &= \langle 1, t, t^2\rangle \times \langle 2t, 1, 0\rangle = \langle -t^2, 2t^3, 1 - 2t^2\rangle\end{aligned}$$

and

$$\begin{aligned}\mathbf{p}'(()t) \times \mathbf{r}(t) + \mathbf{p}(t) \times \mathbf{r}'(t) &= \langle 1 - 2t^2, 2t^3, -t^2\rangle + \langle -t^2, 2t^3, 1 - 2t^2\rangle \\ &= \langle 1 - 3t^2, 4t^3, 1 - 3t^2\rangle\end{aligned}$$

By direct computation of each, we've confirmed that — at least in this case —

$$\frac{d}{dt}\,\mathbf{p}(t) \times \mathbf{r}(t) = \mathbf{p}'(t) \times \mathbf{r}(t) + \mathbf{p}(t) \times \mathbf{r}'(t)$$

This is by no means definitive proof that this relation is always true, but we've also failed to disprove it. ■

---

(8) Evaluate $\displaystyle\int_0^1 \left(\frac{4}{1+t^2}\,\mathbf{j} + \frac{2t}{1+t^2}\,\mathbf{k}\right) dt$.

□ It doesn't really matter to the evaluation of the integral, but this vector function exists only in the yz-plane. The integral is:

$$\begin{aligned}\int_0^1 \left(\frac{4}{1+t^2}\,\mathbf{j} + \frac{2t}{1+t^2}\,\mathbf{k}\right) dt &= \left(4\tan^{-1}(t)\,\mathbf{j} + \ln(1+t^2)\,\mathbf{k}\right)\Big|_0^1 \\ &= 4\tan^{-1}(1)\,\mathbf{j} + \ln(2)\,\mathbf{k} = \pi\,\mathbf{j} + \ln(2)\,\mathbf{k} \quad ■\end{aligned}$$

---

(9) Write the integral that gives the arc length of the vector function from Practice Problem 5 from the point $(0, 2, 2)$ to the point $(15, 17, 5)$.

□ The vector function in question is $\mathbf{r}(t) = \langle t^2 - 1, t^2 + 1, t + 1\rangle$. Note that the points $(0, 2, 2)$ and $(15, 17, 5)$ correspond to $t = 1$ and $t = 4$, respectively. The integrand of the arc length integral will come from components of $\mathbf{r}'(t) = \langle 2t, 2t, 1\rangle$ as follows:

$$\sqrt{(2t)^2 + (2t)^2 + (1)^2} = \sqrt{4t^2 + 4t^2 + 1} = \sqrt{8t^2 + 1}$$

and the arc length integral is

$$L = \int_1^4 \sqrt{8t^2 + 1}\, dt \quad ■$$

B.4.4 *Vector Fields and the Gradient — Practice — Solved*

(1) Find the curl and divergence of $\mathbf{F}(x, y, z) = \langle 0, \cos(xz), -\sin(xy)\rangle$. Is this vector field irrotational or incompressible?

☐ The curl is:

$$\nabla \times \mathbf{F} = \begin{vmatrix} \mathbf{i} & \mathbf{j} & \mathbf{k} \\ \frac{\partial}{\partial x} & \frac{\partial}{\partial y} & \frac{\partial}{\partial z} \\ 0 & \cos(xz) & -\sin(xy) \end{vmatrix}$$
$$= \langle -x\cos(xy) + x\sin(xz), y\cos(xy), -z\sin(xz)\rangle$$

The divergence is

$$\nabla \cdot \mathbf{F} = \left\langle \frac{\partial}{\partial x}, \frac{\partial}{\partial y}, \frac{\partial}{\partial z} \right\rangle \cdot \mathbf{F}$$
$$= \frac{\partial}{\partial x}(0) + \frac{\partial}{\partial y}(\cos(xz)) + \frac{\partial}{\partial z}(-\sin(xy)) = 0$$

Since $\nabla \times \mathbf{F} \neq \mathbf{0}$, the vector field is not irrotational. But $\nabla \times \mathbf{F} = 0$, so it is incompressible. ■

(2) Is this vector field irrotational or incompressible?

$$\mathbf{F}(x, y, z) = \left\langle \frac{x}{x^2 + y^2 + z^2}, \frac{y}{x^2 + y^2 + z^2}, \frac{z}{x^2 + y^2 + z^2} \right\rangle$$

☐ The curl is:

$$\nabla \times \mathbf{F} = \begin{vmatrix} \mathbf{i} & \mathbf{j} & \mathbf{k} \\ \frac{\partial}{\partial x} & \frac{\partial}{\partial y} & \frac{\partial}{\partial z} \\ \frac{x}{x^2+y^2+z^2} & \frac{y}{x^2+y^2+z^2} & \frac{z}{x^2+y^2+z^2} \end{vmatrix} = \langle 0, 0, 0\rangle$$

Since the curl is $\mathbf{0}$, the vector field is irrotational. The divergence is

$$\nabla \cdot \mathbf{F} = \left\langle \frac{\partial}{\partial x}, \frac{\partial}{\partial y}, \frac{\partial}{\partial z} \right\rangle \cdot \mathbf{F}$$
$$= \frac{\partial}{\partial x} \frac{x}{x^2 + y^2 + z^2} + \frac{\partial}{\partial y} \frac{y}{x^2 + y^2 + z^2} + \frac{\partial}{\partial z} \frac{z}{x^2 + y^2 + z^2}$$
$$= \frac{y^2 + z^2 - x^2}{(x^2 + y^2 + z^2)^2} + \frac{x^2 + z^2 - y^2}{(x^2 + y^2 + z^2)^2} + \frac{x^2 + y^2 - z^2}{(x^2 + y^2 + z^2)^2}$$
$$= \frac{x^2 + y^2 + z^2}{(x^2 + y^2 + z^2)^2} = \frac{1}{x^2 + y^2 + z^2}$$

The vector field is not incompressible. ■

(3) Find the gradient of $f(x,y) = y \ln x$, and determine $|\nabla f(1,-3)|$.

☐ First, let's just get the gradient,

$$\nabla f(x,y) = \langle f_x, f_y \rangle = \langle y/x, \ln x \rangle$$

Now note that $|\nabla f(1,-3)|$ is asking for the *magnitude* of the gradient at $(1,-3)$. So,

$$\nabla f(1,-3) = \langle -3, 0 \rangle$$
$$|\nabla f(1,-3)| = \sqrt{(-3)^2 + 0^2} = 3 \quad \blacksquare$$

(4) Assuming s and t are still rectangular coordinates, find the gradient of $g(s,t) = e^{-s} \sin t$.

☐ The lettering doesn't matter; if we're in rectangular coordinates, the gradient will be the vector field composed of the first derivatives with respect to the two variables:

$$\nabla g(s,t) = \langle g_s, g_t \rangle = \langle -e^{-s} \sin t, e^{-s} \cos t \rangle \quad \blacksquare$$

(5) Let f be a scalar function and $\mathbf{F}$ a vector field. Describe why each operation is defined or undefined: (a) $\nabla \times (\nabla \times \mathbf{F})$; (b) $\nabla \cdot (\nabla \cdot \mathbf{F})$; (c) $\nabla f \times \nabla \cdot \mathbf{F}$; (d) $\nabla \cdot (\nabla \times (\nabla f))$.

☐ (a) $\nabla \times (\nabla \times \mathbf{F})$ is OK since $\nabla \times \mathbf{F}$ is a vector function and we find the curl of vector functions.
(b) $\nabla \cdot (\nabla \cdot \mathbf{F})$ is undefined since $\nabla \cdot \mathbf{F}$ is a scalar function and we find the divergence of vector functions.
(c) $\nabla f \times \nabla \cdot \mathbf{F}$ is undefined since $\nabla \cdot \mathbf{F}$ is a scalar function and we find the cross product of two vector functions.
(d) $\nabla \cdot (\nabla \times (\nabla f))$ is OK since each operation produces the type of expression needed by the next one (∇f is a vector function, etc.) $\blacksquare$

(6) Use the partial derivatives found in PP 9 of Sec. 13.5 to construct gradient vectors for the function $z = 3x^2 - 2y^2$ at the points $(1,0)$, $(1,1)$, and $(0,1)$. Create a new version of Fig. B.3 and draw unit vectors in the direction of each gradient vector at the three given points. (The gradient vectors themselves will be too large to fit in the figure.) Do you notice anything interesting about the relationship between the gradient vectors (scaled to unit vectors) and the level curves they start on?

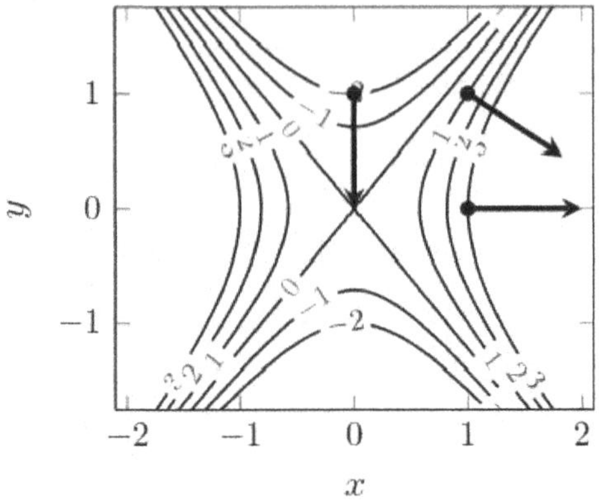

Fig. B.24 Gradients at level curves of $z = 3x^2 - 2y^2$ (w/ PP 6).

□ Since $z_x(1,0) = 6$ and $z_y(1,0) = 0$, then $\nabla(1,0) = \langle 6,0\rangle$. A unit vector in this direction is $\langle 1,0\rangle$.
Since $z_x(1,1) = 6$ and $z_y(1,1) = -4$, then $\nabla(1,1) = \langle 6,-4\rangle$. A unit vector in this direction is $\langle 6/\sqrt{52}, -4/\sqrt{52}\rangle$.

Since $z_x(0,1) = 0$ and $z_y(0,1) = -4$, then $\nabla(0,1) = \langle 0,-4\rangle$. A unit vector in this direction is $\langle 0,-1\rangle$.

These three unit vectors showing the direction of the gradient vectors are displayed in Fig. B.24; it looks like each gradient vector is perpendicular to the level curve at which it originates! ■

B.4.5 *Planes and Tangent Planes — Practice — Solved*

(1) Find the equation of the plane containing the point $(4, 0, 3)$ and with normal vector $\langle 0, 1, 2\rangle$.

☐ We have all the information we need: a point $(x_0, y_0, z_0) = (4, 0, 3)$ and a normal vector $\mathbf{n} = \langle a, b, c\rangle = \langle 0, 1, 2\rangle$ so using the standard form of a plane,

$$\begin{aligned} a(x - x_0) + b(y - y_0) + c(z - z_0) &= 0 \\ 0(x - 4) + 1(y - 0) + 2(z - 3) &= 0 \\ y + 2z &= 6 \quad \blacksquare \end{aligned}$$

(2) Find the equation of the plane containing the points $P(0, 0, 0)$, $Q(2, -4, 6)$ and $R(5, 1, 3)$.

☐ We have our choice of three points on the plane, but we still need a vector perpendicular to this plane. But $\mathbf{PQ} \times \mathbf{PR}$ is such a vector:

$$\begin{aligned} \mathbf{PQ} &= \langle 2, -4, 6\rangle \\ \mathbf{PR} &= \langle 5, 1, 3\rangle \\ \mathbf{PQ} \times \mathbf{PR} &= \langle -18, 24, 22\rangle \end{aligned}$$

So the equation of the plane is (using P as the point),

$$\begin{aligned} -18(x - 0) + (24)(y - 0) + (22)(z - 0) &= 0 \\ -18x + 24y + 22z &= 0 \\ -9x + 12y + 11z &= 0 \quad \blacksquare \end{aligned}$$

(3) Are the planes $2z = 4y - x$ and $3x - 12y + 6z = 1$ parallel, perpendicular, or neither?

☐ From the equation of the plane $2z = 4y - x$, we know its perpendicular vector is $\mathbf{n}_1 = \langle -1, 4, -2\rangle$. From the equation of the plane $3x - 12y + 6z = 1$, we know its perpendicular vector is $\mathbf{n}_2 = \langle 3, -12, 6\rangle$. Since $\mathbf{n}_2 = -3\mathbf{n}_1$, these vectors, and so also the planes, are parallel. ■

(4) Find the equation of the plane tangent to $z = y \ln x$ at $(1, 4, 0)$.

☐ We need:

$$\begin{aligned} f_x &= \frac{y}{x} \to f_x(1, 4) = 4 \\ f_y &= \ln x \to f_y(1, 4) = 0 \end{aligned}$$

so the plane is:

$$\begin{aligned} z - z_0 &= f_x(x_0, y_0)(x - x_0) + f_y(x_0, y_0)(y - y_0) \\ z - 0 &= 4(x - 1) + 0(y - 4) \\ -4x + z &= -4 \quad \blacksquare \end{aligned}$$

---

(5) If $z = 5x^2 + y^2$ and (x, y) changes from $(1, 2)$ to $(1.05, 2.1)$, then what is the resulting dz?

□ We have $dx = 0.05$ and $dy = 0.1$, so

$$\begin{aligned} dz &= f_x(x, y)dx + f_y(x, y)dy = 10x\,dx + 2y\,dy \\ &= 10(1)(0.05) + 2(2)(0.1) = 0.9 \quad \blacksquare \end{aligned}$$

---

(6) Find the equation of the plane containing the point $(-2, 8, 10)$ and perpendicular to the line $x = 1 + t, y = 2t, z = 4 - 3t$.

□ From the equation of the line, a vector parallel to that line is $\langle 1, 2, -3 \rangle$. If the plane is perpendicular to the line, it is perpendicular to this vector, so we can use it as the plane's perpendicular vector:

$$\begin{aligned} 1(x + 2) + 2(y - 8) + (-3)(z - 10) &= 0 \\ x + 2y - 3z &= -16 \quad \blacksquare \end{aligned}$$

---

(7) Find the equation of the plane tangent to $z = e^{x^2 - y^2}$ where $(x, y) = (1, -1)$.

□ We need:

$$\begin{aligned} z_0 = f(x_0, y_0) &= e^0 = 1 \\ f_x = 2xe^{x^2-y^2} &\to f_x(1, -1) = 2e^{1-1} = 2 \\ f_y = -2ye^{x^2-y^2} &\to f_y(1, -1) = 2e^{1-1} = 2 \end{aligned}$$

so the plane is:

$$\begin{aligned} z - z_0 &= f_x(x_0, y_0)(x - x_0) + f_y(x_0, y_0)(y - y_0) \\ z - 1 &= 2(x - 1) + 2(y + 1) \\ 2x + 2y - z &= -1 \quad \blacksquare \end{aligned}$$

---

(8) The length and width of a rectangle are measured as 30cm and 24cm respectively. There is a possible error in measurement of 0.1cm in each direction. Estimate the maximum possible error in the calculated area of the rectangle. (Hint: This is a total differential problem.)

☐ Since $A = LW$, we have $A_L = W$ and $A_W = L$. We are given $L = 30$ and $W = 24$, with $dW = dL = 0.1$. Therefore the total possible error in A is dA, given by

$$\begin{aligned} dA = A_L(L,W)dL + A_W(L,W)dW &= (W)\,dL + (L)\,dW \\ &= 24(0.1) + (30)(0.1) = 5.4 \end{aligned}$$

There is a possible error of $5.4\,cm^2$ in the total area measured. ■

B.4.6 *Directional Derivatives — Practice — Solved*

(1) Find the directional derivative of $f(x,y) = y\ln x$ at (1,-3) in the direction of $\mathbf{u} = \langle -4/5, 3/5\rangle$.

☐ The direction vector $\mathbf{u}$ is already a unit vector, so

$$\begin{aligned} \nabla f(x,y) &= \langle f_x, f_y\rangle = \langle y/x, \ln x\rangle \\ \nabla f(1,-3) &= \langle -3, 0\rangle \\ D_u f(1,-3) &= \nabla f(1,-3)\cdot\mathbf{u} = \langle -3,0\rangle\cdot\left\langle -\frac{4}{5},\frac{3}{5}\right\rangle = \frac{12}{5} \end{aligned}$$ ■

(2) Find the directional derivative of $g(x,y) = e^{-x}\sin y$ at $(0,\pi/3)$ in the direction of $\mathbf{v} = \langle 3,-2\rangle$.

☐ The given direction vector is not a unit vector, so we need a unit vector $\mathbf{u}$ in the direction of $\mathbf{v}$; then, we can proceed as usual:

$$\begin{aligned} \mathbf{u} &= \frac{\mathbf{v}}{|\mathbf{v}|} = \left\langle \frac{3}{\sqrt{13}}, -\frac{2}{\sqrt{13}}\right\rangle \\ \nabla g(x,y) &= \langle g_x, g_y\rangle = \langle -e^{-x}\sin y, e^{-x}\cos y\rangle \\ \nabla g\left(0,\frac{\pi}{3}\right) &= \left\langle -\frac{\sqrt{3}}{2},\frac{1}{2}\right\rangle \\ D_u g(0,\pi/3) = \nabla g(0,\pi,3)\cdot\mathbf{u} &= \left\langle -\frac{\sqrt{3}}{2},\frac{1}{2}\right\rangle\cdot\left\langle \frac{3}{\sqrt{13}}, -\frac{2}{\sqrt{13}}\right\rangle \\ &= \frac{-3\sqrt{3}-2}{2\sqrt{13}} \end{aligned}$$ ■

(3) Find the the directional derivative of $f(x,y) = x^2y^3 - y^4$ at (2,1) in the direction given by $\theta = \pi/4$.

□ We need a unit vector in the given direction; this is

$$\mathbf{u} = \left\langle \cos\frac{\pi}{4}, \sin\frac{\pi}{4} \right\rangle = \langle \frac{1}{\sqrt{2}}, \frac{1}{\sqrt{2}} \rangle$$

So, with $f_x = 2xy^3$ and $f_y = 3x^2y^2 - 4y^3$, we have

$$\nabla f(x,y) = \langle 2xy^3, 3x^2y^2 - 4y^3 \rangle$$
$$\nabla f(2,1) = \langle 4, 8 \rangle$$
$$D_u f(2,1) = \nabla f(2,1) \cdot \mathbf{u} = \langle 4, 8 \rangle \cdot \left\langle \frac{1}{\sqrt{2}}, \frac{1}{\sqrt{2}} \right\rangle = 6\sqrt{2} \quad ■$$

---

(4) Find the maximum rate of change of $f(p,q) = qe^{-p} + pe^{-q}$ at (0,0), and the direction in which this rate of change occurs. (Assume p, q are renamed rectangular coordinates.)

□ We have

$$\nabla f(p,q) = \langle f_p, f_q \rangle = \langle -qe^{-p} + e^{-q}, e^{-p} - pe^{-q} \rangle$$
$$\nabla f(0,0) = \langle 1, 1 \rangle$$
$$|\nabla f(0,0)| = \sqrt{2}$$

So, the maximum rate of change of f at (0,0) is then $|\nabla f(0,0)| = \sqrt{2}$ and this maximum rate occurs in the direction of the gradient, $\langle 1, 1 \rangle$. ■

---

(5) Find the directional derivative of $f(x,y,z) = \sqrt{x + yz}$ at (1,3,1) in the direction of $\mathbf{u} = \langle 2/7, 3/7, 6/7 \rangle$.

□ The direction vector $\mathbf{u}$ is already a unit vector, so

$$\nabla f(x,y,z) = \langle f_x, f_y, f_z \rangle = \left\langle \frac{1}{2\sqrt{x+yz}}, \frac{z}{2\sqrt{x+yz}}, \frac{y}{2\sqrt{x+yz}} \right\rangle$$
$$\nabla f(1,3,1) = \left\langle \frac{1}{4}, \frac{1}{4}, \frac{3}{4} \right\rangle$$
$$D_u f(1,3,1) = \nabla f(1,3,1) \cdot \mathbf{u} = \left\langle \frac{1}{4}, \frac{1}{4}, \frac{3}{4} \right\rangle \cdot \left\langle \frac{2}{7}, \frac{3}{7}, \frac{6}{7} \right\rangle = \frac{23}{28} \quad ■$$

---

(6) Find the directional derivative of $f(x,y,z) = \dfrac{x}{y+z}$ at (4,1,1) in the direction of $\mathbf{v} = \langle 1, 2, 3 \rangle$.

☐ We need a unit vector $\mathbf{u}$ in the direction of $\mathbf{v}$:

$$\mathbf{u} = \frac{\mathbf{v}}{|\mathbf{v}|} = \left\langle \frac{1}{\sqrt{14}}, \frac{2}{\sqrt{14}}, \frac{3}{\sqrt{14}} \right\rangle$$

$$\nabla f(x,y,z) = \langle f_x, f_y, f_z \rangle = \left\langle \frac{1}{y+z}, \frac{-x}{(y+z)^2}, \frac{-x}{(y+z)^2} \right\rangle$$

$$\nabla f(4,1,1) = \left\langle \frac{1}{2}, -1, -1 \right\rangle$$

$$D_u f(4,1,1) = \nabla f(4,1,1) \cdot \mathbf{u} = \left\langle \frac{1}{2}, -1, -1 \right\rangle \cdot \left\langle \frac{1}{\sqrt{14}}, \frac{2}{\sqrt{14}}, \frac{3}{\sqrt{14}} \right\rangle$$
$$= -\frac{9}{2\sqrt{14}} \quad \blacksquare$$

---

(7) Given the 3D temperature function $T(x,y,z) = 20e^{-x^2-y^2-2z^2}$, what is the maximum rate of change of the temperature at the point $(2,-1,2)$, and in what direction does this rate of change occur?

☐ We're going to need the gradient of T at $(2,-1,2)$,

$$T_x = -40xe^{-x^2-y^2-2z^2}$$
$$T_y = -40ye^{-x^2-y^2-2z^2}$$
$$T_z = -80ze^{-x^2-y^2-2z^2}$$
$$\nabla T(x,y,z) = \langle T_x, T_y, T_z \rangle$$
$$= 40e^{-x^2-y^2-2z^2} \langle -x, -y, -2z \rangle$$
$$\nabla T(2,-1,2) = 40e^{-13} \langle -2, 1, -4 \rangle$$

The temperature increases the fastest in the direction of the gradient, so at the point $P(2,-1,2)$ the temperature increases fastest in the direction of $\nabla T(2,-1,2)$, which is in the direction of (removing the icky constant in front since all we need is direction) $\langle -2, 1, -4 \rangle$.

The maximum rate of increase at $P(2,-1,2)$ is the magnitude of the gradient there, which is $40e^{-13}\sqrt{21}$. ■

B.5 Chapter 17: Practice Problem Solutions

B.5.1 *Arc Length Parameterization — Practice — Solved*

(1) Find the arc length of $\mathbf{r}(t) = \langle\sqrt{2}t, e^t, e^{-t}\rangle$ for $0 \le t \le 1$. (Hint: When you create $|\mathbf{r}'(t)| = \sqrt{g(t)}$, the function $g(t)$ will be a perfect square! Really!)

☐ First, build:

$$\begin{aligned}\mathbf{r}'(t) &= \langle\sqrt{2}, e^t, -e^{-t}\rangle \\ |\mathbf{r}'(t)| &= \sqrt{2 + e^{2t} + e^{-2t}} = \sqrt{(e^t + e^{-t})^2} = e^t + e^{-t}\end{aligned}$$

So that

$$L = \int_0^1 |\mathbf{r}'(t)|\, dt = \int_0^1 (e^t + e^{-t})\, dt = e - \frac{1}{e} \quad \blacksquare$$

(2) Find the arc length of $\mathbf{r}(t) = \langle t^2, 2t, \ln t\rangle$ for $1 \le t \le e$. (Hint: When you create $|\mathbf{r}'(t)| = \sqrt{g(t)}$, the function $g(t)$ will be a perfect square! Really!)

☐ First, build:

$$\begin{aligned}\mathbf{r}'(t) &= \left\langle 2t, 2, \frac{1}{t}\right\rangle \\ |\mathbf{r}'(t)| &= \sqrt{4t^2 + 4 + \frac{1}{t^2}} = \sqrt{\left(2t + \frac{1}{t}\right)^2} = 2t + \frac{1}{t}\end{aligned}$$

So that

$$L = \int_1^e |\mathbf{r}'(t)|\, dt = \int_1^e \left(2t + \frac{1}{t}\right) dt = \left(t^2 + \ln t\right)\Big|_1^e = e^2 \quad \blacksquare$$

(3) Find an arc length parameterization of the vector function in You Try It 1. for $t \ge 0$, and determine the location at which the curve accumulates a total arc length of $s = 2$.

☐ Picking up You Try It 1, the arc length of $\mathbf{r}(t) = \langle t^2/2, t^3/3\rangle$ at any t is

$$\begin{aligned}s &= \int_0^t |\mathbf{r}'(\tau)|\, d\tau = \int_0^t \tau\sqrt{1+\tau^2}\, d\tau \\ &= \frac{1}{3}(1+\tau^2)^{3/2}\Big|_0^t = \frac{1}{3}\left[(1+t^2)^{3/2} - 1\right]\end{aligned}$$

Then we turn this inside out to find t in terms of s:

$$\begin{aligned} s &= \frac{1}{3}\left[(1+t^2)^{3/2}-1\right] \\ 3s+1 &= (1+t^2)^{3/2} \\ (3s+1)^{2/3} &= 1+t^2 \\ \sqrt{(3s+1)^{2/3}-1} &= t \end{aligned}$$

Substituting this into the original form of the vector function, we get:

$$\begin{aligned} \mathbf{r}(s) &= \left\langle \frac{1}{2}(\sqrt{(3s+1)^{2/3}-1})^2, \frac{1}{3}(\sqrt{(3s+1)^{2/3}-1})^3 \right\rangle \\ &= \left\langle \frac{1}{2}((3s+1)^{2/3}-1), \frac{1}{3}((3s+1)^{2/3}-1)^{3/2} \right\rangle \end{aligned}$$

This is the arc length parameterized version of the vector function. To now find where we accumulate an arc length of 2, we look for $\mathbf{r}(2)$, which is:

$$\begin{aligned} \mathbf{r}(2) &= \left\langle \frac{1}{2}((3(2)+1)^{2/3}-1), \frac{1}{3}((3(2)+1)^{2/3}-1)^{3/2} \right\rangle \\ &= \left\langle \frac{1}{2}(7^{2/3}-1), \frac{1}{3}(7^{2/3}-1)^{3/2} \right\rangle \quad \blacksquare \end{aligned}$$

---

(4) Revisit the vector function from You Try It 1 and Practice Problem 3, and confirm that $|\mathbf{r}'(s)| = 1$.

□ The arc length parameterized version of this vector function is

$$\mathbf{r}(s) = \langle \frac{1}{2}((3s+1)^{2/3}-1), \frac{1}{3}((3s+1)^{2/3}-1)^{3/2} \rangle$$

from which we get (with simplification behind the scenes),

$$\mathbf{r}'(s) = \left\langle (3s+1)^{-1/3}, (3s+1)^{-1/3}\sqrt{(3s+1)^{2/3}-1} \right\rangle$$

Then,

$$\begin{aligned} |\mathbf{r}'(s)|^2 &= \left((3s+1)^{-1/3}\right)^2 + \left((3s+1)^{-1/3}\sqrt{(3s+1)^{2/3}-1}\right)^2 \\ &= (3s+1)^{-2/3} + (3s+1)^{-2/3}\left((3s+1)^{2/3}-1\right) \\ &= (3s+1)^{-2/3} + 1 - (3s+1)^{-2/3} \\ &= 1 \end{aligned}$$

and since $|\mathbf{r}'(s)|^2 = 1$, we have $|\mathbf{r}'(s)| = 1$. ■

B.5.2 *Contours, Orientation, Pointers — Practice — Solved*

(1) Determine parametric equations and bounds for t ($a \leq t < b$) that will give us an ellipse containing points $(2, 0)$ and $(0, 3)$, oriented positively.

□ By symmetry of the ellipse, two other notable points on the ellipse will be $(-2, 0)$ and $(0, -3)$. The component functions will be a combination of $\sin t$ and $\cos t$ with coefficients of 2 and 3. For positive orientation, we must go around the circle in a counterclockwise direction. If we set $\mathbf{r}(t) = \langle 2\cos t, 3\sin t\rangle$, then at $t = 0$, we are at $(2, 0)$. At $t = \pi/2$, we are at $(0, 3)$. At $t = \pi$, we are at $(-2, 0)$. At $t = 3\pi/2$, we are at $(0, -3)$. And at $t = 2\pi$, we are back at $(2, 0)$. We have traversed the circle via

$$(2, 0) \to (0, 3) \to (-2, 0) \to (0, -3) \to (2, 0)$$

which is counterclockwise (positively oriented). Altogether, it looks like $\mathbf{r}(t) = \langle 2\cos t, 3\sin t\rangle$ for $0 \leq t < 2\pi$ does the job. ■

(2) Determine parametric equations and bounds for t ($a \leq t < b$) that will give an ellipse containing points $(4, 0)$ and $(0, 3)$, oriented negatively.

□ By symmetry of the ellipse, two other notable points on the ellipse will be $(-4, 0)$ and $(0, -3)$. The two given points do not have to be traversed in the order given, we just have to use them both. The component functions will be a combination of $\sin t$ and $\cos t$ with coefficients of 3 and 4. For negative orientation, we must go around the circle in a clockwise direction. If we set $\mathbf{r}(t) = \langle 4\sin t, 3\cos t\rangle$, then at $t = 0$, we are at $(0, 3)$. At $t = \pi/2$, we are at $(4, 0)$. At $t = \pi$, we are at $(0, -3)$. At $t = 3\pi/2$, we are at $(-4, 0)$. And at $t = 2\pi$, we are back at $(0, 3)$. We have traversed the circle via

$$(0, 3) \to (4, 0) \to (0, -3) \to (-4, 0) \to (0, 3)$$

which is clockwise (negatively oriented). Altogether, it looks like $\mathbf{r}(t) = \langle 4\sin t, 3\cos t\rangle$ for $0 \leq t < 2\pi$ does the job. ■

(3) Find $\mathbf{T}$ and $\mathbf{N}$ for the contour $\mathbf{r}(t) = \langle 2t^2, t\rangle$ at $(2, 1)$.

□ Although the full 3D version of $\mathbf{N}(t)$ in (17.13) would apply, we can can use the (simpler) 2D equation (17.12) for this 2D problem.

The ingredients for these equations include:

$$\begin{aligned} \mathbf{r}'(t) &= \langle x'(t), y'(t)\rangle = \langle 4t, 1\rangle \\ \mathbf{r}'(t)^{\perp} &= \langle y'(t), -x'(t)\rangle = \langle 1, -4t\rangle \\ ds &= \sqrt{(x'(t))^2 + (y'(t))^2} = \sqrt{(4t)^2 + (1)^2} = \sqrt{16t^2 + 1} \end{aligned}$$

Note that the point of interest $(2, 1)$ is found by the vector $\langle 2, 1\rangle$, which is generated by $t = 1$. And so,

$$\begin{aligned} \mathbf{r}'(1) &= \langle 4, 1\rangle \\ \mathbf{r}'(1)^{\perp} &= \langle 1, -4\rangle \\ ds(1) &= \sqrt{17} \end{aligned}$$

Then,

$$\begin{aligned} \mathbf{T}(1) &= \frac{\mathbf{r}'(1)}{ds(1)} = \frac{1}{\sqrt{17}}\langle 4, 1\rangle \\ \mathbf{N}(1) &= \frac{\mathbf{r}'(1)^{\perp}}{ds(1)} = \frac{1}{\sqrt{17}}\langle 1, -4\rangle \end{aligned}$$

For some quality assurance, we can confirm that $|\mathbf{T}(1)| = 1$, $|\mathbf{N}(1)| = 1$, and $\mathbf{T}(1) \cdot \mathbf{N}(1) = 0$. ■

---

(4) Find the unit tangent vector $\mathbf{T}$ and unit normal vector $\mathbf{N}$ for $\mathbf{r}(t) = \langle 2e^{-t}, e^{-2t}, t\rangle$ at $t = 0$. Can you demonstrate some quality assurance about your results?

□ The ingredients for finding $\mathbf{T}(0)$ and $\mathbf{N}(0)$ via (17.11) and (17.13) include:

$$\begin{aligned} \mathbf{r}'(t) &= \langle -2e^{-t}, -2e^{-2t}, 1\rangle \\ ds = |\mathbf{r}'(t)| &= \sqrt{4e^{-2t} + 4e^{-4t} + 1} = \sqrt{(2e^{-2t} + 1)^2} = 2e^{-2t} + 1 \\ \mathbf{T}(t) &= \frac{\mathbf{r}'(t)}{ds} = \frac{1}{2e^{-2t} + 1}\langle -2e^{-t}, -2e^{-2t}, 1\rangle \end{aligned}$$

Now we dive into the worst derivative,

$$\begin{aligned} \mathbf{T}'(t) = {} & 4e^{-2t}(2e^{-2t} + 1)^{-2}\langle -2e^{-t}, -2e^{-2t}, 1\rangle \\ & + (2e^{-2t} + 1)^{-1}\langle 2e^{-t}, 4e^{-2t}, 0\rangle \end{aligned}$$

Fortunately, we don't need to simplify it until after plugging in $t = 0$. But reeling in all the necessary results so far,

$$\begin{aligned}\mathbf{T}(0) &= \frac{1}{3}\langle -2, -2, 1\rangle \\ \mathbf{T}'(0) &= 4(3)^{-2}\langle -2, -2, 1\rangle + \frac{1}{3}\langle 2, 4, 0\rangle \\ &= \left\langle -\frac{8}{9}, -\frac{8}{9}, \frac{4}{9}\right\rangle + \left\langle \frac{6}{9}, \frac{12}{9}, 0\right\rangle \\ &= \left\langle -\frac{2}{9}, \frac{4}{9}, \frac{4}{9}\right\rangle \\ |\mathbf{T}'(()0)| &= \sqrt{\frac{4}{81} + \frac{16}{81} + \frac{16}{81}} = \sqrt{\frac{36}{81}} = \frac{2}{3}\end{aligned}$$

and so,

$$\mathbf{N}(0) = \frac{\mathbf{T}'(()0)}{|\mathbf{T}'(()0)|} = \frac{3}{2}\left\langle -\frac{2}{9}, \frac{4}{9}, \frac{4}{9}\right\rangle = \left\langle -\frac{1}{3}, \frac{2}{3}, \frac{2}{3}\right\rangle$$

Altogether, we have

$$\mathbf{T}(0) = \left\langle -\frac{2}{3}, -\frac{2}{3}, \frac{1}{3}\right\rangle \qquad \text{and} \qquad \mathbf{N}(0) = \left\langle -\frac{1}{3}, \frac{2}{3}, \frac{2}{3}\right\rangle$$

As quality assurance, we can confirm that $|\mathbf{T}(0)| = 1$, $|\mathbf{N}(0)| = 1$, and $\mathbf{T}(0) \cdot \mathbf{N}(0) = 0$. ■

---

(5) Find the unit normal vector $\mathbf{N}$ for $\mathbf{r}(t) = (4t^{3/2}/3)\mathbf{i} + t^2\mathbf{j} + t\mathbf{k}$ at $t = 1$. (Hint: Has $\mathbf{T}$ been calculated elsewhere?)

□ We found $\mathbf{T}(1)$ in YTI 4. We used $\mathbf{r}'(t) = 2\sqrt{t}\mathbf{i} + 2t\mathbf{j} + 1\mathbf{k}$, and from this, we can get

$$|\mathbf{r}'(t)| = \sqrt{4t + 4t^2 + 1} = \sqrt{(2t+1)^2} = |2t+1| = 2t+1$$

So a general expression for $\mathbf{T}(t)$ (which wasn't needed in YTI 4 but is required now for $\mathbf{N}(t)$) is

$$\mathbf{T}(t) = \frac{\mathbf{r}'(t)}{|\mathbf{r}'(t)|} = \frac{1}{2t+1}(2\sqrt{t}\mathbf{i} + 2t\mathbf{j} + 1\mathbf{k})$$

Then

$$\begin{aligned}\mathbf{T}'(t) &= \left(\frac{d}{dt}\frac{1}{2t+1}\right)(2\sqrt{t}\mathbf{i} + 2t\mathbf{j} + 1\mathbf{k}) + \frac{1}{2t+1}\frac{d}{dt}(2\sqrt{t}\mathbf{i} + 2t\mathbf{j} + 1\mathbf{k}) \\ &= \left(\frac{-2}{(2t+1)^2}\right)(2\sqrt{t}\mathbf{i} + 2t\mathbf{j} + 1\mathbf{k}) + \frac{1}{2t+1}\left(\frac{1}{\sqrt{t}}\mathbf{i} + 2\mathbf{j} + 0\mathbf{k}\right)\end{aligned}$$

so

$$\begin{aligned}\mathbf{T}'(1) &= -\frac{2}{9}(2\mathbf{i}+2\mathbf{j}+1\mathbf{k})+\frac{1}{3}(\mathbf{i}+2\mathbf{j}+0\mathbf{k})\\ &= -\frac{4}{9}\mathbf{i}+\frac{4}{9}\mathbf{j}-\frac{2}{9}\mathbf{k}+\frac{1}{3}\mathbf{i}+\frac{2}{3}\mathbf{j}\\ &= -\frac{1}{9}\mathbf{i}+\frac{2}{9}\mathbf{j}-\frac{2}{9}\mathbf{k}\end{aligned}$$

and

$$|\mathbf{T}'(1)| = \sqrt{\frac{1}{81}+\frac{4}{81}+\frac{4}{81}} = \sqrt{\frac{9}{81}} = \frac{1}{3}$$

Finally, we get

$$\mathbf{N}(1) = \frac{\mathbf{T}'(1)}{|\mathbf{T}'(1)|} = \frac{3}{1}\left(-\frac{1}{9}\mathbf{i}+\frac{2}{9}\mathbf{j}-\frac{2}{9}\mathbf{k}\right)$$

or cleanly,

$$\mathbf{N}(1) = -\frac{1}{3}\mathbf{i}+\frac{2}{3}\mathbf{j}-\frac{2}{3}\mathbf{k} \quad \blacksquare$$

---

(6) Find the unit tangent vector $\mathbf{T}$ and unit normal vector $\mathbf{N}$ for $\mathbf{r}(t) = \langle \cos t + t\sin t, \sin t - t\cos t, 1\rangle$ at the point $(1, 0, 1)$.

□ Note that the given point corresponds to $t = 0$. Let's pull out the first two components of $\mathbf{r}(t)$ for derivatives:

$$\frac{d}{dt}(\cos t + t\sin t) = -\sin t + \sin t + t\cos t = t\cos t$$
$$\frac{d}{dt}(\sin t - t\cos t) = \cos t - \cos t + t\sin t = t\sin t$$

Nice! Then putting these back together,

$$\begin{aligned}\mathbf{r}'(t) &= \langle t\cos t, t\sin t, 0\rangle\\ ds = |\mathbf{r}'(t)| &= \sqrt{t^2\cos^2 t + t^2\sin^2 t} = t\\ \mathbf{T}(t) &= \frac{\mathbf{r}'(t)}{ds} = \langle \cos t, \sin t, 0\rangle\end{aligned}$$

That's not so bad! Carrying on, then,

$$\begin{aligned}\mathbf{T}'(t) &= \langle -\sin t, \cos t, 0\rangle\\ |\mathbf{T}'(t)| &= \sqrt{\sin^2 t + \cos^2 t + 0} = 1\\ \mathbf{N}(t) &= \frac{\mathbf{T}'(t)}{|\mathbf{T}'(t)|} = \langle -\sin t, \cos t, 0\rangle\end{aligned}$$

Then at $t = 0$ specifically,

$$\mathbf{T}(0) = \langle 1, 0, 0\rangle \text{ and } \qquad \mathbf{N}(0) = \langle 0, 1, 0\rangle \quad \blacksquare$$

B.5.3 *The Fresnet–Serret Frame — Practice — Solved*

(1) Find the unit binormal vector **B** for $\mathbf{r}(t) = \langle e^t, e^t \sin t, e^t \cos t\rangle$ at the point (1,0,1). (Hint: Have **T** and **N** been calculated elsewhere?)

☐ In YTI 6 of Sec. 17.2, we found $\mathbf{T}(t)$ and **N** for this function and point, as:

$$\mathbf{T}(0) = \frac{1}{\sqrt{3}}\langle 1, 1, 1\rangle \qquad \text{and} \qquad \mathbf{N}(0) = \frac{1}{\sqrt{2}}\langle 0, 1, -1\rangle$$

Therefore by (17.14),

$$\mathbf{B}(0) = \mathbf{T}(0) \times \mathbf{N}(0) = \langle -\frac{\sqrt{6}}{3}, \frac{\sqrt{6}}{6}, \frac{\sqrt{6}}{6}\rangle \quad \blacksquare$$

(2) Find the unit binormal vector **B** for $\mathbf{r}(t) = \langle \cos t + t \sin t, \sin t - t \cos t, 1\rangle$ at the point $(1, 0, 1)$. (Hint: Have **T** and **N** been calculated elsewhere?)

In PP 6 of Sec. 17.2, we found $\mathbf{T}(t)$ and **N** for this function and point, as:

$$\mathbf{T}(0) = \langle 1, 0, 0\rangle \text{ and } \qquad \mathbf{N}(0) = \langle 0, 1, 0\rangle$$

Given those, we can probably take a crazy guess that $\mathbf{B}(\mathbf{0}) = \langle 0, 0, 1\rangle$, and when we double check using (17.14),

$$\mathbf{B}(0) = \mathbf{T}(0) \times \mathbf{N}(0) = \langle 0, 0, 1\rangle \quad \blacksquare$$

(3) Determine the complete *TNB*-frame for $\mathbf{r}(t) = \langle e^t, e^t \sin t, e^t \cos t\rangle$ at the point $(e^\pi, 0, e^\pi)$. (Hint: A previously solved problem will be very useful.)

☐ This same vector function appeared in YTI 6 of Sec. 17.2, with the point $t = 0$. The point given here corresponds to $t = \pi$. However, we can still use a lot of the work. From that YTI 6, we learned

$$\mathbf{T}(t) = \frac{1}{\sqrt{3}}\langle 1, \cos t + \sin t, \cos t - \sin t\rangle$$

and so here,

$$\mathbf{T}(\pi) = \frac{1}{\sqrt{3}}\langle 1, (-1) + 0, (-1) - 0\rangle = \frac{1}{\sqrt{3}}\langle 1, -1, -1\rangle$$

Also from that YTI 6, we had

$$\mathbf{T}'(t) = \frac{1}{\sqrt{3}}\langle 0, -\sin t + \cos t, -\sin t - \cos t\rangle$$

and so here,

$$\mathbf{T}'(\pi) = \frac{1}{\sqrt{3}}\langle 0, 0 + (-1), 0 - (-1)\rangle = \frac{1}{\sqrt{3}}\langle 0, -1, 1\rangle$$

from which we find $|\mathbf{T}'(\pi)| = \sqrt{2}/\sqrt{3}$. Therefore,

$$\mathbf{N}(\pi) = \frac{\mathbf{T}'(\pi)}{|\mathbf{T}'(\pi)|} = \frac{1}{\sqrt{2}}\langle 0, -1, 1\rangle$$

From (17.14),

$$\begin{aligned}\mathbf{B}(\pi) = \mathbf{T}(\pi) \times \mathbf{N}(\pi) &= \frac{1}{\sqrt{3}}\langle 1, -1, -1\rangle \times \frac{1}{\sqrt{2}}\langle 0, -1, 1\rangle \\ &= \left\langle -\frac{\sqrt{2}}{\sqrt{3}}, -\frac{1}{\sqrt{6}}, -\frac{1}{\sqrt{6}}\right\rangle\end{aligned}$$

Summarizing, the complete TNB-frame is:

$$\mathbf{T}(\pi) = \left\langle \frac{1}{\sqrt{3}}, -\frac{1}{\sqrt{3}}, -\frac{1}{\sqrt{3}}\right\rangle$$

$$\mathbf{N}(\pi) = \left\langle 0, -\frac{1}{\sqrt{2}}, \frac{1}{\sqrt{2}}\right\rangle$$

$$\mathbf{B}(\pi) = \left\langle -\frac{\sqrt{2}}{\sqrt{3}}, -\frac{1}{\sqrt{6}}, -\frac{1}{\sqrt{6}}\right\rangle$$

Quality assurance: each of $\mathbf{T}(\pi), \mathbf{N}(\pi), \mathbf{B}(\pi)$ is a unit vector, and all three are mutually perpendicular. ■

---

(4) Find the velocity and acceleration functions for the particle moving according to $\mathbf{r}(t) = \langle t^2, \ln t, t\rangle$. Does the particle move at a constant speed?

□ For the position function $\mathbf{r}(t) = \langle t^2, \ln t, t\rangle$, the velocity function is

$$\mathbf{v}(t) = \mathbf{r}'(t) = \left\langle 2t, \frac{1}{t}, 1\right\rangle$$

and acceleration is

$$\mathbf{a}(t) = \mathbf{v}'(t) = \left\langle 2, -\frac{1}{t^2}, 0\right\rangle$$

Since speed is the magnitude of velocity,

$$s = |\mathbf{v}(t)| = \sqrt{4t^2 + \frac{1}{t^2} + 1}$$

which, regardless of simplification, remains a function of t — and so this particle does not move at constant speed. ■

(5) Find the velocity and positions functions for a particle moving with acceleration $\mathbf{a}(t) = t\mathbf{i} + t^2\mathbf{j} + \cos 2t\mathbf{k}$, if we know that $\mathbf{v}(0) = \mathbf{i} + \mathbf{k}$ and $\mathbf{r}(0) = \mathbf{j}$.

□ We have $\mathbf{a}(t) = t\mathbf{i}+t^2\mathbf{j}+\cos 2t\mathbf{k} = \langle t, t^2, \cos 2t\rangle$ with initial conditions $\mathbf{v}(0) = \mathbf{i} + \mathbf{k} = \langle 1, 0, 1\rangle$ and $\mathbf{r}(0) = \mathbf{j} = \langle 0, 1, 0\rangle$. The velocity and acceleration functions are then:

$$\begin{aligned}
\mathbf{v}(t) = \int \mathbf{a}(t)\,dt &= \left\langle \frac{1}{2}t^2 + c_1, \frac{1}{3}t^3 + c_2, \frac{1}{2}\sin 2t + c_3 \right\rangle \\
\mathbf{v}(0) = \langle c_1, c_2, c_3\rangle &= \langle 1, 0, 1\rangle \\
\to \mathbf{v}(t) &= \left\langle \frac{1}{2}t^2 + 1, \frac{1}{3}t^3, \frac{1}{2}\sin 2t + 1 \right\rangle \\
\mathbf{r}(t) = \int \mathbf{v}(t)\,dt &= \left\langle \frac{1}{6}t^3 + t + c_1, \frac{1}{12}t^4 + c_2, -\frac{1}{4}\cos 2t + t + c_3 \right\rangle \\
\mathbf{r}(0) &= \langle c_1, c_2, -\frac{1}{4} + c_3\rangle = \langle 0, 1, 0\rangle \\
\to c_1 = 0, c_2 = 1, \text{ and } \ c_3 = \frac{1}{4} & \\
\to \mathbf{r}(t) &= \left\langle \frac{1}{6}t^3 + t, \frac{1}{12}t^4 + 1, -\frac{1}{4}\cos 2t + t + \frac{1}{4} \right\rangle
\end{aligned}$$

In original notation,

$$\mathbf{v}(t) = \left(\frac{1}{2}t^2 + 1\right)\mathbf{i} + \left(\frac{1}{3}t^3\right)\mathbf{j} + \left(\frac{1}{2}\sin 2t + 1\right)\mathbf{k}$$

$$\mathbf{r}(t) = \left(\frac{1}{6}t^3 + t\right)\mathbf{i} + \left(\frac{1}{12}t^4 + 1\right)\mathbf{j} + \left(-\frac{1}{4}\cos 2t + t + \frac{1}{4}\right)\mathbf{k} \quad \blacksquare$$

(6) Find the tangential and normal components of acceleration for $\mathbf{r}(t) = t\mathbf{i} + t^2\mathbf{j} + 3t\mathbf{k}$.

□ To find the tangential component of acceleration for the given $\mathbf{r}(t)$, we need:

$$\begin{aligned}
\mathbf{r}'(t) &= \mathbf{i} + 2t\mathbf{j} + 3\mathbf{k} \\
|\mathbf{r}'(t)| &= \sqrt{1 + 4t^2 + 9} = \sqrt{4t^2 + 10} \\
\mathbf{r}''(t) &= 2\mathbf{j} \\
\mathbf{r}'(t) \cdot \mathbf{r}''(t) &= 4t
\end{aligned}$$

so then

$$a_T = \frac{\mathbf{r}'(t) \cdot \mathbf{r}''(t)}{|\mathbf{r}'(t)|} = \frac{4t}{\sqrt{4t^2 + 10}}$$

To find the normal component of acceleration, we need $|\mathbf{a}(t)|^2$. Note that $\mathbf{a}(t) = \mathbf{r}''(t)$, which is found above; so,

$$\mathbf{a}(t) = 2\mathbf{j}$$
$$|\mathbf{a}(t)|^2 = 4$$

so that

$$a_N = \sqrt{|\mathbf{a}(t)|^2 - a_T^2} = \sqrt{4 - \left(\frac{4t}{\sqrt{4t^2+10}}\right)^2} = \sqrt{4 - \frac{16t^2}{4t^2+10}}$$
$$= \sqrt{\frac{16t^2+40}{4t^2+10} - \frac{16t^2}{4t^2+10}} = \sqrt{\frac{40}{4t^2+10}} = \frac{2\sqrt{5}}{\sqrt{2t^2+5}} \quad \blacksquare$$

(7) Find the curvature and torsion of $\mathbf{r}(t) = \langle e^t \cos t, e^t \sin t, t\rangle$ at the point (1,0,0).

□ For the curvature of $\mathbf{r}(t) = \langle e^t \cos t, e^t \sin t, t\rangle$ at (1,0,0), i.e. at $t = 0$, we need:

$$\mathbf{r}'(t) = \langle e^t(\cos t - \sin t), e^t(\cos t + \sin t), 1\rangle$$
$$\mathbf{r}''(t) = \langle -2e^t \sin t, 2e^t \cos t, 0\rangle$$
$$\mathbf{r}'(0) = \langle 1, 1, 1\rangle$$
$$|\mathbf{r}'(0)| = \sqrt{3}$$
$$\mathbf{r}''(0) = \langle 0, 2, 0\rangle$$
$$\mathbf{r}'(0) \times \mathbf{r}''(0) = \langle -2, 0, 2\rangle$$
$$|\mathbf{r}'(0) \times \mathbf{r}''(0)| = 2\sqrt{2}$$

so that

$$\kappa = \frac{|\mathbf{r}'(0) \times \mathbf{r}''(0)|}{|\mathbf{r}'(0)|^3} = \frac{2\sqrt{2}}{(\sqrt{3})^3} = \frac{2\sqrt{6}}{9}$$

For torsion, in addition to $\mathbf{r}'(()0)$ and $\mathbf{r}''(()0)$, we need $\mathbf{r}'''(0)$. The full third derivative is

$$\mathbf{r}'''(t) = \langle -2e^t(\sin t + \cos t), 2e^t(\cos t - \sin t), 0\rangle$$

so that $\mathbf{r}'''(0) = \langle -2, 2, 0\rangle$. So, using information computed for the curvature,

$$(\mathbf{r}'(0) \times \mathbf{r}''(0)) \cdot \mathbf{r}'''(0) = \langle -2, 0, 2\rangle \cdot \langle -2, 2, 0\rangle = 4$$

Also,

$$|\mathbf{r}'(0) \times \mathbf{r}''(0)|^2 = (2\sqrt{2})^2 = 8$$

Together, by (17.21),

$$\tau(0) = \frac{(\mathbf{r}'(0) \times \mathbf{r}''(0)) \cdot \mathbf{r}'''(0)}{|\mathbf{r}'(0) \times \mathbf{r}''(0)|^2} = \frac{4}{8} = \frac{1}{2} \quad \blacksquare$$

B.5.4 *Lagrange Multipliers — Practice — Solved*

(1) Find the maximum value of $f(x, y) = x^2 + y^3/3$ that can be achieved by points on the ellipse $2x^2 + y^2 = 6$, and the point(s) at which it occurs.

□ The objective function is $f(x, y) = x^2 + y^3/3$, and its gradient is $\nabla f(x, y) = \langle 2x, y^2 \rangle$.

The constraint function is $g(x, y) = 2x^2 + y^2$, and its gradient is $\nabla g(x, y) = \langle 4x, 2y \rangle$.

The equation of proportionality of these gradients is $\nabla f(x, y) = \lambda \nabla g(x, y)$; this equation, along with the constraint itself, leads to the system of equations,

$$\begin{aligned} 2x &= \lambda(4x) \\ y^2 &= \lambda(2y) \\ 2x^2 + y^2 &= 6 \end{aligned}$$

Simplified, we get

$$\begin{aligned} \frac{1}{2} &= \lambda \\ y &= 2\lambda \\ 2x^2 + y^2 &= 6 \end{aligned}$$

and now we already know $\lambda = 1/2$. With that value, the second equation gives $y = 1$. Then the third equation gives $x = \pm\sqrt{5/2}$. The original simplification of the first equation required that $x = 0$ was no longer under consideration; if $x = 0$, another solution to this system comes with $y = \pm\sqrt{6}$. Similarly, the original simplification of the second equation required that $y = 0$ was no longer under consideration; if $y = 0$, then another solution is $x = \pm\sqrt{3}$. The solutions for λ in the latter two cases are irrelevant. Altogether, we have the following (parts of) solutions to the system of equations, which provide locations of extremes of $f(x, y)$ subject to $g(x, y) = 6$:

$$\left(-\sqrt{\frac{5}{2}}, 1\right), \left(\sqrt{\frac{5}{2}}, 1\right), (0, \sqrt{6}), (0, -\sqrt{6}), (\sqrt{3}, 0), (-\sqrt{3}, 0)$$

The absolute max is located by finding $f(x, y)$ at each candidate point:

$$f\left(-\sqrt{\frac{5}{2}}, 1\right) = \frac{17}{6} \qquad f\left(-\sqrt{\frac{5}{2}}, 1\right) = \frac{17}{6}$$
$$f(0, \sqrt{6}) = 2\sqrt{6} \qquad f(0, -\sqrt{6}) = -2\sqrt{6}$$
$$f(\sqrt{3}, 0) = 3 \qquad f(-\sqrt{3}, 0) = 3$$

The maximum value found is $f(0, \sqrt{6}) = 2\sqrt{6}$. ■

---

(2) Find the maximum value of $f(x, y, z) = x + 2y^2 + 2z$ that can be achieved by points on the sphere $x^2 + y^2 + z^2 = 4$, and the point(s) at which it occurs.

□ The objective function is $f(x, y, z) = x + 2y^2 + 2z$, and its gradient is $\nabla f(x, y, z) = \langle 1, 4y, 2 \rangle$.

The constraint function is $g(x, y, z) = x^2 + y^2 + z^2 = 4$, and its gradient is $\nabla g(x, y, z) = \langle 2x, 2y, 2z \rangle$.

The equation of proportionality of these gradients is $\nabla f(x, y, z) = \lambda \nabla g(x, y, z)$; this equation, along with the constraint itself, leads to the system of equations,

$$\begin{aligned} 1 &= \lambda(2x) \\ 4y &= \lambda(2y) \\ 2 &= \lambda(2z) \\ x^2 + y^2 + z^2 &= 4 \end{aligned}$$

or

$$\begin{aligned} 1 &= 2\lambda x \\ 2y &= \lambda y \\ 1 &= \lambda z \\ x^2 + y^2 + z^2 &= 4 \end{aligned}$$

Tracking down these solutions would be simple but tedious, so let's have technology do it. From Wolfram Alpha, the solutions are:

$$\left(\frac{1}{4}, -\frac{\sqrt{59}}{4}, \frac{1}{2}\right) \quad \left(\frac{1}{4}, \frac{\sqrt{59}}{4}, \frac{1}{2}\right) \quad \left(-\frac{2}{\sqrt{5}}, 0, -\frac{4}{\sqrt{5}}\right) \quad \left(\frac{2}{\sqrt{5}}, 0, \frac{4}{\sqrt{5}}\right)$$

The absolute max is located by finding $f(x, y, z)$ at each candidate point:

$$f\left(\frac{1}{4}, -\frac{\sqrt{59}}{4}, \frac{1}{2}\right) = \frac{69}{8} \qquad f\left(\frac{1}{4}, \frac{\sqrt{59}}{4}, \frac{1}{2}\right) = \frac{69}{8}$$

$$f\left(-\frac{2}{\sqrt{5}}, 0, -\frac{4}{\sqrt{5}}\right) = -2\sqrt{5} \qquad f\left(\frac{2}{\sqrt{5}}, 0, \frac{4}{\sqrt{5}}\right) = 2\sqrt{5}$$

The maximum value of 69/8 is found at two points, $(1/4, -\sqrt{59}/4, 1/2)$ and $(1/4, \sqrt{59}/4, 1/2)$. ■

---

(3) To generate Fig. 13.21, I had to find the minimum and maximum values of $z = 1 + xy$ subject to the constraint $(x-3)^2 + (y-2)^2 = 0.25$. What were these values, and at what points did they occur?

□ With $f(x, y) = 1 + xy$ and $g(x, y) = (x-3)^2 + (y-2)^2$, the relation $\nabla f(x, y) = \lambda \nabla g(x, y)$ and the constraint leads to the system of equations

$$\begin{aligned} y &= 2\lambda(x-3) \\ x &= 2\lambda(y-2) \\ (x-3)^2 + (y-2)^2 &= 0.25 \end{aligned}$$

By Wolfram Alpha, we get the solutions $(x, y) \approx (2.753, 1.565)$, for which $f(x, y) = 5.31$, and $(x, y) \approx (3.295, 2.404)$, for which $f(x, y) = 8.92$. These provide the minimum and maximum values of $f(x, y)$ subject to the given constraint, and the locations at which they occur. ■

---

(4) Find the point on the ellipse $x^2 + 2y^2 = 4$ that is closest to the point $(1, 1)$.

□ We want to minimize the distance to $(x, y) = (1, 1)$ for points satisfying $x^2 + 2y^2 = 4$. As in earlier optimization problems, it suffices to minimize the square of the distance, $D^2 = (x-1)^2 + (y-1)^2$. This expression for D^2 is our objective function, and the constraint is the left side of the equation of the ellipse. The gradients of each are:

$$\begin{aligned} \nabla D^2(x, y) &= \langle 2(x-1), 2(y-1) \rangle \\ \nabla g(x, y) &= \langle 2x, 4y \rangle \end{aligned}$$

Setting the proportion $\nabla D^2(x, y) = \lambda \nabla g(x, y)$ and introducing the constraint, we get the system:

$$\begin{aligned} 2(x-1) &= \lambda(2x) \\ 2(y-1) &= \lambda(4y) \\ x^2+2y^2 &= 4 \end{aligned}$$

or

$$\begin{aligned} x-1 &= \lambda x \\ y-1 &= 2\lambda y \\ x^2+2y^2 &= 4 \end{aligned}$$

The solutions are, using Wolfram Alpha, $(x,y) \approx (1.086, 1.188)$ and $(x,y) \approx (-1.879, -0.484)$. (The exact versions are too nasty for human consumption.) Intuition suggests that the point with the positive coordinates is the one closest to $(1,1)$, but let's compute D^2 for each point just to be sure:

$$\begin{aligned} D^2\,(1.086, 1.188) &\approx (1.806-1)^2 + (1.188-1)^2 \approx 0.685 \\ D^2\,(-1.879, -0.484) &\approx (-1.879-1)^2 + (-0.484-1)^2 \approx 10.5 \end{aligned}$$

Therefore, $(x,y) \approx (1.086, 1.188)$ is the point on the ellipse $x^2+2y^2=4$ that is closest to the point $(1,1)$. ■

---

(5) What is the largest volume of possible of a cylinder which is sized according to $\pi r^2 + h = 20$, and what are the dimensions which give that volume?

□ The volume of a cylinder is $V = \pi r^2 h$, and this is our objective function to be maximized. The constraint is $\pi r^2 + h = 20$. The gradients are $\nabla V(r,h) = \langle 2\pi r h, \pi r^2 \rangle$ and $\nabla g(r,h) = \langle 2\pi r, 1 \rangle$. Setting the proportion $\nabla V(r,h) = \lambda \nabla g(r,h)$ and introducing the constraint, we get the system:

$$\begin{aligned} 2\pi r h &= \lambda(2\pi r) \\ \pi r^2 &= \lambda(1) \\ \pi r^2 + h &= 20 \end{aligned}$$

or

$$\begin{aligned} rh &= \lambda r \\ \pi r^2 &= \lambda \\ \pi r^2 + h &= 20 \end{aligned}$$

The solutions are, using Wolfram Alpha,

$$(r,h)=\left(-\sqrt{\frac{10}{\pi}},10\right)\qquad (r,h)=\left(\sqrt{\frac{10}{\pi}},10\right)\qquad (r,h)=(0,20)$$

The first solution is irrelevant as a negative r is not appropriate. The third solution clearly gives the minimum volume (V = 0). So the maximum volume occurs at $(r,h)=(\sqrt{10/\pi},10)$, and that volume is

$$V_{max}=\pi\left(\sqrt{\frac{10}{\pi}}\right)^2(10)=100 \quad \blacksquare$$

---

(6) Find the maximum value of $f(x,y,z,w)=2x-3y+4z+w$ determined by points on the "hypersphere" $x^2+y^2+z^2+w^2=3$.

☐ The gradient of the objective function is $\nabla f=\langle 2,-3,4,1\rangle$, and the gradient of the objective function is $\nabla g=\langle 2x,2y,2z,2w\rangle$. The relation $\nabla f=\lambda\nabla g$ along with the constraint then give this system of equations:

$$\begin{aligned} 2 &= \lambda(2x)\\ -3 &= \lambda(2y)\\ 4 &= \lambda(2z)\\ 1 &= \lambda(2w)\\ x^2+y^2+z^2+w^2 &= 3 \end{aligned}$$

This system would be simple to solve by hand, but why spoil a good thing? Wolfram Alpha gives two solution sets,

$$(x,y,z,w)=\left(-\sqrt{\frac{2}{5}},\frac{3}{\sqrt{10}},-2\sqrt{\frac{2}{5}},-\frac{1}{\sqrt{10}}\right),\ \left(\sqrt{\frac{2}{5}},-\frac{3}{\sqrt{10}},2\sqrt{\frac{2}{5}},\frac{1}{\sqrt{10}}\right)$$

and at these points we have

$$f\left(-\sqrt{\frac{2}{5}},\frac{3}{\sqrt{10}},-2\sqrt{\frac{2}{5}},-\frac{1}{\sqrt{10}}\right)=-3\sqrt{10}$$

$$f\left(\sqrt{\frac{2}{5}},-\frac{3}{\sqrt{10}},2\sqrt{\frac{2}{5}},\frac{1}{\sqrt{10}}\right)=3\sqrt{10}$$

and so the maximum value of $f(x,y,z,w)$ subject to $x^2+y^2+z^2+w^2=3$ is $3\sqrt{10}$. ■

(7) Find the absolute extremes of $f(x, y) = xy(1-x-y) = xy - x^2y - xy^2$ anywhere on and inside the unit circle. (Compare to Practice Problem 2 in Sec. 14.2.)

□ First, note that our constraint here is $x^2 + y^2 \leq 1$.

In Practice Problem 2 of Sec. 14.2, we found that this function has a local maximum at $(1/3, 1/3)$, which does satisfy the constraint $x^2 + y^2 \leq 1$. (There were saddle points as well, but those won't be absolute extremes.)

We'll use Lagrange Multipliers to examine the boundary of the unit circle for other possible extremes. Since $\nabla f(x, y) = \langle y - 2xy - y^2, x - x^2 - 2xy\rangle$ and $\nabla g(x, y) = \langle 2x, 2y\rangle$, then the relation $\nabla f(x, y) = \lambda \nabla g(x, y)$ and the constraint itself gives this system of equations:

$$\begin{gathered} y - 2xy - y^2 = \lambda(2x) \\ x - x^2 - 2xy = \lambda(2y) \\ x^2 + y^2 = 1 \end{gathered}$$

Wolfram Alpha reports six solutions (x, y) to this system:

$$\left(-\frac{1}{6}(1+\sqrt{17}), \frac{1}{6}(\sqrt{17}-1)\right), \left(\frac{1}{6}(\sqrt{17}-1), -\frac{1}{6}(1+\sqrt{17})\right),$$

$$(0,1), (1,0), \left(-\frac{1}{\sqrt{2}}, -\frac{1}{\sqrt{2}}\right), \left(\frac{1}{\sqrt{2}}, \frac{1}{\sqrt{2}}\right)$$

To locate extremes, we now have seven points to compare (the critical point from the interior of the unit circle and these six points from the boundary of the unit circle). The function values are (with some approximation):

$$f\left(\frac{1}{3}, \frac{1}{3}\right) = \frac{1}{27}$$

$$f\left(-\frac{1}{6}(1+\sqrt{17}), \frac{1}{6}(\sqrt{17}-1)\right) = -\frac{16}{27} \approx -0.59$$

$$f\left(\frac{1}{6}(\sqrt{17}-1), -\frac{1}{6}(1+\sqrt{17})\right) = -\frac{16}{27} \approx -0.59$$

$$f(0,1) = 0$$
$$f(1,0) = 0$$
$$f\left(-\frac{1}{\sqrt{2}}, -\frac{1}{\sqrt{2}}\right) = \frac{1+\sqrt{2}}{2} \approx 1.21$$
$$f\left(\frac{1}{\sqrt{2}}, \frac{1}{\sqrt{2}}\right) = \frac{1-\sqrt{2}}{2} \approx -0.21$$

The absolute maximum of $f(x, y)$ subject to the given constraint occurs at one point, and is

$$f\left(-\frac{1}{\sqrt{2}}, -\frac{1}{\sqrt{2}}\right) = \frac{1+\sqrt{2}}{2}$$

The absolute minimum of $f(x, y)$ subject to the given constraint occurs at two points, and is

$$f\left(-\frac{1}{6}(1+\sqrt{17}), \frac{1}{6}(\sqrt{17}-1)\right)$$
$$= f\left(\frac{1}{6}(\sqrt{17}-1), -\frac{1}{6}(1+\sqrt{17})\right) - \frac{16}{27} \approx -0.59 \quad \blacksquare$$

B.5.5 *Parametric Surfaces — Practice — Solved*

(1) Figure 15.29 shows the paraboloid $x = 5 - y^2 - z^2$ for $0 \leq x \leq 5$. Provide the parametric equations for this surface and give the appropriate bounds on the parameters.

□ This is basically YTI 4 flipped on its side. The parametric equations are

$$\begin{cases} x = & 5 - s \\ y = & \sqrt{s}\cos t \\ z = & \sqrt{s}\sin t \end{cases} \quad \text{with} \quad 0 \leq s \leq 5 \quad ; \quad 0 \leq t < 2\pi$$

Or in vector form (with the same bounds),

$$\mathbf{r}(s, t) = \langle 5 - s, \sqrt{s}\sin t, \sqrt{s}\sin t\rangle$$

Note that these expressions aren't unique; for example, we could swap the equations for y and z. ■

(2) Figure C.9 shows the paraboloid $z = 2x^2 + y^2$ for $0 \leq z \leq 10$. Provide the parametric equations for this surface and give the appropriate bounds on the parameters.

□ The cross sections of this surface are ellipses, not circles. I'll show you the parametric equations first, then explain two ways to get to them. They are

$$\begin{cases} x = \frac{1}{\sqrt{2}}\sqrt{s}\cos t \\ y = \sqrt{s}\cos t \\ z = s \end{cases} \quad \text{with} \quad 0 \leq s \leq 10 \;\; ; \;\; 0 \leq t < 2\pi$$

One way to get there is in a reverse-engineering sort of way, where we recognize that with these equations, we have $x^2 = (s/2)\cos^2 t$ and so we'd have to multiply x^2 by 2 in order to combine it with $y^2 = s\sin^2 t$ and then build $2x^2 + y^2 = z$. The second way to do it is more "by the book"; we rewrite the surface as $z = x^2/(1/\sqrt{2})^2 + y^2$, and then when we consult Appendix A.2, we have a match to $z = x^2/a^2 + y^2/b^2$ with $a = 1/\sqrt{2}$ and $b = 1$. Those values can be plugged in to the templates for the parametric equations to yield our result. So either way, we get there! ■

---

(3) A quadric surface has the following parametric (vector) equations. Identify this surface by giving its "regular" expression in rectangular coordinates.

$$\mathbf{r}(s,t) = \langle \sqrt{9-s}\cos t, \sqrt{9-s}\sin t, s\rangle \quad \text{for} \quad 0 \leq s \leq 9, 0 \leq t < 2\pi$$

□ With $x = \sqrt{9-s}\cos t$ and $y = \sqrt{9-s}\sin t$, we have $x^2 + y^2 = 9 - s$; since $z = s$, then, this becomes $x^2 + y^2 = 9 - z$ or $z = 9 - x^2 - y^2$. This is consistent with given bounds as being an inverted paraboloid $z = 9 - x^2 - y^2$ above the xy-plane. ■

---

(4) A quadric surface has the parametric equations $x = 1 + s + t, y = -s + t, z = 2s$, where s and t can be any real numbers. Identify this surface by giving its "regular" expression in rectangular coordinates.

□ Since all three equations are linear, this is going to be a plane! If we bundle the equations as $\searrow(s,t) = \langle s+t, 1-s+t, 2s\rangle$ then we can expand to isolate s and t:

$$\searrow(s,t) = \langle 1,0,0\rangle + s\langle 1,-1,2\rangle + t\langle 1,1,0\rangle$$

From this, we see a point on the plane is located by $\mathbf{r}_0 = \langle 1,0,0\rangle$, and two vectors in the plane are $\langle 1,-1,2\rangle$ and $\langle 1,1,0\rangle$. A vector perpendicular to the plane is then

$$\langle 1,-1,2\rangle \times \langle 1,1,0\rangle = \langle -2,2,2\rangle$$

We can now form the rectangular version of this plane as

$$\begin{aligned} a(x-x_0)+b(y-y_0)+c(z-z_0)&=0 \\ -2(x-1)+2(y-0)+2(z-0)&=0 \\ z&=x-y-1 \end{aligned}$$

or any other alternate but equivalent form. ■

---

(5) A quadric surface has the parametric equations $x = s\cos t, y = s^2, z = s\sin t$, for $0 \le s \le \pi$ and $0 \le t \le \pi$. Identify this surface by giving its "regular" expression in rectangular coordinates and appropriate bounds on those coordinates.

□ If we assign $x = s\cos t$, $y = s$, $z = s\sin t$, then x and z together are related by $x^2 + z^2 = s^2$. But since $y = s^2$, we have $y = x^2 + z^2$. So this is a portion of paraboloid opening around the y-axis, from $y = 0$ to $y = \pi^2$. ■

---

(6) Figure C.20 shows the hyperboloid $x^2 + 3y^2 - z^2 = 1$ between $z = -1$ and $z = 4$. Provide the parametric equations for this surface and give the appropriate bounds on the parameters.

□ Appendix A.2 gives a recipe for the parametric equations of a hyperboloid of one sheet if the equation looks like:

$$\frac{x^2}{a^2}+\frac{y^2}{b^2}-\frac{z^2}{c^2}=1$$

We can rewrite the hyperboloid $x^2 + 3y^2 - z^2 = 1$ as:

$$\frac{x^2}{1^2}+\frac{y^2}{(1/\sqrt{3})^2}-\frac{z^2}{1^2}=1$$

to get a match to $a = 1$, $b = 1/\sqrt{3}$, and $c = 1$. Then by the template given in the Appendix, the parametric equations for this surface will be

$$\begin{cases} x = \sqrt{1+s^2}\cos t \\ y = \frac{1}{\sqrt{3}}\sqrt{1+s^2}\sin t \\ z = s \end{cases} \quad \text{with} \quad -1 \le s \le 4 \ ; \quad 0 \le t < 2\pi \quad ■$$

B.6 Chapter 18: Practice Problem Solutions

B.6.1 *Line Integrals — Practice — Solved*

---

(1) Evaluate $\int_C ye^x\,ds$, where C is the line from $(1,2)$ to $(4,7)$.

□ This is a line integral with respect to arc length ds, so we'll need the arc length version of the line integral. We can describe the contour C as

$$x = 1 + 3t \qquad y = 2 + 5t \qquad \text{for } 0 \le t \le 1$$

so that

$$\sqrt{[x'(t)]^2 + [y'(t)]^2} = \sqrt{(3)^2 + (5)^2} = \sqrt{34}$$

and

$$\begin{aligned}\int_C ye^x\,ds &= \int_a^b f(x(t), y(t))\sqrt{[x'(t)]^2 + [y'(t)]^2}\,dt \\ &= \int_0^1 (2+5t)e^{1+3t}(\sqrt{34})\,dt = \frac{\sqrt{34}e}{9}(16e^3 - 1) \quad \blacksquare\end{aligned}$$

---

(2) Evaluate $\int_C \sin x\,dx + \cos y\,dy$ where C is the top half of $x^2 + y^2 = 1$ from $(1,0)$ to $(-1,0)$ joined to the line from from $(-1,0)$ to $(-2,3)$.

□ Note that this line integral does not involve the arc length parameter, so we'll just convert everything directly to parametric form. The contour C has two parts: the top half of $x^2 + y^2 = 1$ from $(1,0)$ to $(-1,0)$ (call this C_1) and then the line from $(-1,0)$ to $(-2,3)$ (call this C_2). The contours are

$$\begin{aligned} C_1 :\ & x = \cos t \quad y = \sin t \quad \text{for} \quad 0 \le t \le \pi \\ & dx = -\sin t\,dt \quad dy = \cos t\,dt \\ C_2 :\ & x = -1 - t \quad y = 3t \quad \text{for} \quad 0 \le t \le 1 \\ & dx = -dt \quad dy = 3\,dt \end{aligned}$$

The integrals are:

$$\int_{C_1} \sin x\,dx + \cos y\,dy = \int_0^\pi \sin(\cos t)(-\sin t\,dt) + \cos(\sin t)(\cos t\,dt) = 0$$

$$\begin{aligned}\int_{C_2} \sin x\,dx + \cos y\,dy &= \int_0^1 \sin(-1-t)(-dt) + \cos(3t)(3\,dt) \\ &= \sin 3 - \cos 2 + \cos 1\end{aligned}$$

Putting these together,

$$\int_C \sin x\,dx + \cos y\,dy = \int_{C_1} \sin x\,dx + \cos y\,dy + \int_{C_2} \sin x\,dx + \cos y\,dy$$
$$= \sin 3 - \cos 2 + \cos 1 \quad \blacksquare$$

---

(3) Evaluate $\int_C \mathbf{F}\cdot d\mathbf{r}$ where $\mathbf{F}(x,y,z) = \langle yz, xz, xy\rangle$ and $\mathbf{r}(t) = \langle t, t^2, t^3\rangle$ for $0 \le t \le 2$.

☐ From $\mathbf{r}(t)$ we have

$$x = t \to dx = dt$$
$$y = t^2 \to dy = 2t\,dt$$
$$z = t^3 \to dz = 3t^2\,dt$$

so that

$$\int_C \mathbf{F}\cdot d\mathbf{r} = \int_C yz\,dx + xz\,dy + xy\,dz$$
$$= \int_0^2 (t^2)(t^3)(dt) + (t)(t^3)(2t\,dt) + t(t^2)(3t^2\,dt)$$
$$= \int_0^2 (6t^5)\,dt = 64 \quad \blacksquare$$

---

(4) Evaluate $\int_C x^2 z\,ds$, where C is the line segment from $(0,6,-1)$ to $(4,1,5)$.

☐ Since this is a line integral with respect to arc length ds, we 'll use the arc length version of the line integral. We can describe the contour C as

$$x = 4t \quad y = 6 - 5t \quad z = -1 + 6t \qquad \text{for } 0 \le t \le 1$$

so that

$$\sqrt{[x'(t)]^2 + [y'(t)]^2 + [z'(t)]^2} = \sqrt{(4)^2 + (-5)^2 + (6)^2} = \sqrt{77}$$

and

$$\int_C x^2 z\,ds = \int_a^b f(x(t), y(t), z(t))\sqrt{[x'(t)]^2 + [y'(t)]^2 + [z'(t)]^2}\,dt$$
$$= \int_0^1 (4t)^2(-1+6t)(\sqrt{77})\,dt = \frac{56\sqrt{77}}{3} \quad \blacksquare$$

(5) Evaluate $\int_C z\,dx + x\,dy + y\,dz$ along the contour C that is given by

$$x = t^2 \quad y = t^3 \quad z = t^2 \quad \text{for} \quad 0 \le t \le 1$$

□ Since this is a line integral with respect to dx, dy and dz we can construct the parametric version of the line integral directly. From the parametric description of the line integral, we get $dx = 2t\,dt$, $dy = 3t^2\,dt$ and $dz = 2t\,dt$, and so

$$\begin{aligned}\int_C z\,dx + x\,dy + y\,dz &= \int_0^1 t^2(2t\,dt) + (t^2)(3t^2 dt) + (t^3)(2t\,dt) \\ &= \int_0^1 (2t^3 + 5t^4)\,dt = \frac{3}{2} \quad \blacksquare\end{aligned}$$

(6) Find the work done by $\mathbf{F}(x, y, z) = \langle z, y, -x \rangle$ around the contour $\mathbf{r}(t) = \langle t, \sin t, \cos t \rangle$ for $0 \le t \le \pi$.

□ We want to evaluate $\int_C \mathbf{F} \cdot d\mathbf{r}$. From $\mathbf{r}(t)$ we have

$$\begin{aligned} x = t &\to dx = dt \\ y = \sin t &\to dy = \cos t\,dt \\ z = \cos t &\to dz = -\sin t\,dt \end{aligned}$$

so that

$$\begin{aligned}\int_C \mathbf{F} \cdot d\mathbf{r} &= \int_C z\,dx + y\,dy - x\,dz \\ &= \int_0^\pi (\cos t\,dt) + (\sin t)(\cos t\,dt) - t(-\sin t\,dt) \\ &= \int_0^\pi (\cos t + \sin t \cos + t \sin t)\,dt = \pi\end{aligned}$$

(7) *(Bonus! Following up Sec. 15.3 ...)* We expect $\int_C f(x, y)\,ds = 0$ for which of the following combinations of function $f(x, y)$ and path of integration C?

I1) $f(x, y) = x^2 y$ and C is the upper half of $2x^2 + 3y^2 = 6$?
I2) $f(x, y) = x^2 y$ and C is the right half of $2x^2 + 3y^2 = 6$?
I3) $f(x, y) = xy^2$ and C is the on $y = x^2 - 1$ from $(-1, 0)$ to $(1, 0)$?

□ (I1) is not zero, but (I2) and (I3) are. In (I1), y is positive everywhere on the contour C, and so we are integrating an expression $(x^2 y)$

which is always positive. In (I2), the positive contributions from y are balanced by equal negative contributions; similarly, contributions from x in xy^2 are balanced due to the symmetry of the parabolic contour C around the y-axis. ■

B.6.2 *Conservative Vector Fields — Practice — Solved*

(1) Determine if the vector field $\mathbf{F}(x,y) = \langle x^3 + 4xy, 4xy - y^3 \rangle$ is conservative. If it is, find a potential function for it.

□ Matching to the form $\mathbf{F}(x,y) = \langle P(x,y), Q(x,y) \rangle$, we have

$$P = x^3 + 4xy \to \frac{\partial P}{\partial y} = 4x$$
$$Q = 4xy - y^3 \to \frac{\partial Q}{\partial x} = 4y$$

Since $\partial P/\partial y \neq \partial Q/\partial x$ then the vector field is not conservative, and there is no function $f(x,y)$ such that $\nabla f = \mathbf{F}$. ■

(2) Determine if the vector field $\mathbf{F}(x,y) = \langle e^y, xe^y \rangle$ is conservative. If it is, find a potential function for it.

□ Matching to the form $\mathbf{F}(x,y) = \langle P(x,y), Q(x,y) \rangle$, we have

$$P = e^y \to \frac{\partial P}{\partial y} = e^y$$
$$Q = xe^y \to \frac{\partial Q}{\partial x} = e^y$$

Since $\partial P/\partial y = \partial Q/\partial x$ then the vector field is conservative. So there is a function $f(x,y)$ such that $\nabla f = \mathbf{F}$.

We know from $\mathbf{F}$ that for this function, $f_x = e^y$ and $f_y = xe^y$. Based on f_x, we know that at worst, $f(x,y) = xe^y + g(y)$ where $g(y)$ is some unknown function of y.

With $f(x,y)$ in this form, $f_y = xe^y + g'(y)$. But since we know $f_y = xe^y$ we have that $g'(y) = 0$ and $g(y) = C$. So, $f(x,y) = xe^y + C$. ■

(3) Determine if the vector field $\mathbf{F}(x,y,z) = \langle 3z^2, \cos y, 2xz \rangle$ is conservative. If it is, find a potential function for it.

☐ Since $\nabla \times \mathbf{F} = \langle 0, 4z, 0 \rangle$ which is not $\mathbf{0}$, then $\mathbf{F}$ is not conservative, and so there is no scalar function f such that $\mathbf{F} = \nabla f$. ■

---

(4) Evaluate $\displaystyle\int_C \mathbf{F} \cdot d\mathbf{r}$ where

$$\mathbf{F}(x, y) = \left\langle \frac{y^2}{1 + x^2}, 2y \tan^{-1}(x) \right\rangle$$

and C is the curve given by $\mathbf{r}(t) = \langle t^2, 2t \rangle$ for $0 \leq t \leq 1$.

☐ A quick check will show that $\mathbf{F}$ is conservative. Therefore, we can find a potential function for it and evaluate the integral using the Fundamental Theorem for Line Integrals. The potential function for this vector field is a function $f(x, y)$ such that $f_x = y^2/(1 + x^2)$ and $f_y = 2y \tan^{-1}(x)$. Based on f_x, we know that at worst, $f(x, y) = y^2 \tan^{-1}(x) + g(y)$ where $g(y)$ is some unknown function of y.

With $f(x, y)$ in this form, $f_y = 2y \tan^{-1}(x) + g'(y)$. But since we know $f_y = 2y \tan^{-1}(x)$ we have that $g'(y) = 0$ and $g(y) = C$. Let's choose $C = 0$ so that

$$f(x, y) = y^2 \tan^{-1}(x)$$

Next we need $\int_C \mathbf{F} \cdot d\mathbf{r}$ where $\mathbf{r}(t) = \langle t^2, 2t \rangle$ for $0 \leq t \leq 1$. The values of $\mathbf{r}(t)$ and $f(x, y)$ at the endpoints of this curve, i.e. at $t = 0$ and $t = 1$, are

$$\begin{aligned} \mathbf{r}(1) &= \langle 1, 2 \rangle \rightarrow f(1, 2) = 2^2 \tan^{-1}(1) = \pi \\ \mathbf{r}(0) &= \langle 0, 0 \rangle \rightarrow f(0, 0) = 0 \end{aligned}$$

Then by the Fundamental Theorem for Line Integrals, we have

$$\int_C \mathbf{F} \cdot d\mathbf{r} = f(1, 2) - f(0, 0) = \pi \quad ■$$

---

(5) Determine if the vector field $\mathbf{F}(x, y) = \langle 1 + 2xy + \ln x, x^2 \rangle$ is conservative. If it is, find a potential function for it.

☐ Matching to the form $\mathbf{F}(x, y) = \langle P(x, y), Q(x, y) \rangle$, we have

$$\begin{aligned} P = 1 + 2xy + \ln x &\rightarrow \frac{\partial P}{\partial y} = 2x \\ Q = x^2 &\rightarrow \frac{\partial Q}{\partial x} = 2x \end{aligned}$$

Since $\partial P/\partial y = \partial Q/\partial x$ then the vector field is conservative. So there is a function $f(x, y)$ such that $\nabla f = \mathbf{F}$.

We know from $\mathbf{F}$ that for this function, $f_x = 1 + 2xy + \ln x$ and $f_y = x^2$. Based on f_x, we know that at worst,

$$f(x, y) = x + x^2y + x(\ln x - 1) + g(y) = x^2y + x\ln x + g(y)$$

where $g(y)$ is some unknown function of y. With $f(x, y)$ in this form, $f_y = x^2 + g'(y)$. But since we know $f_y = x^2$ we have that $g'(y) = 0$ and $g(y) = C$. So,

$$f(x, y) = x^2y + x\ln x + C \quad \blacksquare$$

---

(6) Determine if the vector field $\mathbf{F}(x, y, z) = \langle e^z, 1, xe^z\rangle$ is conservative. If it is, find a potential function for it.

□ Since $\nabla \times \mathbf{F} = \mathbf{0}$ then $\mathbf{F}$ is conservative, and so there is a scalar function f such that $\mathbf{F} = \nabla f$. Since that's true, we know $f_x = e^z$, $f_y = 1$, and $f_z = xe^z$. Without any further ado, it's pretty easy to figure out that $f(x, y, z) = xe^z + y + C$. ■

---

(7) Use the Fundamental Theorem for Line Integrals to evaluate $\int_C \mathbf{F} \cdot d\mathbf{r}$ where $\mathbf{F}(x, y, z) = \langle 2xz + y^2, 2xy, x^2 + 3z^2\rangle$, and C is the curve given by $\mathbf{r}(t) = \langle t^2, t + 1, z\rangle$ for $0 \le t \le 1$.

□ The potential function for this vector field is a function $f(x, y, z)$ such that $f_x = 2xz + y^2$, $f_y = 2xy$ and $f_z = x^2 + 3z^2$. We could go through the routine of finding $f(x, y, z)$ based on these derivatives, but I think this one's straightforward enough to deduce that

$$f(x, y, z) = x^2z + xy^2 + z^3$$

Next we need $\int_C \mathbf{F} \cdot d\mathbf{r}$ where C is the contour given by $x = t^2$, $y = t+1$, and $z = 2t - 1$ for $0 \le t \le 1$. The endpoints of this curve and values of $f(x, y, z)$ there are

$$(x(0), y(0), z(0)) = (0, 1, -1) \rightarrow f(0, 1, -1) = 0 + 0 + (-1)^3 = -1$$
$$(x(1), y(1), z(1)) = (1, 2, 1) \rightarrow f(1, 2, 1) = 1 + 4 + 1 = 6$$

Then by the Fundamental Theorem for Line Integrals, we have

$$\int_C \mathbf{F} \cdot d\mathbf{r} = f(1, 2, 1) - f(0, , 1, -1) = 7 \quad \blacksquare$$

B.6.3 *Surface Integrals — Practice — Solved*

(1) Find $\iint_S xy\,dS$ where the surface S is the triangular region with vertices P(1,0,0), Q(0,2,0) and R(0,0,2).

□ This surface is part of a plane. Which one? Well, you can construct **PQ** × **PR** to get a normal vector and build the equation of the plane that way. Or you can examine the points to see they satisfy $2x+y+z=2$. Either way, we have the plane $z=2-2x-y$ so that $g_x=-2$ and $g_y=-1$. The region D is the intersection of this plane with the xy-plane, which is $2x+y=2$ (found by setting $z=0$). So using the scalar function version of the surface integral,

$$\begin{aligned}\iint_S xy\,dS &= \iint_D xy\sqrt{g_x^2+g_y^2+1}\,dA\\ &= \int_0^1\int_0^{-2x+2} xy\sqrt{(-2)^2+(-1)^2+1}\,dydx\\ &= \sqrt{6}\int_0^1\int_0^{-2x+2} xy\,dydx = \frac{\sqrt{6}}{6} \quad \blacksquare\end{aligned}$$

(2) Find the surface area of the portion of the paraboloid $z=4-x^2-y^2$ above the xy-plane.

□ The paraboloid intersects the xy-plane in the circle $x^2+y^2=4$. So the surface area of the paraboloid above the xy-plane is the surface area of the paraboloid above the region D that is the circle $x^2+y^2=4$. So, with $z=4-x^2-y^2$, we have $g_x=-2x$ and $g_y=-2y$, and

$$\sqrt{g_x^2+g_y^2+1}=\sqrt{4x^2+4y^2+1}$$

But note that the region and expression to be integrated are suitable for polar coordinates; D is described $0\le r\le 2$ and $0\le\theta\le 2\pi$, and $\sqrt{4x^2+4y^2+1}=\sqrt{4r^2+1}$. So the surface area is given by

$$\begin{aligned}A_S &= \iint_S (1)\,dS = \iint_D (1)\sqrt{g_x^2+g_y^2+1}\,dA\\ &= \int_0^{2\pi}\int_0^2 \sqrt{4r^2+1}\cdot rdrd\theta = \int_0^{2\pi}\left(\frac{1}{12}(4r^2+1)^{3/2}\Big|_0^2\right)d\theta\\ &= \int_0^{2\pi}\frac{1}{12}(17^{3/2}-1)d\theta = \frac{\pi}{6}(17^{3/2}-1) \quad \blacksquare\end{aligned}$$

(3) Find $\iint_S \mathbf{F} \cdot d\mathbf{S}$ where $\mathbf{F} = \langle xy, 4x^2, yz \rangle$ and the surface S is $z = xe^y$ over $0 \le x \le 1$ and $0 \le y \le 1$, oriented positively.

☐ Matching the surface to the form $z = g(x, y)$ we have $g = xe^y$, so

$$\frac{\partial g}{\partial x} = e^y \quad ; \quad \frac{\partial g}{\partial y} = xe^y$$

Matching to the form $\mathbf{F} = \langle P, Q, R \rangle$, we have

$$P = xy \quad ; \quad Q = 4x^2 \quad ; \quad R = yz = y(xe^y)$$

So by the vector version of the surface integral,

$$\begin{aligned}\iint_S \mathbf{F} \cdot d\mathbf{S} &= \iint_D \left(-P\frac{\partial g}{\partial x} - Q\frac{\partial g}{\partial y} + R \right) dA \\ &= \iint_D \left(-xy(e^y) - (4x^2)(xe^y) + y(xe^y) \right) dA \\ &= \int_0^1 \int_0^1 \left(-4x^3 e^y \right) dydx = 1 - e \quad \blacksquare\end{aligned}$$

(4) Find $\iint_S yz\, dS$ where the surface S is the part of the plane $x+y+z = 1$ in the first octant.

☐ Since $z = 1 - x - y$ we have $g_x = -1$ and $g_y = -1$. The region D is the intersection of this plane with the xy-plane, which is $x + y = 1$ (found by setting $z = 0$). So using the scalar function version of the surface integral,

$$\begin{aligned}\iint_S yz\, dS &= \iint_D yz\sqrt{g_x^2 + g_y^2 + 1}\, dA \\ &= \int_0^1 \int_0^{-x+1} y(1 - x - y)\sqrt{(-1)^2 + (-1)^2 + 1}\, dydx \\ &= \sqrt{3} \int_0^1 \int_0^{-x+1} y(1 - x - y)\, dydx = \frac{\sqrt{3}}{24} \quad \blacksquare\end{aligned}$$

(5) Find the surface area of the part of the hyperbolic paraboloid $z = y^2 - x^2$ between the cylinders $x^2 + y^2 = 1$ and $x^2 + y^2 = 4$.

☐ This piece of the surface is over the region in the xy-plane between the circles $x^2 + y^2 = 1$ and $x^2 + y^2 = 4$. With $z = y^2 - x^2$, we get $g_x(x, y) = -2x$ and $g_y(x, y) = 2y$, so

$$\sqrt{g_x(x, y)^2 + g_y(x, y)^2 + 1} = \sqrt{4x^2 + 4y^2 + 1}$$

The region and expression to be integrated are suitable for polar coordinates; D is described $1 \leq r \leq 2$ and $0 \leq \theta \leq 2\pi$, and $\sqrt{4x^2+4y^2+1} = \sqrt{4r^2+1}$. So the surface area is given by

$$\begin{aligned} A_S &= \iint_S (1)\,dS = \iint_D (1)\sqrt{g_x^2+g_y^2+1}\,dA \\ &= \int_0^{2\pi}\int_1^2 \sqrt{4r^2+1}\cdot r\,dr\,d\theta = \int_0^{2\pi}\left(\frac{1}{12}(4r^2+1)^{3/2}\Big|_1^2\right)d\theta \\ &= \int_0^{2\pi}\frac{1}{12}(17^{3/2}-5^{3/2})d\theta = \frac{\pi}{6}(17^{3/2}-5^{3/2}) \quad \blacksquare \end{aligned}$$

---

(6) Find $\iint_S \mathbf{F}\cdot d\mathbf{S}$ where $\mathbf{F} = \langle x, y, z^4\rangle$ and the surface S is $z = \sqrt{x^2+y^2}$ under $z = 1$ oriented negatively.

□ Matching the surface to the form $z = g(x, y)$ we have $g = \sqrt{x^2+y^2}$ and

$$\frac{\partial g}{\partial x} = \frac{x}{\sqrt{x^2+y^2}} \quad ; \quad \frac{\partial g}{\partial y} = \frac{x}{\sqrt{x^2+y^2}}$$

Matching to the form $\mathbf{F} = \langle P, Q, R\rangle$, we have

$$P = x \quad ; \quad Q = y \quad ; \quad R = z^4 = (x^2+y^2)^2$$

Since the surface is oriented negatively we will change the sign on our integral to account for the orientation. Polar coordinates will be useful; the boundary of the domain is $1 = \sqrt{x^2+y^2}$ or $x^2+y^2 = 1$:

$$\begin{aligned} \iint_S \mathbf{F}\cdot d\mathbf{S} &= -\iint_D \left(-P\frac{\partial g}{\partial x} - Q\frac{\partial g}{\partial y} + R\right) dA \\ &= -\iint D\left(-x\cdot\frac{x}{\sqrt{x^2+y^2}} - y\cdot\frac{y}{\sqrt{x^2+y^2}} + (x^2+y^2)^2\right) dA \\ &= \iint D\left(\sqrt{x^2+y^2} - (x^2+y^2)^2\right) dA \\ &= \int_0^{2\pi}\int_0^1 (r - r^4)\,r\,dr\,d\theta = \frac{\pi}{3} \quad \blacksquare \end{aligned}$$

---

(7) *(Bonus! Following up Sec. 15.3 ...)* We expect $\iint Sf(x, y, z)\,dS =$ 0 for which of the following combinations of function $f(x, y, z)$ and surfaces of integration S?

I1) $f(x,y) = e^{x^2+y^2+z^2}$ and S is the upper half of the unit sphere
I2) $f(x,y) = ze^{x^2+y^2}$ and S is the right half of the unit sphere?
I3) $f(x,y) = \sin(x)\cos(yz)$ and S is the paraboloid $z = x^2 + y^2$ from $z = 0$ to $z = 1$?

□ (I1) is not zero, but (I2) and (I3) are. The integrand of (I1) is always positive, whereas the integrands in (I2) and (I3) cooperate with the symmetric regions of integration to cause every positive value of the integrand function to have an equal but opposite partner. Also note that this is about the only intuition we can have about (I2) and (I3), since they would be near impossible to evaluate by hand. ■

B.6.4 *Green's Theorem — Practice Problems — Solutions*

---

(1) Find $\oint_C x^2y^2\,dx + 4xy^3\,dy$ where C is the boundary of the triangle with corners traversed in order (0,0) to (1,3) to (0,3) and back to (0,0).

□ Matching the integral to the form $\oint_C Pdx + Qdy$, we have

$$P = x^2y^2 \to \frac{\partial P}{\partial y} = 2x^2y$$

$$Q = 4xy^3 \to \frac{\partial Q}{\partial x} = 4y^3$$

Note that C is the area between the lines $y = 3x$ and $y = 3$ from $x = 0$ to $x = 1$; this is closed and positively oriented, so Green's Theorem applies and (you don't need to see all the trivial steps of evaluation, right?)...

$$\oint_C x^2y^2\,dx + 4xy^3\,dy = \iint_D \left(\frac{\partial Q}{\partial x} - \frac{\partial P}{\partial y}\right) dA$$

$$= \int_0^1 \int_{3x}^3 (4y^3 - 2x^2y)\,dydx = \frac{318}{5} \quad ■$$

---

(2) Find $\oint_C \mathbf{F} \cdot d\mathbf{r}$ where $\mathbf{F}(x,y) = \langle y^2\cos x, x^2 + 2y\sin x\rangle$ and C is the triangle from (0,0) to (2,6) to (2,0) and back to (0,0).

□ Matching the vector field to the form $\mathbf{F}(x,y) = \langle P(x,y), Q(x,y)\rangle$, we have

$$P = y^2 \cos x \to \frac{\partial P}{\partial y} = 2y \cos x$$

$$Q = x^2 + 2y \sin x \to \frac{\partial Q}{\partial x} = 2x + 2y \cos x$$

Note that C bounds the region under $y = 3x$ from $x = 0$ to $x = 2$. C is closed but negatively oriented (clockwise) — see Fig. B.25. Green's Theorem applies, but we have to change the sign to account for the orientation:

$$\oint_C \mathbf{F} \cdot d\mathbf{r} = \oint_C P\,dx + Q\,dy = -\iint_D \left(\frac{\partial Q}{\partial x} - \frac{\partial P}{\partial y} \right) dA$$
$$= -\iint_D (2x)\,dA = -\int_0^2 \int_0^{3x} (2x)\,dydx = -16 \quad \blacksquare$$

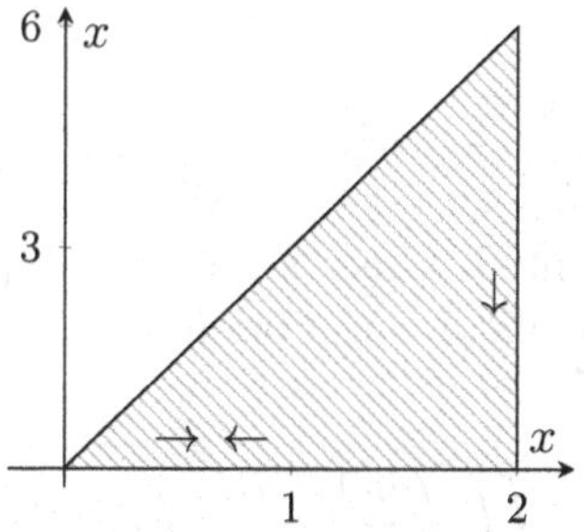

Fig. B.25 The triangle from $(0, 0)$ to $(2, 6)$ to $(2, 0)$ and back.

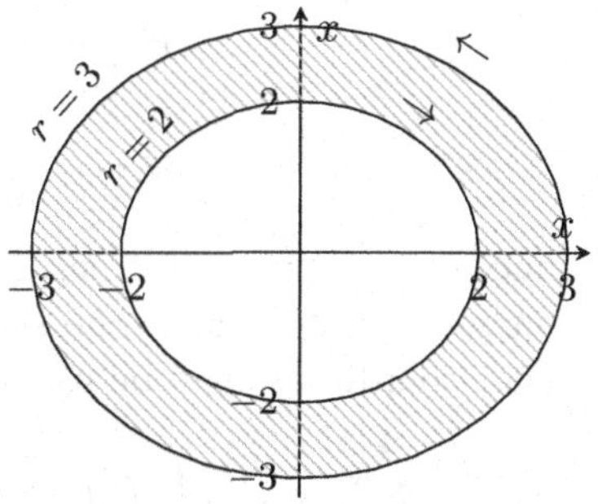

Fig. B.26 Region between $r = 2$ and $r = 3$.

---

(3) Find $\oint_C xe^{-2x}\,dx + (x^4 + 2x^2y^2)\,dy$ where C is the boundary of the region inside the circles $x^2 + y^2 = 4$ and $x^2 + y^2 = 9$.

☐ Matching the integral to the form $\int_C P\,dx + Q\,dy$, we have

$$P = xe^{-2x} \to \frac{\partial P}{\partial y} = 0$$

$$Q = x^4 + 2x^2y^2 \to \frac{\partial Q}{\partial x} = 4x^3 + 4xy^2$$

Note that C not simple, but Green's Theorem still applies; we can easily describe the region D inside C using polar coordinates — see

Fig. B.26 — so

$$\begin{aligned}\oint_C xe^{-2x}\,dx+(x^4+2x^2y^2)\,dy &= \iint_D \left(\frac{\partial Q}{\partial x}-\frac{\partial P}{\partial y}\right) dA\\ &= \iint_D 4x(x^2+y^2)\,dA = \int_0^{2\pi}\int_2^3 4r\cos\theta(r^2)\,rdrd\theta\\ &= \int_0^{2\pi}\int_2^3 4r^3\cos\theta\,rdrd\theta = 0 \quad \blacksquare\end{aligned}$$

---

(4) Find $\oint_C \mathbf{F}\cdot d\mathbf{r}$ where $\mathbf{F}(x,y)=\langle e^x+x^2y, e^y-xy^2\rangle$ and C is the clockwise perimeter of $x^2+y^2=25$.

□ Matching the vector field to the form $\mathbf{F}(x,y)=\langle P(x,y),Q(x,y)\rangle$, we have

$$P=e^x+x^2y\to\frac{\partial P}{\partial y}=x^2$$

$$Q=e^y-xy^2\to\frac{\partial Q}{\partial x}=-y^2$$

Since C is clockwise it is negatively oriented. Green's Theorem applies, but we have to change the sign to account for the orientation. Also, polar coordinates will be useful:

$$\begin{aligned}\oint_C \mathbf{F}\cdot d\mathbf{r} = \oint_C P\,dx+Q\,dy &= -\iint_D \left(\frac{\partial Q}{\partial x}-\frac{\partial P}{\partial y}\right) dA\\ &= -\iint_D \left(-y^2-x^2\right)\,dA = \iint_D \left(x^2+y^2\right)\,dA\\ &= \int_0^{2\pi}\int_0^5 (r^2)\,r\,drd\theta = \frac{625\pi}{2} \quad \blacksquare\end{aligned}$$

B.6.5 *The Divergence Theorem — Practice — Solved*

---

(1) Find $\iint_S \mathbf{F}\cdot d\mathbf{S}$ for the vector field $\mathbf{F}=\langle x^2z^3, 2xyz^3, xz^4\rangle$ where S is the surface of the region E that is the rectangular box $-1\le x\le 1$, $-2\le y\le 2$, $-3\le z\le 3$.

□ The divergence of $\mathbf{F}$ is:

$$\nabla\cdot\mathbf{F}=\frac{\partial}{\partial x}(x^2z^3)+\frac{\partial}{\partial y}(2xyz^3)+\frac{\partial}{\partial z}(xz^4)=2xz^3+2xz^3+4xz^3=8xz^3$$

Then by the Divergence Theorem,

$$\iint_S \mathbf{F} \cdot d\mathbf{S} = \iiint_E \nabla \cdot \mathbf{F}(x, y, z)\, dV$$
$$= \int_{-1}^{1} \int_{-2}^{2} \int_{-3}^{3} (8xz^3)\, dz dy dx = 0 \quad \blacksquare$$

---

(2) Evaluate $\iint_S \mathbf{F} \cdot d\mathbf{S}$ for $\mathbf{F}(x, y, z) = \langle xy, (y^2 + e^{xz^2}), \sin(xy)\rangle$ and S is the surface of the region bounded by $z = 1 - x^2$, $z = 0$, $y = 0$, $y = 2$.

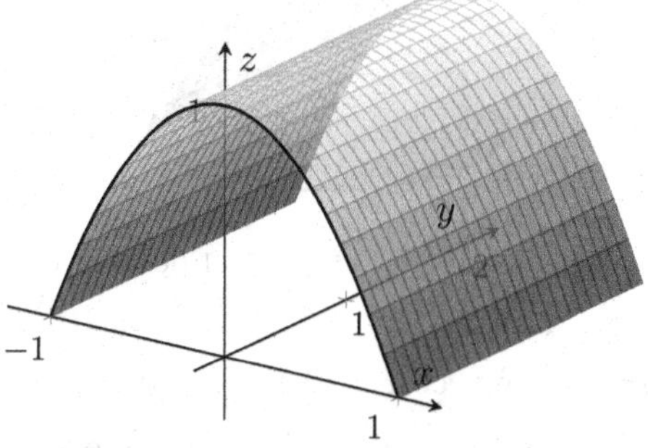

Fig. B.27 The parabolic cylinder $z = 1 - x^2$ for $0 \le y \le 2$, $z \ge 0$.

☐ The divergence of **F** is:

$$\nabla \cdot \mathbf{F} = \frac{\partial}{\partial x}(xy) + \frac{\partial}{\partial y}(y^2 + e^{xz^2}) + \frac{\partial}{\partial z}(\sin(xy)) = y + 2y + 0 = 3y$$

The region of integration is under the upper half of the parabolic cylinder $z = 1 - x^2$. It looks like a quonset hut — see Fig. B.27. Since we want the region above $z = 0$, our x values are restricted between -1 and 1. Bounds on y are given explicitly. By the Divergence Theorem, then,

$$\iint_S \mathbf{F} \cdot d\mathbf{S} = \iiint_E \nabla \cdot \mathbf{F}\, dV = \int_{-1}^{1} \int_{0}^{2} \int_{0}^{1-x^2} (3y)\, dz dy dx = 8 \quad \blacksquare$$

---

(3) Find $\iint_S \mathbf{F} \cdot d\mathbf{S}$ for the vector field $\mathbf{F} = \langle x^3 y, -x^2y^2, -x^2yz\rangle$ where S is the surface of the hyperboloid $x^2 + y^2 - z^2 = 1$ between $z = -2$ and $z = 2$.

☐ The surface S is shown in Fig. B.28. The divergence of **F** is:

$$\nabla \cdot \mathbf{F} = \frac{\partial}{\partial x}(x^3 y) + \frac{\partial}{\partial y}(-x^2y^2) + \frac{\partial}{\partial z}(-x^2yz) = 3x^2y - 2x^2y - x^2y = 0$$

Then by the Divergence Theorem,

$$\iint_S \mathbf{F} \cdot d\mathbf{S} = \iiint_E \nabla \cdot \mathbf{F}\, dV = \iiint_E (0) dz dy dx = 0 \quad \blacksquare$$

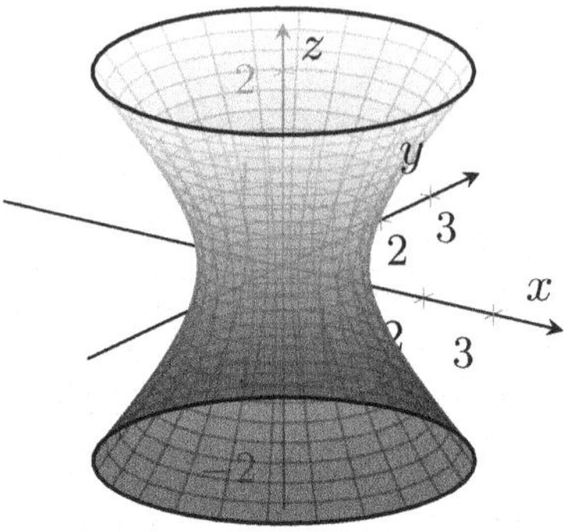

Fig. B.28 The hyperboloid $x^2 + y^2 - z^2 = 1$ between $z = -2$ and $z = 2$.

---

(4) Find $\iint_S \mathbf{F} \cdot d\mathbf{S}$ for the vector field $\mathbf{F} = \langle x^2y, xy^2, 2xyz \rangle$ where S is the surface of the tetrahedron formed by the plane $x + 2y + z = 2$ in the first octant.

□ The divergence of $\mathbf{F}$ is:

$$\nabla \cdot \mathbf{F} = \frac{\partial}{\partial x}(x^2y) + \frac{\partial}{\partial y}(xy^2) + \frac{\partial}{\partial z}(2xyz) = 2xy + 2xy + 2xy = 6xy$$

For the Divergence Theorem, note that the domain underlying the tetrahedron in the xy-plane is between the line $x + 2y = 2$ and the axes, so

$$\begin{aligned} \iint_S \mathbf{F} \cdot d\mathbf{S} &= \iiint_E \nabla \cdot \mathbf{F}\, dV \\ &= \int_0^2 \int_0^{-x/2+1} \int_0^{2-x-2y} (6xy)\, dz dy dx = \frac{2}{5} \quad \blacksquare \end{aligned}$$

B.6.6 *Stokes' Theorem — Practice — Solved*

---

(1) Compute $\oint_C \mathbf{F} \cdot d\mathbf{r}$ for the vector field $\mathbf{F} = \langle yz, 2xz, e^{xy} \rangle$ where C is the boundary of the cylinder $x^2 + y^2 = 16$ at $z = 5$.

□ Note that the surface itself is simply the portion of $z = 5$ inside the cylinder, so that the cylinder only defined the perimeter of the surface, which is a circle of radius 4. We'll use Stokes' Theorem to convert this to a surface integral involving $\nabla \times \mathbf{F}$, which is

$$\nabla \times \mathbf{F} = \langle xe^{xy} - 2x, -ye^{xy} + y, z\rangle$$

which in turn is converted to a double integral in the xy-plane. Equation (18.4) wraps that all up.

Matching the surface to the form $z = g(x, y) = 5$ we have $g_x = g_y = 0$. Matching the curl, $\nabla \times \mathbf{F} = \langle P, Q, R\rangle$, we have

$$P = xe^{xy} - 2x \quad ; \quad Q = -ye^{xy} + y \quad ; \quad R = z = 5$$

So by (18.4), we have

$$\begin{aligned}\oint_C \mathbf{F} \cdot d\mathbf{r} &= \iint_S \nabla \times \mathbf{F} \cdot d\mathbf{S} = \iint_D \left(-P\frac{\partial g}{\partial x} - Q\frac{\partial g}{\partial y} + R\right) dA \\ &= \iint_D (0 + 0 + 5)\ dA = \int_0^{2\pi} \int_0^4 5r dr d\theta = 80\pi \quad \blacksquare\end{aligned}$$

---

(2) Compute $\displaystyle\iint_S (\nabla \times \mathbf{F}) \cdot d\mathbf{S}$ for the vector field $\mathbf{F} = \langle yz, xz, xy\rangle$ where S is the paraboloid $z = 9 - x^2 - y^2$ above $z = 5$.

□ The boundary curve ∂S of this surface is the circle $x^2 + y^2 = 4$ (with $z = 5$) and can be expressed as

$$\begin{aligned} x &= 2\cos t && \to dx = -2\sin t\, dt \\ y &= 2\sin t && \to dy = 2\cos t\, dt \\ z &= 5 && \to dz = 0 \end{aligned}$$

for $0 \le t \le 2\pi$. By Stokes' Theorem we convert the surface integral into a line integral:

$$\begin{aligned}\iint_S (\nabla \times \mathbf{F}) \cdot d\mathbf{S} &= \oint_{\partial S} \mathbf{F} \cdot d\mathbf{r} = \oint_{\partial S} (yz)dx + (xz)dy + (xy)dz \\ &= \int_0^{2\pi} (2\sin t)(5)(-2\sin t dt) + (2\cos t)(5)(2\cos t dt) + 0 \\ &= \int_0^{2\pi} (20\cos^2 t - 20\sin^2 t)dt = 0 \quad \blacksquare\end{aligned}$$

(3) Compute $\int_C \mathbf{F} \cdot d\mathbf{r}$ for the vector field $\mathbf{F} = \langle e^{-x}, e^x, e^z \rangle$ where C is the boundary of the plane $2x + y + 2z = 2$ in the first octant, traversed counterclockwise.

□ Remember that this is asking us to evaluate $\oint_C \mathbf{F} \cdot d\mathbf{r}$. We'll use Stokes' Theorem via (18.4) to convert this to a double integral involving the $\nabla \times \mathbf{F}$, which is $\nabla \times \mathbf{F} = \langle 0, 0, e^x \rangle$ (details omitted).

Matching the surface to the form $z = g(x, y)$ we have $g(x, y) = 1 - x - y/2$, so that $g_x = -1$ and $g_y = -1/2$. Matching the curl to the form $\nabla \times \mathbf{F} = \langle P, Q, R \rangle$, we have

$$P = 0 \quad ; \quad Q = 0 \quad ; \quad R = e^x$$

So by (18.4),

$$\begin{aligned}
\oint_C \mathbf{F} \cdot d\mathbf{r} = \iint_S \nabla \times \mathbf{F} \cdot d\mathbf{S} &= \iint_D \left(-P\frac{\partial g}{\partial x} - Q\frac{\partial g}{\partial y} + R \right) dA \\
&= \iint_D (0 + 0 + e^x) \, dA \\
&= \int_0^1 \int_0^{-2x+2} (e^x) \, dy dx = 2e - 4 \quad \blacksquare
\end{aligned}$$

(4) Find the work done by the vector field $\mathbf{F} = \langle x^2 y^3 z, \sin(xyz), xyz \rangle$ around the bounding contour of the cone $y^2 = x^2 + z^2$ between $y = 0$ and $y = 3$ with normal vectors oriented outwards.

□ The boundary curve ∂S of this surface is the circle $x^2 + z^2 = 9$ and, to be positively oriented, can be expressed as

$$\begin{aligned}
x &= 3\cos t &&\to dx = -3\sin t \, dt \\
z &= 3\sin t &&\to dz = 3\cos t \, dt \\
y &= 3 &&\to dy = 0
\end{aligned}$$

for $0 \leq t \leq 2\pi$. By Stokes' Theorem,

$$\begin{aligned}
\iint_S (\nabla \times \mathbf{F}) \cdot d\mathbf{S} &= \oint_{\partial S} \mathbf{F} \cdot d\mathbf{r} \\
&= \oint_{\partial S} (x^2 y^3 z) \, dx + \sin(xyz) \, dy + (xyz) \, dz
\end{aligned}$$

Preparing all terms with their parametric forms,

$$\begin{aligned}
(x^2y^3z)\,dx &= (3\cos t)^2(3)^3(3\sin t)(-3\sin t\,dt)\\
&= (-3^7)\cos^2 t\sin^2 t\\
\sin(xyz)\,dy &= 0\\
(xyz)\,dz &= (3\cos t)(3\sin t)(3)(3\cos t\,dt)\\
&= 3^4\cos^2 t\sin t
\end{aligned}$$

Together,

$$\begin{aligned}
\iint_S (\nabla\times\mathbf{F})\cdot d\mathbf{S} &= \int_0^{2\pi}((-3^7)\cos^2 t\sin^2 t + 3^4\cos^2 t\sin t)\,dt\\
&= -\frac{3^7\pi}{4} \quad \blacksquare
\end{aligned}$$

Appendix C

Solutions to All Challenge Problems

C.1 Chapter 13: Challenge Problem Solutions

C.1.1 *Life in Three Dimensions — Challenge — Solved*

(1) What is described by the expression $x = y$ in 2D and 3D?

□ In 2D, this is a line. In 3D, it's an infinite plane; start with the line $x = y$ in the xy-plane, then allow all z-coordinates, since z-coordinates are unrestricted by the expression. The line $x = y$ stretches into a vertical plane, bisecting the octants it passes through. ■

(2) What is described by the expression $x^2 + y^2 = 1$ in 2D and 3D?

□ In 2D, the expression $x^2 + y^2 = 1$ describes the unit circle. In 3D, it describes a vertical cylinder. The z-coordinate is unrestricted, so take the circle $x^2 + y^2 = 1$ in the xy-plane and let it expand infinitely in the z-directions. Voila, a cylinder. ■

(3) Is the triangle defined by the points $A(1, 2, -3)$, $B(3, 4, -2)$, and $C(3, -2, 1)$ an isoceles triangle?

□ The triangle connecting these points has sides:

$$\begin{aligned}
|AB| &= \sqrt{4 + 4 + 1} = 3 \\
|BC| &= \sqrt{0 + 36 + 9} = \sqrt{45} \\
|AC| &= \sqrt{4 + 16 + 16} = \sqrt{36}
\end{aligned}$$

The triangle is not isoceles since no two lengths are the same. But, note that $|BC|^2 = |AB|^2 + |AC|^2$, so this is actually a right triangle! ■

C.1.2 *Multivariable Functions — Challenge — Solved*

(1) Describe the domain and range of $f(x, y) = \sqrt{y-x}\ln(y+x)$; include a sketch of the domain.

□ From the square root part, we must have $y - x \geq 0$. From the natural log part, we need $y + x > 0$. Together, we need $y \geq x$ and $y > -x$. This is the V-shaped region above the two lines $y = x$ and $y = -x$ in the xy-plane, including the line $y = x$ but not $y = -x$. Figure C.1 shows this region in $\mathbb{R}^2$. ■

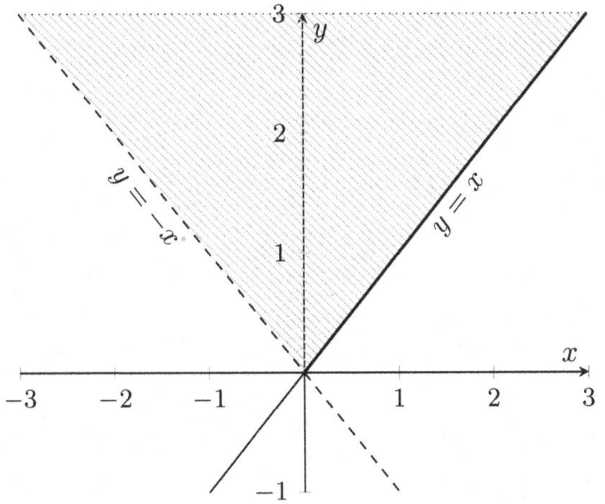

Fig. C.1 The domain of $f(x, y) = \sqrt{y-x}\ln(y+x)$.

(2) Give TWO different sets of parametric equations that produce the line segment starting at $(1, 1, 2)$ and ending at $(-2, 0, 2)$.

□ Let's start with $x = 1 - 3t, y = 1 - t, z = 2$ for $0 \leq t \leq 1$. Next, how about we set the range of t to $0 \leq t \leq 3$, and then we can have $x = 1 - t, y = 1 - \dfrac{t}{3}, z = 2$. ■

(3) Do the lines $x = t, y = t, z = t$ and $x = 2 - s$, $y = -1 + 2s$, $z = (s + 1)/2$ share a point?

□ We can make the x and y coordinates match by forcing $t = 2 - s$ and $t = -1 + 2s$, i.e. $2 - s = -1 + 2s$, i.e. $s = 1$ and so also $t = 1$. Passing these values to the z-coordinates we get (1) $z = 1$ and (2)

$z = (1+1)/2 = 1$. So the z-coordinate matches, too! The values $t = 1$ and $s = 1$ both give the point $(1, 1, 1)$, and this is the point the lines share. ■

C.1.3 *3D Surfaces — Challenge — Solved*

(1) A plane shares x and y intercepts with the plane $2x + y - 3z - 4 = 0$ but has its own z intercept of $(0, 0, -2)$. What is the equation of this plane?

□ The plane $2x + y - 3z - 4 = 0$ has an x intercept of $(2, 0, 0)$ and a y intercept of $(0, 4, 0)$, so we're looking for the plane going through the points $(2, 0, 0)$, $(0, 4, 0)$, and $(0, 0, -2)$. Plugging each point into the standard form for a plane, we get

$$\begin{aligned} 2a + d &= 0 \\ 4b + d &= 0 \\ -2c + d &= 0 \end{aligned}$$

Choosing the free parameter $d = -4$, we get $a = 2$, $b = 1$ and $c = -2$, so the equation of this plane is $2x + y - 2z - 4 = 0$, which can be rewritten as $z = x + y/2 - 2$. ■

(2) Identify the surface $4x^2 + 4y^2 + 4z^2 - 8x + 16y = 1$. Give at least two pieces of identifying information that distinguishes this surface from others of the same type.

□ Completing the square on the given equation (factor first!),

$$\begin{aligned} 4x^2 + 4y^2 + 4z^2 - 8x + 16y &= 1 \\ 4(x^2 - 2x + (-1)^2) + 4(y^2 + 4y + (2)^2) + 4z^2 &= 1 + 4(-1)^2 + 4(2)^2 \\ 4(x^2 - 2x + 1) + 4(y^2 + 4y + 4) + 4z^2 &= 1 + 4 + 16 \\ 4(x-1)^2 + 4(y+2)^2 + 4z^2 &= 21 \\ (x-1)^2 + (y+2)^2 + z^2 &= \frac{21}{4} \end{aligned}$$

So this is a sphere with center $(1, -2, 0)$ and radius $\sqrt{21}/2$. ■

(3) Identify the surface $25y^2 + z^2 = 100 + 4x^2$ using traces. Provide at least one trace parallel to each each of the xy-, yz-, and xz-planes.

□ In the coordinate planes $x = 0$, $y = 0$, and $z = 0$, we get the traces $25y^2 + z^2 = 100$, $z^2 - 4x^2 = 100$, and $25y^2 - 4x^2 = 100$, which are an ellipse, and two hyperbolas respectively. The trace in any plane $x = c$ looks like $25y^2 + z^2 = 100 + 4c^2$ which is an ellipse. The trace in any plane $y = c$ is $z^2 - 4x^2 = 100 - 25c^2$ which is a hyperbola. The trace in any plane $z = c$ looks like $25y^2 - 4x^2 = 100 - c^2$ which is also a hyperbola. This is a hyperboloid of one sheet. ■

---

(4) *(Bonus! This problem will connect to others in later sections.)* Find the level surface of $w = 2x^2 + y^2 + z$ corresponding to $w = 4$, and present that level surface as a contour plot showing the values $z = 0, 1, 2, 3$. (Thus, from a 4D hypersurface, we generate a 3D level surface, and for that we show several 2D level curves!)

□ The level surface of $w = 2x^2 + y^2 + z$ corresponding to $w = 4$ is the collection of points (x, y, z) in $\mathbb{R}^3$ that yield $2x^2 + y^2 + z = 4$. We can rewrite this relation as $z = 4 - 2x^2 - y^2$, which is an inverted paraboloid with vertex $(0, 0, 4)$. To display this surface as a contour plot as specified, using the values $z = 0, 1, 2, 3$, we collect the following four contours (level curves):

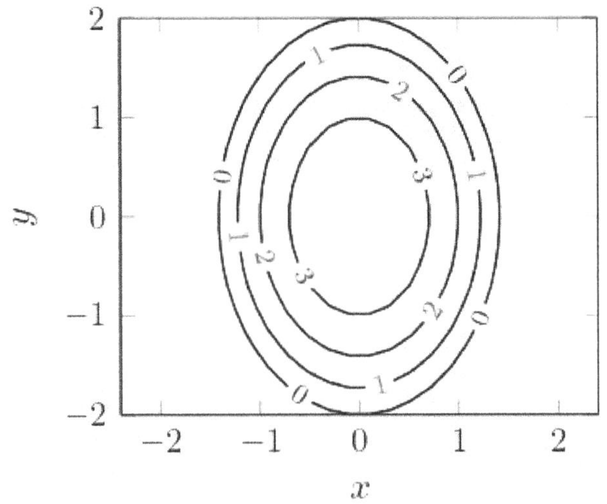

Fig. C.2 Some level curves for $z = 4 - 2x^2 - y^2$ (w/ CP 4).

- $0 = 4 - 2x^2 - y^2$, i.e. $2x^2 + y^2 = 4$
- $1 = 4 - 2x^2 - y^2$, i.e. $2x^2 + y^2 = 3$
- $2 = 4 - 2x^2 - y^2$, i.e. $2x^2 + y^2 = 2$
- $3 = 4 - 2x^2 - y^2$, i.e. $2x^2 + y^2 = 1$

Each contour (level curve) is an ellipse, as shown in Fig. C.2. ■

C.1.4 *Limits of Multivariable Functions — Challenge — Solved*

---

(1) Investigate the limit $\lim\limits_{(x,y)\to(0,0)} \dfrac{6x^3y}{2x^4+y^4}$.

□ Along the path $x = 0$,

$$\lim_{(x,y)\to(0,0)} \frac{6x^3y}{2x^4+y^4} = \lim_{(x,y)\to(0,0)} \frac{0}{y^4} = 0$$

Along the path $y = x$,

$$\lim_{(x,y)\to(0,0)} \frac{6x^3y}{2x^4+y^4} = \lim_{(x,y)\to(0,0)} \frac{6x^4}{3x^4} = 2$$

So we get two different limits on two different paths, and the limit does not exist. ■

---

(2) Investigate the limit $\lim\limits_{(x,y)\to(0,0)} \dfrac{x^4-y^4}{x^2+y^2}$.

□ With some factoring, we get:

$$\begin{aligned}\lim_{(x,y)\to(0,0)} \frac{x^4-y^4}{x^2+y^2} &= \lim_{(x,y)\to(0,0)} \frac{(x^2-y^2)(x^2+y^2)}{x^2+y^2}\\ &= \lim_{(x,y)\to(0,0)} (x^2-y^2) = 0\end{aligned}$$ ■

---

(3) Where is the function $f(x,y,z) = \sqrt{1-x^2-y^2-z^2}$ continuous?

□ The domain of this function is all points such that $1-x^2-y^2-z^2 \geq 0$, i.e. everywhere $x^2+y^2+z^2 \leq 1$. This is everywhere inside and on the unit sphere. Within this domain, all points (x,y,z) are points of continuity. ■

C.1.5 *Partial Derivatives — Challenge — Solved*

(1) Find both first order derivatives of $f(s,t) = \dfrac{st}{s^2+t^2}$.

☐ Using the quotient rule, we get

$$\frac{\partial f}{\partial s} = \frac{t(s^2+t^2) - st(2s)}{(s^2+t^2)^2} = \frac{t^3 - ts^2}{(s^2+t^2)^2} = \frac{t(t^2-s^2)}{(s^2+t^2)^2}$$

$$\frac{\partial f}{\partial t} = \frac{s(s^2+t^2) - st(2t)}{(s^2+t^2)^2} = \frac{s^3 - st^2}{(s^2+t^2)^2} = \frac{s(s^2-t^2)}{(s^2+t^2)^2} \quad \blacksquare$$

(2) Find all second order derivatives of $u = e^{-s}\sin t$.

☐ Starting with first derivatives,

$$\frac{\partial u}{\partial s} = -e^{-s}\sin t \qquad \text{and} \qquad \frac{\partial u}{\partial t} = e^{-s}\cos t$$

so that

$$\frac{\partial^2 u}{\partial s^2} = e^{-s}\sin t \quad ; \quad \frac{\partial^2 u}{\partial t \partial s} = \frac{\partial^2 u}{\partial s \partial t} = -e^{-s}\cos t \quad ;$$

$$\frac{\partial^2 u}{\partial t^2} = -e^{-s}\sin t \quad \blacksquare$$

(3) Does $u = \ln\sqrt{x^2+y^2}$ satisfy Laplace's Equation?

☐ We have

$$u_x = \frac{x}{x^2+y^2} \quad \to \quad u_{xx} = \frac{y^2-x^2}{(x^2+y^2)^2}$$

$$u_y = \frac{y}{x^2+y^2} \quad \to \quad u_{xx} = \frac{x^2-y^2}{(x^2+y^2)^2}$$

so

$$u_{xx} + u_{yy} = \frac{y^2-x^2}{(x^2+y^2)^2} + \frac{x^2-y^2}{(x^2+y^2)^2} = 0$$

and the function DOES satisfy Laplace's equation. ■

(4) *(Bonus! This problem connects to CP 4 in Sec. 13.3 and will be used later, too.)* For $w = 2x^2 + y^2 + z$, compute $w_x(1,1,1)$, $w_y(1,1,1)$, and $w_z(1,1,1)$. In CP 4 of Sec. 13.3, you presented some level curves associated with this level surface.[4] Can you find where the point $(1,1,1)$ is indicated on the plot of those level curves?

[4]You can also see those level curves in the solutions for that section.

☐ Since $w_x = 4x$, then $w_x(1,1,1) = 4$. Since $w_y = 2y$, then $w_y(1,1,1) = 2$. Since $w_z = 1$, then $w_z(1,1,1) = 1$.

The point $(1,1,1)$ falls on the level surface for $w = 4$, since $w(1,1,1) = 4$. In the 2D diagram of level curves representing this surface, we should see the point $(1,1,1)$ represented on the level curve for $z = 1$, simply because the z-coordinate of this point is 1. But also, the level surface for $w = 4$ is the surface $2x^2 + y^2 + z = 4$, i.e. $z = 4 - 2x^2 - y^2$ — and when $x = 1$ and $y = 1$, we find $z = 1$. It's all consistent. The point representing $(1,1,1)$ has been added to the diagram from CP 4 of Sec. 13.3, as new Fig. C.3. ■

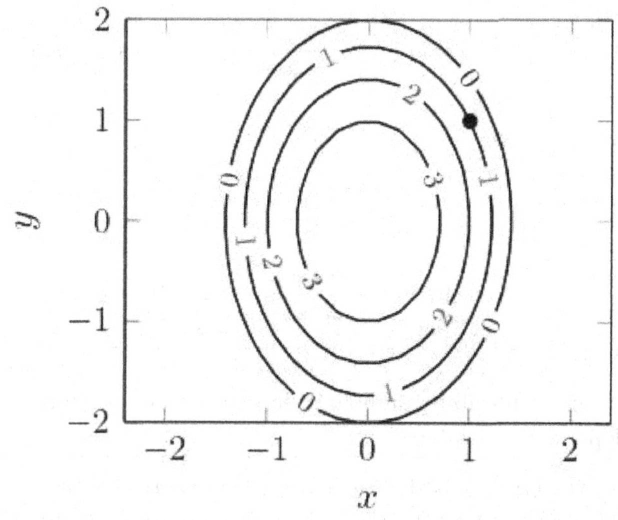

Fig. C.3 Some level curves for $z = 4 - 2x^2 - y^2$ (w/ CP 4).

C.2 Chapter 14: Challenge Problem Solutions

C.2.1 *The Chain Rule — Challenge — Solved*

(1) If $z = f(\alpha(s,t), \beta(s,t)) = \ln \alpha \cos \beta$ with $\alpha(s,t) = 2s + t$ and $\beta(s,t) = s - 2t$, find the first derivatives of z with respect to the independent variables s and t. Write your final expressions in terms of s and t.

□ We have that z is ultimately a function of s and t, going through the intermediate variables α and β. So,

$$\begin{aligned}\frac{\partial z}{\partial s} &= \frac{\partial z}{\partial \alpha}\frac{\partial \alpha}{\partial s} + \frac{\partial z}{\partial \beta}\frac{\partial \beta}{\partial s} \\ &= \frac{\cos\beta}{\alpha}\cdot(2) - \ln\alpha\sin\beta\cdot(1) \\ &= \frac{2\cos\beta}{\alpha} - \ln\alpha\sin\beta = \frac{2\cos(s-2t)}{2s+t} - \ln(2s+t)\sin(s-2t)\end{aligned}$$

$$\begin{aligned}\frac{\partial z}{\partial t} &= \frac{\partial z}{\partial \alpha}\frac{\partial \alpha}{\partial t} + \frac{\partial z}{\partial \beta}\frac{\partial \beta}{\partial t} \\ &= \frac{\cos\beta}{\alpha}\cdot(1) - \ln\alpha\sin\beta\cdot(-2) \\ &= \frac{\cos\beta}{\alpha} + 2\ln\alpha\sin\beta = \frac{\cos(s-2t)}{2s+t} + 2\ln(2s+t)\sin(s-2t) \quad \blacksquare\end{aligned}$$

(2) If $w = w(s,t)$ with $s = s(x,y,z,p)$, and $t = t(x,y,z,p)$, write chain rule expressions for all possible first derivatives of w with respect to the four independent variables.

□ There will be four first derivatives of w; these derivatives are with respect to x, y, z, p going through the intermediate variables of s, t. The chain rule formulations are:

$$\begin{aligned}\frac{\partial w}{\partial x} &= \frac{\partial w}{\partial s}\frac{\partial s}{\partial x} + \frac{\partial w}{\partial t}\frac{\partial t}{\partial x} \\ \frac{\partial w}{\partial y} &= \frac{\partial w}{\partial s}\frac{\partial s}{\partial y} + \frac{\partial w}{\partial t}\frac{\partial t}{\partial y} \\ \frac{\partial w}{\partial z} &= \frac{\partial w}{\partial s}\frac{\partial s}{\partial z} + \frac{\partial w}{\partial t}\frac{\partial t}{\partial z} \\ \frac{\partial w}{\partial p} &= \frac{\partial w}{\partial s}\frac{\partial s}{\partial p} + \frac{\partial w}{\partial t}\frac{\partial t}{\partial p} \quad \blacksquare\end{aligned}$$

---

(3) If $T = \cos(x^2 + y^2 + z^2)$ where $x = u + v$, $y = 2u - v$ and $z = 3uv^2$, find the values of all possible first derivatives of T with respect to u and v when $u = 1, v = 2$.

□ There are two first partials with respect to u, v, going through the intermediate variables x, y, z. We want them for $u = 1, v = 2$. First, note that for these values we have

$$x = u + v = 3$$
$$y = 2u - v = 0$$
$$z = 3uv^2 = 12$$

Then with the growing list of values $u = 1, v = 2, x = 3, y = 0, z = 12$, we have

$$\frac{\partial T}{\partial x} = -2x\sin(x^2 + y^2 + z^2) = -6\sin 153$$
$$\frac{\partial T}{\partial y} = -2y\sin(x^2 + y^2 + z^2) = 0$$
$$\frac{\partial T}{\partial z} = -2z\sin(x^2 + y^2 + z^2) = -24\sin 153$$

and also

$$\frac{\partial x}{\partial u} = 1 \qquad \frac{\partial y}{\partial u} = 2 \qquad \frac{\partial z}{\partial u} = 3v^2 = 12$$
$$\frac{\partial x}{\partial v} = 1 \qquad \frac{\partial y}{\partial v} = -1 \qquad \frac{\partial z}{\partial v} = 6uv = 12$$

So finally with the chain rule,

$$\begin{aligned}\frac{\partial T}{\partial u} &= \frac{\partial T}{\partial x}\frac{\partial x}{\partial u} + \frac{\partial T}{\partial y}\frac{\partial y}{\partial u} + \frac{\partial T}{\partial z}\frac{\partial z}{\partial u}\\ &= (-6\sin 153)(1) + 0 + (-24\sin 153)(12) = -294\sin 153\\ \frac{\partial T}{\partial v} &= \frac{\partial T}{\partial x}\frac{\partial x}{\partial v} + \frac{\partial T}{\partial y}\frac{\partial y}{\partial v} + \frac{\partial T}{\partial z}\frac{\partial z}{\partial v}\\ &= (-6\sin 153)(1) + 0 + (-24\sin 153)(12) = -294\sin 153 \quad \blacksquare\end{aligned}$$

C.2.2 *Optimization — Challenge — Solved*

(1) Find and characterize the critical points of $f(x,y) = 8xy(x+y)+\sqrt{17}$.

☐ Writing out $f(x,y) = 8x^2y + 8xy^2 + \sqrt{17}$, we have that

$$f_x = 16xy + 8y^2 = 8y(2x+y)$$
$$f_y = 8x^2 + 16xy = 8x(x+2y)$$

Remember we are looking for when $f_x = f_y = 0$. Starting with f_x, we see $f_x = 0$ only at $y = 0$ or when $y = -2x$.

- When $y = 0$ then $f_y = 8x^2$, and so f_y will also be zero only when $x = 0$. So we've found that both $f_x = 0$ AND $f_y = 0$ when $y = 0$ AND $x = 0$, so $(0,0)$ is a critical point.
- When $y = -2x$ then $f_y = 8x(x - 4x) = -24x^2$, and so again f_y will also be zero only when $x = 0$. Given that $y = -2x$ here, we've again found that both $f_x = 0$ AND $f_y = 0$ when $y = 0$. Looks like $(0,0)$ is our only critical point!

Note that if you started looking at when $f_y = 0$ (at $x = 0$ or $x = -2y$) you'd come to the same conclusion.

Now let's get ready for the second derivative test, via Useful Fact 14.1, to categorize this critical point of $(x,y) = (0,0)$.

$$f_{xx} = 16y \to f_{xx}(0,0) = 0$$
$$f_{yy} = 16x \to f_{yy}(0,0) = 0$$
$$f_{xy} = 16(x+y) \to f_{xy}(0,0) = 0$$
$$D(0,0) = f_{xx}(0,0)f_{yy}(0,0) - [f_{xy}(0,0)]^2 = 0$$

Since $D(0,0) = 0$ the second derivative test is inconclusive. Uh oh! Hey, maybe that's why this is a challenge problem — not to see if you can handle really complicated derivative calculations, but to see if you remember the fundamental meaning of a saddle point. Note that $f(0,0)$ itself is $\sqrt{17}$. The value of $f(x,y)$ is also $\sqrt{17}$ along either of the x- or y-axes, and also along the line $y = -x$. Will $f(x,y)$ wobble above and below $\sqrt{17}$ otherwise? Of course. Note that, for example,

$$f(0.1, 0.1) > \sqrt{17}$$
$$f(0.1, 0.05) < \sqrt{17}$$

Our critical point is a saddle point. ■

(2) Find the absolute extremes of $f(x, y) = 3x^2 + 2xy + y^2$ over the domain D with $-2 \leq x \leq 2, 0 \leq y \leq 3$.

□ Are there extremes over the interior of D? With $f_x = 6x + 2y$ and $f_y = 2x + 2y$, we have $f_x = f_y = 0$ at (0,0) which is indeed in the interior of D.

Let edge L1 be the left boundary of D, where $x = -2$. Here, the function reduces to $f(-2, y) = 12 - 4y + y^2$. There is a Calc I style critical point here at $y = 2$, which leads us to the possible extreme of $f(x, y)$ at $(-2, 2)$.

Let edge L2 be the right boundary of D, where $x = 2$. Here, the function reduces to $f(2, y) = 12 + 4y + y^2$. Now $f'(2, y) = 4 + 2y$, which is zero where $y = -2$. This might lead to a possible extreme of $f(x, y)$ at $(x, y) = (2, -2)$, but that's outside D, so we don't care about it.

Let edge L3 be the lower boundary of D, where $y = 0$. Here, the function reduces to $f(x, 0) = 3x^2$. There is a Calc I style critical point here at $x = 0$, which leads us to a possible extreme of $f(x, y)$ at $(0, 0)$, but we already know about that one.

Let edge L4 be the upper boundary of D, where $y = 3$. Here, the function reduces to $f(x, 3) = 3x^2 + 6x + 9$. There is a Calc I style critical point here at $x = -1$, which leads us to the possible extreme of $f(x, y)$ at $(-1, 3)$.

At the critical points we've identified on the interior, we get

$$f(0,0) = 0 \quad ; \quad f(-2,2) = 8 \quad ; \quad f(-1,3) = 6$$

At the vertices of our region,

$$f(-2,0) = 12 \;\; ; \;\; f(-2,3) = 9 \;\; ; \;\; f(2,0) = 12 \;\; ; \;\; f(2,3) = 33$$

Comparing the critical points and vertices, we see an absolute maximum of 33 at (2,3), and absolute minimum of 0 at (0,0). ■

(3) Identify the coordinates of the point on the plane $2x - y + z = 16$ that is closest to the origin.

□ Translated, we are asking for the point on the plane at which the distance to $(0, 0, 0)$ is minimized. As in other examples, we'll make life easier by minimizing d^2 (since whatever minimizes d^2 also minimizes d). The distance between (0,0,0) and any point in the universe at all is given by:

$$d^2 = (x - 0)^2 + (y - 0)^2 + (z - 0)^2 = x^2 + y^2 + z^2$$

This is our objective function. Our constraint is that we're only interested in points on the plane $2x - y + z = 16$, i.e. $z = 16 - 2x + y$, which reduces our objective function to:

$$d^2 = x^2 + y^2 + (16 - 2x + y)^2$$

We can now treat the right hand side as the function $f(x, y)$ we want to minimize, and start our search for a critical point with $f_x = 0$. Note that

$$f_x = 2x + 2(16 - 2x + y)(-2) = 10x - 4y - 64$$

Then $10x - 4y - 64 = 0$ gives that $f_x = 0$ when $4y = 10x - 64$.

Now seeing that $f_x = 0$ when $4y = 10x - 64$, we look for where $f_y = 0$, too:

$$\begin{aligned} -4x + 4y + 32 &= 0 \\ -4x + (10x - 64) + 32 &= 0 \\ 6x - 32 &= 0 \\ x &= \frac{16}{3} \end{aligned}$$

Passing that back to $4y = 10x - 64$,

$$\begin{aligned} 4y &= 10\left(\frac{16}{3}\right) - 64 \\ y &= \frac{40}{3} - 16 = -\frac{8}{3} \end{aligned}$$

And so putting our values together, we see that the distance from the plane to the origin is minimized at the point with $(x, y) = (16/3, -8/3)$.

For the full set of coordinates, we need the z-coordinate, too, and that comes from the equation of the plane:

$$z = 16 - 2x + y = 16 - 2\left(\frac{16}{3}\right) - \frac{8}{3} = \frac{8}{3}$$

and so the full coordinates of the point closest to the origin are

$$(x, y, z) = \left(\frac{16}{3}, -\frac{8}{3}, \frac{8}{3}\right)$$

One technical detail should be addressed, and that is: how do we know we MINIMIZED the distance? One argument would be that there is no max distance between the given plane and the origin. Or, we could dive into the second derivative test:

$$\begin{aligned}
f_x = 2x + 2(16 - 2x + y)(-2) &= 10x - 4y - 64 \\
f_y = 2y + 2(16 - 2x + y) &= -4x + 4y + 32 \\
f_{xx} &= 10 \\
f_{yy} &= 4 \\
f_{xy} &= -4 \\
D(x, y) = f_{xx}f_{yy} - [f(x, y)]^2 &= 24
\end{aligned}$$

Since $D(x, y) > 0$ and $f_{xx} > 0$ everywhere, any critical point is a local minimum. ■

C.2.3 *Double Integrals — Challenge — Solved*

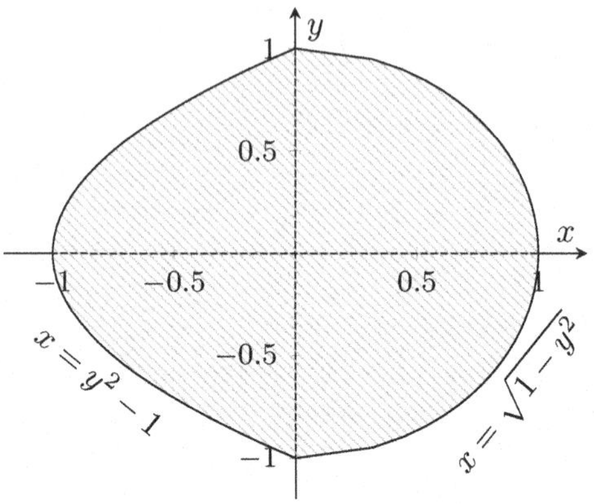

Fig. C.4 Region of integration for CP 1.

---

(1) Evaluate $\iint_D (1)\, dA$ where D is the region to the right of the parabola $x = y^2 - 1$ and to the left of the semicircle $x = \sqrt{1 - y^2}$. Once you find the value of the integral, state what geometric measure you just calculated.

☐ Note that $x = y^2 - 1$ is a parabola opening sideways to the right from the vertext of $(-1, 0)$. The semicircle opens to the left from $(1, 0)$. This region is shown in Fig. C.4. Since the boundaries of D are given in terms of x, we can set the inner limits of integration to be these two bounds, that is we'll have $y^2 - 1 \le x \le \sqrt{1 - y^2}$. For the outer (constant) bounds on y, we have to find where the two curves intersect:

$$\begin{aligned} y^2 - 1 &= \sqrt{1 - y^2} \\ y^4 - 2y^2 + 1 &= 1 - y^2 \\ y^4 - y^2 &= 0 \\ y^2(y^2 - 1) &= 0 \end{aligned}$$

This happens where $y = -1, 0, 1$. These correspond to $x = 0, -1, 0$ respectively. The intersection $(-1, 0)$ is where the FULL circle on the right might intersect the vertex of the parabola, but that is outside of D. We're interested in the min and max values of y, where the two curves criss-cross as they intersect the y-axis at $y = -1, 1$. And so the full integration is:

$$\iint_D (1)\, dA = \int_{-1}^{1} \int_{y^2-1}^{\sqrt{1-y^2}} (1)\, dx dy$$

The inner integral would not be hard, but the outer integral would require trigonometric substitution because of the $\sqrt{1-y^2}$ term. So let's hand this off to a CAS to get:

$$\iint_D (1)\, dA = \int_{-1}^{1} \int_{y^2-1}^{\sqrt{1-y^2}} (1)\, dx dy = \frac{\pi}{2} + \frac{4}{3}$$

We've computed the area in between the paraboloid and the semicircle. ■

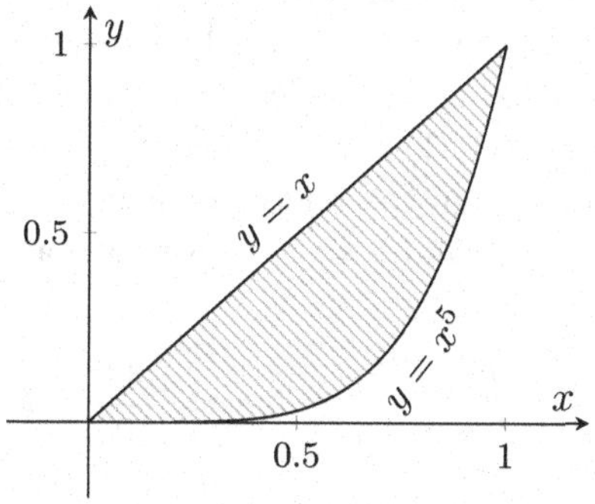

Fig. C.5 Region of integration for CP 2.

---

(2) Find the volume under the surface $z = 2x + y^2$ over the region in the first quadrant bounded by $y = x^5$ and $y = x$.

□ Note that $y = x^5$ and $y = x$ will intersect at $(0, 0)$ and $(1, 1)$ in the first quadrant. This region is shown in Fig. C.5. The line $y = x$ is the "higher" curve and therefore provides the upper limit of integration:

$$\iint_D (x^2 + 3y^2)\, dA = \int_0^1 \int_{x^5}^{x} (2x + y^2)\, dy dx = \frac{149}{336} \quad ■$$

---

(3) Reverse the order of integration of, and then evaluate, the following integral

$$\int_{-1}^{0} \int_{-\sqrt{y+1}}^{\sqrt{y+1}} y^2\, dx dy$$

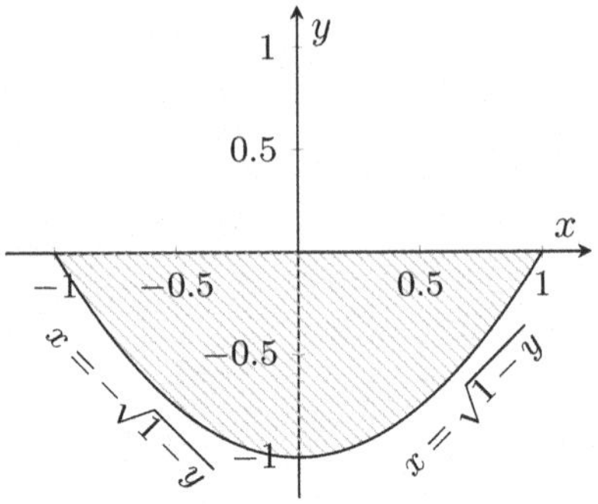

Fig. C.6 $x = \pm\sqrt{1-y}$ aka $y = x^2 - 1$, for CP 3.

□ Note that the limits, along with the current ordering, tell us we're looking at a domain from $x = -\sqrt{y+1}$ to $x = \sqrt{y+1}$ for $-1 \leq y \leq 0$. This means we're looking at the region between the two halves of the paraboloid $y = x^2 - 1$, below the x-axis. This region is shown in Fig. C.6. We can redesign the region as being between $y = x^2 - 1$ and $y = 0$ between $x = -1$ and $x = 1$, so we have an alternate version of the integral:

$$\int_{-1}^{1} \int_{x^2-1}^{0} y^2 \, dydx = \frac{32}{105} \quad \blacksquare$$

C.2.4 *Triple Integrals — Challenge — Solved*

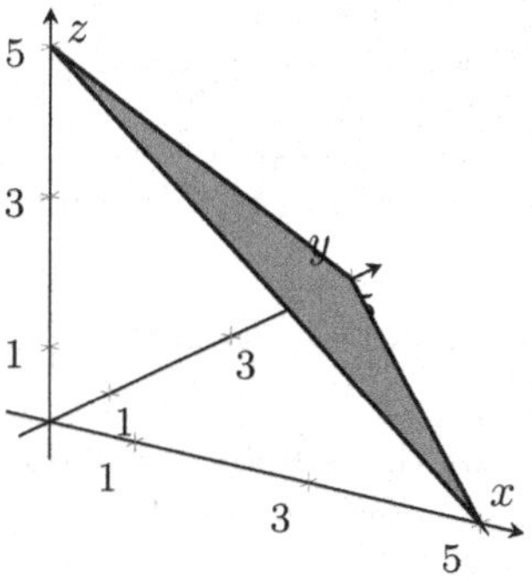

Fig. C.7 Region in the first octant bounded by $x + y + z = 5$ (w/ CP 1).

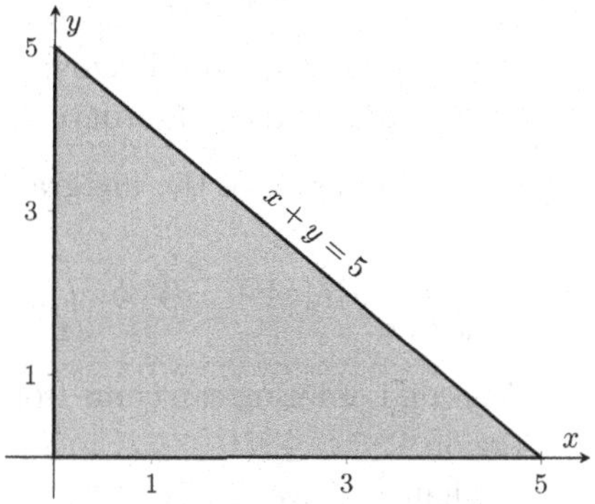

Fig. C.8 2D region under $x+y+z=5$ (w/ CP 1).

(1) Given a 3D region of integration E and a function $f(x, y, z)$ defined in that region, the average value of f over E is given by

$$f_{avg} = \frac{1}{V(E)} \iiint_E f(x, y, z)\, dV$$

where $V(E)$ is the volume of E. Find the average value of $f(x, y, z) = xz + 5z + 10$ over the region in the first octant between the plane $x + y + z = 5$ and the coordinate planes.

□ The 3D solid described here is shown in Fig. C.7, and the resulting 2D region in the xy-plane is in Fig. C.8. The bounds of E can be given as $0 \le x \le 5$, $0 \le y \le 5 - x$, and $0 \le z \le 5 - x - y$. So the volume of E is

$$\int_0^5 \int_0^{5-x} \int_0^{5-x-y} (1)\, dzdydz = \frac{125}{6}$$

Then the integration of f over E results in:

$$\int_0^5 \int_0^{5-x} \int_0^{5-x-y} (xz + 5z + 10)\, dzdydz = \frac{4375}{12}$$

Then the average of f over E is

$$f_{avg} = \frac{1}{V(E)} \iiint_E f(x, y, z)\, dV = \frac{6}{125} \cdot \frac{4375}{12} = \frac{35}{2} \quad \blacksquare$$

(2) Evaluate $\iiint_E y\ln(x) + z\,dV$ where E is defined as the region $E = \{(x,y,z) : 1 \le x \le e; 0 \le y \le \ln(x); 0 \le z \le 1\}$ using TWO equivalent orderings of integration. Obviously, you should get the same value from the integral with each ordering.

☐ One ordering of the integral comes directly from the description:

$$\iiint_E y\ln(x) + z\,dV = \int_1^e \int_0^{\ln(x)} \int_0^1 y\ln(x) + z\,dzdydx = \frac{7}{2} - e$$

A second ordering can come from reordering the bounds of x and y: The region $1 \le x \le e; 0 \le y \le \ln(x)$ is also known as $0 \le y \le 1, e^y \le x \le e$, so that we have

$$\iiint_E y\ln(x) + z\,dV = \int_0^1 \int_{e^y}^e \int_0^1 y\ln(x) + z\,dzdxdy = \frac{7}{2} - e$$

There are other arrangements, but these are the most direct. ■

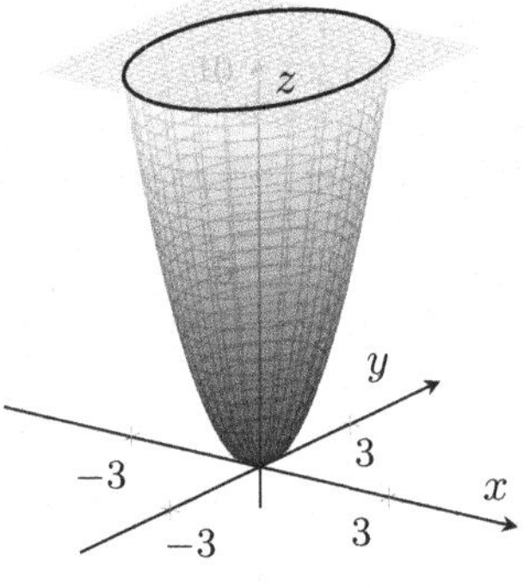

Fig. C.9 Region between $z = 2x^2 + y^2$, $z = 10$ (w/ CP 3).

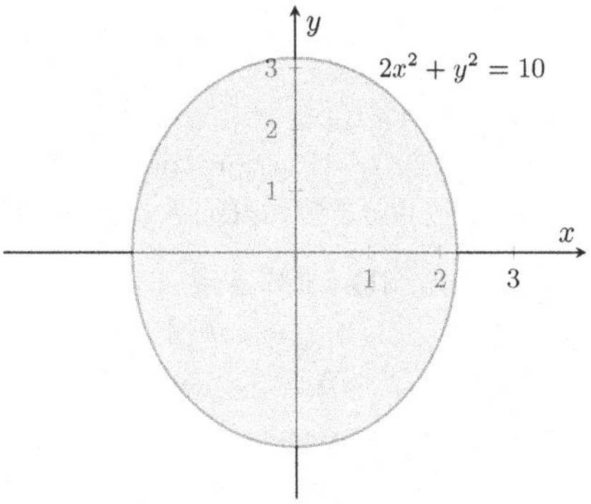

Fig. C.10 2D region from $z = 2x^2 + y^2$ vs $z = 10$ (w/ CP 3).

(3) Find the volume of region between the elliptic paraboloid $z = 2x^2 + y^2$ and the plane $z = 10$. (Use of tech for evaluation is highly recommended!)

☐ The 3D solid described here is shown in Fig. C.9, and the resulting 2D region in the xy-plane is in Fig. C.10. The paraboloid opens upward around the z-axis, so the "floor" of this solid is the paraboloid itself and the "roof" is the plane $z = 10$. This gives limits on z: $2x^2 + y^2 \leq z \leq 10$. With z pinned down, we get limits on y and z by exploring the region of the xy-plane used by this solid.

The intersection of the paraboloid with the plane $z = 10$ is $2x^2 + y^2 = 10$. This is an ellipse whose long axis is in the y direction. We can solve for y as $-\sqrt{10 - 2x^2} \leq y \leq \sqrt{10 - 2x^2}$. The limits on x come from the intersection of the ellipse with the x-axis: $2x^2 + (0)^2 = 10$ gives $x = \pm\sqrt{5}$.

We must order limits of integration so that variables are progressively eliminated; the final limits must be constants. So, z must go first (inner), then y (middle), and finally x (outer): So,

$$V = \iiint_E dV = \int_{-\sqrt{5}}^{\sqrt{5}} \int_{-\sqrt{10-2x^2}}^{\sqrt{10-2x^2}} \int_{2x^2+y^2}^{10} dz dy dx = 25\pi\sqrt{(2)} \quad \blacksquare$$

C.3 Chapter 15: Challenge Problem Solutions

C.3.1 *Double Ints in Polar Coords — Challenge — Solved*

(1) Given a 2D region R and a function f defined in that region, the average value of f over R is given by

$$f_{avg} = \frac{1}{A(R)} \iint_R f\, dA$$

where $A(R)$ is the area of R. Find the average value of $f(x, y) = \sqrt{x^2 + y^2}$ over the region bounded by $r = 3\sin 2\theta$ between $\theta = 0$ and $\theta = \pi/2$.

□ The area of the region R is given in polar coordinates by

$$A(R) = \iint_R (1)\, dA = \int_0^{\pi/2} \int_0^{3\sin 2\theta} (1)\, r\, dr d\theta = \frac{9\pi}{8}$$

Converted to polar coordinates, the function f is $f = r$, and so the integral of the function over the region is

$$\iint_R f\, dA = \int_0^{\pi/2} \int_0^{3\sin 2\theta} (r)\, r\, dr d\theta = 6$$

And so

$$f_{avg} = \frac{1}{A(R)} \iint_R f\, dA = \frac{8}{9\pi} \cdot 6 = \frac{16}{3\pi} \quad \blacksquare$$

(2) Use a double integral in polar coordinates to find the volume between the paraboloid $z = 16 - 2x^2 - 2y^2$ and the plane $z = 2$.

□ Figures C.11 and C.12 show the full 3D region as well as the corresponding polar region in the 2D plane. The paraboloid intersects the plane $z = 2$ where $16 - 2x^2 - 2y^2 = 2$, i.e. where $x^2 + y^2 = 7$. So the volume between the paraboloid and the plane is below the paraboloid (which in polar coordinates is $z = 16 - 2r^2$) and above the plane $z = 2$; the volume sits over the region $0 \leq r \leq \sqrt{7}$ and $0 \leq \theta \leq 2\pi$. So,

$$\iint_R [(16 - 2r^2) - 2]\, dA = \int_0^{2\pi} \int_0^{\sqrt{7}} (14 - 2r^2) \cdot r\, dr d\theta = 49\pi \quad \blacksquare$$

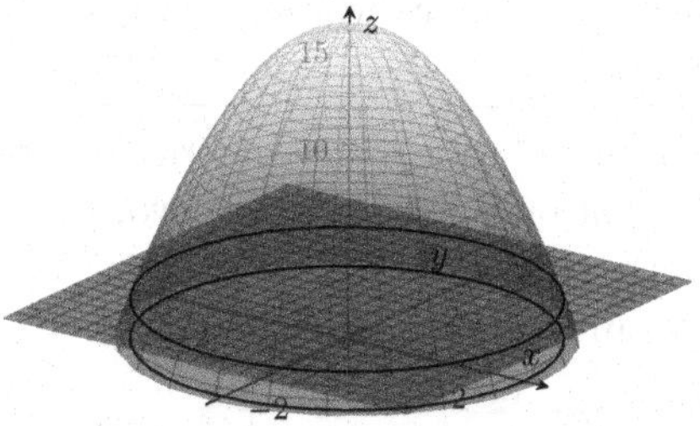

Fig. C.11 Under the paraboloid $z = 16 - 2x^2 - 2y^2$, above the plane $z = 2$.

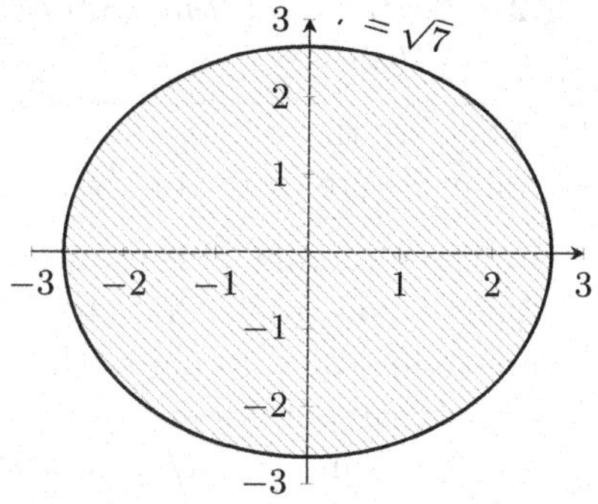

Fig. C.12 Intersection of $z = 16 - 2x^2 - 2y^2$ and $z = 2$.

---

(3) Convert the following integral into polar coordinates and then evaluate:

$$\int_0^3 \int_{-\sqrt{9-y^2}}^{\sqrt{9-y^2}} (x^2 + y^2)^2 \, dx dy$$

□ The limits of integration describe the upper half of a circle of radius 3 (since y goes from 0 to 3) — which is the region $0 \leq r \leq 3$ and $0 \leq \theta \leq \pi$. The function being integrated is, in polar form, $(r^2)^2$ or r^4. And so, we have

$$\int_0^3 \int_{-\sqrt{9-y^2}}^{\sqrt{9-y^2}} (x^2 + y^2)^2 \, dx dy = \int_0^{\pi} \int_0^3 (r^4) \cdot r dr d\theta = \frac{243\pi}{2} \quad ■$$

C.3.2 *Cylindrical and Spherical Coords — Challenge — Solved*

---

(1) What are the cylindrical coordinates of the point whose spherical coordinates are $(\rho, \theta, \phi) = (2, 3\pi/4, \pi/6)$?

□ The θ coordinate is the same in both cylindrical and spherical coordinates, so we don't have to convert that one. For the other two, there are complicated formulas that directly convert between spherical and cylindrical, but it's just as well to go through rectangular. With $\rho = 2$, $\theta = 3\pi/4$, and $\phi = \pi/6$, we have

$$x = \rho\cos\theta\sin\phi = 2\cos\frac{3\pi}{4}\sin\frac{\pi}{6} = -\frac{1}{\sqrt{2}}$$
$$y = \rho\sin\theta\sin\phi = 2\sin\frac{3\pi}{4}\sin\frac{\pi}{6} = +\frac{1}{\sqrt{2}}$$
$$z = \rho\cos\phi = 2\cos\frac{\pi}{6} = \sqrt{3}$$

Now note that the cylindrical z is also the rectangular z, so $z = \sqrt{3}$. And $r = \sqrt{x^2 + y^2} = 1$. So the cylindrical coordinates of the given point are

$$(r, \theta, z) = (1, \frac{3\pi}{4}, \sqrt{3})$$

A bit of double checking can be done by noting the original spherical coordinates put the point over the second quadrant, where $x < 0, y > 0, z > 0$ and everything that follows is consistent with that. ■

---

(2) Suppose a three-dimensional region is described with the following bounds in rectangular coordinates:

- The full 3D region is bounded below by the paraboloid $z = x^2 + y^2$ and above by the cone $z = \sqrt{x^2 + y^2}$
- In the xy-plane, the region covers the right half of the unit circle, thus we have $0 \le x \le \sqrt{1 - y^2}$ and $-1 \le y \le 1$.

Describe the bounds of this same region in cylindrical coordinates by giving the bounds of r, θ, and z.

□ The two surfaces around this region are shown in Fig. C.13. We convert the equations of the surfaces to polar coordinates,

$$z = x^2 + y^2 \quad\to\quad z = r^2 \text{ (lower surface)}$$
$$z = \sqrt{x^2 + y^2} \quad\to\quad z = \sqrt{r^2} = r \text{ (upper surface)}$$

Since we're restricted to the right half of the unit circle in the xy-plane, we select the ranges $-\pi/2 \leq \theta \leq \pi/2$ and $0 \leq r \leq 1$. ■

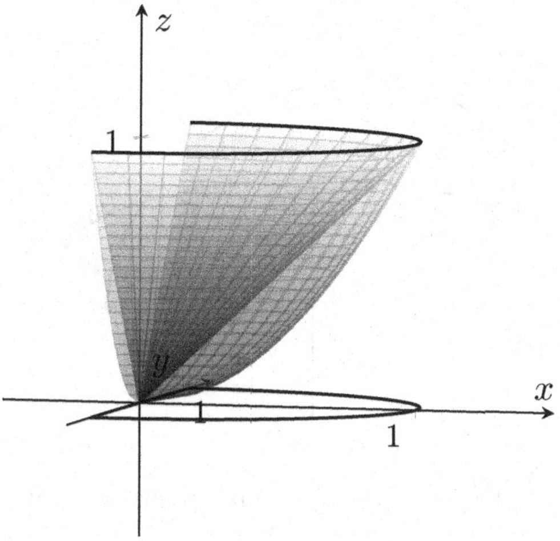

Fig. C.13 Above $z = r^2$, below $z = r$, $-\pi/2 \leq \theta \leq \pi/2$.

---

(3) Suppose a three-dimensional region is described with the following bounds in rectangular coordinates:

- The full 3D region is underneath the hemisphere $z = \sqrt{4 - x^2 - y^2}$ and above the xy-plane
- In the xy-plane, the region covers the upper half of a circle of radius 2 centered at the origin, thus we have $-2 \leq x \leq 2$ and $0 \leq y \leq \sqrt{4 - x^2}$.

Describe the bounds of this same region in spherical coordinates by giving the bounds of ρ, θ, and ϕ.

□ Note that the given hemisphere has a radius of 2. (The portion of this hemisphere as described is shown in Fig. C.14. So, in spherical coordinates, the full region under the hemisphere $z = \sqrt{4 - x^2 - y^2}$ and above the xy-plane is given by

$$0 \leq \rho \leq 2 \qquad 0 \leq \theta \leq 2\pi \qquad 0 \leq \phi \leq \frac{\pi}{2}$$

Again, that's the entire region under the hemisphere. The second restriction above, though, says we only want half of that whole zone (the half extending along the positive y-axis). This cuts our range of θ in half. And so altogether, we have

$$0 \leq \rho \leq 2 \qquad 0 \leq \theta \leq \pi \qquad 0 \leq \phi \leq \frac{\pi}{2} \quad \blacksquare$$

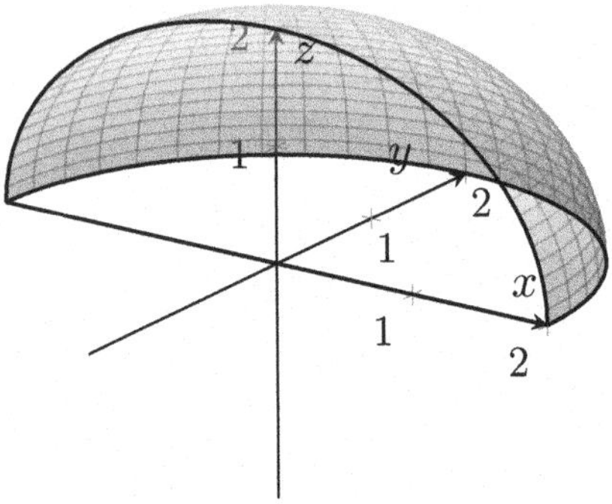

Fig. C.14 Inside $\rho = 2$, above xy-plane, forward of xz-plane.

C.3.3 *Triple Ints in Cyl. & Spher. Coords — Challenge — Solved*

(1) Let's pretend the corn silo pictured in Fig. C.15 is bounded on its sides by the cylinder $x^2 + y^2 = 225$ and above by the (inverted) cone $z = 50 - \sqrt{x^2 + y^2}/9$. Construct a triple integral in cylindrical coordinates that would give the volume of this silo, and compute it.

Fig. C.15 Corn silo with cylinder and cone (with CP 1).

□ We can describe the region of integration as being inside the cylinder $x^2 + y^2 = 225$ and below the (inverted) cone $z = 50 - \sqrt{x^2 + y^2}/9$. Because of the cylindrical body of the region, cylindrical coordinates is the way to go. Note that in cylindrical coordinates, the "roof" of the silo is $z = 50 - r/9$, and so the entire region is bounded in the z direction with $0 \le z \le 50 - r/9$. The equation of the bounding cylinder, in cylindrical coordinates, is $r = 15$. So in the horizontal directions, we have $0 \le r \le 15$ and $0 \le \theta \le 2\pi$. In all, the volume is

$$V = \iiint_E (1)\, dV = \int_0^{2\pi} \int_0^{15} \int_0^{50-r/9} (1)\, r\, dz dr d\theta = 11{,}000\pi$$

and that volume is in $11,000\pi$ cubic units, where units are whatever is built into the original equations! ■

(2) Evaluate $\iiint_E xyz, dV$, where E is the region between the spheres $\rho = 1$ and $\rho = 3$ and above the cone $\phi = 2\pi/3$.

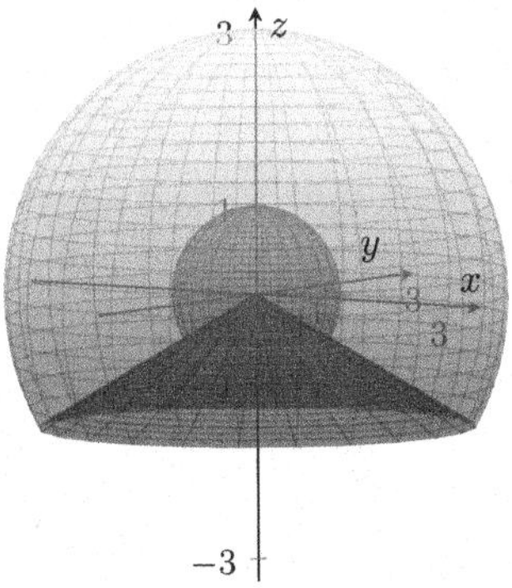

Fig. C.16 Above $\phi = 2\pi/3$, between $\rho = 1$, $\rho = 3$.

□ The given region E of integration is between the spheres $\rho = 1$ and $\rho = 3$ and above the cone $\phi = 2\pi/3$. Bounds on ρ are given by the spheres. The region above the cone is $0 \leq \phi \leq 2\pi/3$ (remember, ϕ is measures from the top down). And, since there are no other restrictions, we allow $0 \leq \theta \leq 2\pi$. The region E is shown in Fig. C.16. Also, converting the given function,

$$xyz = (\rho \sin\phi \cos\theta)(\rho \sin\phi \sin\theta)(\rho \cos\phi) = \rho^3 \sin^2\phi \cos\phi \sin\theta \cos\theta$$

The volume element in spherical coordinates is $dV = \rho^2 \sin\phi \, d\rho d\phi d\theta$. So, the triple integral becomes

$$\int_0^{2\pi} \int_0^{2\pi/3} \int_1^3 (\rho^3 \sin^2\phi \cos\phi \sin\theta \cos\theta)\rho^2 \sin\phi \, d\rho d\phi d\theta = 0$$

Note that 0 is a perfectly valid result, since this integral does NOT compute a volume! ■

(3) Evaluate the following integral by first converting to an appropriate coordinate system; explain the geometric meaning of your final value:

$$\int_{-4}^{4} \int_0^{\sqrt{16-y^2}} \int_{-\sqrt{16-x^2-y^2}}^{5} (1) \, dz dx dy$$

□ From the given limits of integration on z, we can see that the "floor" of this region is the lower half of the sphere $x^2 + y^2 + z^2 = 16$, and the "roof" is the plane $z = 5$. From the limits on x and y, we see we are using up the upper half of a circle of radius 4. So imagine an invisible cylinder of radius 4 going from the rim of the lower hemisphere up to the plane $z = 5$; that cylinder forms the outer "walls" of this region. The region is shown in Fig. C.17. Clearly (I hope it's clear!), cylindrical coordinates look the most appropriate, and we have

$$0 \leq r \leq 4 \qquad 0 \leq \theta \leq \pi \qquad -\sqrt{16 - r^2} \leq z \leq 5$$

and then the requested integral is

$$\int_{-4}^{4} \int_{0}^{\sqrt{16-y^2}} \int_{-\sqrt{16-x^2-y^2}}^{5} (1)\, dz dx dy = \int_{0}^{\pi} \int_{0}^{4} \int_{-\sqrt{16-r^2}}^{5} (1)\, r\, dz dr d\theta$$
$$= \frac{184\pi}{3}$$

Because we integrated the function $f = 1$ over the region of integration, we just computed the volume of this hemisphere / cylinder thingy. ■

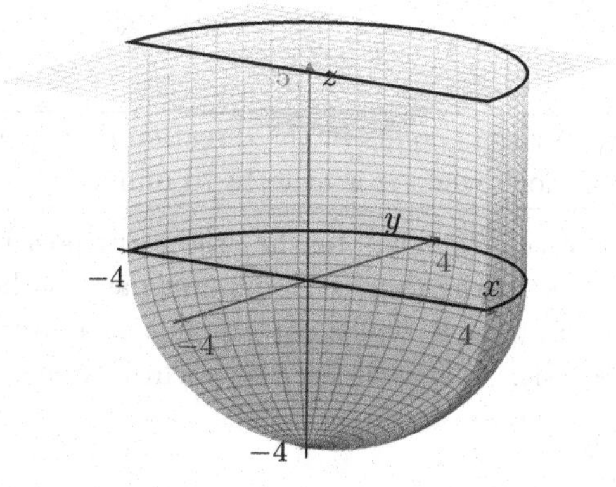

Fig. C.17 Region between $z = -\sqrt{16 - r^2}$ and $z = 5$, for $y \geq 0$.

C.4 Chapter 16: Challenge Problem Solutions

C.4.1 *Vector Basics — Challenge — Solved*

(1) Find a vector of length 3 in the opposite direction of $\mathbf{v} = \langle 1, -1, 2 \rangle$.

□ The vector $\mathbf{v} = \langle 1, -1, 2 \rangle$ has length $|\mathbf{v}| = \sqrt{1+1+4} = \sqrt{6}$, so a unit vector in the same direction is

$$\frac{\mathbf{v}}{|\mathbf{v}|} = \left\langle \frac{1}{\sqrt{6}}, -\frac{1}{\sqrt{6}}, \frac{2}{\sqrt{6}} \right\rangle$$

A vector of length 3 in the SAME direction as $\langle 1, -1, 2 \rangle$ is then

$$3\frac{\mathbf{v}}{|\mathbf{v}|} = \left\langle \frac{3}{\sqrt{6}}, -\frac{3}{\sqrt{6}}, \frac{6}{\sqrt{6}} \right\rangle$$

so a vector of length 3 in the OPPOSITE direction is:

$$\left\langle -\frac{3}{\sqrt{6}}, \frac{3}{\sqrt{6}}, -\frac{6}{\sqrt{6}} \right\rangle$$

If you like your radicals simplified, this vector is also known as

$$\left\langle -\frac{\sqrt{6}}{2}, \frac{\sqrt{6}}{2}, -\sqrt{6} \right\rangle \quad \blacksquare$$

(2) Find a unit vector that points in the direction of a minute hand on an analog 12-hour clock when it is exactly 10 minutes past the hour.

□ The center of the clock will be the origin; the positive x-axis points towards 3 o-clock. If the time is 10 minutes past the hour, the minute hand is pointing at the 2, which corresponds to an angle of $\pi/6$ above the positive x-axis. (Do you see why each 5 minute interval on the clock corresponds to an angle of $\pi/6$?).

If we represent the minute hand itself as a vector $\mathbf{M}$ with length r, its components are

$$\mathbf{M} = \langle x, y \rangle = \left\langle r\cos\frac{\pi}{6}, r\sin\frac{\pi}{6} \right\rangle = \left\langle \frac{\sqrt{3}r}{2}, \frac{r}{2} \right\rangle$$

Now we don't know what r is, but since we're after a unit vector in the direction of the minute hand, why not just set $r = 1$? Then our minute hand is itself the unit vector, and its components are $\langle \sqrt{3}/2, 1/2 \rangle$. $\blacksquare$

(3) If $\mathbf{v}$ is any vector $\langle v_1, v_2, v_3\rangle$ and $\mathbf{w} = 5\mathbf{v}$, use the definition of magnitude (length) to prove that the length of $\mathbf{w}$ is always 5 times the length of $\mathbf{v}$. (Sure, we can say "Well, duh, of course it is!", but can you prove it in the mathematical court of law?)

□ Note that since $\mathbf{v} = \langle v_1, v_2, v_3\rangle$, then $\mathbf{w} = \langle 5v_1, 5v_2, 5v_3\rangle$. Then since, generically, $|\mathbf{w}| = \sqrt{w_1^2 + w_2^2 + w_3^2}$, we have

$$|\mathbf{w}| = \sqrt{25v_1^2 + 25v_2^2 + 25v_3^2} = \sqrt{25(v_1^2 + v_2^2 + v_3^2)} = 5\sqrt{v_1^2 + v_2^2 + v_3^2}$$

But, the right most square root expression is just $|\mathbf{v}|$ itself, and so

$$|\mathbf{w}| = 5|\mathbf{v}| \quad \blacksquare$$

C.4.2 *Dot and Cross Products — Challenge — Solved*

(1) If $\mathbf{a} = \langle 1, 6, -2\rangle$ and $\mathbf{b} = \langle 2, -3, 1\rangle$, what are $\text{comp}_{\mathbf{a}}\mathbf{b}$ and $\text{proj}_{\mathbf{a}}\mathbf{b}$?

□ We have $\mathbf{a} \cdot \mathbf{b} = -18$ and $|\mathbf{a}| = \sqrt{41}$ so that

$$\text{comp}_{\mathbf{a}}\mathbf{b} = \frac{\mathbf{a} \cdot \mathbf{b}}{|\mathbf{a}|} = \frac{-18}{\sqrt{41}}$$

with which we get:

$$\begin{aligned}\text{proj}_{\mathbf{a}}\mathbf{b} &= \left(\frac{\mathbf{a} \cdot \mathbf{b}}{|\mathbf{a}|}\right)\left(\frac{\mathbf{a}}{|\mathbf{a}|}\right) = \left(\frac{-18}{\sqrt{41}}\right)\left(\frac{1}{\sqrt{41}}\right)\langle 1, 6, -2\rangle \\ &= -\frac{18}{41}\langle 1, 6, -2\rangle\end{aligned}$$

If you like to distribute constants, this is $\langle -18/41, -108/41, 36/41\rangle$. ■

(2) Find a unit vector orthogonal to both $\mathbf{i} - 2\mathbf{j}$ and $\mathbf{i} + \mathbf{k}$.

□ Let's find just any vector orthogonal to both, then scale it to a unit vector. A vector orthogonal to both of these vectors will come from their cross product. We can find that $\langle 1, -2, 0\rangle \times \langle 1, 0, 1\rangle = \langle -2, -1, 2\rangle$; the length of this vector is 3. Therefore a unit vector orthogonal to the two given vectors is $\langle -2/3, -1/3, 2/3\rangle$. ■

(3) Let $\mathbf{v} = \langle p, q, r\rangle$ be any vector in $\mathbf{R}^3$ and let $\mathbf{w}$ be any scalar multiple of $\mathbf{v}$, i.e. $w = c\mathbf{v}$. The cross product $\mathbf{v} \times \mathbf{w}$ will always have the same result; find that result, and show how you know what it is.

□ The value of $\mathbf{v} \times (c\mathbf{v})$ will always be $\mathbf{0}$. Here's how we know: We have

$$\mathbf{v} \times (c\mathbf{v}) = \begin{vmatrix} \mathbf{i} & \mathbf{j} & \mathbf{k} \\ p & q & r \\ cp & cq & cr \end{vmatrix} = \begin{vmatrix} q & r \\ cq & cr \end{vmatrix} \mathbf{i} - \begin{vmatrix} p & r \\ cp & cr \end{vmatrix} \mathbf{j} + \begin{vmatrix} p & q \\ cp & cq \end{vmatrix} \mathbf{k}$$
$$= (cqr - cqr)\mathbf{i} - (cpr - cpr)\mathbf{j} + (cpq - cpq)\mathbf{k}$$
$$= 0\mathbf{i} + 0\mathbf{j} + 0\mathbf{k} \quad \blacksquare$$

C.4.3 *Vector Functions — Challenge — Solved*

---

(1) Suppose object A is moving along the path $\mathbf{r}_1(t) = \langle t, t^2, t^3 \rangle$ and object B is moving along the path $\mathbf{r}_2(s) = \langle 1 + 2s, 1 + 6s, 1 + 14s \rangle$.

(a) Find a vector that points from object A to object B at $t = s = 2$.
(b) How far apart are the objects at that instant?
(c) Are there any points that the two paths share? If so, find them. If not, say how you know.

□ (a) When $t = s = 2$, object A is at $\mathbf{r}_1(2) = \langle 2, 4, 8 \rangle$ and object B is at $\mathbf{r}_2(2) = \langle 5, 13, 29 \rangle$. So, a vector that points from A to B is $\mathbf{r}_2(2) - \mathbf{r}_1(2) = \langle 3, 9, 21 \rangle$.

(b) The distance between the two objects at this instant is $|\langle 3, 9, 21 \rangle| = \sqrt{531} = 3\sqrt{59}$.

(c) There are two points the paths share. One shared point is somewhat "obvious". Note that $\langle 1, 1, 1 \rangle$ is a vector / point on both paths, and it happens at $t = 1$ and $s = 0$. By observation, you *might* also notice that $t = 2$ and $s = 1/2$ yield the same location of $\langle 2, 4, 8 \rangle$. But there's a systematic way to search, too:

We can take the coordinates one at a time and see what it takes to make them equal. To make the x-coordinates the same, we'd need $t = 1 + 2s$. To make the y-coordinates the same, we'd need $t^2 = 1 + 6s$. To make these two things happen simultaneously, we need $(1 + 2s)^2 = 1 + 6s$.

This works when

$$\begin{aligned} 1+4s+4s^2 &= 1+6s \\ -2s+4s^2 &= 0 \\ -2s(1-2s) &= 0 \\ s=0\ ;\ s &= \frac{1}{2} \end{aligned}$$

When $s = 0$, we have $t = 1 + 2s = 1$; at these values, (as we already knew) the curves both hit $\langle 1, 1, 1\rangle$. But then also when $s = 1/2$ we have $t = 1 + 2s = 2$, and with these values, both curves hit $\langle 2, 4, 8\rangle$. So there are two points where the curves intersect, $\langle 1, 1, 1\rangle$ and $\langle 2, 4, 8\rangle$.

■

---

(2) Consider the vector curve $\mathbf{r}(t) = \langle 2\cos t, 2t/\pi, 2\sin t\rangle$.

(a) Find the vector equation of the line tangent to the vector curve at the location given by $\langle 0, 1, 2\rangle$.

(b) Write the integral that would give the total arc length of the curve from $\langle 2, 0, 0\rangle$ to $\langle 0, 1, 2\rangle$.

□ (a) Note that the location $\langle 0, 1, 2\rangle$ happens when $t = \pi/2$. For the tangent line here, we need:

$$\begin{aligned} \mathbf{r}'(t) &= \langle -2\sin t, \frac{2}{\pi}, 2\cos t\rangle \\ \mathbf{r}'\left(\frac{\pi}{2}\right) &= \langle -2, \frac{2}{\pi}, 0\rangle \end{aligned}$$

So we have an initial vector $\mathbf{r}_0 = \langle 0, 1, 2\rangle$ and a parallel vector $\mathbf{r}'(\pi/2)$, and the tangent line is formed as:

$$\mathbf{r}_0 + t\,\mathbf{r}'\left(\frac{\pi}{2}\right) = \langle 0, 1, 2\rangle + t\langle -2, \frac{2}{\pi}, 0\rangle = \langle -2t, 1 + \frac{2t}{\pi}, 2\rangle$$

(b) For the arc length from $\langle 2, 0, 0\rangle$ to $\langle 0, 1, 2\rangle$, note that $\langle 2, 0, 0\rangle$ corresponds to $t = 0$, and $\langle 0, 1, 2\rangle$ corresponds to $t = \pi/2$. Then to set up the arc length integral, we identify

$$f(t) = 2\cos t \quad , \quad g(t) = \frac{2}{\pi}t \quad , \quad h(t) = 2\sin t$$

from which we then get

$$f'(t) = -2\sin t \quad , \quad g'(t) = \frac{2}{\pi} \quad , \quad h'(t) = 2\cos t$$

so that

$$\sqrt{[f'(t)]^2+[g'(t)]^2+[h'(t)]^2} = \sqrt{4\sin^2 t + \frac{4}{\pi^2} + 4\cos^2 t}$$
$$= 2\sqrt{1+\frac{1}{\pi^2}} = 2\sqrt{\frac{\pi^2+1}{\pi^2}} = \frac{2}{\pi}\sqrt{\pi^2+1}$$

and so the arc length integral is

$$L = \int_a^b \sqrt{[f'(t)]^2+[g'(t)]^2+[h'(t)]^2}\,dt = \frac{2}{\pi}\int_0^{\pi/2} \sqrt{\pi^2+1}\,dt \quad ■$$

---

(3) If tangent lines to a vector function $\mathbf{r}(t)$ can be found as

$$\left\langle \frac{1}{1+t^2}, \frac{1}{2\sqrt{t}}, e^{-t} \right\rangle$$

and the vector function passes through the point $(pi/4, 1, 1-1/e)$, what is the vector function $\mathbf{r}(t)$?

□ The vector equation giving the tangent lines is the derivative of the actual vector function. So, we know that

$$\mathbf{r}'(t) = \left\langle \frac{1}{1+t^2}, \frac{1}{2\sqrt{t}}, e^{-t} \right\rangle$$

and then

$$\mathbf{r}(t) = \int \mathbf{r}'(t)\,dt = \left\langle \tan^{-1}(t), \sqrt{t}, -e^{-t} \right\rangle + \mathbf{C}$$

Now note that the first two coordinates of the point $(\pi/4, 1, 1-1/e)$ occur for $t=1$. However, at $t=1$, the final component would only be $-1/e$, so we need the extra 1 to come from the arbitrary vector $\mathbf{C}$. With $\mathbf{C} = \langle 0,0,1\rangle$, we have

$$\mathbf{r}(t) = \left\langle \tan^{-1}(t), \sqrt{t}, 1-e^{-t} \right\rangle$$

If that loose argument about $\mathbf{C}$ doesn't work for you, try this: we can say that we recognize

$$\mathbf{r}(1) = \left(\frac{\pi}{4}, 1, 1-\frac{1}{e}\right)$$

That is,

$$\mathbf{r}(1) = \left\langle \tan^{-1}(1), \sqrt{1}, -e^{-1} \right\rangle + \mathbf{C} = \left(\frac{\pi}{4}, 1, 1-\frac{1}{e}\right)$$
$$\left\langle \frac{\pi}{4}, 1, -\frac{1}{e} \right\rangle + \mathbf{C} = \left(\frac{\pi}{4}, 1, 1-\frac{1}{e}\right)$$
$$\mathbf{C} = \langle 0,0,1\rangle \quad ■$$

(4) (Bonus Time in *the Pit!*) Can you apply the technique used to prove Useful Fact 16.10 (in *the Pit!*) to prove the expression which Practice Problem 8 suggests might be true?

☐ All cross products shown here are done "behind the scenes." You should recreate any you don't trust! Given two vector functions $\mathbf{p}(t) = \langle p_1(t), p_2(t), p_3(t)\rangle$ and $\mathbf{r}(t) = \langle r_1(t), r_2(t), r_3(t)\rangle$ in $\mathbb{R}^3$, their cross product is (hiding the (t) dependence for brevity),

$$\mathbf{p}(t) \times \mathbf{r}(t) = \langle p_2r_3 - p_3r_2, p_3r_1 - p_1r_3, p_1r_2 - p_2r_1\rangle$$

so that

$$\frac{d}{dt}\mathbf{p} \times \mathbf{r} = \langle p_2'r_3 + p_2r_3' - p_3'r_2' - p_3r_2', p_3'r_1 + p_3r_1' - p_1'r_3' - p_1r_3', \tag{C.1}$$

$$p_1'r_2 + p_1r_2' - p_2'r_1' - p_2r_1'\rangle \tag{C.2}$$

For comparison, we'll compute $\mathbf{p}'(t) \times \mathbf{r}(t)$ and $\mathbf{p}(t) \times \mathbf{r}'(t)$ separately, find their derivatives, and add them up.

$$\begin{aligned}\mathbf{p}' \times \mathbf{r} &= \langle p_1', p_2', p_3'\rangle \times \langle r_1, r_2, r_3\rangle \\ &= \langle p_2'r_3 - p_3'r_2', p_3'r_1 - p_1'r_3', p_1'r_2 - p_2'r_1'\rangle \\ \mathbf{p} \times \mathbf{r}' &= \langle p_1, p_2, p_3\rangle \times \langle r_1', r_2', r_3'\rangle \\ &= \langle p_2r_3' - p_3r_2', p_3r_1' - p_1r_3', p_1r_2' - p_2r_1'\rangle\end{aligned}$$

Adding,

$$\mathbf{p}' \times \mathbf{r} + \mathbf{p} \times \mathbf{r}' = \langle p_2'r_3 + p_2r_3' - p_3'r_2' - p_3r_2', p_3'r_1 + p_3r_1' - p_1'r_3' - p_1r_3', \tag{C.3}$$

$$p_1'r_2 + p_1r_2' - p_2'r_1' - p_2r_1'\rangle \tag{C.4}$$

Comparison of (C.1) and (C.3) confirms the identity

$$\frac{d}{dt}\mathbf{p}(t) \times \mathbf{r}(t) = \mathbf{p}'(t) \times \mathbf{r}(t) + \mathbf{p}(t) \times \mathbf{r}'(t) \quad \blacksquare$$

C.4.4 *Vector Fields and the Gradient — Challenge — Solved*

(1) Find the curl and divergence of $\mathbf{F}(x, y, z) = \langle xyz, x^2y^2z^2, y^2z^3 \rangle$. (Simplify each as much as possible.)

☐ The curl is:

$$\begin{aligned}\nabla \times \mathbf{F} &= \begin{vmatrix} \mathbf{i} & \mathbf{j} & \mathbf{k} \\ \frac{\partial}{\partial x} & \frac{\partial}{\partial y} & \frac{\partial}{\partial z} \\ xyz & x^2y^2z^2 & y^2z^3 \end{vmatrix} \\ &= (2yz^3 - 2x^2y^2z)\mathbf{i} - (0 - xy)\mathbf{j} + (2xy^2z^2 - xz)\mathbf{k} \\ &= 2yz(z^2 - x^2y)\mathbf{i} + xy\mathbf{j} + xz(2y^2z - 1)\mathbf{k}\end{aligned}$$

The divergence is

$$\begin{aligned}\nabla \cdot \mathbf{F} &= \left\langle \frac{\partial}{\partial x}, \frac{\partial}{\partial y}, \frac{\partial}{\partial z} \right\rangle \cdot \langle xyz, x^2y^2z^2, y^2z^3 \rangle \\ &= \frac{\partial}{\partial x}(xyz) + \frac{\partial}{\partial y}(x^2y^2z^2) + \frac{\partial}{\partial z}(y^2z^3) \\ &= yz + 2x^2yz^2 + 3y^2z^2 = yz(1 + 2x^2z + 3yz) \quad \blacksquare\end{aligned}$$

(2) Some of you may be familiar with the idea that the gravitational force due to an object is inversely proportional to the square of the distance between the object and the point of interest,

$$F = \frac{c}{r^2} = \frac{c}{x^2 + y^2}$$

where c is a constant containing several other constants mushed together, and r is the distance from the large body (presuming the object is at the coordinate origin. Find the gradient of this function. (Optional: In your expression for the gradient, try to introduce r anywhere you see an equivalent expression in x and y.)

☐

$$\begin{aligned}\nabla f &= \left\langle \frac{\partial}{\partial x} \frac{c}{x^2 + y^2}, \frac{\partial}{\partial y} \frac{c}{x^2 + y^2} \right\rangle = \left\langle -\frac{2cx}{(x^2 + y^2)^2}, \frac{-2cy}{x^2 + y^2} \right\rangle \\ &= -2c \left\langle \frac{x}{(x^2 + y^2)^2}, \frac{y}{(x^2 + y^2)^2} \right\rangle\end{aligned}$$

We could then reintroduce r:

$$\nabla f = -2c \left\langle \frac{x}{r^4}, \frac{y}{r^4} \right\rangle \quad \blacksquare$$

(3) This is a problem for those of you who like puzzlers. Make up a couple of simple scalar functions $f(x, y, z)$. Find the gradient of each function, then find the curl of each gradient. What do you get? Make a conjecture as to what you'll *always* get for $\nabla \times (\nabla f)$, the curl of a gradient. Demonstrate why your conjecture will be true no matter what scalar function $f(x, y, z)$ you start with.

□ The conjecture should be that $\nabla \times (\nabla f) = \mathbf{0}$. Let's see why. $\nabla f = \langle f_x, f_y, f_z \rangle$, so that $\nabla \times (\nabla f)$ is given by

$$\begin{vmatrix} \mathbf{i} & \mathbf{j} & \mathbf{k} \\ \frac{\partial}{\partial x} & \frac{\partial}{\partial y} & \frac{\partial}{\partial z} \\ f_x & f_y & f_z \end{vmatrix} = \langle f_{zy} - f_{yz}, -(f_{zx} - f_{xz}), f_{yx} - f_{xy} \rangle$$

But since the components of f have continuous derivatives, we know the mixed second order partials are all the same: $f_{zy} = fyz$, $f_{zx} = f_{xz}$ and $f_{yx} - f_{xy}$. Therefore each component of the cross product is 0, and $\nabla \times (\nabla f) = \mathbf{0}$. Hooray! ■

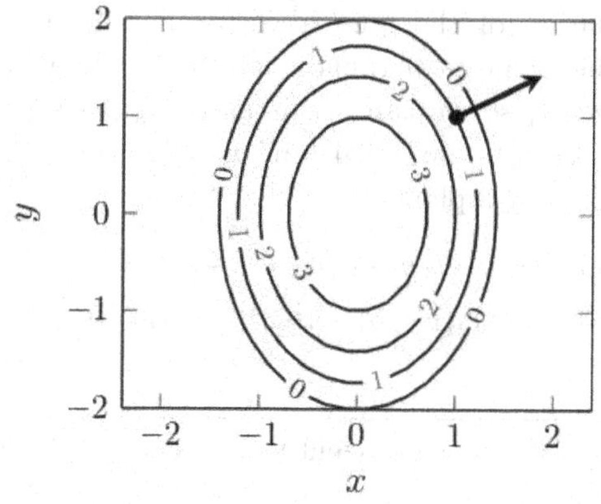

Fig. C.18 A gradient vector and level curves for $z = 4 - 2x^2 - y^2$ (w/ CP 4).

(4) *(Bonus! This problem has been developing in CP 4 of Secs. 13.3 and 13.5.)* Use the partial derivatives found in CP 4 of Sec. 13.5 to construct gradient vectors for the function $w = 2x^2 + y^2 + z$ at the point $(1, 1, 1)$. Draw the projection of this gradient vector into a new version of Fig. C.3. Do you notice anything interesting?

□ Since $w_x(1,1,1) = 4$, $w_y(1,1,1) = 2$, $w_z(1,1,1) = 1$, then $\nabla(1,1,1) = \langle 4, 2, 1\rangle$. A unit vector in this direction is $\langle 4/\sqrt{21}, 2/\sqrt{21}, 1/\sqrt{21}\rangle$. Projected into the xy-plane, this vector is $\langle 4/\sqrt{21}, 2/\sqrt{21}, 1/\sqrt{21}\rangle$; Fig. C.18 shows this vector originating from the point $(1, 1)$; it looks like this gradient vector is perpendicular to the level curve at which it originates. FFT: We can't tell directly from the plot, but what do you think the odds are that this gradient vector is actually perpendicular to the level *surface* depicted in Fig. C.18? ■

C.4.5 *Planes and Tangent Planes — Challenge — Solved*

(1) Find the equation of the plane containing the point $(3, -5, 4)$ and perpendicular to the line $x = 1 + 2t, y = 1 - t, z = 4 - 3t$.

□ From the equation of the line, a vector parallel to that line is $\langle 2, -1, -3\rangle$. If the plane is perpendicular to the line, it is perpendicular to this vector, so we can use it as the plane's perpendicular vector. Using $(x_0, y_0, z_0) = (3, -5, 4)$ and $\mathbf{n} = \langle a, b, c\rangle = \langle 2, -1, -3\rangle$ in the general equation for a plane:

$$\begin{aligned} a(x - x_0) + b(y - y_0) + c(z - z_0) &= 0 \\ 2(x - 3) + (-1)(y + 5) + (-3)(z - 4) &= 0 \\ 2x - y - 3z &= -1 \quad ■ \end{aligned}$$

(2) Find the equation of the plane tangent to $z = \sqrt{x^2 - y^2}$ where $(x, y) = (2, -1)$.

□ We need:

$$\begin{aligned} z_0 = f(x_0, y_0) &= \sqrt{(2)^2 - (-1)^2} = \sqrt{3} \\ f_x = \frac{x}{\sqrt{x^2 - y^2}} &\to f_x(2, -1) = \frac{2}{\sqrt{3}} \\ f_y = -\frac{y}{\sqrt{x^2 - y^2}} &\to f_y(2, -1) = \frac{1}{\sqrt{3}} \end{aligned}$$

so the plane is:

$$\begin{aligned} z - z_0 &= f_x(x_0, y_0)(x - x_0) + f_y(x_0, y_0)(y - y_0) \\ z - \sqrt{3} &= \frac{2}{\sqrt{3}}(x - 2) + \frac{1}{\sqrt{3}}(y + 1) \\ \sqrt{3}z - 3 &= 2(x - 2) + 1(y + 1) \\ \sqrt{3}z &= 2x + y \\ z &= \frac{2}{\sqrt{3}}x + \frac{1}{\sqrt{3}}y \quad \blacksquare \end{aligned}$$

---

(3) The radius and height of a cylinder are measured as 30cm and 24cm respectively. There is a possible error in measurement of 0.1cm in each direction. Estimate the maximum possible error in the calculated volume of the cylinder.

□ Since $V(r, h) = \pi r^2 h$, we have $V_h = \pi r^2$ and $V_r = 2\pi rh$. We are given $r = 30$ and $h = 24$, with $dr = dh = 0.1$. Therefore the total possible error in V is dV, given by

$$\begin{aligned} dV &= V_r(r, h)dr + V_h(r, h)dh = (2\pi rh)\,dr + (\pi r^2)\,dh \\ &= (2\pi \cdot 30 \cdot 24)(0.1) + (\pi 30^2)(0.1) = 234\pi \end{aligned}$$

There is a possible error of $234\pi\, cm^3$ in the total volume measured. That's a lot! ■

C.4.6 *Directional Derivatives — Challenge — Solved*

(1) Find the directional derivative of $f(x, y, z) = \sqrt{xy + z}$ at (1,3,1) in the direction of $\mathbf{w} = \langle 3, 2, 6\rangle$.

□ We need a unit vector $\mathbf{u}$ in the direction of $\mathbf{w}$. Since $|\mathbf{w}| = 7$, that unit vector would be

$$\mathbf{u} = \left\langle \frac{3}{7}, \frac{2}{7}, \frac{6}{7} \right\rangle$$

Then we have

$$\nabla f(x, y, z) = \langle f_x, f_y, f_z \rangle = \left\langle \frac{y}{2\sqrt{xy+z}}, \frac{x}{2\sqrt{xy+z}}, \frac{1}{2\sqrt{xy+z}} \right\rangle$$

$$\nabla f(1, 3, 1) = \left\langle \frac{3}{2(2)}, \frac{1}{2(2)}, \frac{1}{2(2)} \right\rangle = \left\langle \frac{3}{4}, \frac{1}{4}, \frac{1}{4} \right\rangle$$

$$D_u f(1, 3, 1) = \nabla f(1, 3, 1) \cdot \mathbf{u} = \left\langle \frac{3}{4}, \frac{1}{4}, \frac{1}{4} \right\rangle \cdot \left\langle \frac{3}{7}, \frac{2}{7}, \frac{6}{7} \right\rangle = \frac{17}{28} \quad \blacksquare$$

(2) Find the directional derivative of $f(x, y, z) = y/(x+z)$ at the point $P = (1, 4, 1)$ in the direction of $\mathbf{v} = \langle 1, 2, 1\rangle$. If S is the surface represented by the graph of f, and a bug standing at P started walking on S in the direction of $\mathbf{v}$, would the bug be walking uphill or downhill?

□ We need a unit vector $\mathbf{u}$ in the direction of $\mathbf{v}$. Since $|\mathbf{v}| = \sqrt{6}$, that unit vector would be

$$\mathbf{u} = \left\langle \frac{1}{\sqrt{6}}, \frac{2}{\sqrt{6}}, \frac{1}{\sqrt{6}} \right\rangle$$

Then we have

$$\nabla f(x, y, z) = \langle f_x, f_y, f_z \rangle = \left\langle -\frac{y}{(x+z)^2}, \frac{1}{x+z}, -\frac{y}{(x+z)^2} \right\rangle$$

$$\nabla f(1, 4, 1) = \left\langle -1, \frac{1}{2}, -1 \right\rangle$$

and so

$$\begin{aligned} D_u f(1, 4, 1) &= \nabla f(1, 4, 1) \cdot \mathbf{u} \\ &= \left\langle -1, \frac{1}{2}, -1 \right\rangle \cdot \left\langle \frac{1}{\sqrt{6}}, \frac{2}{\sqrt{6}}, \frac{1}{\sqrt{6}} \right\rangle = -\frac{1}{\sqrt{6}} \end{aligned}$$

Since this directional derivative is negative, then the ant at P walking the the direction of $\mathbf{v}$ would be going downhill. ■

(3) Given the 3D temperature function $T(x, y, z) = 5e^{-x^2-y^2-2z^2}$, what is the (simplified) maximum rate of change of the temperature at the point $(1/\sqrt{2}, 0, 1/\sqrt{2})$, and in what direction does this rate of change occur?

□ We're going to need the gradient of T at P:

$$\begin{aligned} T_x &= -10xe^{-x^2-y^2-2z^2} \\ T_y &= -10ye^{-x^2-y^2-2z^2} \\ T_z &= -20ze^{-x^2-y^2-2z^2} \\ \nabla T(x, y, z) &= \langle T_x, T_y, T_z \rangle \\ &= 10e^{-x^2-y^2-2z^2} \langle -x, -y, -2z \rangle \\ \nabla T\left(\frac{1}{\sqrt{2}}, 0, \frac{1}{\sqrt{2}}\right) &= 10e^{-3/2} \left\langle -\frac{1}{\sqrt{2}}, 0, -\frac{2}{\sqrt{2}} \right\rangle \end{aligned}$$

The temperature increases the fastest in the direction of the gradient, so at the point P the temperature increases fastest in the direction of $\nabla T(P)$. Since all we want is a direction, we could report this as the messy $\nabla T(P)$, or the slightly tidier $\langle -1/\sqrt{2}, 0, -\sqrt{2} \rangle$, or even $\langle -1, 0, -2 \rangle$. (Make sure you know why the latter gives the same direction as the others!)

The maximum rate of increase at P is the magnitude of the gradient there, which is

$$\begin{aligned} 10e^{-3/2} \left(\left(-\frac{1}{\sqrt{2}} \right)^2 + 0 + (\sqrt{2})^2 \right)^{1/2} &= 10e^{-3/2} \left(\frac{1}{2} + 2 \right)^{1/2} \\ &= 10e^{-3/2} \cdot \sqrt{\frac{5}{2}} = 10\sqrt{\frac{5}{2e^3}} \quad \blacksquare \end{aligned}$$

C.5 Chapter 17: Challenge Problem Solutions

C.5.1 *Arc Length Parameterization — Challenge — Solved*

(1) Find the arc length of of $\mathbf{r}(t) = \langle t^2,\ \sin t - t\cos t,\ \cos t + t\sin t\rangle$ for $0 \le t \le \pi$.

☐ First, build:

$$\mathbf{r}'(t) = \langle 2t, t\sin t, t\cos t\rangle$$
$$|\mathbf{r}'(t)| = \sqrt{4t^2 + t^2\sin^2 t + t^2\cos^2 t} = \sqrt{5t^2} = \sqrt{5}t$$

So that

$$L = \int_0^{\pi} |\mathbf{r}'(t)|\, dt = \int_0^{\pi} \sqrt{5}t\, dt = \frac{\sqrt{5}\pi^2}{2} \quad \blacksquare$$

(2) Find the arc length (use computational aid to estimate it, if needed) of the segment of $y = x^3$ from $(-1,-1)$ to $(1,1)$. (Hint: Can you parameterize this curve?)

☐ To put this curve in the context of a vector function, let's think of $y = x^3$ from $(-1,-1)$ to $(1,1)$ as $\mathbf{r}(t) = \langle t, t^3\rangle$ for $-1 \le t \le 1$. Then we can form the following expressions:

$$\mathbf{r}'(t) = \langle 1, 3t^2\rangle$$
$$|\mathbf{r}'(t)| = \sqrt{1 + 9t^4}$$

So that

$$L = \int_{-1}^{1} |\mathbf{r}'(t)|\, dt = \int_{-1}^{1} \sqrt{1+9t^4}\, dt \approx 3.096$$

That integral cannot be done by hand, so I got a bit of help from a CAS. ■

(3) Find an arc length parameterization of $\mathbf{r}(t) = \langle \sin 2t,\ 2t^{3/2}/3,\ \cos 2t\rangle$ for $t \ge 0$, and determine the location at which the curve accumulates a total arc length of $s = 1$.

☐ For this vector function,

$$\mathbf{r}'(t) = \langle 2\cos 2t, \sqrt{t}, -2\sin 2t\rangle$$
$$|\mathbf{r}'(t)| = \sqrt{4\cos^2 2t + t + 4\sin^2 2t} = \sqrt{4+t}$$

so that the arc length parameter is:

$$s = \int_0^t |\mathbf{r}'(\tau)|\, d\tau = \int_0^t \sqrt{4+\tau}\, d\tau = \frac{2}{3}(4+\tau)^{3/2}\Big|_0^t$$
$$= \frac{2}{3}\left[(4+t)^{3/2} - 4^{3/2}\right] = \frac{2}{3}\left[(4+t)^{3/2} - 8\right]$$

Then we turn this inside-out to find t in terms of s:

$$\frac{3}{2}s + 8 = (4+t)^{3/2}$$
$$\left(\frac{3}{2}s + 8\right)^{2/3} = 4 + t$$
$$\left(\frac{3}{2}s + 8\right)^{2/3} - 4 = t$$

We need to substitute this into the original form of the vector function. So take a deep breath, and accept the fact that what was once simply $\mathbf{r}(t) = \langle \sin 2t, 2t^{3/2}/3, \cos 2t \rangle$ is now $\mathbf{r}(s) = \langle x(s), y(s), z(s) \rangle$, where

$$x(s) = \sin 2\left[\left(\frac{3}{2}s + 8\right)^{2/3} - 4\right]$$
$$y(s) = 2\left[\left(\frac{3}{2}s + 8\right)^{2/3} - 4\right]^{3/2}$$
$$z(s) = \cos 2\left[\left(\frac{3}{2}s + 8\right)^{2/3} - 4\right]$$

Well, that's something, isn't it? To find where we accumulate an arc length of $s = 1$, we can

(a) Find the value of t that corresponds to $s = 1$, and plug that into $\mathbf{r}(t)$, or ...
(b) Plug $s = 1$ into $\mathbf{r}(s)$

Either way, it's a mess, and we get:

$$\mathbf{r}(1) = \left\langle \sin\left[2\left(\frac{19}{2}\right)^{2/3} - 8\right], 2\left[\left(\frac{19}{2}\right)^{2/3} - 4\right]^{3/2},\right.$$
$$\left. \cos\left[2\left(\frac{19}{2}\right)^{2/3} - 8\right]\right\rangle \quad \blacksquare$$

C.5.2 *Contours, Orientation, Pointers — Challenge — Solved*

---

(1) The orbital path of a comet follows the vector function $\mathbf{r}(t) = \langle \cosh t, \sinh t \rangle$. Two aliens riding on the comet decide to jump off before the comet gets close to Earth, because right now Earth is a pretty dumb place to be. The both leap at $t = \ln 2$; one leaps off in the direction of $\mathbf{T}$, and one leaps off in the direction of $\mathbf{N}$. If they fly away in straight lines, find the vector equations of those lines. (This identity might be useful: $\sinh^2 t + \cosh^2 t = \cosh(2t)$.)

☐ Let's start calculating with (17.11) and (17.12), to build up to $\mathbf{T}(\ln 2)$:

$$\begin{aligned}
\mathbf{r}'(t) &= \langle \sinh t, \cosh t \rangle \\
ds = |\mathbf{r}'(t)| &= \sqrt{\sinh^2 t + \cosh^2 t} = \sqrt{\cosh(2t)} \\
\mathbf{T}(t) &= \frac{\mathbf{r}'(t)}{ds} = \frac{1}{\sqrt{\cosh(2t)}} \langle \sinh t, \cosh t \rangle \\
\mathbf{T}(\ln 2) &= \frac{1}{\sqrt{\cosh(2\ln 2)}} \langle \sinh \ln 2, \cosh \ln 2 \rangle \\
&= \frac{1}{\sqrt{17/8}} \left\langle \frac{3}{4}, \frac{5}{4} \right\rangle = \sqrt{\frac{8}{17}} \left\langle \frac{3}{4}, \frac{5}{4} \right\rangle
\end{aligned}$$

Going further, all the way to $\mathbf{N}(\ln 2)$,

$$\begin{aligned}
\mathbf{r}'(t)^{\perp} &= \langle y'(t), -x'(t) \rangle = \langle \cosh t, -\sinh t \rangle \\
\mathbf{N}(t) &= \frac{\mathbf{r}'(t)^{\perp}}{ds} = \frac{1}{\sqrt{\cosh(2t)}} \langle \cosh t, -\sinh t \rangle \\
\mathbf{N}(\ln 2) &= \frac{1}{\sqrt{\cosh(2\ln 2)}} \langle \cosh \ln 2, -\sinh \ln 2 \rangle \\
&= \frac{1}{\sqrt{17/8}} \left\langle \frac{5}{4}, -\frac{3}{4} \right\rangle = \sqrt{\frac{8}{17}} \left\langle \frac{5}{4}, -\frac{3}{4} \right\rangle
\end{aligned}$$

Since the point marked by $t = \ln 2$ is $\mathbf{r}(\ln 2) = \langle 5/4, 3/4 \rangle$, then the vector equations of the lines along $\mathbf{T}$ and $\mathbf{N}$ at $t = \ln 2$ are:

$$\begin{aligned}
\mathbf{r}(t)_T &= \left\langle \frac{5}{4}, \frac{3}{4} \right\rangle + p \cdot \sqrt{\frac{8}{17}} \left\langle \frac{3}{4}, \frac{5}{4} \right\rangle \\
&= \left\langle \frac{5}{4}, \frac{3}{4} \right\rangle + \frac{2\sqrt{2}p}{\sqrt{17}} \left\langle \frac{3}{4}, \frac{5}{4} \right\rangle
\end{aligned}$$

$$\begin{aligned}\mathbf{r}(t)_N &= \left\langle \frac{5}{4}, \frac{3}{4} \right\rangle + p \cdot \sqrt{\frac{8}{17}} \left\langle \frac{5}{4}, -\frac{3}{4} \right\rangle \\ &= \left\langle \frac{5}{4}, \frac{3}{4} \right\rangle + \frac{2\sqrt{2}p}{\sqrt{17}} \left\langle \frac{5}{4}, -\frac{3}{4} \right\rangle\end{aligned}$$

(note that p is chosen arbitrarily as the parameter). ■

---

(2) Determine **T** and **N** for $\mathbf{r}(t) = \langle t^2, 2t^3/3, t \rangle$ at the point $(1, 2/3, 1)$. Can you demonstrate some quality assurance about your results?

□ Note that the given point corresponds to $t = 1$. Here we go into a long string of calculations; as always, be alert to where we can jump out of general formulas into numerical computations at the specific point. Some derivatives will have been found behind the scenes. To start with, we have

$$\begin{aligned}\mathbf{r}'(t) &= \langle 2t, 2t^2, 1 \rangle \\ ds = |\mathbf{r}'(t)| &= \sqrt{4t^2 + 4t^4 + 1} = \sqrt{(2t^2+1)^2} = 2t^2 + 1 \\ \mathbf{T}(t) &= \frac{\mathbf{r}'(t)}{ds} = \left\langle \frac{2t}{2t^2+1}, \frac{2t^2}{2t^2+1}, \frac{1}{2t^2+1} \right\rangle\end{aligned}$$

At $t = 1$ specifically,

$$\mathbf{T}(1) = \left\langle \frac{2}{3}, \frac{2}{3}, \frac{1}{3} \right\rangle$$

Now back to the general form,

$$\mathbf{T}'(t) = \left\langle \frac{2(1-2t^2}{(2t^2+1)^2}, \frac{4t}{(2t^2+1)^2}, \frac{-4t}{(2t^2+1)^2} \right\rangle$$

and again at $t = 1$ specifically,

$$\mathbf{T}'(1) = \left\langle -\frac{2}{9}, \frac{4}{9}, -\frac{4}{9} \right\rangle \quad \text{and} \quad |\mathbf{T}'(1)| = \frac{2}{3}$$

So finally,

$$\mathbf{N}(1) = \frac{\mathbf{T}'(1)}{|\mathbf{T}'(1)|} = \left\langle -\frac{1}{3}, \frac{2}{3}, -\frac{2}{3} \right\rangle$$

Collecting our results, we have

$$\mathbf{T}(1) = \left\langle \frac{2}{3}, \frac{2}{3}, \frac{1}{3} \right\rangle \qquad \text{and} \qquad \mathbf{N}(1) = \left\langle -\frac{1}{3}, \frac{2}{3}, -\frac{2}{3} \right\rangle$$

As a double-check, we can confirm that $|\mathbf{T}(1)| = 1$, $|\mathbf{N}(1)| = 1$, and $\mathbf{T}(1) \cdot \mathbf{N}(0) = 0$. ■

(3) A track where horses race is in the shape of the ellipse $\mathbf{r}(t) = \langle 3\cos t, 2\sin t\rangle$ (for $0 \le t \le 2\pi$). The coordinate system is centered at the beer tent in the center of the lawn inside the racetrack, and the axes split the ellipse into four quadrants, as this equation suggests. One particularly moody horse decides enough is enough, and when he reaches the spot on the track marked by $t = \pi/4$, he breaks off the track and runs away in a direction perfectly tangent to his original path on the track. So:

(a) What are the coordinates at which he crosses the y-axis and escapes to freedom?

(b) If a gust of wind adds an acceleration vector of $\langle 0, 10\rangle$ to aid the horse, what is the component of that acceleration vector in this tangent direction of escape?

☐ From the original vector function, we have

$$\mathbf{r}\left(\frac{\pi}{4}\right) = \left\langle \frac{3}{\sqrt{2}}, \frac{2}{\sqrt{2}} \right\rangle$$

Then, more generally,

$$\begin{aligned} \mathbf{r}'(t) &= \langle -3\sin t, 2\cos t\rangle \\ ds = |\mathbf{r}'(t)| &= \sqrt{9\sin^2 t + 4\cos^2 t} \\ \mathbf{T}(t) &= \frac{\mathbf{r}'(t)}{ds} = \frac{1}{\sqrt{9\sin^2 t + 4\cos^2 t}}\langle -3\sin t, 2\cos t\rangle \end{aligned}$$

At $t = \pi/4$ specifically,

$$\mathbf{T}\left(\frac{\pi}{4}\right) = \frac{\sqrt{2}}{\sqrt{13}}\left\langle -\frac{3}{\sqrt{2}}, \frac{2}{\sqrt{2}} \right\rangle = \left\langle -\frac{3}{\sqrt{13}}, \frac{2}{\sqrt{13}} \right\rangle$$

(a) To answer the first question, we need the equation of the tangent line at the point $\mathbf{r}(\pi/4)$. This point along with the tangent vector $\mathbf{T}\left(\frac{\pi}{4}\right)$ are the information we need to form the (vector) equation of this line. Because the parameter t is already in use for the curve, and s represents arc length parameter specifically, let's use p as the parameter for the line

$$L(p) = \mathbf{r}_0 + p\mathbf{r}_1 = \left\langle \frac{3}{\sqrt{2}}, \frac{2}{\sqrt{2}} \right\rangle + p \cdot \left\langle -\frac{3}{\sqrt{13}}, \frac{2}{\sqrt{13}} \right\rangle$$

We can untangle the specific equations for the x and y coordinates of this line as:

$$x = \frac{3}{\sqrt{2}} - p \cdot \frac{3}{\sqrt{2}}$$
$$y = \frac{2}{\sqrt{2}} + p \cdot \frac{2}{\sqrt{2}}$$

The horse running on this line crosses the y-axis when $x = 0$; the parameter p which makes this happen is $p = 1$. And so at $p = 1$, the y-coordinate is $y = 4/\sqrt{2}$. Altogether, the horse crosses the y-axis at

$$(x, y) = \left(0, \frac{4}{\sqrt{2}}\right)$$

(b) If an acceleration vector is $\vec{a} = \langle 0, 10\rangle$, then the component of that vector along $\mathbf{T}$ is (remember that $|\mathbf{T}| = 1$):

$$a_T = \frac{\mathbf{a} \cdot \mathbf{T}}{|\mathbf{T}|} = \langle 0, 10\rangle \cdot \left\langle -\frac{3}{\sqrt{13}}, \frac{2}{\sqrt{13}} \right\rangle = \frac{20}{\sqrt{13}} \quad \blacksquare$$

C.5.3 *The Fresnet–Serret Frame — Challenge — Solved*

(1) The shuttlecraft *Galileo* is lifting off from the planet Taurus 2 along the path given by $\mathbf{r}(t) = \langle t^5/5, \sqrt{2}t^3/3, t\rangle$ in the galactic coordinate system. When at the point marked by $t = 1$, the shuttle launches a probe in the direction of the binormal vector $\mathbf{B}(1)$. If the probe travels in a straight line, it will cross through two of three galactic coordinate planes (xy, xz, or yz). Which two planes will it cross, and at which galactic coordinates?

□ We need to generate the Fresnet–Serret frame for $\mathbf{r}(t)$, which starts with:

$$\mathbf{r}'(t) = \left\langle t^4, \sqrt{2}t^2, 1 \right\rangle$$
$$|\mathbf{r}'(t)| = \sqrt{(t^4)^2 + (\sqrt{2}t^2)^2 + 1^2} = \sqrt{(t^4)^2 + 2t^4 + 1}$$
$$= \sqrt{(t^4+1)^2} = t^4 + 1$$
$$\mathbf{T}(t) = \frac{\mathbf{r}'(t)}{|\mathbf{r}'(t)|} = \frac{1}{t^4+1}\langle t^4, \sqrt{2}t^2, 1\rangle$$
$$\mathbf{T}'t = -\frac{4t^3}{(t^4+1)^2}\langle t^4, \sqrt{2}t^2, 1\rangle + \frac{1}{t^4+1}\langle 4t^3, 2\sqrt{2}t, 0\rangle$$

Then at $t = 1$,

$$\mathbf{T}(1) = \frac{1}{2}\langle 1, \sqrt{2}, 1\rangle = \left\langle \frac{1}{2}, \frac{1}{\sqrt{2}}, \frac{1}{2}\right\rangle$$
$$\mathbf{T}'(1) = \frac{4}{4}\langle 1, \sqrt{2}, 1\rangle + \frac{1}{2}\langle 4, 2\sqrt{2}, 0\rangle = \langle 1, 0, -1\rangle$$
$$|\mathbf{T}'(1)| = \sqrt{2}$$
$$\mathbf{N}(1) = \frac{\mathbf{T}'1}{|\mathbf{T}'1|} = \left\langle \frac{1}{\sqrt{2}}, 0, -\frac{1}{\sqrt{2}}\right\rangle$$

Then

$$\mathbf{B}(1) = \mathbf{T}(1) \times \mathbf{N}(1) = \left\langle \frac{1}{2}, \frac{\sqrt{2}}{2}, \frac{1}{2}\right\rangle \times \left\langle \frac{1}{\sqrt{2}}, 0, -\frac{1}{\sqrt{2}}\right\rangle$$
$$= \left\langle -\frac{1}{2}, \frac{1}{\sqrt{2}}, -\frac{1}{2}\right\rangle$$

Summarizing,

$$\mathbf{T}(1) = \left\langle \frac{1}{2}, \frac{1}{\sqrt{2}}, \frac{1}{2}\right\rangle$$
$$\mathbf{N}(1) = \left\langle \frac{1}{\sqrt{2}}, 0, -\frac{1}{\sqrt{2}}\right\rangle$$
$$\mathbf{B}(1) = \left\langle -\frac{1}{2}, \frac{1}{\sqrt{2}}, -\frac{1}{2}\right\rangle$$

The point of departure is $\mathbf{r}(1) = \langle 1/5, \sqrt{2}/3, 1\rangle$, and so the vector equation of the line directed by $\mathbf{B}(1)$ is (with parameter p):

$$\mathbf{r}_B(p) = \left\langle \frac{1}{5}, \frac{\sqrt{2}}{3}, 1\right\rangle + p \cdot \left\langle -\frac{1}{2}, \frac{1}{\sqrt{2}}, -\frac{1}{2}\right\rangle$$

The line $\mathbf{r}_B(p)$ hits a coordinate plane whenever $x = 0$, $y = 0$, or $z = 0$, and these happen at:

$$x = 0 : \rightarrow \frac{1}{5} + p \cdot \left(-\frac{1}{2}\right) \rightarrow p = \frac{2}{5}$$
$$y = 0 : \rightarrow \frac{\sqrt{2}}{3} + p \cdot \frac{1}{\sqrt{2}} = 0 \rightarrow p < 0 \text{ so nevermind}$$
$$z = 0 : \rightarrow 1 + p \cdot \left(-\frac{1}{2}\right) = 0 \rightarrow p = 2$$

And so, the probe launched in the direction of **B** crosses the galactic yz-plane at a parameter value $p = 2/5$; this corresponds to the point

$$\begin{aligned}\mathbf{r}_B\left(\frac{2}{5}\right) &= \left\langle \frac{1}{5}, \frac{\sqrt{2}}{3}, 1 \right\rangle + \frac{2}{5} \cdot \left\langle -\frac{1}{2}, \frac{1}{\sqrt{2}}, -\frac{1}{2} \right\rangle \\ &= \left\langle \frac{1}{5}, \frac{\sqrt{2}}{3}, 1 \right\rangle + \left\langle -\frac{1}{5}, \frac{\sqrt{2}}{5}, -\frac{1}{5} \right\rangle = \left\langle 0, \frac{8\sqrt{2}}{15}, \frac{4}{5} \right\rangle\end{aligned}$$

It then crosses the galactic xy-plane at a parameter value $p = 2$; this corresponds to the point

$$\mathbf{r}_B(2) = \left\langle \frac{1}{5}, \frac{\sqrt{2}}{3}, 1 \right\rangle + 2 \cdot \left\langle -\frac{1}{2}, \frac{1}{\sqrt{2}}, -\frac{1}{2} \right\rangle = \left\langle -\frac{4}{5}, \frac{4\sqrt{2}}{3}, 0 \right\rangle \quad \blacksquare$$

(2) Find the tangential and normal components of acceleration for $\mathbf{r}(t) = \langle e^{-t}\cos t, e^{-t}\sin t \rangle$ at $t = \pi/4$. (Hint: How can you rig up a cross product involving vectors that are only two-dimensional?)

☐ To find the tangential component of acceleration we need:

$$\begin{aligned}\mathbf{r}'(t) &= \langle -e^{-t}(\cos t + \sin t), e^{-t}(\cos t - \sin t) \rangle \\ |\mathbf{r}'(t)| &= \ldots = \sqrt{2}e^{-t} \\ \mathbf{r}''(t) &= \langle 2e^{-t}\sin t, -2e^{-t}\cos t \rangle \\ \mathbf{r}'(t) \cdot \mathbf{r}''(t) &= -2e^{-2t}\end{aligned}$$

so then

$$a_T = \frac{\mathbf{r}'(t) \cdot \mathbf{r}''(t)}{|\mathbf{r}'(t)|} = \frac{-2e^{-2t}}{\sqrt{2}e^{-t}} = -\sqrt{2}e^{-t}$$

and at $t = \pi/4$, we get

$$a_T\left(\frac{\pi}{4}\right) = -\sqrt{2}e^{-\pi/4}$$

To find the normal component of acceleration, we need $|\mathbf{a}(t)|^2$. Note that $\mathbf{a}(t) = \mathbf{r}''(t)$, which is found above; so,

$$\begin{aligned}\mathbf{a}(t) &= \langle 2e^{-t}\sin t, -2e^{-t}\cos t \rangle \\ |\mathbf{a}(t)|^2 &= 4e^{-2t}\end{aligned}$$

so that

$$\begin{aligned}a_N &= \sqrt{|\mathbf{a}(t)|^2 - a_T^2} = \sqrt{4e^{-2t} - (-\sqrt{2}e^{-t})^2} \\ &= \sqrt{4e^{-2t} - 2e^{-2t}} = \sqrt{2e^{-2t}} = \sqrt{2}e^{-t}\end{aligned}$$

so at $t = \pi/4$, we get

$$a_N\left(\frac{\pi}{4}\right) = \sqrt{2}e^{-\pi/4} \quad \blacksquare$$

(3) Show that the curvature of a circle of radius a is $\kappa = 1/a$. What is the torsion anywhere on this circle?

☐ A circle of radius a can described by the vector function $\mathbf{r}(t) = \langle a\cos t, a\sin t, 0\rangle$. We want to show that the curvature at any point is $\kappa = 1/a$. To compute the curvature at any point, we need to find

$$\kappa = \frac{|\mathbf{r}'(t) \times \mathbf{r}''(t)|}{|\mathbf{r}'(t)|^3}$$

So we need several items:

$$\mathbf{r}'(t) = \langle -a\sin t, a\cos t, 0\rangle$$
$$|\mathbf{r}'(t)| = \sqrt{a^2\sin^2 t + a^2\cos^2 t + 0} = a$$
$$\mathbf{r}''(t) = \langle -a\cos t, -a\sin t, 0\rangle$$
$$\mathbf{r}'(t) \times \mathbf{r}''(t) = \langle 0, 0, a^2\sin^2 t + a^2\cos^2 t\rangle \text{ details begind the scenes}$$
$$|\mathbf{r}'(0) \times \mathbf{r}''(0)| = a^2$$

so that

$$\kappa = \frac{|\mathbf{r}'(t) \times \mathbf{r}''(t)|}{|\mathbf{r}'(t)|^3} = \frac{a^2}{(a)^3} = \frac{1}{a}$$

The torsion function for this curve is $\tau(t) = 0$, because this is a two-dimensional curve. ■

(4) (Bonus Time in *the Pit*!) Can you derive the relation (17.15) in Useful Fact 17.2?

☐ Suppose we have a scalar multiple of a vector function, i.e.

$$\mathbf{v}(t) = \beta(t)\langle f(t), g(t), h(t)\rangle$$

We are going to prove that

$$\frac{d}{dt}\beta(t)\mathbf{v}(t) = \beta'(t)\mathbf{v}(t) + \beta(t)\mathbf{v}'(t)$$

Starting with $\mathbf{v}(t) = \beta(t)\langle f(t), g(t), h(t)\rangle$, we can multiply the scalar function $\beta(t)$ into each component:

$$\mathbf{v}(t) = \langle \beta(t)f(t), \beta(t)g(t), \beta(t)h(t)\rangle$$

When finding $\mathbf{v}'(t)$, then, the regular old scalar function product rule applies to each component function:

$$\mathbf{v}'(t) = \langle \beta'(t)f(t)+\beta(t)f'(t), \beta'(t)g(t)+\beta(t)g'(t), \beta'(t)h(t)+\beta(t)h'(t)\rangle$$

Let's separate the terms involving $\beta(t)$ and $\beta'(t)$:

$$\mathbf{v}'(t) = \langle \beta'(t)f(t), \beta'(t)g(t), \beta'(t)h(t)\rangle + \langle \beta(t)f'(t), \beta(t)g'(t), \beta(t)h'(t)\rangle$$
$$= \beta'(t)\langle f(t), g(t), h(t)\rangle + \beta(t)\langle f'(t), g'(t), h'(t)\rangle$$

and this can be tidied up as:

$$\mathbf{v}'(t) = \beta'(t)\mathbf{v}(t) + \beta(t)\mathbf{v}'(t) \quad ■$$

C.5.4 *Lagrange Multipliers — Challenge — Solved*

(1) The galactic company that manufactures phasers (p), pulse rifles (r), and communicator badges (b) for the United Federation of Planets earns a profit of 10 quatloos for every phaser sold, 20 quatloos for every pulse rifle, and 5 quatloos for every communicator badge. The logistics of their operations require that the combined number of these items made each day is held strictly to $2p^2 + r^2 + 4b^2 = 10000$. How many of each item should they make each day to maximize profit?

□ The objective function is $f(p, r, b) = 10p + 20r + 5b$, and its gradient is $\nabla f(p, r, b) = \langle 10, 20, 5\rangle$.

The constraint function is $g(p, r, b) = 2p^2 + r^2 + 4b^2$, and its gradient is $\nabla g(p, r, b) = \langle 4p, 2q, 8r\rangle$.

The equation of proportionality of these gradients is $\nabla f(p, r, b) = \lambda \nabla g(p, r, b)$; this equation, along with the constraint itself, leads to the system of equations,

$$\begin{aligned} 10 &= 4p\lambda \\ 20 &= 2r\lambda \\ 5 &= 8b\lambda \\ 2p^2 + r^2 + 4b^2 &= 10000 \end{aligned}$$

Wolfram Alpha reports two solution sets for these equations, but one has negative values for one or more variables, which is irrelevant in the context of the problem. The only suitable solution is, in exact form,

$$p = \frac{200}{\sqrt{73}} \quad , \quad r = \frac{80}{\sqrt{73}} \quad , \quad b = \frac{50}{\sqrt{73}}$$

This is one of those times when the exact forms of the solution values aren't as useful as the approximations (rounded to the nearest integer):

$$p \approx 23 \quad , \quad r \approx 94 \quad , \quad b \approx 6$$

So they should make 23 phasers, 94 pulse rifles, and 6 communicator badges per day to maximize profit. ■

(2) Find the maximum value of $f(x, y, z) = xy + 2yz - 3x + 3z$ that can be achieved anywhere on the unit sphere.

□ The objective function is $f(x, y, z) = xy + 2yz - 3x + 3z$, and its gradient is $\nabla f(x, y, z) = \langle y - 3, x + 2z, 2y + 3\rangle$.

The constraint function is $g(x, y, z) = x^2 + y^2 + z^2$ (from the equation of the unit circle), and its gradient is $\nabla g(x, y, z) = \langle 2x, 2y, 2z\rangle$.

The equation of proportionality of these gradients is $\nabla f(x, y, z) = \lambda \nabla g(x, y, z)$; this equation, along with the constraint itself ($x^2 + y^2 + z^2 = 1$, the unit circle), leads to the system of equations,

$$\begin{aligned} y - 3 &= \lambda(2x) \\ x + 2z &= \lambda(2y) \\ 2y + 3 &= \lambda(2z) \\ x^2 + y^2 + z^2 &= 1 \end{aligned}$$

Wolfram Alpha reports two sets of real solutions to this system in approximate form, because the exact forms are *nasty*. These approximate solutions are:

$$(x, y, z) = (0.624, 0.198, -0.756) \ ; \ (x, y, z) = (-0.624, 0.198, 0.756)$$

The values of the objective function at these points will provide the min and max of the objective function.

$$\begin{aligned} f\,(0.624, 0.198, -0.756) &\approx -4.316 \\ f\,(-0.624, 0.198, 0.756) &\approx 4.316 \end{aligned}$$

So, the maximum value of $f(x, y, z)$ that can be achieved on the unit sphere is about 4.316. ■

---

(3) Find the point on the plane $2x + 3y + 4z = 12$ that is closest to the point $(5, 5, 5)$.

□ We want to minimize the distance to $(x, y, z) = (5, 5, 5)$ for points satisfying $2x + 3y + 4z = 12$. It suffices to minimize the square of the distance, $D^2 = (x - 5)^2 + (y - 5)^2 + (z - 5)^2$. This expression for D^2 is our objective function, and the constraint is the left side of the equation of the plane. The gradients of each are:

$$\begin{aligned} \nabla D^2(x, y, z) &= \langle 2(x - 5), 2(y - 5), 2(z - 5)\rangle \\ \nabla g(x, y, z) &= \langle 2, 3, 4\rangle \end{aligned}$$

Setting the proportion $\nabla D^2(x, y, z) = \lambda \nabla g(x, y, z)$ and introducing the constraint, we get the system:

$$\begin{aligned} 2(x-5) &= 2\lambda \\ 2(y-5) &= 3\lambda \\ 2(z-5) &= 4\lambda \\ 2x + 3y + 4z &= 12 \end{aligned}$$

The solution is, using Wolfram Alpha, $(x, y, z) = (79/29, 46/29, 13/29)$. This is the point on the given plane closest to the point $(5, 5, 5)$. ■

C.5.5 *Parametric Surfaces — Challenge — Solved*

(1) Identify the parametric surface $\mathbf{r}(s, t) = \langle 3\sin t\cos s, 3\sin t\sin s, 3\cos t\rangle$ for $0 \le s \le \pi/2$ and $0 \le t \le \pi/2$.

□ If we break out the individual equations, we have

$$\begin{aligned} x &= 3\sin t\cos s \\ y &= 3\sin t\sin s \\ z &= 3\cos t \end{aligned}$$

The first two form $x^2 + y^2 = 9\sin^2 t$. Then with the third, we get $x^2 + y^2 + z^2 = 9$. This is a sphere of radius 3. But, the parameters s and t were restricted. The parameter s acts as the polar angle in spherical coordinates, and t acts as the azimuthal angle (you can tell by comparing $z = 3\cos t$ to $z = 3\cos\phi$, where the latter is the direct conversion equation between spherical and rectangular coordinates).

Together, the restrictions $0 \le s \le \pi/2$ and $0 \le t \le \pi/2$ keep us in the first octant. So, this is a sphere of radius 3, centered at the origin, in the first octant. ■

(2) Give parametric equations for the portion of the plane $z = 2x + y$ that lies inside the cylinder $x^2 + y^2 = 1$.

□ We have to decide: do we focus on the plane first, or the cylinder? The cylinder is just the lateral boundary; the cross section is a circle of radius 1. It seems easiest set up parametric equations so that we are filling x and y coordinates from within the given circle, and then set z

according to the equation of the plane. Since the unit circle itself can be presented as $x = \cos t$ and $y = \sin t$ for $0 \le t \le 2\pi$, then we can also fill the interior of the circle (in the horizontal plane) with $x = s\cos t$ and $y = s\sin t$ for $0 \le s \le 1$ and $0 \le t \le 2\pi$. Then we can just directly build $z = 2x + y$ as $z = 2s\cos t + s\sin t$. Together,

$$\begin{cases} x = s\cos t \\ y = s\sin t \\ z = 2s\cos t + s\sin t \end{cases} \quad \text{with} \quad 0 \le s \le 1 \;\; ; \;\; 0 \le t < 2\pi$$

Note this is not unique; another perfectly good set of parametric equations is:

$$\begin{cases} x = s\sin t \\ y = t\cos t \\ z = 2s\sin t + s\cos t \end{cases} \quad \text{with} \quad 0 \le s \le 1 \;\; ; \;\; 0 \le t < 2\pi \quad \blacksquare$$

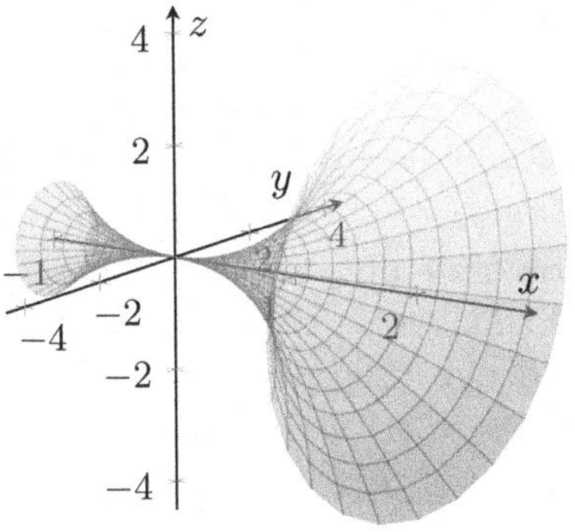

Fig. C.19 $f(x) = x^2$ revolved around the x-axis on $[-1, 2]$.

(3) Do you remember surfaces of revolution? Those can be drawn parametrically, too! A figure in Chapter 9 (Volume 1), duplicated here as Fig. C.19, shows the surface generated when the curve $f(x) = x^2$ is revolved around the x-axis on the interval $[-1, 2]$. Give parametric equations for this surface of revolution.

□ In this trip down memory lane, you should recall that the solid of revolution will have circular cross sections for each x in $[-1, 2]$. The radius of each circle is $f(x)$. And so, we must form parametric equations that sweep along the x-axis from $x = -1$ to $x = 2$ and form a circle of radius x^2 in the yz-direction at each x. Our parametric equations can be:

$$\begin{cases} x = s \\ y = s^2 \cos t \\ z = s^2 \sin t \end{cases} \quad \text{with} \quad -1 \leq s \leq 2 \;\; ; \;\; 0 \leq t < 2\pi \quad \blacksquare$$

C.6 Chapter 18: Challenge Problem Solutions

C.6.1 *Line Integrals — Challenge — Solved*

---

(1) Evaluate $\int_C (x-4)(z-5)\,ds$, where C is the line segment from $(4,1,5)$ to $(0,6,-1)$.

□ Since this is a line integral with respect to arc length ds, we 'll use the arc length version of the line integral. We need ANY parametric description of the line segment, and the most basic one using $0 \leq t \leq 1$ is:

$$x = 4 - 4t \quad y = 1 + 5t \quad z = 5 - 6t \quad \text{for } 0 \leq t \leq 1$$

so that

$$\sqrt{[x'(t)]^2 + [y'(t)]^2 + [z'(t)]^2} = \sqrt{(-4)^2 + (5)^2 + (-6)^2} = \sqrt{77}$$

and

$$\int_C (x-4)(z-5)\,ds$$

$$= \int_a^b f(x(t), y(t), z(t))\sqrt{[x'(t)]^2 + [y'(t)]^2 + [z'(t)]^2}\,dt$$

$$= \int_0^1 (-4t)(-6t)(\sqrt{77})\,dt = 24\sqrt{77}\int_0^1 t^2\,dt = 8\sqrt{77} \quad \blacksquare$$

---

(2) Evaluate $\int_C y\,dx + z\,dy + x\,dz$ along the contour C that is given by

$$x = t^3 \quad y = t^2 \quad z = t \quad \text{for} \quad 0 \leq t \leq 1$$

□ Since this is a line integral with respect to dx, dy and dz we can construct the parametric version of the line integral directly. From the parametric description of the line integral, we get $dx = 3t^2\,dt$, $dy = 2t\,dt$ and $dz = dt$, and so

$$\int_C y\,dx + z\,dy + x\,dz = \int_0^1 t^2(3t^2\,dt) + (t)(2t\,dt) + (t^3)(dt)$$

$$= \int_0^1 (3t^4 + 2t^2 + t^3)\,dt = \frac{91}{60} \quad \blacksquare$$

(3) Find the work done by $\mathbf{F}(x,y,z) = \langle -y, z^2, x\rangle$ around the contour $\mathbf{r}(t) = \langle \sin t, t, \cos t\rangle$ for $0 \le t \le 2\pi$.

☐ We want to evaluate $\int_C \mathbf{F} \cdot d\mathbf{r}$. From $\mathbf{r}(t)$ we have

$$\begin{aligned} x &= \sin t \rightarrow dx = \cos t\, dt \\ y &= t \rightarrow dy = dt \\ z &= \cos t \rightarrow dz = -\sin t\, dt \end{aligned}$$

so that

$$\begin{aligned} \int_C \mathbf{F} \cdot d\mathbf{r} &= \int_C -y\, dx + z^2\, dy + x\, dz \\ &= \int_0^{2\pi} (-t)(\cos t\, dt) + (\cos t)^2 (dt) + (\sin t)(-\sin t\, dt) \\ &= \int_0^{2\pi} (-t\cos t + \cos^2 t - \sin^2 t)\, dt = 0 \quad \blacksquare \end{aligned}$$

(4) *(Bonus! Following up Sec. 15.3 ...)* We expect $\displaystyle\int_C \mathbf{F}(x,y) \cdot d\mathbf{r} = 0$ for which of the following combinations of vector field $\mathbf{F}(x,y)$ and oriented path of integration C?

I4) $\mathbf{F}(x,y) = \langle y^3, x^3\rangle$ and C is the circle of radius 2, oriented counterclockwise

I5) $\mathbf{F}(x,y) = \langle x/y, x+y\rangle$ and C is the cardioid $r = 2+\cos\theta$, oriented counterclockwise

I6) $\mathbf{F}(x,y) = \langle (x+y)^2, x^2y^2\rangle$ and C follows $y = 1/(x^2+1)$ from $x = -2$ to $x = 2$.

☐ Integrals (I4) and (I5) are zero, but (I6) is not. In (I6), the components of $d\mathbf{r}$ are non-negative. The components of $\mathbf{F}$ are also non-negative. So we are integrating a dot product which is always non-negative. The result cannot be zero. In (I5) and (I6) the symmetry of the contour C cooperates with the positive and negative contributions of x and y for perfect balance. Draw some pictures! Sketch C, then draw several representatives of $d\mathbf{r}$ along C. Also draw some representative samples of $\mathbf{F}$. While this does not offer conclusive proof of the integrals' values being zero, we are only making *predictions* about the behavior of the integrals.

Note that in (I4), C is a simple closed contour and the integral is zero, yet $\mathbf{F}(x, y)$ is not conservative. This gives a chance to reinforce what you can and can't conclude from statements such as, "If $\mathbf{F}$ is conservative and has continuous first derivatives, then $\int_C \mathbf{F}(x, y) \cdot d\mathbf{r} = 0$ along any piecewise-smooth closed contour C." We cannot say the reverse, that if the same integral is zero, then $\mathbf{F}$ must have been conservative. ■

C.6.2 *Conservative Vector Fields — Challenge — Solved*

(1) Determine if the vector field $\mathbf{F}(x, y) = \langle y^2 + e^x + xe^x, 2xy\rangle$ is conservative. If it is, find a potential function for it.

☐ Matching to the form $\mathbf{F}(x, y) = \langle P(x, y), Q(x, y)\rangle$, we have

$$P = y^2 + xe^x + e^x \to \frac{\partial P}{\partial y} = 2y$$

$$Q = 2xy \to \frac{\partial Q}{\partial x} = 2y$$

Since $\partial P/\partial y = \partial Q/\partial x$ then the vector field is conservative. So there is a function $f(x, y)$ such that $\nabla f = \mathbf{F}$.

We know from $\mathbf{F}$ that for this function, $f_x = y^2 + e^x + xe^x$ and $f_y = 2xy$. Based on f_y, we know that at worst,

$$f(x, y) = xy^2 + g(x)$$

where $g(x)$ is some unknown function of x. With $f(x, y)$ in this form, we see that $f_x = y^2 + g'(x)$. But from the gradient vector field, we also know $f_x = y^2 + e^x + xe^x$ So the mystery function must satisfy $g'(x) = e^x + xe^x$. Through integration, this gives that $g(x) = xe^x + C$. So altogether,

$$f(x, y) = xy^2 + xe^x + C \quad ■$$

(2) Determine if the vector field $\mathbf{F}(x, y, z) = \langle 3x^2, -\cos(y), 2xz\rangle$ is conservative. If it is, find a potential function for it.

☐ Investigating $\nabla \times \mathbf{F}$, we have

$$\nabla \times \mathbf{F} = \begin{vmatrix} \mathbf{i} & \mathbf{j} & \mathbf{k} \\ \frac{\partial}{\partial x} & \frac{\partial}{\partial y} & \frac{\partial}{\partial z} \\ 3x^2 & -\cos(y) & 2xz \end{vmatrix} = (0)\mathbf{i} - (2z)\mathbf{j} + (0)\mathbf{k} \neq \mathbf{0}$$

Since $\nabla \times \mathbf{F} \neq \mathbf{0}$, then $\mathbf{F}$ is NOT conservative, and so there will not be a scalar function f such that $\mathbf{F} = \nabla f$. ■

---

(3) Use the Fundamental Theorem for Line Integrals to evaluate $\int_C \mathbf{F} \cdot d\mathbf{r}$ where $\mathbf{F}(x, y, z) = \langle 10x + 3y + yz, 3x + 20y + xz, xy\rangle$ and C is the curve given by

$$\mathbf{r}(t) = \left\langle t^{5/2} - 1, \sqrt{t+3}, \sin\left(\frac{\pi}{2}t\right)\right\rangle \quad \text{for} \quad 0 \leq t \leq 1$$

□ The potential function for this vector field is a function $f(x, y, z)$ such that

$$f_x = 10x + 3y + yz \quad , \quad f_y = 3x + 20y + xz \quad , \quad f_z = xy$$

We could go through a systematic routine of finding $f(x, y, z)$ based on these derivatives, but I think this one's straightforward enough to deduce directly that

$$f(x, y, z) = 5x^2 + 3xy + 10y^2 + xyz$$

In preparation for the Fundamental Theorem of Line Integrals, note that the starting and ending points of the contour C are indicated by the vectors:

$$\mathbf{r}(0) = \langle -1, \sqrt{3}, 0\rangle$$
$$\mathbf{r}(1) = \langle 0, 2, 1\rangle$$

The values of $f(x, y, z)$ at these endpoints are:

$$\begin{aligned} f(-1, \sqrt{3}, 0) &= 5(-1)^2 + 3(-1)(\sqrt{3}) + 10(\sqrt{3})^2 + (-1)(\sqrt{3})(0) \\ &= 35 - 3\sqrt{3} \\ f(0, 2, 1) &= 5(0)^2 + 3(0)(2) + 10(2)^2 + (0)(2)(1) = 40 \end{aligned}$$

Then by the Fundamental Theorem for Line Integrals, we have

$$\int_C \mathbf{F} \cdot d\mathbf{r} = f(0, 2, 1) - f(-1, \sqrt{3}, 0) = 40 - (35 - 3\sqrt{3}) = 5 + 3\sqrt{3}$$ ■

C.6.3 *Surface Integrals — Challenge — Solved*

(1) Find $\iint_S y^2(3-z)\,dS$ where the surface S is the part of the plane $x+y+z=3$ in the first octant.

□ The upper bound of our eventual region of integration D is the intersection of the plane $z=3-x-y$ with the xy-plane, which is $x+y=3$ (found by setting $z=0$). In the first octant, then, the bounds of D are $0\le x\le 3$ and $0\le y\le 3-x$. Remember that to evaluate the scalar function version of the surface integral, we will need to eliminate any z from the integrand; here, the integrand is $y^2(3-z)$, which on the surface $z=3-x-y$ becomes $y^2(3-(3-x-y))$ or $y^2(x+y)$. Also note that the equation of the surface identifies $z=g(x,y)=3-x-y$, so that $g_x=-1$ and $g_y=-1$. And so altogether:

$$\begin{aligned}\iint_S y^2(3-z)\,dS &= \iint_D y^2(3-z)\sqrt{g_x^2+g_y^2+1}\,dA\\ &= \int_0^3\int_0^{3-x} y^2(x+y)\sqrt{(-1)^2+(-1)^2+1}\,dydx\\ &= \sqrt{3}\int_0^3\int_0^{3-x} y^2(x+y)\,dydx = \frac{81\sqrt{3}}{5} \quad ■\end{aligned}$$

(2) Find the surface area of the portion of the surface $z=4-x^2-y^2$ over the region between $x^2+y^2=2$, $x^2+y^2=9$, and $y=0$.

□ The portion of the surface as described is over the region in the xy-plane between the upper half circles $x^2+y^2=2$ and $x^2+y^2=9$, so when we set up our integral, polar coordinates may be useful. With $z=g(x,y)=4-x^2-y^2$, we get $g_x(x,y)=-2x$ and $g_y(x,y)=-2y$, so

$$\sqrt{g_x(x,y)^2+g_y(x,y)^2+1}=\sqrt{4x^2+4y^2+1}$$

When we start setting up our integral as

$$A_S=\iint_S (1)\,dS=\iint_D (1)\sqrt{g_x^2+g_y^2+1}\,dA=\iint_D \sqrt{4x^2+4y^2+1}\,dA$$

we see that, yep, polar coordinates are going to be best because our region of integration involves upper semicircles ($0\le\theta\le\pi$) of radius $\sqrt{2}$ and 3, *and* the integrand converts to $\sqrt{4r^2+1}$. Altogether, then

$$\begin{aligned}A_S &= \iint_S (1)\,dS = \iint_D (1)\sqrt{g_x^2+g_y^2+1}\,dA\\ &= \int_0^{\pi}\int_{\sqrt{2}}^{3}\sqrt{4r^2+1}\cdot r\,drd\theta = \left(\frac{37\sqrt{37}}{12}-\frac{9}{4}\right)\pi \quad ■\end{aligned}$$

(3) Find $\iint_S \mathbf{F} \cdot d\mathbf{S}$ where $\mathbf{F} = \langle x, y, z^2 \rangle$ and the surface S is the inverted cone $z = 4 - \sqrt{x^2 + y^2}$ above the xy-plane (oriented positively).

□ The surface is identified as $z = g(x, y) = 4 - \sqrt{x^2 + y^2}$, so

$$\frac{\partial g}{\partial x} = -\frac{x}{\sqrt{x^2 + y^2}} \quad ; \quad \frac{\partial g}{\partial y} = -\frac{y}{\sqrt{x^2 + y^2}}$$

Matching the vector field to the form $\mathbf{F} = \langle P, Q, R \rangle$ and incorporating the equation of the surface for z, we have

$$P = x \quad ; \quad Q = y \quad ; \quad R = z^2 = (4 - \sqrt{x^2 + y^2})^2$$

Let's start putting together our integral:

$$\begin{aligned}
\iint_S \mathbf{F} \cdot d\mathbf{S} &= \iint_D \left(-P\frac{\partial g}{\partial x} - Q\frac{\partial g}{\partial y} + R \right) dA \\
&= \iint_D \left(-x \cdot \frac{-x}{\sqrt{x^2 + y^2}} - y \cdot \frac{-y}{\sqrt{x^2 + y^2}} + (4 - \sqrt{x^2 + y^2})^2 \right) dA \\
&= \iint_D \left(\frac{x^2}{\sqrt{x^2 + y^2}} + \frac{y^2}{\sqrt{x^2 + y^2}} + (4 - \sqrt{x^2 + y^2})^2 \right) dA \\
&= \iint_D \left(\frac{x^2 + y^2}{\sqrt{x^2 + y^2}} + (4 - \sqrt{x^2 + y^2})^2 \right) dA \\
&= \iint_D \left(\sqrt{x^2 + y^2} + (4 - \sqrt{x^2 + y^2})^2 \right) dA
\end{aligned}$$

OK, it looks like we're ready for polar coordinates; our region of integration D is the intersection of the surface with the xy-plane: $4 - \sqrt{x^2 + y^2} = 0$ becomes $x^2 + y^2 = 16$, and that's a full circle of radius 4, with bounds $0 \le r \le 4$, $0 \le \theta \le 2\pi$. Altogether,

$$\iint_S \mathbf{F} \cdot d\mathbf{S} = \int_0^{2\pi} \int_0^4 \left(r + (4 - r)^2 \right) r \, dr d\theta = \frac{256\pi}{3} \quad \blacksquare$$

(4) *(Bonus! Following up Sec. 15.3 ...)* We expect $\iint S\mathbf{F}(x, y, z) \cdot d\mathbf{S} = 0$ for which of the following combinations of vector field $\mathbf{F}(x, y, z)$ and oriented surface S?

I4) $\mathbf{F}(x, y, z) = \langle xy, xyz, 0 \rangle$ and S is the inverted paraboloid $z = 4 - x^2 - y^2$ oriented outwards

I5) $\mathbf{F}(x,y,z) = \langle x+y+z, e^{-xyz}, \cos(x)\sin(y)\rangle$ and S is the upwards oriented plane $z=6$ over $\{(x,y) : -\pi \le x \le \pi, -\pi \le y \le \pi\}$

I6) $\mathbf{F}(x,y,z) = \langle x,y,z\rangle$ and S is the unit sphere oriented outwards

---

□ All three integrals (I4)–(I6) yield zero. (I5) is deceptive — it looks awful, but helps stress the importance of the relationship $d\mathbf{S} = \mathbf{n}dS$, where $\mathbf{n}$ is normal to the surface, and here we can use $\mathbf{n} = \langle 1,0,0\rangle$ — so $\mathbf{F}\cdot\mathbf{n}$ is actually very simple. ■

C.6.4 *Green's Theorem — Challenge — Solved*

---

(1) Find $\displaystyle\oint_C x^2e^{-3x}\,dx + \frac{2}{3}(x^2+y^2)^3\,dy$ where C is the boundary of the region inside the circles $x^2+y^2=2$ and $x^2+y^2=9$.

□ Matching the integral to the form $\int_C P\,dx + Q\,dy$, we have

$$P = x^2e^{-3x} \to \frac{\partial P}{\partial y} = 0$$

$$Q = \frac{2}{3}(x^2+y^2)^3 \to \frac{\partial Q}{\partial x} = 4x(x^2+y^2)^2$$

Anticipating use of Green's Theorem, we can easily describe the region D inside C using polar coordinates, so

$$\oint_C x^2e^{-3x}\,dx + \frac{2}{3}(x^2+y^2)^3\,dy = \iint_D \left(\frac{\partial Q}{\partial x} - \frac{\partial P}{\partial y}\right)\,dA$$

$$= \iint_D 4x(x^2+y^2)^2\,dA = \int_0^{2\pi}\int_{\sqrt{2}}^{3} 4(r\cos\theta)(r^2)^2 r\,drd\theta$$

$$= \int_0^{2\pi}\int_{\sqrt{2}}^{3} 4r^6\cos\theta\,drd\theta = 0 \quad ■$$

---

(2) Find $\displaystyle\oint_C \mathbf{F}\cdot d\mathbf{r}$ where

$$\mathbf{F}(x,y) = \langle e^{-x} + \frac{1}{2}x^3y^2, \sin(y) - \frac{1}{2}x^2y^3\rangle$$

and C is the perimeter of $x^2+y^2=16$, oriented clockwise.

☐ Matching the vector field to the form $\mathbf{F}(x,y) = \langle P(x,y), Q(x,y)\rangle$, we have

$$P = e^{-x} + +\frac{1}{2}x^3y^2 \to \frac{\partial P}{\partial y} = x^3y$$
$$Q = \sin(y) - \frac{1}{2}x^2y^3 \to \frac{\partial Q}{\partial x} = -xy^3$$

Since C is clockwise it is negatively oriented. Green's Theorem applies, but we have to change the sign to account for the orientation. Also, polar coordinates will be useful:

$$\begin{aligned}\oint_C \mathbf{F}\cdot d\mathbf{r} &= \oint_C P\,dx + Q\,dy = -\iint_D \left(\frac{\partial Q}{\partial x} - \frac{\partial P}{\partial y}\right) dA \\ &= -\iint_D \left(-xy^3 - x^3y\right) dA = \iint_D xy\left(x^2+y^2\right) dA \\ &= \int_0^{2\pi}\int_0^4 (r\cos\theta)(r\sin\theta)(r^2)r\,drd\theta \\ &= \int_0^{2\pi}\int_0^4 r^5\cos\theta\sin\theta\,drd\theta = 0 \quad \blacksquare\end{aligned}$$

---

(3) The integral $\oint_C \mathbf{F}\cdot d\mathbf{r}$ gives the *circulation* of the vector field $\mathbf{F}$ around the contour C. Use Green's Theorem to find the circulation of $\mathbf{F}(x,y) = \langle y, -x\rangle$ around a circle $x^2+y^2 = a^2$ (where a is the constant radius of the circle). Take the boundary as being oriented positively.

☐ Matching the integral to the form $\int_C P\,dx + Q\,dy$, we have

$$P = y \to \frac{\partial P}{\partial y} = 1$$
$$Q = -x \to \frac{\partial Q}{\partial x} = -1$$

Anticipating use of Green's Theorem, we can easily describe the region D inside the circle using polar coordinates, so

$$\begin{aligned}\oint_C \mathbf{F}\cdot d\mathbf{r} &= \oint_C y\,dx - x\,dy = \iint_D \left(\frac{\partial Q}{\partial x} - \frac{\partial P}{\partial y}\right) dA \\ &= \iint_D (-1-1)\,dA = \int_0^{2\pi}\int_0^a (-2)r\,drd\theta = -2\pi a^2 \quad \blacksquare\end{aligned}$$

C.6.5 *The Divergence Theorem — Challenge — Solved*

(1) Find $\iint_S \mathbf{F} \cdot d\mathbf{S}$ for the vector field $\mathbf{F} = \langle -x^3y + yz, x^2y^2 + e^x, x^2yz \rangle$ where S is the surface of the hyperboloid $x^2 + 3y^2 - z^2 = 1$ between $z = -1$ and $z = 4$ — see Fig. C.20.

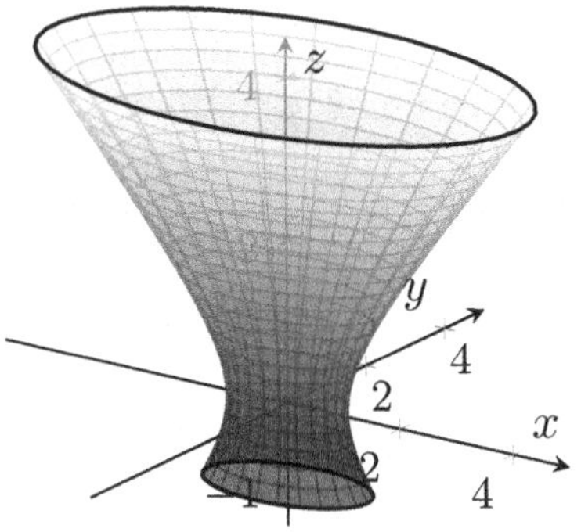

Fig. C.20 The hyperboloid $x^2 + 3y^2 - z^2 = 1$ between $z = -1$ and $z = 4$.

☐ The divergence of $\mathbf{F}$ is:

$$\nabla \cdot \mathbf{F} = \frac{\partial}{\partial x}(-x^3y + yz) + \frac{\partial}{\partial y}(x^2y^2 + e^x) + \frac{\partial}{\partial z}(x^2yz)$$
$$= -3x^2y + 2x^2y + x^2y = 0$$

So by the divergence theorem,

$$\iint_S \mathbf{F} \cdot d\mathbf{S} = \iiint_E \nabla \cdot \mathbf{F} \, dV = \iiint_E (0)\, dzdydx = 0 \quad \blacksquare$$

(2) Find $\iint_S \mathbf{F} \cdot d\mathbf{S}$ for the vector field $\mathbf{F} = \langle x^2yz + e^y, xy^2z + \sin(xz), \pi - xyz^2 \rangle$ where S is the surface of the tetrahedron formed by the plane $2x + 2y + z = 4$ in the first octant — see Fig. C.21.

☐ The divergence of $\mathbf{F}$ is:

$$\nabla \cdot \mathbf{F} = \frac{\partial}{\partial x}(x^2yz + e^y) + \frac{\partial}{\partial y}(xy^2z + \sin(xz)) + \frac{\partial}{\partial z}(\pi - xyz^2)$$
$$= 2xyz + 2xyz - 2xyz = 2xyz$$

For the divergence theorem, note that the domain underlying the tetrahedron in the xy-plane is between the line $2x+2y = 4$ (a.k.a. $y = 2-x$)

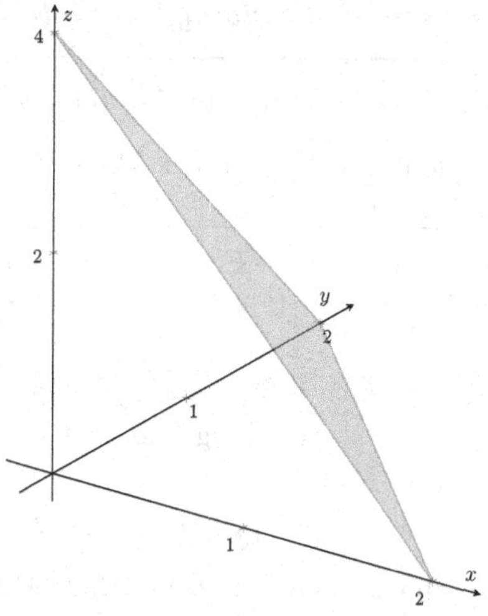

Fig. C.21 The plane $2x + 2y + z = 4$ in the first octant.

and the axes, so bounds are $0 \leq y \leq 2 - x$ and $0 \leq x \leq 2$.

$$\iint_S \mathbf{F} \cdot d\mathbf{S} = \iiint_E \nabla \cdot \mathbf{F}\, dV = \int_0^2 \int_0^{2-x} \int_0^{4-2x-2y} (2xyz)\, dzdydx = \frac{32}{45}$$

■

---

(3) As noted, the integral $\iint_S \mathbf{F} \cdot d\mathbf{S}$ gives the *flux* of the vector field $\mathbf{F}$ across the surface S. Use the Divergence Theorem to find the flux of a vector field $\mathbf{F}(x, y, z) = \langle bz - cy, cx - az, ay - bx \rangle$ across ANY smooth closed surface S in $\mathbf{R}^3$ (where a, b, c are constants).

□ Since

$$\nabla \cdot \mathbf{F} = \frac{\partial}{\partial x}(bz - cy) + \frac{\partial}{\partial y}(cx - az) + \frac{\partial}{\partial z}(ay - bx) = 0$$

then

$$\iint_S \mathbf{F} \cdot d\mathbf{S} = \iiint_E (0)\, dV = 0$$

for any smooth closed surface S and its interior E. ■

C.6.6 *Stokes' Theorem — Challenge — Solved*

(1) Compute $\int_C \mathbf{F}\cdot d\mathbf{r}$ for the vector field $\mathbf{F} = \langle (1+y)z, (1+z)x, (1+x)y\rangle$ where C is the boundary of the plane $2x+2y+z=4$ in the first octant, traversed counterclockwise.

□ We'll use Stokes' Theorem via Eq. (18.4) to convert this to a double integral involving $\nabla\times\mathbf{F}$, which is $\nabla\times\mathbf{F} = \langle 1,1,1\rangle$ (details omitted).

The equation of the surface is $z = g(x,y) = 4-2x-2y$, so we have $g_x = -2$ and $g_y = -2$. Matching the curl to the form $\nabla\times\mathbf{F} = \langle P,Q,R\rangle = \langle 1,1,1\rangle$, we have

$$P=1 \quad ; \quad Q=1 \quad ; \quad R=1$$

The domain D in the xy-plane underneath C is bounded by $x=0$, $y=0$, and the line $2x+2y=4$ (a.k.a. $y=2-x$). So by (18.4),

$$\begin{aligned}\oint_C \mathbf{F}\cdot d\mathbf{r} = \iint_S \nabla\times\mathbf{F}\cdot d\mathbf{S} &= \iint_D \left(-P\frac{\partial g}{\partial x} - Q\frac{\partial g}{\partial y} + R\right) dA \\ &= \iint_D (-(1)(-2)-(1)(-2)+1)\, dA \\ &= \int_0^2\int_0^{2-x} (5)\, dydx = 10 \quad \blacksquare\end{aligned}$$

(2) Compute $\iint_S (\nabla\times\mathbf{F})\cdot d\mathbf{S}$ for the vector field $\mathbf{F} = \langle z^2, -3xy, x^3y^3\rangle$ where S is the top part of the inverted paraboloid $z = 5-x^2-y^2$ above the plane $z=1$, oriented positively.

□ The boundary curve ∂S of this surface is the circle $x^2+y^2=4$ (at $z=1$) and would usually be expressed (with positive orientation) as

$$\begin{aligned} x &= 2\cos t & &\rightarrow\ dx = -2\sin t\, dt \\ y &= 2\sin t & &\rightarrow\ dy = 2\cos t\, dt \\ z &= 1 & &\rightarrow\ dz = 0 \end{aligned}$$

for $0\le t\le 2\pi$. By Stokes' Theorem, then,

$$\iint_S (\nabla\times\mathbf{F})\cdot d\mathbf{S} = \oint_{\partial S}\mathbf{F}\cdot d\mathbf{r} = \oint_{\partial S}(z^2)\,dx + (-3xy)\,dy + (x^3y^3)\,dz$$

Because $dz=0$, we can drop the third term in the integrand, and have:

$$\cdots = \int_0^{2\pi} (1)^2(-2\sin t\, dt) + (-3)(2\cos t)(2\sin t)(2\cos t\, dt) + 0$$

And this cleans up pretty nicely:

$$\cdots = \int_0^{2\pi} (-2\sin t - 24\cos^2 t \sin t)\,dt = 0 \quad \blacksquare$$

---

(3) The integral $\oint_C \mathbf{F} \cdot d\mathbf{r}$ gives the *circulation* of the vector field $\mathbf{F}$ around the contour C. Let g be the function $g(x, y, z) = xe^z \sin(y)$ and let $\mathbf{F}(x, y, z) = \nabla g$. What is the total circulation due to $\mathbf{F}$ around the contour C, which is the intersection of the plane $x + y + z = 5$ and the cylinder $x^2 + y^2 = 11$, oriented positively? (Hint: There's an easy way, and there's a hard way...)

□ According to Stokes' Theorem, we want to set up the conversion

$$\oint_C \mathbf{F} \cdot d\mathbf{r} = \iint_S \nabla \times \mathbf{F} \cdot d\mathbf{S}$$

for the given vector field $\mathbf{F}$ and contour C. But since the vector field $\mathbf{F}$ is explicitly defined as the gradient of a scalar function, then we know that $\mathbf{F}$ is a conservative vector field — and therefore its curl is zero. With $\nabla \times \mathbf{F} = \mathbf{0}$, then, the right hand integral is zero, meaning that the left hand integral — the circulation — is also zero. ■

Index

Printed in the United States
by Baker & Taylor Publisher Services

Printed in the United States
by Baker & Taylor Publisher Services